Sensitivity Analysis

Sensitivity Analysis

Edited by

Andrea Saltelli
European Commission, Joint Research Centre, Italy

Karen Chan
European Commission, Joint Research Centre, Italy

E. Marian Scott
University of Glasgow, UK

JOHN WILEY & SONS, LTD

Chichester • New York • Weinheim • Brisbane • Singapore • Toronto

This paperback edition first published 2008

Registered office
John Wiley & Sons Ltd, The Atrium, Southern Gate, Chichester, West Sussex, PO19 8SQ, United Kingdom

For details of our global editorial offices, for customer services and for information about how to apply for permission to reuse the copyright material in this book please see our website at www.wiley.com.

A catalogue record for this book is available from the British Library.

ISBN: 978-0-470-74382-9 (P/B)

Contents

Editors' Preface

Our intentions in preparing this handbook were to try and bring to the attention of the wider modelling community the many different sensitivity analysis (SA) methodologies that have been developed and, through real-world applications, to demonstrate their usefulness in the scientific challenges we face in model development and use and model corroboration.

The handbook contains 22 chapters written by 25 contributors from 10 countries, and is divided into four sections. In the first section, a general introduction and overview of the field is given (Chapters 1 and 2). This is followed by a methodological section, which includes chapters on the most widely used SA techniques (Chapters 4–6), followed by some less widely used approaches (Chapters 7–11). The third section then introduces a series of applications from a wide variety of applied sciences, including ecology, chemistry, physics, economics and policy-making (Chapters 12–20). Finally, the last section of the handbook includes two chapters (21 and 22) that take a more general view of SA, and indeed the activity of modelling, and try to describe the challenges remaining. A brief description of some SA software and their capabilities is provided in the Appendix.

A draft version of this volume was used in the Summer School on Sensitivity Analysis held in Venice in the summer of 1999, and all the students who provided a constructive feedback are gratefully acknowledged. Special thanks go to Dr Sharon Clutton of Wiley for following our work, as well as to all contributors for the time and effort they put into this work.

A. Saltelli
K. Chan
E. M. Scott
June 2000

Contributors

Dr Ömer Faruk Alış
Program in Applied and Computational Mathematics
Princeton University
Princeton
NJ 08544
USA

Terry Andres
AECL Whiteshell Laboratories
Pinawa, Manitoba
ROE 1LO
Canada

Dr Jeff Arnold
Atmospheric Sciences Modelling Division
Air Resources Laboratory
National Oceanic and Atmospheric Admin (NOAA)
79 Alexander Drive, Building 4201
Research Triangle Park
NC 27709
USA

Professor M. B. Beck
Warnell School of Forest Resources
University of Georgia
Athens
Georgia 30602-2152
USA

Dr Marco G. Beghi
Istituto Nazionale per la Fisica della Materia — Dipartimento di Ingegneria Nucleare
Politecnico di Milano
Via Ponzio 34/3
20133 Milano
Italy

Dr C. E. Bottani
Istituto Nazionale per la Fisica della Materia — Dipartimento di Ingegneria Nucleare
Politecnico di Milano
Via Ponzio 34/3
20133 Milano
Italy

Dr Francesca Campolongo
Institute for Systems, Informatics and Safety
EC Joint Research Centre
TP 361
21020 Ispra (VA)
Italy

Professor Jeff D. Cawlfield
Department of Geological & Petroleum Engineering
University of Missouri at Rolla
129 McNutt Hall
Rolla
MO 65401-0249
USA

Dr Karen Chan
Institute for Systems, Informatics and Safety
EC Joint Research Centre
TP 361
21020 Ispra (VA)
Italy

Dr Jining Chen
Department of Environmental System Engineering
TsingHua University
Beijing
PR China

Professor Roger M. Cooke
Department of Mathematics
Delft University of Technology
Mekelweg 4
PO Box 5031
2600 GA Delft
The Netherlands

Dr Freddie J. Davis
Amarillo National Research Center
600 S. Tyler, Suite 800
Amarillo
TX 79101
USA

Dr Robin L. Dennis
Atmospheric Sciences Modelling Division
Air Resources Laboratory
National Oceanic and Atmospheric Admin (NOAA)
79 Alexander Drive, Building 4201
Research Triangle Park
NC 27709, USA

Raoul Depoutot
Statistical Office of the European Commission (EUROSTAT)
Bat. BECH
5 Rue Alphonse Weicker
Luxembourg

Professor David Draper
Statistics Group
Department of Mathematical Sciences
University of Bath
Claverton Down
Bath BA2 7AY
UK

Professor Jon C. Helton
Department of Mathematics
Arizona State University
Tempe
AZ 85287-1804
USA

Professor David Ríos Insua
Department of Experimental Sciences and Engineering
Universidad Rey Juan Carlos
28933 Mostoles
Madrid
Spain

Jochen Jesinghaus
Institute for Systems, Informatics and Safety
EC Joint Research Centre
TP 361
21020 Ispra (VA)
Italy

Professor Jack Kleijnen
Department of Information Systems
Center for Economic Research
School of Management and Economics
Tilburg University, PO Box 90153
5000 LE Tilburg
The Netherlands

Professor Jacinto Martín
School of Engineering
Politecnic University of Madrid
10071 Cáceres
Spain

Dr Jan M. van Noortwijk
HKV Consultants
PO Box 2120
Lelystad
8203 AC
The Netherlands

Dr Rosanna Pastorelli
Intituto Nazionale per la Fisica della Materia — Dipartimento di Ingegneria Nucleare
Politecnico di Milano
Via Ponzio 34/3
20133 Milano
Italy

Dr Christophe Planas
Institute for Systems, Informatics and Safety
EC Joint Research Centre
TP 361
21020 Ispra (VA)
Italy

Pedro Prado
CIEMAT
Dpto. Impacto Ambiental de la Energia
Avda. Complutense, 22
28040 - Madrid
Spain

Dr Maila Puolamaa
Statistical Office of the European Commission (EUROSTAT)
Bat. BECH
5 Rue Alphonse Weicker
Luxembourg

Professor Herschel Rabitz
Department of Chemistry
Princeton University
Princeton
NJ 08544
USA

Dr Fabrizio Ruggeri
Consiglio Nazionale della Ricerche, Istituto per la Applicazioni della Matematica e dell'Informatica
Via Ampere 56
I-20131 Milano
Italy

Andrea Saltelli
Institute for Systems, Informatics and Safety
EC Joint Research Centre
TP 361
21020 Ispra (VA)
Italy

Dr E. Marian Scott
Department of Statistics
Mathematics Building
University Gardens
University of Glasgow
Glasgow G12 8QW
UK

Professor Ilya M. Sobol'
Institute for Mathematical Modelling
Russian Academy of Sciences
4 Miusskaya Square
Moscow 125047
Russia

Tine Sørensen
Department of Biostatistics
University of Copenhagen
Blegdamsvej 3
2200 København N
Denmark

Dr Stefano Tarantola
Institute for Systems, Informatics and Safety
EC Joint Research Centre
TP 361
21020 Ispra (VA)
Italy

Dr Gail S. Tonnesen
Atmospheric Sciences Modelling Division
Air Resources Laboratory
National Oceanic and Atmospheric Admin (NOAA)
79 Alexander Drive, Building 4201
Research Triangle Park
NC 27709, USA

Dr Tamas Turányi
Department of Physical Chemistry
Eötvös University (ELTE)
H-1518 Budapest-112
PO Box 32
Hungary
and
Chemical Research Center
H-1525 Budapest
PO Box 17
Hungary

Dr José-Manuel Zaldívar Comenges
Institute for Systems, Informatics and Safety
EC Joint Research Centre
TP 280
21020 Ispra (VA)
Italy

That is what we meant by science. That both question and answer are tied up with uncertainty, and that they are painful. But that there is no way around them. And that you hide nothing; instead, everything is brought out into the open.

Peter Høeg, *Borderliners*

— Sensitivity analysis for modellers?
— Would you go to an orthopaedist who didn't use X-ray?

Jean Marie Furbringer

I

Introduction

1

What is Sensitivity Analysis?

Andrea Saltelli

European Commission, Joint Research Centre, Ispra, Italy

1.1 INTRODUCTION

Sensitivity analysis (SA) is the study of how the variation in the output of a model (numerical or otherwise) can be apportioned, qualitatively or quantitatively, to different sources of variation, and of how the given model depends upon the information fed into it. On this basis, we contend that SA is a prerequisite for model building in any setting, be it diagnostic or prognostic, and in any field where models are used.

Models are developed to approximate or mimic systems and processes of different natures (e.g. physical, environmental, social, or economic), and of varying complexity. Many processes are so complex that physical experimentation is too time-consuming, too expensive, or even impossible. As a result, to explore systems and processes, investigators often turn to mathematical or computational models.

A mathematical model is defined by a series of equations, input factors, parameters, and variables aimed to characterize the process being investigated. Input is subject to many sources of uncertainty including errors of measurement, absence of information and poor or partial understanding of the driving forces and mechanisms. This imposes a limit on our confidence in the response or output of the model. Further, models may have to cope with the natural intrinsic variability of the system, such as the occurrence of stochastic events.

Good modelling practice requires that the modeller provide an evaluation of the confidence in the model, possibly assessing the uncertainties associated with the modelling process and with the outcome of the model itself.

Originally, SA was created to deal simply with uncertainties in the input variables and model parameters. Over the course of time, the ideas have been extended to incorporate model conceptual uncertainty, i.e. uncertainty in model structures, assumptions, and specifications. As a whole, SA is used to increase the confidence in the model and its predictions, by providing an understanding of how the model response variables respond to changes in the inputs, be they data used to calibrate it, model structures, or factors, i.e. the model-independent variables. SA is thus closely linked to uncertainty analysis (UA),

Sensitivity Analysis. Edited by A. Saltelli *et al.*

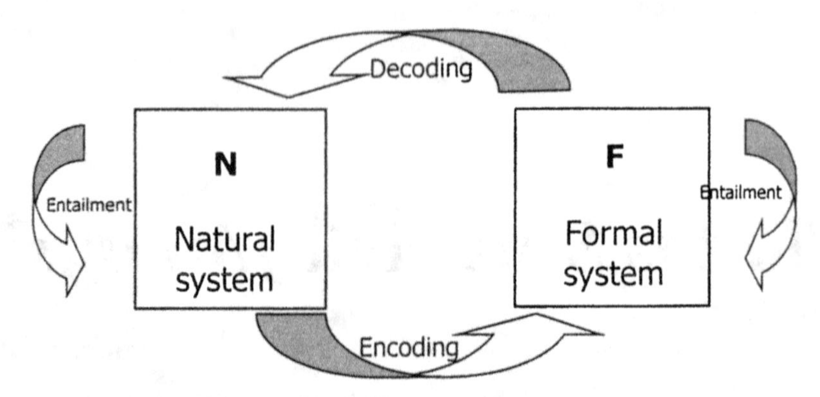

Figure 1.1 Formalization of the modelling process taken from Rosen (1991).

which aims to quantify the overall uncertainty associated with the response as a result of uncertainties in the model input. In this book, we shall, however, focus on SA.

All of the above might be summarized by the following loose definition of SA:

Definition *Sensitivity analysis studies the relationships between information flowing in and out of the model*

For the definition we have taken inspiration from Rosen's (1991) formalisation of the modelling activity (Figure 1.1), where the link between a model (driven by a formal entailment structure) and the world (driven by different classes of causality) is the process of 'encoding' (from world to model) and 'decoding' (from model to world). Encoding and decoding are not themselves 'entailed' by anything, i.e. they are the object of the modeller's craftsmanship. Yet those two activities are the essence and the purpose of the modelling process, i.e. one writes a model in the hope that the decoding operation will provide insight into the world. This is only possible, in our view, if the uncertainty in the information provided by the model (the substance of use for the decoding exercise) is carefully apportioned to the uncertainty associated with the encoding process.

In the remaining part of this introductory chapter, we shall briefly describe why, how and when to perform SA.

1.2 AN EXAMPLE

Having compromised ourselves with a definition of SA, we may now proceed to offer an example that illustrates different ways of measuring sensitivity. We start by highlighting the importance of what one uses to *measure* sensitivity.

Consider a model for a dry-cleaning bill,

$$C = \sum_i C_i, \tag{1.1}$$

where the total cost on the bill, C, equalling the sum of the individual costs of the items laundered, C_i, may be strongly influenced by an individual item (the cost of a fur present in the selection of items, say). One measure of sensitivity may tell us that the *variation* in the total cost C is equally sensitive to all items in the sample. This would be the case if the sensitivity of an item were taken to be the derivative of the total bill with respect to the cost of a single item in the bill, i.e.

$$S_i = \frac{\partial C}{\partial C_i}. \tag{1.2}$$

Here the derivative is computed at a point $C^0 = (C_1^0, C_2^0, \ldots)$, i.e. a point where all the C_i are fixed to some reference value C_i^0 (the term 'nominal value' is often used). In other words, the quantity S_i is the local sensitivity index measuring the effect on C of perturbing C_i around a reference or central value C^0. In this case, S_i is equal to one for all items.

Another choice of sensitivity measure would explore what happens to the total cost if all the items in the sample are allowed a finite variation in cost. The derivative above could be normalized by the mean of output (the total cost) and input (the cost of the item). The sensitivity index would then measure the effect on C of perturbing C_i by a fixed fraction of C_i's reference value (C_i's mean value, in this case), i.e.

$$S_i = \frac{\partial C}{\partial C_i} \frac{C_i^0}{C^0}, \tag{1.3}$$

and the cost of the most expensive item, i.e. the fur, would be *measured* as the most influential factor. Alternatively, the sensitivity index could measure the effect on C of perturbing C_i by a fixed fraction of C_i's standard deviation, i.e.

$$S_i = \frac{\partial C}{\partial C_i} \frac{\mathrm{std}(C_i)}{\mathrm{std}(C)}. \tag{1.4}$$

If the standard deviations were equal, all items would again be judged to be equally important in determining the variation of the bill. If, more realistically, the standard deviations were different, e.g. higher standard deviations were associated with higher mean values, the cost of the fur would again be identified as the most influential factor.

SA is not concerned with what causes the output of the model to be what it is, but what the sources of variation in that output are. Using the measures defined by (1.4), the cost of the fur is found to be the predominant factor only if it drives most of the variation in the total cost. If we use the sensitivity measures defined by (1.2) then the cost of the fur is as important as all other factors, even if it contributes predominantly to the total cost. The purpose of this discussion on the measure is to highlight that the type of measure employed, selected on the basis of the context or use one desires to make of SA, has a direct consequence on the outcome of the analysis. Different measures have different uses and applications, and a universal recipe for measuring sensitivity does not exists. Good practice should instead be our goal, and this is the subject of the present volume.

1.3 WHY CARRY OUT A SENSITIVITY ANALYSIS?

In the context of numerical modelling, SA means very different things to different people (compare the reviews by Turanyi (1990a), Janssen *et al.* (1990), Helton (1993), and Goldsmith (1998)). For a reliability engineer, SA could be the process of moving or changing components in the design of a plant to investigate how a fault tree analysis for that plant would change. For a chemist, SA could be the analysis of the strength of the relation between kinetic or thermodynamic inputs and measurable outputs of a reaction system. For a software engineer, SA could be related to the robustness and reliability of the software with respect to different assumptions. For an economist, the task of SA could be to appraise how stable the estimated parameters of a model (customarily derived via regression) are with respect to all factors that were excluded from the regression, thus ascertaining whether parameter estimation is robust or fragile. For a developer of expert systems, it is important to measure sensitivity with respect to the quantiles of the 'prior' distributions. For a statistician, involved in statistical modelling, sensitivity analysis is mostly known and practised under the heading of 'robustness analysis'. Statisticians are mostly interested in 'distributional robustness', intended as insensitivity with respect to small deviations from the assumptions about the underlying distribution assumed for the data (Huber, 1981a).

These different types of analyses have in common the aim to investigate how a given computational model responds to variations in its inputs. Modellers conduct SA to determine:

(a) if a model resembles the system or processes under study;
(b) the factors that mostly contribute to the output variability and that require additional research to strengthen the knowledge base;
(c) the model parameters (or parts of the model itself) that are insignificant, and that can be eliminated from the final model;
(d) if there is some region in the space of input factors for which the model variation is maximum;
(e) the optimal regions within the space of the factors for use in a subsequent calibration study;
(f) if and which (group of) factors interact with each other.

Under (a), the model does not properly reflect the processes involved if it exhibits strong dependence on supposedly non-influential factors or if the range of model predictions is not a sound one. In this case, SA highlights the need to revise the model structure. It often happens that the model turns out to be highly tuned to a specific value of a factor, up to the point that necessary changes, e.g. resulting from new evidence, lead to unacceptable variation in the model predictions. When this happens, it is likely that in order to optimize the simulation, some parameter values have been chosen incorrectly. This reflects lack of conceptual understanding of the role of the parameters in the system.

Under (b), SA can assist the modeller in deciding whether the parameter estimates are sufficiently precise for the model to give reliable predictions. If not, further work can be directed towards improved estimation of those parameters that give rise to the greatest uncertainty in model predictions. If the model sensitivity seems congruent with (i.e. does not contradict) our understanding of the system being modelled, SA will open up the possibility of improving the model by prioritizing measurement of the most influential factors. In this way, the impacts of measurement errors on computational results can be minimized.

Under (c), we mean insignificant in the sense of 'not affecting the variation of the output'; this concept is taken up at the end of the book, where we discuss the 'relevance' of models. According to some investigators, when the model is used in a case of conflicting stakes (e.g. siting a facility or licensing a practice), the model should not be more complex than needed, and factors/processes that are insignificant should be removed.

As far as (e) is concerned, we stress the need for 'global' optimization. One should investigate the space of the factors in its entirety, and not just around some nominal points (see also the Monte Carlo filtering approach mentioned in **Chapters 2** and **21**).

Point (f) is an important technicality: often factors have combined effects that cannot be reduced to the sum of the individual ones. This is relevant, since the presence of an interaction has implications for all of the above points (calibration, determination of critical points, etc.).

1.4 HOW TO PERFORM SENSITIVITY ANALYSIS?

Anticipating here a notion offered in **Chapter 2**, i.e. that of a sampling-based sensitivity analysis, we try to draft a flow chart for the SA process. A sampling-based SA is one in which the model is executed repeatedly for combinations of values sampled from the distribution (assumed known) of the input factors. The following steps can be identified:

1. Design the experiment (identify what question the model should answer) and determine which of its input factors should concern the analysis.
2. Assign probability density functions or ranges of variation to each input factor.
3. Generate an input vector/matrix through an appropriate design.
4. Evaluate the model, thus creating an output distribution for the response of interest.
5. Assess the influences or relative importance of each input factor on the output variable(s).

The basic steps in a SA are illustrated in Figure 1.2, the process starting in the upper left corner of the diagram.

One starts by defining one or a series of candidate models to answer the question considered, and selecting factors for the analysis (Which factors shall I include? Do I vary all factors?). The output variable(s) are also selected at this time. Distributions must be defined for each uncertain input factor, where 'constant' is just a particular case. When using derivatives as a sensitivity measure (Equation (1.2) above), one apparently does not need this

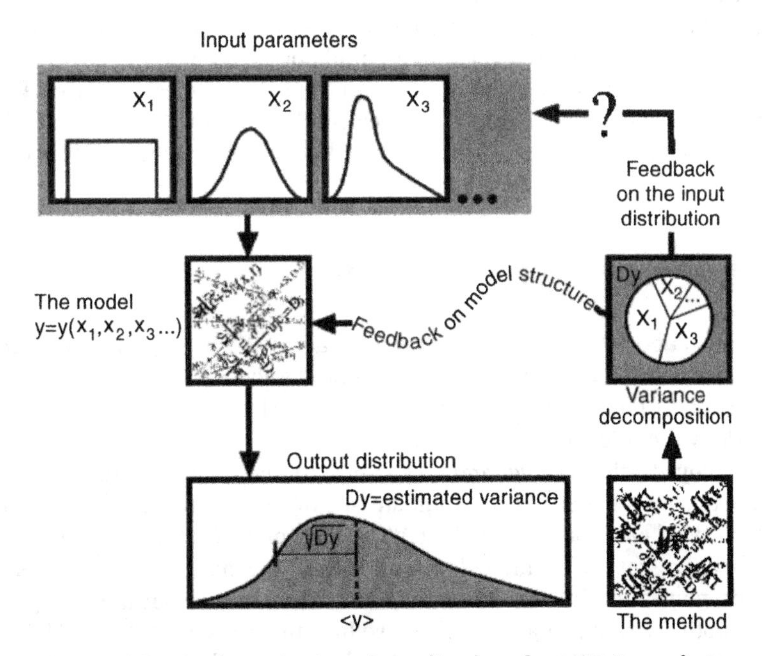

Figure 1.2 A schematic view of sampling-based sensitivity analysis.

step. Nevertheless, when derivatives are evaluated via incremental ratios, a uniform distribution is *de facto* assumed over the small interval selected for the evaluation.

As a rule, SA is performed on a set of factors. A factor could be:

- an input datum (e.g. an uncertain exchange rate);
- the distribution parameter describing some aleatory process (e.g. an unknown frequency rate for earthquakes in a given area);
- a trigger factor, whose value drives the selection of alternative mechanisms (e.g. a linear versus a nonlinear process) or scenarios (different greenhouse gas emission policies).

Factors can be varied one-at-a-time (OAT), all the other factors being held constant. When using random sampling, or experimental designs, several factors can be varied at once. If a correlation structure is specified for the input factors, realizations can be generated from the multivariate input distribution using *ad hoc* sampling procedures.

The sample is fed through the model, i.e. the model is run repeatedly for each realization to obtain an output sample for the variable(s) of interest. This can be used to build an empirical probability distribution for the response variable(s). Means, standard deviations, confidence bounds, cumulative distribution functions, etc. can then be estimated.

So far we have implemented an uncertainty analysis, i.e. we have quantified the variation in the model response. The next step, sensitivity analysis, is the apportioning of the uncertainty according to source (i.e. to the factors).

A possible representation of the results (shown in Figure 1.2) would be a pie chart that partitions the variance of the output. This variance decomposition allows the identification of the influential factors, forming a basis for all the modelling investigations described in the previous section.

As far as the implementation of the analysis is concerned, different formal approaches can be followed, and a very large number of techniques are available in the literature (see the review articles quoted above). SA articles can be found in a large number of journals, and a number of special issues and proceedings are available (JSCS, 1997; RESS, 1997; SAMO, 1998; CPC, 1999). The present volume aims to cover a wide range of techniques, as well as to provide a reference handbook and guide to sensitivity analysis as a discipline. Thus, we attempt a classification of the existing methods (see **Chapter 2**). This has been somehow complicated by the wealth of applications that can be found for SA, and the consequent fact that SA has many different meanings to different people, as discussed.

1.5 GOALS OF SENSITIVITY ANALYSIS

We have mentioned above why one should perform a sensitivity analysis, including model development, verification, calibration, model identification, and mechanism reduction. Without pretension of being exhaustive, we now try to elaborate some of these concepts, and indicate briefly the role SA has to play.

SA can be employed prior to a *calibration* exercise to investigate the tuning importance of each parameter, i.e. to identify a candidate set of important factors for calibration, since the difficulty of calibrating models against field or laboratory data increases with the number of processes to be modelled (and hence the number of parameters to be estimated). This may allow a dimensionality reduction of the parameter space where the calibration/optimization is made. SA can also help to ensure that the problem is not an ill-conditioned one. Quantitative SA methods (i.e. those that tell how much more important one factor is than another, see **Chapter 2**) can be appropriate when both the model inputs and the available data are affected by uncertainties. The question answered is what factors are allowed to be calibrated and at what confidence, given the data and their uncertainty.

While SA was originally created to deal with the uncertainties in the input factors, recent developments have seen some of the ideas being extended to incorporate structural uncertainty as mentioned above. In this way, SA also touches on the difficult problem of *model quality*[1] and is an important element of judgement for the corroboration, or falsification, of the scientific hypotheses embedded in a model. SA can be used to ensure that the response of the model to its input factors can be accounted for, that the model does not exhibit strong dependence on supposedly non-influential factors and that the range of model predictions is a sensible one.

SA can be an effective tool for *model identification*. By pinpointing experimental conditions in which the ability to discriminate among the various models is a maximum, SA can identify the most appropriate model structures and specifications competing to describe available evidence.

This is closely related to *mechanism reduction*, determining a subset of input factors accounting for (most of) the output variance. This enables the insignificant factors to be identified and eliminated from the final model. In this way, irrelevant parts of the model can be dropped, or a simpler model can be built or extracted from a more complex one (*model lumping*).

The above points have some epistemological implications concerning the 'relevance' of a model (see also **Chapter 21**). It has been argued that often the complexity of models largely exceeds the actual 'requirements'. The view of Oreskes *et al.* (1994) is that models should be heuristic constructs built for a task. They should not be more complex than they need to be. Following this reasoning, a model would be 'relevant' when its input factors actually cause variation in the model response that is the object of the analysis. Model 'irrelevance' would flag a bad model, a model used out of context, or a model unable to provide the answer being sought.

Another possible goal for SA is to determine if there is some region in the space of input parameters for which the model variation is maximum or divergent. This is useful in control theory, where one might also be interested in the initiation of chaotic behavior for some combinations of model parameters.

In the field of risk analysis, the goal of SA is to identify risk-governing parameters. This may help the analyst, especially when he/she has knowledge on the factors that is not explicitly coded in the problem. For example, two factors are given the same uncertainty in input, but the analyst is more confident in (or scared of) one of them. If the dreaded factor turns out to be the risk-governing one, mitigating actions can be taken.

In general, SA can assist the modeller in deciding whether model performance is sufficient to the task, and when this is not the case, to provide guidance as to where to invest to solve the problem.

---

[1] Judgements of merit on the quality of a model when applied to a given task are often made on the basis of processes known as verification and validation. Lately the meaning attached to these terms has been very much the subject of debate (see Sheng *et al.* (1993) for a classical system analysis approach, and Konikov and Bredehoeft (1992) and Oreskes *et al.* (1994) for a more radical criticism of these concepts). We shall often use the term 'corroboration', proposed by Oreskes *et al.*, to mean the reinforcement of an hypothesis by means of non-contradiction between the prediction of a model and the evidence, and the term 'falsification' of a hypotheses for the opposite occurrence. We shall try to avoid the terms 'verification' and 'validation' altogether.

1.6 PROPERTIES OF VARIOUS TYPES OF SENSITIVITY ANALYSIS TECHNIQUES

The choice of which SA method to adopt is difficult, since each technique has strengths and weaknesses. Such a choice depends on the problem the investigator is trying to address, on the characteristics of the model under study, and also on the computational cost that the investigator can afford.

One possible way of grouping sensitivity analysis methods is into three classes: *screening methods*, *local SA methods* and *global SA methods*. This distinction is somewhat arbitrary, since screening tests also can be viewed as either local or global. Further, the first class is characterized with respect to its use (screening), while the other two are characterized with respect to how they treat factors.

1.6.1 Screening

In dealing with models that are computationally expensive to evaluate and have a large number of input parameters, *screening experiments* can be used to identify the parameter subset that controls most of the output variability (with low computational effort). This is based on the experience that often only a few of the input parameters have a significant effect on the model output. As a drawback, these 'economical' methods tend to provide qualitative sensitivity measures, i.e. they rank the input factors in order of importance, but do not quantify how much more important a given factor is than another. In contrast, a quantitative method would give, for example, the exact percentage of the total output variance that each factor (or group of factors) is accounting for. There is clearly a trade-off between computational cost and information.

1.6.2 Local SA

Local SA concentrates on the local impact of the factors on the model. Local SA is usually carried out by computing partial derivatives of the output functions with respect to the input variables. In order to compute the derivatives numerically, the input parameters are allowed to vary within a small interval of fractional variation around a nominal value.[2] The interval is usually the same for all of the variables and is not related to our degree of knowledge of the variables. Present-day computational tools for local SA allow large numbers of sensitivity coefficients to be computed simultaneously. This is often used to solve a so-called 'inverse problem', e.g. how to back-calculate the value of kinetic constants for reactions whose output are not directly measurable, based on the concentration of end-products further down the reaction pathway (Rabitz, 1989). One can see local SA as a particular case of one-factor-at-a-time (OAT) approach, since when one factor is varied, all others are held constant.

Local methods are less helpful when SA is used to compare the effect of various factors on the output, as in this case the relative uncertainty of each input should be weighted. This can be achieved by some kind of differential analysis, where an incremental ratio is considered. As an example, in Capaldo and Pandis (1997), the target sensitivity measure is

$$S_i = \frac{x_i}{y}\frac{\partial y}{\partial x_i},$$

[2]This is not true if a 'direct' solution method is used, see Chapter 5, p. 83.

i.e. the effect on the relative variation of y of perturbing x_i by a fixed fraction of x_i's central value. S_i is estimated by

$$\hat{S}_i = \frac{\ln y - \ln y_b}{\ln x_i - \ln x_{ib}},$$

where the subscript b indicates a baseline value and x_i a generic input factor. Each of the x_i is then given a different variation between the baseline and a 'sensitivity test value'. This approach is practicable when the variation around a baseline of the input variables is small, since in general it allows the input–output relationship to be assumed linear. Sometimes, the range of variation is taken as identical for all the variables (e.g. $\pm 5\%$ of the nominal value in Falls *et al.*, 1989), and the relative importance of the input parameters is thus assessed. We should nevertheless advise against this practice unless the model is known in advance to be linear. It has been recognized in the literature for a number of years (Cukier *et al.*, 1973) that when the model is nonlinear and various input variables are affected by uncertainties of different orders of magnitude, a global SA method should be used.

1.6.3 Global SA

Global SA techniques have been discussed by Cukier *et al.* (1978), Iman and Helton (1988), Sobol' (1990b), Helton *et al.* (1991), and Saltelli and Homma (1992), among others. For a review of applications of global variance-based sensitivity analysis techniques, see Saltelli *et al.*, 2000. Global SA apportions the output uncertainty to the uncertainty in the input factors, described typically by probability distribution functions that cover the factors' ranges of existence. The ranges are valuable, since they represent our knowledge or lack of it with respect to the model and its parameterization. More specifically, we should define global methods by the following two properties:

First global property: The inclusion of influence of scale and shape *The sensitivity estimates of individual factors incorporate the effect of the range and the shape of their probability density functions.*

Second global property: Multidimensional averaging *The sensitivity estimates of individual factors are evaluated varying all other factors as well.*

A global SA technique thus incorporates the influence of the whole range of variation and the form of the probability density function of the input. A global method evaluates the effect of x_i while all other $x_j, j \neq i$, are varied as well. In contrast, the local perturbative approach is based on partial derivatives, the effect of the variation of the input factor x_i when all other $x_j, j \neq i$, are kept constant at their central (nominal) value.

1.7 CHOICE OF METHODS

An important property to consider when choosing which SA technique to employ, is the following:

Model independence property *The level of additivity or linearity of the model does not influence the accuracy of the method.*

Not all SA methods are model-independent. A Monte Carlo analysis coupled with a linear regression can provide useful insight into the relative influence of factors when these relate linearly to the output of interest. Differential analysis can be used with caution when the range of variation of the factors is small. Some methods for SA are fairly inexpensive but perform poorly for nonlinear, non-additive models. The choice of the method for a particular application is discussed in detail in **Chapter 2**.

Further important properties of SA methods will be introduced in **Chapter 8** when discussing the variance-based approaches. These touch on the capacity of a method to account for factor-to-factor interactions, the possibility to group factors into sets and to treat each set as a factor, and computational efficiency.

1.8 ABOUT THE CHAPTERS AHEAD

The book is divided into *Introduction, Methods, Applications*, and *Conclusions*, and is not meant to be read from start to finish. The reader should feel free to skip those sections not immediately relevant, and jump to the parts in which he/she is interested or which are relevant for the specific problem at hand. The hurried reader could simply read **Chapter 2** for the most appropriate tool to solve a problem, and jump to the chapter where the tool is described.

Part I: Introduction

Following this introductory chapter is **Chapter 2**: *Hitchhiker's Guide to Sensitivity Analysis*, where, as the title suggests, we offer a sort of simplified tourist guide to SA. It is composed of a summary description of the various methods with elements of evaluation and comparison, suggesting the most appropriate methods for different settings. This chapter is useful for the reader needing brief descriptions and a pointer to find the chapter most useful for his/her problem. It includes several different analytical functions used as test models throughout the book.

Part II: Methods

This section is devoted to more detailed descriptions of the SA methodologies. In **Chapter 3** we review briefly the theory of experimental design. In **Chapter 4**, we take a look at *Screening Methods*, including the methods of Morris, Cotter, and others. **Chapter 5** offers a description of *Local Methods*, including the adjoint method and the use of functional sensitivities. In **Chapter 6**: *Sampling-Based Methods*, a description is given of the various Monte Carlo-based sensitivity analyses. These include scatterplots, correlation, and regression analysis. Also described are the sampling techniques on which these are based. **Chapter 7**: *Reliability Algorithms: FORM and SORM* deals with sensitivity analysis for risk analysis, with a thorough description of the FORM and SORM methods. **Chapter 8** introduces *Variance-Based methods*, including Sobol' indices, FAST and extended FAST. **Chapter 9**: *Managing the Tyranny of Parameters in Mathematical Modelling of Physical Systems* presents a generalized model decomposition scheme (high-dimensional model replacement, HDMR) whose application to SA appears promising. **Chapter 10** gives a review of *Bayesian Sensitivity Analysis*. **Chapter 11** describes how *Graphical Methods*, such as cob-webs, can be employed.

Part III: Applications

The third section is more practically oriented. If the reader has a specific problem at hand, this section can give useful hints about what to use and how to implement the analysis. **Chapter 12** is an introduction to the Applications Section. It describes the modelling process from formulation of objectives, model building, development, and use. The following chapters give examples of specific uses. **Chapter 13** shows how to use SA to treats model and scenario uncertainty. **Chapter 14** is an *application of UA, SA to modelling time series*, and also includes a relevant application to model uncertainty. **Chapter 15** is an example of a dataless precalibration analysis in solid state physics (how SA can be used prior to calibration). **Chapter 16** is an application of the FORM method to a hydrogeological problem. **Chapter 17** offers one-at-a-time (OAT) and global analyses for interpreting air quality model predictions, with an interesting study of the interplay between system uncertainty and regulatory targets. **Chapter 18** is a comparison of *different SA methods on a chemical reaction model.* **Chapter 19** is an *application to logistic equations and population dynamics*; both **Chapters 18** and **19** show applications of SA in the model building process. **Chapter 20** deals with SA used to assess the quality of models used for environmental policy, and is interesting in its way of discriminating among uncertainties in the data versus uncertainty in the indicator building process.

Part IV: Conclusions

The final section rounds off the book by considering the epistemological aspects of SA. **Chapter 21**: *Assuring the Quality of Models Designed for Predictive Tasks* is a discussion of the scientific method in the context of model validation, including a description of generalized sensitivity analysis (GSA). The final **Chapter 22** is dedicated to the *Fortune and Future of SA*, and tries to probe some of the issues touched in the present introduction, such as model transparency and model relevance. It also offers an epistemic perspective of SA as an ingredient of the modelling process.

The Appendix

This aims to give an overview of some specialized SA software. It does not provide in-depth information about the software, but gives a brief description and a guide to where and how to obtain it.

2

Hitchhiker's Guide to Sensitivity Analysis

Francesca Campolongo, Andrea Saltelli, Tine Sørensen and Stefano Tarantola

European Commission, Joint Research Centre, Ispra, Italy

2.1 INTRODUCTION

2.1.1 Purpose

This chapter gives the reader a first overview of the various methods that are currently used in sensitivity analysis (SA). At the same time, it tries to suggest which methods should be used in which settings or applications, based on the pooled experience of the various authors of this volume. The reader with an application at hand will hence be able to focus his/her reading on the chapters more relevant to his/her task. A description of the tasks and methods is given first. Suggestions on different applications for different methods are given in the second part of the chapter.

2.1.2 Notation and Terminology

Let us assume that we are studying a system involving a vector of k input factors $\mathbf{x} = (x_1, x_2, \ldots, x_k)$ and an output variable y. In practice, the input factors are affected by several kinds of heterogeneous uncertainties that reflect our imperfect knowledge of the system. Hence it is convenient for the purpose of sensitivity analysis to treat them as random variables with assumed probability distributions. The vector $\mathbf{x}$ can be seen as a realization of a random vector $\mathbf{X}$, characterized by a joint probability density function (p.d.f.) $p(\mathbf{X}) = p(X_1, X_2, \ldots, X_k)$, assumed to be known (we use the convention that capital letters denote random variables and lower case letters correspond to their realizations). The output y can then also be seen as a realization of a random variable Y, and the relationship between the input factors and the output under study can be represented by a

Sensitivity Analysis. Edited by A. Saltelli *et al.*

mathematical construction $f(\cdot)$ such that

$$Y = f(X_1, X_2, \ldots, X_k) = f(\mathbf{X}). \tag{2.1}$$

Hence, the output has its own p.d.f., which can be estimated in different ways. Throughout this chapter, we shall assume that a single output is observed, but multiple outputs could be encountered in practical problems.

In our SA context, *factors* include both variables and parameters, where variables are measured quantities (e.g. production of CO_2) or quantities varied by the experimentalist (e.g. in mechanical physics, the force imposed on the system), and parameters are quantities internal to the model (e.g. quantities to be estimated in a calibration, such as a reaction rate in a chemical system).

Summary statistics of the output can be computed from the γth moment of Y, given by

$$\langle Y^{(\gamma)} \rangle = \int_{\Omega^k} f^{\gamma}(\mathbf{X}) p(\mathbf{X}) d\mathbf{X}, \tag{2.2}$$

where Ω^k is the k-dimensional space of input factors. This integral gives the starting point to compute sensitivity measures of various kinds.

In the following, when describing the various SA methodologies, we shall use the term *computational cost* as an indicator of efficiency. The computational cost will be given in terms of number of *model evaluations* needed for the analysis.

2.1.3 Categorization of Sensitivity Analysis

Different sensitivity analysis strategies may be applied, depending on the setting. Here we identify three main settings:

- *factor screening*, where the task is to identify influential factors in a system with many factors;
- *local SA*, where the emphasis is on the local (point) impact of the factors on the model; local SA involves partial derivatives, and is analytical;
- *global SA*, where the emphasis is on apportioning the output uncertainty to the uncertainty in the input factors; global SA typically takes a sampling approach, and the uncertainty range given in the input reflects our imperfect knowledge of the model factors.

Factor screening may be useful as a first step when dealing with a model containing a large number of factors. Often, only a few of the input parameters and groupings of parameters have a significant effect on the model output. Screening experiments can be used to identify, with low computational effort, the subset of factors that controls most of the output variability. The most common screening designs are described in Chapter 4.

Local SA (see Chapter 5) is usually carried out by computing partial derivatives of the output functions with respect to the input factors. In order to compute the derivative numerically, the input factors are varied within a small interval around a nominal value. The interval is usually the same for all of the factors, and is not related to our degree of knowledge of the factors.

The local sensitivity approach is practicable when the variation around the midpoint of the input factors is small; in general, the input–output relationship is assumed to be linear. Furthermore, if the range of variation is the same for all the factors (e.g. $\pm 10\%$ of the nominal value) then the relative importance of the input factors can be determined.

One shortcoming of the linear sensitivity approach is that it is not possible to assess effectively the impact of possible differences in the scale of variation of the input factors (unless the model itself is linear). When significant uncertainty exists in the input factors, the linear sensitivities alone are not likely to provide a reliable estimator of the output uncertainty in the model. When the model is nonlinear and various input variables are affected by uncertainties of different orders of magnitude, a global sensitivity method should be used.

In *global SA*, the aim is to apportion the uncertainty in the output variable to the uncertainty in each input factor. Distributions for each factor provide the input for the analysis. These distributions are valuable, since they represent our knowledge (or lack of it) with respect to the model and its parameterization. A SA experiment is usually considered to be global when (a) all the parameters are varied simultaneously and (b) the sensitivity is measured over the entire range of each input parameter (see Chapter 1).

In the following subsections, we shall give some essential elements of the various methods. Techniques have been further subdivided among the following classes:

- screening designs;
- differential analysis (mostly local);
- Monte Carlo analysis (global);
- response surface methodology (global);
- Fourier amplitude sensitivity test (FAST, global).

Given the large number of settings where SA is applied and the large number of methodological approaches, the classification presented here is only speculative, and its only purpose is that of giving an ordered presentation of methods. Alternative taxonomies can be thought of, such as one based on methods being either analytic or Monte Carlo to start with, and local or global as a further subdivision.

2.2 SCREENING DESIGNS

Screening designs (see Chapter 4) are preliminary numerical experiments whose purpose is to isolate the most important factors from amongst a large number that may affect a particular model response.

Typical screening designs are *one-at-a-time* (OAT) *experiments* in which the impact of changing the values of each factor is evaluated in turn (Daniel, 1958, 1973). The experiment, which uses the 'standard' values, is defined as the *control experiment*. For each factor, two extreme values are then selected by the analyst (normally the control values are 'midway' between the two extremes). The magnitudes of residuals, defined as the difference between the perturbed experimental results and the control, are compared in order to evaluate factors to which the model is significantly sensitive.

One limitation of the OAT experiments is that they allow only the evaluation of the main effects (the effects of the input factors without including their mutual interactions, i.e. the first-order effects). The use of *factorial experimentation* (Box *et al.*, 1978b) allows not only for the evaluation of main effects but also for interactions. In a factorial experiment approach, all factors are perturbed simultaneously to one of their possible values, called 'levels', and all the possible combinations of values are covered. The computational cost of the numerical experiment is then l^k, where k is the number of factors and l is the number of levels.

When the cost of a full factorial experiment becomes too high, because the number of factors is too large or the model evaluation is time-consuming, a useful alternative is given by the *fractional factorial* (FF) experiment (Box *et al.*, 1978b). In a FF experiment, the number of runs required is reduced, since only some of the possible interactions among factors are examined. The desired main effects and the interaction effects are calculated in exactly the same way as in the full experiment. The loss of information about some of the higher-order interactions is balanced out by the reduced cost of the experiment.

Many of the screening methods described in the literature rely on strict assumptions about the nature or absence of interactions between factors. One exception is that of Cotter (1979). Cotter's method does not require prior assumptions about interactions, and its results are hence easier to interpret. This design is called the *systematic fractional replicate design*, and it requires $2k + 2$ trials for a k-factor experiment. Although computationally efficient, this method has the major disadvantage of a lack of precision.

In models with several uncertain factors, a screening method requiring fewer simulations than factors might be needed. One such method is the *iterated fractional factorial design* (IFFD). IFFD prescribes a set of statistical experiments that can be used to estimate the additive effects (i.e. the main effect), quadratic effects, and two-way interactions of influential factors. The number of simulations required can be much smaller than the total number of factors examined. IFFD works better when the effects of a few highly influential factors dominate the model output.

A different approach to identify the important inputs in a computational model whose evaluations are expensive, or where the number of input factors is large, was proposed by Morris (1991). This approach is intended specifically for those cases in which the number of experimental runs has to be proportional to the number of input factors, and there are no simplifying assumptions about the form of the model. The Morris method is described in Chapter 4; examples of its application can be found in Chapters 18 and 19.

2.3 DIFFERENTIAL ANALYSIS

Differential analysis (see Chapter 5) is often used in conjunction with the kinetics of a spatially homogeneous reaction system. It can be modelled by an initial-value ODE problem of the form

$$\frac{dc}{dt} = f(\mathbf{c}, \mathbf{k}), \qquad \mathbf{c}(0) = \mathbf{c}^0, \tag{2.3}$$

where $\mathbf{c}$ is the n-vector of concentrations, $\mathbf{k}$ is the k-vector of system parameters, and t represents time. The parameters may include rate coefficients. Initial conditions are not considered in the vector $\mathbf{k}$.

A Taylor series is used to approximate the model (2.3). Once constructed, this series is used as a surrogate for the original model in analytical uncertainty and sensitivity studies.

A differential analysis involves four steps. In the first step, base values and ranges are selected for each input factor. In the second step, a Taylor series approximation to the output $\mathbf{c}$ is developed around the base values of the inputs. In the third step, variance propagation techniques are used to estimate the uncertainty in $\mathbf{c}$ in terms of its expected value and its variance. In the final step, the Taylor series approximations are used to estimate the importance of individual input factors.

One problem arising in a differential analysis is the determination of an appropriate order for the Taylor series approximation. Estimates for expected value and variance, in the third step, vary according to the order of approximation.

In the final step, there are different ways of measuring factor importance. The expressions giving expected value and variance of the output $\mathbf{c}$ may provide estimates for the fractional contribution to the variance. Normalized first–order partial derivatives, in the Taylor series approximation, can measure the effect on the response that results from perturbing k_i by a fixed fraction of its base value (for a short review, see Helton, 1993).

The greatest effort in a differential analysis is the determination of the partial derivatives in the Taylor series approximation. Usually only the first-order partial derivatives, called the *first-order local sensitivity coefficients*, are computed and studied. They constitute the *sensitivity matrix* **S**, which represents a linear approximation of the dependence of the solutions on parameter changes.

The matrix **S** can be obtained by differentiation if the analytical solution of the ODEs (2.3) is known. Unfortunately this is seldom (if ever) the case. Such simple systems are rarely met, and numerical methods have to be applied.

A number of specialized techniques have been developed to facilitate the calculation of these derivatives. These include:

- the *brute force method*, or indirect method, which uses finite-difference approximations;
- the *method of Miller and Frenklach* (1983), based on approximations by empirical models of the solution of system (2.3) in a parameter region at time t;
- the so-called *direct methods*, based on the equation

$$\frac{d}{dt}\frac{\partial \mathbf{c}}{\partial k_i} = \mathbf{J}(t)\frac{\partial \mathbf{c}}{\partial k_i} + \frac{\partial \mathbf{f}(t)}{\partial k_i}, \tag{2.4}$$

  where $\mathbf{J}(t) = \partial \mathbf{f}/\partial \mathbf{c}$, and the initial condition for $\partial \mathbf{c}/\partial k_i$ is a zero vector, which is derived by differentiation of (2.3) with respect to k_j;
- the *Green function method*, also called the variational method;
- the *polynomial approximation method*, elaborated by Hwang (1985), which transforms the sensitivity differential equations (2.4) into a set of algebraic ones.

As discussed in Chapter 1, the characteristics of each method may be different for different problems. Thus, the choice of which method to adopt depends on the problem with which an investigator is dealing.

If the system under consideration is not a spatially homogeneous constant-parameter system (e.g. the one given in (2.3)), but is a system in which the parameters are also functions of time and space, other SA methods are more appropriate. In this case, the appropriate sensitivity analysis is based on their perturbation by another function using the principles of nonlinear functional analysis. Dickinson and Gelinas (1976) were the first to tackle the problem of parameter functions, and introduced a sensitivity measure depending on the perturbing function (see Turanyi, 1990a). Since this measure could only be calculated by a procedure similar to the brute force method, an unambiguous method that could be calculated by more sophisticated techniques was sought. The sensitivity measure meeting those requirements was named *sensitivity density* (Demiralp and Rabitz, 1981). A description of the *sensitivity density* measures and of their properties can be found in Chapter 5.

A generalized sensitivity density has also been introduced as an appropriate sensitivity measure for an inhomogeneous reaction system. Functional derivatives for the study of reaction–diffusion systems were first computed by Koda (1982) and Rabitz and co-workers (Demiralp and Rabitz, 1981; see Chapter 5 for further references). These measures indicate the effect on the concentration of species A at a given time t and location $\mathbf{x}$ of a variation of another species B at different time t' and location x', e.g.

$$\frac{\partial[\mathrm{A}]_{t,\mathbf{x}}}{\partial[\mathrm{B}]_{t',\mathbf{x}'}}, \qquad t' < t. \tag{2.5}$$

The sensitivity matrix $\mathbf{S}$ containing the local sensitivities $\partial\mathbf{c}/\partial k_i$ provides information on the importance of the parameters. In particular, the local sensitivities provide information on the effect of small changes of parameters from the nominal values. The order of importance that can be deduced from local sensitivities is called order of *tuning* importance (Turanyi, 1990a).

If a parameter is eliminated from the model (i.e. its value is set to zero), it may or may not cause large changes in the model results. In chemical models, a frequent observation is that deleting a parameter with negligible tuning importance sometimes causes large changes in the model results, while elimination of a high-tuning-importance parameter causes much smaller changes. Therefore a different order of importance can be created: the order of importance that is determined by setting the parameter to zero and rerunning the model. This is called the order of *reduction* importance (Turanyi, 1990a).

In reality, the influence (on the output) due to individual parameters is usually not as relevant as the influence due to a group of parameters. These parameter groups cause functional connections between the sensitivity coefficients, and they can be identified by inspection and comparison of the elements of the sensitivity matrix. However, more sophisticated methods are available to determine the influence of parameter groups. For instance, *principal component analysis* (PCA) (Vajda *et al.*, 1985) can be used to extract meaningful kinetic information from linear sensitivity coefficients computed for several species of a reaction system at several time points. PCA can be very useful in mechanism reduction. Sometimes the elimination of the reactions one by one may cause significant changes in the solution, while the elimination of reaction pairs has no significant consequences. PCA can be used to identify groups of reactions that can be eliminated, this method having the ability to reveal those parts of the mechanism that consist of strongly interacting reactions and to indicate their importance within the system.

2.4 MONTE CARLO ANALYSIS

Monte Carlo (MC) analysis (see Chapter 6) is based on performing multiple evaluations with randomly selected model input, and then using the results of these evaluations to determine both uncertainty in model predictions and apportioning to the input factors their contribution to this uncertainty. In general, the analysis involves five steps: the selection of ranges and distributions for each X_i; generation of a sample from the ranges and distributions specified in the first step; evaluation of the model for each element of the sample; uncertainty analysis; and sensitivity analysis. A special case of MC analysis involves using quasi-random numbers, where the sample generation in the second step is not random at all.

2.4.1 Selection of Probability Distribution Functions

In the first step, in the absence of information about ranges and distributions for the input variables, a crude characterization may be adequate, especially if the analysis is primarily exploratory. Uniform or log-uniform distributions may be assumed and physical plausibility arguments might be used to establish the ranges. Sensitivity analysis results generally depend more on the selected ranges than on the assigned distributions. However, distributional assumptions can have an impact on the estimated distributions for output variables, so care must be used in developing distributions for the input variables when particular interest is placed on the estimation of distributions (or quantiles) of the output variables.

If appropriate data are available, it may be possible to estimate distributions and distribution parameters for the data with formal statistical procedures. See Chatfield (1993) for a critical review of the process of model selection and calibration.

Unfortunately, most parameters used in modelling are not amenable to statistical estimation for several reasons. One is a problem of scale (parameters have been estimated over a time scale or on a physical scale that is much shorter than the one needed by the model). Some variables may not be observable. Some factors represent the occurrences of rare events (i.e. scenario probabilities). Finally, often models use lumped parameters, which have little resemblance to reality. There is a growing interest in the use of expert elicitation techniques to tackle the problem of defining the input distributions (Cooke, 1991).

2.4.2 Sampling

The second step in the MC analysis involves the selection of a sample from the distributions developed in the first step. Various sampling procedures are used in MC studies (Chapter 6). Among these are:

- random sampling;
- stratified sampling (including Latin hypercube sampling);
- quasi-random sampling.

McKay *et al.* (1979) have reviewed the first two methods. A description of the third method can be found in Sobol' (1990b), which is also a good 'primer' in MC simulation.

When two or more variables are correlated, it is necessary to introduce the appropriate correlation structure into the sample. A method for correlated variables is also presented in Chapter 6.

Random sampling

In *random sampling*, a sample $(\mathbf{x}_1, \mathbf{x}_2, \ldots, \mathbf{x}_N)$ of the desired dimension N is generated from the joint distribution of the input variables or, when these are independent, from their marginal distributions. Random sampling is also referred to as pseudo-random, since the random numbers are machine-generated by a deterministic process and not random *stricto sensu*. From the statistical point of view, random sampling has advantages, since it produces unbiased estimates of the mean and the variance of Y.

Stratified sampling

The purpose of *stratified sampling* is to achieve a better coverage of the sample space of the input factors. Let the sample space S of the input vector $\mathbf{X}$ be partitioned into I disjoint strata $S_1, \ldots, S_I$. Represent the size of each S_i, $i = 1, \ldots, I$, as the probability that $\mathbf{X}$ is in S_i, i.e. as $p_i = P(\mathbf{X} \in S_i)$. Obtain a random sample $\mathbf{x}_h$, $h = 1, \ldots, n_i$, from S_i, where $\sum_{i=1}^{I} n_i = N$. In particular, when $I = 1$, the result is a random sample over the entire sample space.

Latin hypercube sampling

Latin hypercube sampling (LHS) may be considered a particular case of stratified sampling. The range of each input factor X_j, $j = 1, \ldots, k$, is divided into N intervals of equal marginal probability $1/N$, and one observation of each input factor is made in each interval. Thus, there are N non-overlapping realizations for each of the k input factors. One of the realizations on X_1 is randomly selected (each observation is equally likely to be selected), matched with a randomly selected realization of X_2, and so on up till X_k. These collectively constitute the first sample, $\mathbf{x}_1$. One of the remaining realizations on X_1 is then matched at random with one of the remaining observations on X_2, and so on, to get $\mathbf{x}_2$. A similar procedure is followed for $\mathbf{x}_3, \ldots, \mathbf{x}_N$, which exhausts the observations and results in a Latin hypercube sample. The method has the advantage of ensuring that the input factor has all portions of its distribution represented by input values.

LHS performs better than the previous two sampling strategies when the output is dominated by only a few components of the input factors. The method ensures that each of these components is represented in a fully stratified manner, no matter which component might turn out to be important. McKay *et al.* (1979) proved that when the output y is a monotonic function of each of its arguments, LHS is better than random sampling for estimating the mean and the population distribution function. Stein (1987) proved that, asymptotically, LHS is better than random sampling in that it provides an estimator (of the expectation of the output function) with lower variance. In particular, Stein showed that, the closer the output function is to being additive in its input variables, the greater is the reduction in variance.

A method for producing LHS samples when the components of the input variables are statistically dependent is also given in Stein (1987).

Although, in some cases, LHS was shown to be preferable to simple random sampling, there are still examples in which, dealing with non-additive, non-monotonic functions, the performance of LHS is equivalent to or worse than the performance of simple random sampling (see below). Hence, the superiority of this method is still under discussion.

Quasi-random sampling

Several *quasi-random sequences* are available, and are reviewed in Bratley and Fox (1988). We limit our discussion to Sobol' LP_τ sequences (Sobol', 1967, 1976, 1990a), since in our experience they have yielded good results in sensitivity analysis studies. The algorithm proposed by Sobol' generates quasi-random numbers that are characterized by an enhanced convergence rate. The crude Monte Carlo rate $N^{1/2}$ of stochastic convergence is in some cases replaced by a convergence rate of $N^{-1+\varepsilon}$, with an arbitrary $\varepsilon > 0$.

A description of how these numbers are generated is given in Bratley and Fox (1988). A comparative study of the performance of quasi-random numbers and crude Monte Carlo and LHS sampling is made by Homma and Saltelli (1995). On the non-monotonic test function employed by those authors, the advantage of using the Sobol' method is evident. The performances of crude Monte Carlo and LHS were very similar (and inferior to the performance of the quasi-random sampling). This is not surprising, since, as pointed out in the previous section, the advantage of using LHS rather than simple Monte Carlo has been proved only for monotonic functions (see also Chapter 8).

Correlation control

Control of correlations between variables within a sample is extremely important and difficult, because the imposed correlations have to be consistent with the proposed variable distributions. Iman and Conover (1982) proposed a method of controlling the correlation structure in random and Latin hypercube samples. The method is described in Chapter 6.

2.4.3 Evaluation of the Model

The third step in a Monte Carlo analysis is the evaluation of the model for each of the sample elements. This step is conceptually simple: each element is supplied to the model as input, creating a sequence of results of the form $y_i = f(\mathbf{x}_i)$, $i = 1, \ldots, N$, to be used in the uncertainty and sensitivity analysis.

2.4.4 Uncertainty Analysis

The fourth step, *uncertainty analysis*, is straightforward. The expected value and variance for the output variable y are estimated by

$$\hat{E}(Y) = \frac{1}{N}\sum_{i=1}^{N} y_i, \tag{2.6}$$

$$\hat{V}(Y) = \frac{1}{N-1}\sum_{i=1}^{N} [y_i - \hat{E}(Y)]^2 \tag{2.7}$$

if random sampling or LHS is used, while for stratified sampling the factors $1/N$ and $1/(N-1)$ must be replaced by weights w_i, $i = 1, \ldots, N$, to reflect the probability and number of observations associated with the individual stratum. McKay *et al.* (1979) showed that under various conditions, LHS results in more stable estimates of the mean than random sampling. Both estimates are unbiased for random sampling, while for LHS the estimated mean is unbiased but the estimated variance is known to be biased.

2.4.5 Sensitivity Analysis

The final step is sensitivity analysis, to apportion the variation in the output to the different sources of variation in the system. Many techniques can be used, yielding different measures of sensitivity, and details are given in the following subsections.

Scatterplots and correlation coefficients

Scatterplots are one of the most intuitive and straightforward techniques for sensitivity analysis. These are plots of the output variable y_i against x_{ij} for each input factor x_j, where $j = 1, \ldots, k$, and $i = 1, \ldots, N$ are the model simulations. Scatterplots may reveal relationships between model inputs and model predictions, such as nonlinear relationships, and thresholds (Helton, 1993). They can be considered global measures of importance, and are model-independent. One disadvantage of the method is the need to generate and examine a large number of plots, at least one per input factor, possibly multiplied by the number of time points if the output is time-dependent. Furthermore, scatterplots offer a qualitative measure of sensitivity, since the relative importance of variables can be estimated but not quantified (see also Chapter 11 on graphical methods).

Another simple measure of sensitivity is given by the *Pearson product moment correlation coefficient (PEAR)*, which is the usual linear correlation coefficient computed on the points

(x_{ij}, y_i), $i = 1, \ldots, N$. The correlation r_{x_jy} between the input variable X_j and the output Y is defined by

$$r_{x_jy} = \frac{\sum_{i=1}^{N}(x_{ij} - \bar{x}_j)(y_i - \bar{y})}{\left[\sum_{i=1}^{N}(x_{ij} - \bar{x}_j)^2\right]^{1/2}\left[\sum_{i=1}^{N}(y_i - \bar{y})^2\right]^{1/2}}, \tag{2.8}$$

where

$$\bar{y} = \sum_i \frac{y_i}{N}, \qquad \bar{x}_j = \sum_i \frac{x_{ij}}{N}.$$

The coefficient r_{x_jy} provides a measure of the linear relationship between X_j and Y. For nonlinear models, the *Spearman coefficient (SPEA)* is preferred as a measure of correlation. It is computed by using the ranks of both y and x_j instead of the raw values (Conover, 1980). The basic assumptions underlying the Spearman coefficient are:

(a) both the x_{ij} and the y_i are random samples from their respective populations;
(b) the measurement scale of both variables is at least ordinal.

Regression analysis

More quantitative measures of sensitivity are based on *regression analysis*. A multivariate sample of the input $\mathbf{x}$ is generated by some sampling strategy (dimension $N \times k$), and the corresponding sequence of N output values is computed using the model under analysis. If a linear regression model is being sought, it takes the form

$$y_i = b_0 + \sum_j b_j x_{ij} + \varepsilon_i, \qquad j = 1, 2, \ldots, k, \tag{2.9}$$

where the b_j are the regression coefficients that must be determined and ε_i is the error (residual) due to the approximation. One common way of determining the coefficients b_j is to use least-squares analysis (Draper and Smith, 1981).

Once the b_j have been computed, they can be used to indicate the importance of individual input variables x_j with respect to the uncertainty in the output y. In fact, assuming that $\mathbf{b}$ has been computed, the regression model can be rewritten as

$$\frac{y - \bar{y}}{\hat{s}} = \sum_j \frac{b_j \hat{s}_j}{\hat{s}} \frac{x_j - \bar{x}_j}{\hat{s}_j} \tag{2.10}$$

where

$$\bar{y} = \sum_i \frac{y_i}{N}, \qquad \bar{x}_j = \sum_i \frac{x_{ij}}{N}, \tag{2.11}$$

$$\hat{s} = \left[\sum_i \frac{(y_i - \bar{y})^2}{N-1}\right]^{1/2}, \qquad \hat{s}_j = \left[\sum_i \frac{(x_{ij} - \bar{x}_j)^2}{N-1}\right]^{1/2}. \tag{2.12}$$

The coefficients $b_j\hat{s}_j/\hat{s}$ are called *standardized regression coefficients* (SRCs). These can be used for sensitivity analysis (when the x_j are independent), since they quantify the effect of varying each input variable away from its mean by a fixed fraction of its variance while maintaining all other variables at their expected values.

When using the SRCs, it is also important to consider the model coefficient of determination,

$$R_y^2 = \frac{\sum_{i=1}^{N} (\hat{y}_i - \bar{y})^2}{\sum_{i=1}^{N} (y_i - \bar{y})^2}, \tag{2.13}$$

where $\hat{y}_i$ denotes the estimate of y_i obtained from the regression model. R_y^2 provides a measure of how well the linear regression model can reproduce the actual output y. R_y^2 represents the fraction of the variance of the output explained by the regression model. The closer R_y^2 is to unity, the better is the model performance. The validity of the SRCs as measures of sensitivity is conditional on the degree to which the regression model fits the data, i.e. to R_y^2.

To determine the adequacy of a regression model, the *predicted error sum of squares* (*PRESS*) can be used. PRESS is constructed as follows. The ith observation is removed from the N available observations, and a regression model is constructed on the $N - 1$ observations. For the removed observation y_i, the value $\hat{y}_i(k)$ is estimated by using the new regression model. The PRESS statistic (Allen, 1971) for the regression model with k variables is computed as

$$\text{PRESS}_k = \sum_{i=1}^{N} [y_i - \hat{y}_i(k)]^2. \tag{2.14}$$

The preferred regression model is the one with the smallest PRESS value. The use of PRESS is essential when regression models are used for sensitivity analysis purposes, given that such analysis often involves many input variables affected by considerable uncertainty.

A non-parametric regression technique that has been applied for sensitivity analysis is the projection pursuit method. A description of this method, which involves regressing Y on arbitrary functions of the factors (rather than on factors directly), is given in Chapter 13.

Correlation measures

Another interesting measure of importance is given by *partial correlation coefficients* (*PCCs*). These coefficients are based on the concepts of correlation and partial correlation. The partial correlation coefficient between the output variable Y and the input variable X_j is obtained from the use of a sequence of regression models. First the following two models are constructed:

$$\hat{Y} = b_0 + \sum_{h \neq j} b_h x_h, \qquad \hat{X}_j = c_0 + \sum_{h \neq j} c_h x_h. \tag{2.15}$$

Then the results of these two regressions are used to define the new variables $Y - \hat{Y}$ and $X_j - \hat{X}_j$. The partial correlation coefficient between Y and X_j is defined as the correlation coefficient between $Y - \hat{Y}$ and $X_j - \hat{X}_j$ (Helton, 1993). Thus, the PCCs provide a measure of the strength of the linear relationship between two variables after a correction has been made for the linear effects of the other variables in the analysis. In other words, the PCC gives the strength of the correlation between Y and a given input X_j after adjustment for any effect due to correlation between X_j and any of the X_i, $i \neq j$.

Since SRCs measure the effect on the output variable that results from perturbing an input variable by a fixed fraction of its standard deviation, PCCs and SRCs provide related but not identical measures of variable importance. In particular:

- SRCs are sensitive to all input distributions; the SRCs can provide a decomposition of the output variance according to the input factors;
- PCCs provide a measure of variable importance that tends to exclude the effects of other variables.

However, for the case in which the input variables are uncorrelated, the order of variable importance based either on SRCs or PCCs (in their absolute values) is exactly the same.

Rank transformation

Since regression analysis is based on the linear relationships between the explanatory variables and the dependent variable, it often performs poorly when this relationship is nonlinear, yielding a low value of the R_y^2 coefficient computed on the raw values. To avoid the problem of nonlinearity, rank transformations are frequently employed. Ranks can cope with nonlinear (albeit monotonic) relationships between the input–output distributions, allowing the use of linear regression techniques. Rank-transformed statistics are more robust, and provide a useful solution in the presence of long-tailed input and output distributions. However, care must be employed when interpreting the results of analyses based on rank transformations, since any conclusion drawn using ranks does not translate easily to the original model.

The rank transform is a simple procedure, which involves replacing the data with their corresponding ranks. A vector of N output values $\mathbf{y} = (y_1, \ldots, y_N)$ is generated by repeatedly evaluating the model for a set of N sampled vectors $(x_{11}, \ldots, x_{1k}), \ldots, (x_{N1}, \ldots, x_{Nk})$, where k is the number of variables. The observations are then replaced by their corresponding ranks 1 (highest value) to N (lowest), where $R(y_i)$ is the rank assigned to the ith value of y. The model can also be rank-transformed for the independent variables; i.e., for fixed i, each of the independent variables x_{ij}, $j = 1, \ldots, k$, is replaced with its corresponding ranks 1 to N. The usual least-squares regression analysis is then performed on the regression equation expressing $R(Y_i)$ in terms of x_{ij} or $R(x_{ij})$, respectively. If R_y^{*2}, the model coefficient of determination based on the rank-transformed model, is higher than R_y^2, the one based on the raw data, then the *standardized rank regression coefficients* (*SRRCs*) or the *partial rank correlation coefficients* (*PRCCs*) can be used for sensitivity analysis instead of the SRCs or the PCCs, respectively. The ranked variables are often used, because generally the R_y^{*2} values are higher than the R_y^2 values, especially for nonlinear models. The difference between R_y^2 and R_y^{*2} is a useful indicator of the nonlinearity of the model.

The performance of the SRRCs and the PRCCs is shown to be very satisfactory when the model output varies monotonically with each independent variable. However, in the presence of strong non-monotonicity, the results may become dubious (low R_y^{*2}). Another limitation in the use of ranks is that the transformation alters the model being studied, so that the resulting sensitivity measures give us information on a different model. The new model is not only more linear than the original one but also more additive. The relative weight of the first-order terms is increased at the expense of the interactions and higher-order interactions. As a result, the influence of those factors that influence the output by way of interactions may be overlooked in an analysis based on the ranks. This difficulty increases with the dimensionality of the problem, and may lead to the failure of a rank-based sensitivity analysis (see Saltelli *et al.*, 1993; Saltelli and Sobol', 1995).

Two-sample tests

Two-sample test statistics have been employed as measures of sensitivity. In particular, the *Smirnov test*, the *Cramér–von Mises test*, the *Mann–Whitney test*, and the *two-sample t-test statistic* are used.

The application of such tests to sensitivity analysis comes from the idea of partitioning the sample for the factor X_j under consideration into two subsamples according to the quantiles of the output (Y) distribution. If the distribution of X_j in the two subsamples can be shown to be dissimilar then the factor under consideration is considered influential. For instance, the values of the x_{ij} corresponding to the output y_j above the 90th quantile of the $F(Y)$ distribution may constitute one subsample and all the other x_{ij} the other subsample. If the distribution of the two subsamples can be shown to be different based on a two-sample test then X_j is said to be an influential variable.

The tests are described in Conover (1980), and a discussion of their performances in SA is given in Saltelli and Marivoet (1990). In spite of its ingenuity, this application of two-sample tests to SA suffers from a number of shortcomings. First of all, the estimates tend to be poor as far as robustness is concerned (their predictions vary considerably from one sample to the next). They tend to be qualitative. Finally, the results depend upon the choice of the quantile for splitting the sample. We would not recommend such methods unless the application really justified it, e.g. when a given fixed portion of the tail of the output distribution must necessarily be characterized; this could apply in reliability analysis, but even there we would recommend a different method (see Chapter 16).

Stepwise regression analysis

Stepwise regression analysis (Helton, 1993) provides an alternative to constructing a regression model containing all the input variables. A sequence of regression models is constructed using the following steps:

(i) the first regression model contains the most influential (on the output variable) input variable;
(ii) the second model introduces the next most influential input variables (given the one from the previous step);
(iii) the third model introduces a third variable (given the variables from steps (i) and (ii));

and so on, until the point is reached at which subsequent models are unable to increase, meaningfully, the amount of variation in the output variable that can be accounted for.

The model coefficients of determination R_y^2 computed at successive steps of the analysis provide a measure of variable importance by indicating how much of the variation in the dependent variable can be accounted for by all variables selected at each step. Also, the individual SRCs in the individual regression models provide an indication of variable importance. When the input variables are uncorrelated, the size of the coefficient of determination R_y^2 attributable to the individual variables, the absolute values of the SRCs, and the absolute values of the PCCs, are identical. When variables are correlated, care must be used in the interpretation of the results of a regression analysis, since the regression coefficients can change in ways that are basically unrelated to the importance of the individual variables as correlated variables are added to and deleted from the regression model, i.e. the results are conditional on what is already in the model.

A delicate part of this technique consists in deciding when to stop the construction process of the consecutive regression models. Calculation of the regression coefficients associated with the input variables, in order to check whether they are significantly different from zero or not, can be a useful method to adopt. *F*-statistic values are conventionally used to control which variables should be included in the model (see p. 93 of Draper and Smith, 1981). More details about the stepwise regression analysis can be found in Chapter 6 and in Draper and Smith (1981).

Other Monte Carlo-based methods for sensitivity analysis

Fedra *et al.* (1981) present Monte Carlo filtering as an alternative to the concept of local calibration. The analyst should refrain from searching for the optimal solution, but rest with the plausible ones by a process of mapping factors values into the output space, then censoring the input corresponding to unacceptable *Y*s. An evolution of Monte Carlo filtering that has proven useful in hydrology is the Bayesian generalized likelihood estimation (GLUE), due to Beven and Binley (1992); see also Freer *et al.* (1996). In GLUE, many (Monte Carlo-generated) different combinations of factor value and alternative model structures can be compared with the evidence, e.g. a historical record of precipitations and run off for a given basin, and this is used to evaluate the likelihood of any given combination of factor value/model structure. Once a forecast is needed for a new set of forcing data (precipitations), this is made by using all of the previously explored Monte Carlo combinations of hydrological parameters and assumptions, each weighted by its likelihood.

Spear and Hornberger (1980) (see also Hornberger and Spear, 1981) present generalized sensitivity analysis, i.e. MC filtering followed by a partition of the value of the factor into two subsets (the one corresponding to acceptable *Y*s and the one corresponding to unacceptable *Y*s). The Kolmogorov–Smirnov two-sample test is used to identify influential factors. An extension of this approach linked to the use of GLUE is in Freer *et al.* (1996).

Chang and Delleur (1992) use the approach of Hornberger and Spear (1981) (renamed for the occasion as regionalized sensitivity analysis), and add multidimensional calibration to it.

Young *et al.* (1996) use generalized sensitivity analysis as input to data-based mechanistic modelling (DBM).

2.5 MEASURES OF IMPORTANCE

The methods described above are all Monte Carlo-based and involve regression or correlation analysis. Another global sensitivity analysis method, which can be used in conjunction with Monte Carlo, is the *measure of importance*. This is described in the literature under different forms, and is based on the estimation of the following quantity:

$$\frac{\mathrm{Var}_{X_j}[E(Y|X_j = x_j)]}{\mathrm{Var}(Y)}, \tag{2.16}$$

where $E(Y|X_j = x_j)$ denotes the expectation of Y conditional on a fixed value of X_j, and Var_{X_j} stands for variance over all the possible values of X_j. McKay (1995) called the numerator of (2.16) the *variance correlation expectation* (*VCE*) and the ratio the *correlation ratio.*

Iman and Hora (1990) observed that, although mathematically correct, the importance measure lacks robustness, and can be highly influenced by outliers associated with long-tailed input distributions. In Hora and Iman (1986), the measure of importance of a variable X_j is defined as

$$I_j = \sqrt{\mathrm{Var}(Y) - E[\mathrm{Var}(Y|X_j = x_j)]}. \tag{2.17}$$

The quantity $\mathrm{Var}(Y) - E[\mathrm{Var}(Y|X_j = x_j)]$ in (2.17) is identically the *VCE* as defined earlier. This measure was found to be non-robust and highly influenced by outliers.

Iman and Hora (1990) suggested an alternative measure, namely

$$\frac{\mathrm{Var}_{X_j}[E(\log Y|X_j = x_j)]}{\mathrm{Var}(\log Y)}, \tag{2.18}$$

where $E(\log Y|X_j = x_j)$ is estimated using linear regression. Similarly, the rank-based importance measure can be used. These measures have the advantage of robustness, but the conclusions based on $\log Y$ or $R(Y)$ are not easily interpreted back to the original model, as discussed above in the case of the rank regression and correlation measures (Saltelli and Sobol', 1995).

A discussion of the increased robustness of the rank version of sensitivity analysis statistics can be found in Saltelli and Marivoet (1990). Rank versions of the importance measure were tested in Saltelli *et al.* (1993), Homma and Saltelli (1994), and McKay and Beckman (1994).

Another importance measure, introduced by Ishigami and Homma (1989, 1990), is given by the following statistic:

$$\mathrm{HIM}(X_j) = \frac{1}{N}\sum_{i=1}^{N} y_i y_i^j, \tag{2.19}$$

where N is the number of computer simulations, each corresponding to a different set of values for the input factors, y_i is the output for the ith run, and y_i^j is the output generated for the ith run when all the variables but X_j have been resampled. The meaning of the HIM measure is intuitive: if X_j is an influential variable, a high value of y_i will be associated with high values of y_i^j and HIM will be high; for a non-influential variable, y_i and y_i^j will be associated randomly, and HIM will be smaller.

One disadvantage of the HIM measure is the lack of robustness. A possible solution to this problem (tested in Saltelli *et al.*, 1993) lies in replacing the raw values with their ranks.

2.5.1 Sobol' Sensitivity Indices

Sobol' sensitivity indices (Sobol', 1990b, 1993), can be thought of as a generalization of concepts introduced above for HIM and the correlation ratio. Sobol' sensitivity indices and the elegant mathematical theory underlying their construction are described in detail in Chapter 8.

The method proposed by Sobol' is based on the decomposition of the function $f(\mathbf{x}) = f(x_1, x_2, \ldots, x_k)$ into summands of increasing dimensions, so that the variance D of $f(\mathbf{x})$ can itself be decomposed as

$$D = \sum_{j=1}^{k} D_j + \sum_{1 \leqslant i<j \leqslant k} D_{ij} + \cdots + D_{12\ldots k}. \tag{2.20}$$

Each generic term in this decomposition can be computed by straightforward Monte Carlo integration. A first-order sensitivity coefficient is defined as

$$S_j = \frac{D_j}{D}, \tag{2.21}$$

while sensitivity estimates for higher-order terms are given by

$$S_{i_1 \ldots i_s} = D_{i_1 \ldots i_s}/D, \tag{2.22}$$

with the obvious property that the sum over all the combinations of indices of the sensitivity estimates is one.

The sensitivity indices have a natural interpretation since they represent the fraction of the total variance of $f(\mathbf{x})$ that is due to any individual factor or combination of factors. One limitation of the Sobol' sensitivity indices is their high computational cost. When dealing with models that are nonlinear, the computation of higher-order sensitivity indices is essential. Unfortunately, one separate sample of size N is needed to compute each $S_{i_1 \ldots i_s}$. Given that the number of summations in (2.20) is 2^k, then $N \times 2^k$ model evaluations are to be computed. If the number of variables is large, such a cost is prohibitive. For this reason, Homma and Saltelli (1996) used the Sobol' method to compute a total index that estimates the total contribution to the variance of $f(\mathbf{x})$ that is due to a certain input variable X. The *total effect term* $\mathrm{TS}(j)$ is defined as the sum of all the terms where at least one of the indices $i_1, \ldots, i_s$ is equal to j. The new approach reduces the number of model evaluations to $N \times (k+1)$ while still providing a measure of the higher-order effects. Results of the analysis can be easily displayed in an intuitive graphical way, by normalizing each $\mathrm{TS}(j)$ by the sum of the $\mathrm{TS}(j)$, $j = 1, \ldots, k$. Usually this sum is larger than one, unless the model is perfectly additive, but the indices can be normalized. The total normalized indices can be plotted in the form of a pie chart, giving the fraction of variance accounted for by each input. When the output is a time series, the chart may take the form of a time-dependent diagram bounded between 0 and 1 (see Saltelli *et al.*, 1998).

Recent investigations have also shown that the computational efficiency of the Sobol' method may be improved by using a sampling strategy that is called Winding Stairs. The Winding Stairs sampling strategy is described in Chapter 8.

For complex models with several compartments or submodels, or simply with sets of factors pertaining to different logical levels, Sobol' indices can be computed efficiently for the set, rather than for the factor (Chapter 8).

2.5.2 FAST (Fourier Amplitude Sensitivity Test)

The Fourier amplitude sensitivity test (FAST) offers an alternative approach to compute exactly the same indices as the Sobol' method, i.e. first-order S_J, higher-order indices $S_{i_1 \cdots i_s}$, and total effects $\mathrm{TS}(j)$. For instance, both S_Js (the one from FAST and that from Sobol') are estimates of the same quantity

$$\frac{\mathrm{Var}_{X_j}[E(Y|X_j)]}{\mathrm{Var}(y)}.$$

FAST was created in the 1970s by Cukier *et al.* (1973, 1978) and developed further by Koda *et al.* (1979) and McRae *et al.* (1982), and extended by Saltelli *et al.* (1999b). FAST represents one of the most elegant methods for sensitivity analysis, and works for monotonic and non-monotonic models alike.

The basis of the FAST approach is a transformation that converts a multidimensional integral over all the uncertain model inputs to a one-dimensional integral, via a search curve that scans the whole parameter space. The scanning is done so that each axis of the factor space is explored with a different frequency. A Fourier decomposition is used to obtain the fractional contribution of the individual input factors to the variance of the model prediction.

Classic FAST can be used to compute first-order terms S_J, while extended FAST (Saltelli *et al.*, 1999b) can be used to compute the total indices $\mathrm{TS}(j)$. In both cases, FAST is computationally more efficient than Sobol', although enhanced Sobol' are being presently tested (Chapter 8). FAST has also the advantage that the same set of model evaluations can be used to compute both S_J and $\mathrm{TS}(j)$ for a given factor X.

2.6 RESPONSE SURFACE METHODOLOGY

This procedure is based on developing a response surface approximation to the model under consideration. This approximation is then used as a surrogate for the original model in uncertainty and sensitivity analysis. The analysis can be divided into six steps:

- selection of ranges and distributions for each variable X_j;
- development of an experimental design defining the combinations of variable values for which model evaluations will be performed;
- evaluations of the model;
- construction of a response surface approximation to the original model;
- uncertainty analysis;
- sensitivity analysis.

Different types of experimental designs are available to select the design points at which the model will be evaluated. The one ultimately selected will depend on many factors: the number of independent variables under consideration, the possible presence of quadratic or higher-order effects, the possible importance of variable interactions, and the computational effort required to evaluate the model.

Although the sequence of design points for the evaluations seem to have exactly the same attributes as a sample of elements selected for a Monte Carlo study, it is important to highlight the following distinction: the design points are selected in this analysis with a classical experimental design, which ensures that a specified structure exists between the values of individual x_j but allows no probabilistic weight to be assigned to the x_j. In contrast, in a Monte Carlo analysis, it is possible to assign a weight to each element (those elements selected with a probabilistic procedure), which can be useful in the construction of estimated means and variances for the output.

In the fourth step, in order to construct a response surface approximation to the original model, the most frequently used technique is that of least squares. If only first-order terms are considered, the resulting surface has the form

$$y = b_0 + \sum_{j=1}^{k} b_j x_j. \tag{2.23}$$

This surface plays the same role in a response surface methodology as the Taylor series in differential analysis. In the fifth and sixth steps, in order to estimate respectively the uncertainty in Y and the sensitivity of Y to the individual X_j, the expected value and variance of

Y are estimated using approximation (2.23), and the coefficients in b_j are normalized so that the effects of changes in the X_j are apparent.

More details about the response surface methodology can be found in Box and Draper (1987), Kleijnen (1987), and Myers (1971).

2.7 FORM AND SORM

In Chapter 1, we stated that sensitivity analysis is mainly concerned with the variation of the output induced by variation in the input, be it local (as in derivative-based methods) or global (as in e.g. variance-based measures). In some instances, the analyst is not interested in the magnitude of Y (and hence its potential variation) but in the probability of Y exceeding some critical value. The constraint (e.g. $Y - Y_{\text{crit}} \leqslant 0$) determines a hypersurface in the space Ω of the input factors $\mathbf{X}$. The minimum distance between some design point for $\mathbf{X}$ and the hypersurface is the quantity of interest. Let β denote such a minimum distance for some assigned joint distribution of the input $\mathbf{X}$. In these settings, one can chose as sensitivity measure the derivative of β with respect to the input factors. Such a quantity should not be confused with the local derivative of Y with respect to the inputs, since the action of taking the minimum of β over the hyper-space of $\mathbf{X}$ introduces an element of probabilistic weighting. The *first-order reliability method* (FORM) offers such a probabilistic measure (see Chapter 7). It gives an estimate of how much a given input factor may drive the risk (probability of failure) of the system.

2.8 COMPARING DIFFERENT APPROACHES

In the field of global sensitivity analysis based on Monte Carlo sampling, several quantitative comparisons can be found in the literature.

1. Several studies on the performance of sensitivity analysis techniques on different test models were reported by Iman and Helton (1988). The sensitivity analysis techniques under examination included:

- Response surface replacement, used in conjunction with fractional factorial design;
- differential analysis;
- partial rank correlation coefficient in conjunction with LHS.

Results of these studies pointed to the effectiveness of the regression-based non-parametric techniques such as the standardized rank regression coefficient and partial rank correlation coefficient. In particular, the investigators concluded that the non-parametric techniques used in conjunction with LHS are the most robust, being able to cope with model nonlinearity (better than in the case of fractional design) and to scan all the space of the input variables (differential analysis only provides information around a point in the space of the variables).

2. In Saltelli and Marivoet (1990), a number of sensitivity analysis techniques (the Pearson correlation, the Spearman coefficient, PCC, PRCC, SRC, SRRC, the Smirnov test statistic, the Cramér–von Mises test statistic, the Mann–Whitney test statistic and the two-sample t-test statistic) are compared in the case of a nonlinear model response. Hypothesis testing was systematically applied to quantify the degree of confidence in the results given by the various estimators. The main conclusions of that intercomparison exercise can be summarised as follows:

- the relative stability of sensitivity analysis indicators depends upon the sample size. When increasing the sample size, the non-parametric estimators (the Spearman coefficient, PRCC, SRRC, the Smirnov test statistic, the Cramér–von Mises test statistic, and the Mann–Whitney test statistic) tend to converge to its lowest asymptote;
- the disagreement between estimators is significant, and increases when decreasing the sample size;
- the estimators PRCC and SRCC appear to be, in general, the most robust and reliable.

3. In Saltelli *et al.* (1993), another comparison of techniques was carried out for nonlinear non-monotonic models. The first analysis showed that non-parametric tests based on ranks are systematically better than their parametric equivalent based on the raw values. However, the existing non-parametric methods (such as SRRC, PRCC and the Spearman coefficient) are fairly reproducible and accurate when the model output varies linearly or at least monotonically with each independent variable, but their accuracy becomes dubious in the presence of model non-monotonicity. The variance-based measures, in contrast appeared capable of overcoming the difficulties posed by model non-monotonicity, but are relatively expensive to apply. These findings were reinforced by a subsequent study (Homma and Saltelli, 1996) devoted to Sobol' sensitivity indices and to their total effect version. The Sobol' sensitivity indices have been shown to be reliable and accurate. Saltelli and Bolado (1998) have shown that FAST is more efficient computationally (quicker convergence) than Sobol'. The extended FAST (Saltelli *et al.*, 1999b) method allowed a reduction of the computational cost. With this latter method, it is in fact possible to compute simultaneously S_j and $\mathrm{TS}(j)$. Further comparison of computational performances are in Chapter 8.

4. A screening test (IFFD) was compared against SRRC and a variance-based measure in Saltelli *et al.* (1994). The analysis indicated that IFFD predictions are extremely reproducible and that IFFD performs as well as the ranked version of the variance-based measure, and better than SRRC whenever the sample size is large enough. Furthermore, IFFD is more robust than SRRC in that it can detect quadratic effects. Although IFFD is less robust than the variance-based method, since it cannot detect higher-order effects, it has to be kept in mind that the computational effort required to evaluate a variance-based measure is considerably larger than the one needed to compute IFFD.

2.9 ANALYTICAL TEST MODELS

A set of test functions is offered in this section to allow the reader to evaluate the performance of the sensitivity measures introduced in this volume. These test functions are also used as benchmarks in some of the methodological chapters (Part II).

The following three subsections offer analytical test models, classified according to their complexity. For each test model, the distribution functions of the input factors are prescribed. Basic output statistics such as PEAR, SRC, PCC, SPEA, SRRC, and PRCC are displayed in tables, and graphical representations, i.e. 3D and scatter plots, are given for some cases.

Notes

- In this chapter, $X \sim U(a, b)$ has been used to denote a continuous variable X distributed uniformly in the range (a, b). $X \sim DU(N)$ denotes a discrete random variable X uniformly distributed from 1 to N.

- Some statistics are estimated at a sample size 16 000 using SPSS 8.0 software.

2.9.1 Linear Test Problems

Model 1

$$Y = \sum_{j=1}^{k} X_j,$$

where $k = 3$, $X_j \sim U(\bar{x}_j - \sigma_j, \bar{x}_j + \sigma_j)$, $\bar{x}_j = 3^{j-1}$, and $\sigma_j = 0.5\bar{x}_j$. The respective summary statistics for the output and sensitivity coefficients are shown in Tables 2.1 and 2.2.

Table 2.1 Summary statistics of the output for Model 1.

$E(Y)$	$V(Y)$	R_y^2
13	91/12	1

Table 2.2 Summary of sensitivity coefficients for Model 1.

Factors	PEAR (Y,X_j)	PCC (Y,X_j)	$V_{X_j}[E(Y\|X_j)]$	$E_{X_j}[V(Y\|X_j)]$
X_1	$1/\sqrt{91}$	1	1/12	90/12
X_2	$3/\sqrt{91}$	1	9/12	82/12
X_3	$9/\sqrt{91}$	1	81/12	10/12

Model 2

$$Y = X_1 + X_2,$$

with a correlation structure between X_1 and X_2. The joint probability distribution function is

$$p(x_1, x_2) = \begin{cases} 2 & \text{if } 0 \leqslant x_1, x_2 \leqslant 0.5 \text{ or } 0.5 \leqslant x_1, x_2 \leqslant 1, \\ 0 & \text{elsewhere.} \end{cases}$$

The respective summary statistics for the output and sensitivity coefficients are shown in Tables 2.3 and 2.4, and the scatterplot is shown in Figure 2.1.

Table 2.3 Summary statistics of the output for Model 2.

$E(Y)$	$V(Y)$	PEAR (X_1,X_2)	R_y^2
1	7/24	3/4	1

Table 2.4 Summary of sensitivity coefficients for Model 2.

Factor	PEAR (Y,X_j)	PCC	SRC	$V_{X_j}[E(Y\mid X_j)]$	$E_{X_j}[V(Y\mid X_j)]$
X_j	$\sqrt{7/8}$	1	$\sqrt{2/7}$	13/48	1/48

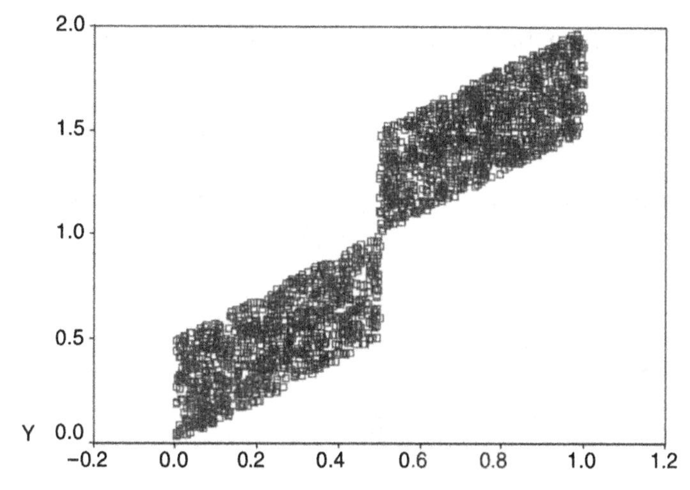

Figure 2.1 Scatterplot for $Y = X_1 + X_2$ versus X_1.

The rank-based statistics are equivalent to non-rank-based ones, since the model is linear.

Model 3

$$Y = \sum_{j=1}^{k} c_j(X_j - \tfrac{1}{2}),$$

where $k = 22$, $X_j \sim U(0,1)$, and $c_j = (j-11)^2$ (source: Sobol', 1993). The respective summary statistics for the output and the sensitivity coefficients are shown in Tables 2.5 and 2.6.

Table 2.5 Summary statistics of the output for Model 3.

$E(Y)$	$V(Y)$	R_y^2
0	$\frac{1}{12} \times 65\,307$	1

Table 2.6 Summary of sensitivity coefficients for Model 3.

	PEAR (Y,X_j)	PCC (Y,X_j)	$V_{X_j}[E(Y\mid X_j)]$	$E_{X_j}[V(Y\mid X_j)]$
X_j	$c_j \Big/ \sqrt{\sum_{j=1}^{k} c_j^2}$	1	$V(X_j) \times c_j^2$	$V(X_j) \times \sum_{i \neq j} c_i^2$

2.9.2 Monotonic Test Problems

Model 4

$$Y = X_1 + X_2^4,$$

where three possible configurations (a), (b) and (c) are considered. They are

(a) $X_j \sim U(0,1)$;
(b) $X_j \sim U(0,3)$;
(c) $X_j \sim U(0,5)$.

The respective summary statistics for the output and sensitivity coefficients are shown in Tables 2.7 and 2.8.

Table 2.7 Summary statistics of the output for Model 4.

Configuration	$E(Y)$	$V(Y)$	$\hat{R}_y^2$	$\hat{R}_y^{2*}$
(a)	7/10	139/900	0.89	0.89
(b)	177/10	46 731/100	0.75	0.96
(c)	255/2	25 × 40 003/36	0.75	0.98

Table 2.8 Summary of sensitivity coefficients for Model 4.

	PEAR (Y,X_j)	PCC (Y,X_j)	$\frac{V_{X_j}[E(Y\mid X_j)]}{V(Y)}$	$\widehat{\text{SPEA}}$ (Y,X_j)	$\widehat{\text{PRCC}}$ (Y,X_j)
Configuration (a)					
X_1	$15/\sqrt{417} \approx 0.73$	$15/\sqrt{273} \approx 0.91$	$75/139 \approx 0.54$	0.76	0.91
X_2	$4 \times \sqrt{3}/\sqrt{139} \approx 0.59$	$\sqrt{3}/2 \approx 0.87$	$64/139 \approx 0.46$	0.55	0.85
Configuration (b)					
X_1	$\sqrt{\frac{1}{15577}} \times 5 \approx 0.04$	$\sqrt{\frac{25}{3913}} \approx 0.08$	$25/15577 \approx 0.0016$	0.16	0.61
X_2	$\sqrt{\frac{1}{15577}} \times 4 \times 3^3 \approx 0.87$	$\frac{9}{\sqrt{108}} \approx 0.87$	$2^6 3^5/15577 \approx 0.9984$	0.96	0.97
Configuration (c)					
X_1	$\sqrt{\frac{3}{40003}} \approx 0.0$	$\sqrt{\frac{3}{10003}} \approx 0.02$	$\frac{3}{40003} \approx 0.0001$	0.08	0.50
X_2	$100 \times \sqrt{\frac{3}{40003}} \approx 0.86$	$\sqrt{3}/2 \approx 0.86$	$\frac{40000}{40003} \approx 0.9999$	0.99	0.99

Model 5

$$Y = \exp\left(\sum_{j=1}^{k} b_j X_j\right) - I_k,$$

where

$$I_k = \prod_{j=1}^{k} \frac{e^{b_j} - 1}{b_j},$$

and $X_j \sim U(0,1)$. Two possible configurations are considered:

(a) $k = 6$ and $b_1 = 1.5, b_2 = \cdots = b_6 = 0.9$;

(b) $k = 20$ and $b_j = \begin{cases} 0.6 & \text{for } 1 \leqslant j \leqslant 10, \\ 0.4 & \text{for } 11 \leqslant j \leqslant 20 \end{cases}$

(source: Sobol' and Levitan, 1999). The respective summary statistics for the output and sensitivity coefficients are shown in Tables 2.9 and 2.10.

Table 2.9 Summary statistics of the output for Model 5.

Configuration	$E(Y)$	$V(Y)$	$\hat{R}_y^2$	$\hat{R}_y^{2*}$
(a)	0	427.2751	0.80	0.97
(b)	0	18 022	0.81	0.96

Table 2.10 Summary of sensitivity coefficients for Model 5.

	PEAR (Y, X_j)	PCC (Y, X_j)	$\frac{V_{X_j}[E(Y \mid X_j)]}{V(Y)}$	$\widehat{\text{SPEA}}$ (Y, X_j)	$\widehat{\text{PRCC}}$ (Y, X_j)
Configuration (a)					
X_1	0.51	0.76	0.2870	0.59	0.96
$X_{j \neq 1}$	0.32	0.58	0.1057	0.35	0.88
Configuration (b)					
$X_{j \leqslant 10}$	0.24	0.47	0.0562	0.26	0.76
$X_{j > 10}$	0.16	0.32	0.0250	0.17	0.61

Model 6

$$Y = \frac{X_2^4}{X_1^2}.$$

Here we consider two possible configurations:

(a) $X_j \sim U(0.9, 1.1)$;

(b) $X_j \sim U(0.5, 1.5)$

(source: Gardner *et al.*, 1981). The second configuration is designed to compare local and global techniques. The non-additive features of the model can be investigated over the assigned range of variation. The respective summary statistics for the output and sensitivity coefficients are shown in Tables 2.11 and 2.12, and a 3D graphical representation, for case (b) is shown in Figure 2.2.

Table 2.11 Summary statistics of the output for Model 6.

Configuration	$E(Y)$	$V(Y)$	$\hat{R}^2_y$	$\hat{R}^{2*}_y$
(a)	$\frac{100}{99}(1.1^5 - 0.9^5)$ $= 1.03$	$\frac{25}{27}(1.1^9 - 0.9^9)(0.9^{-3} - 1.1^{-3})$ $= 0.07$	0.98	0.99
(b)	$0.2 \times (1.5^5 - 0.5^5) \times \left(\frac{1}{0.5} - \frac{1}{1.5}\right)$ $= 2.0166$	$\frac{1}{3}\left(\frac{1}{0.5^3} - \frac{1}{1.5^3}\right) - \left[0.2 \times (1.5^5 - 0.5^5)\left(\frac{1}{0.5} - \frac{1}{1.5}\right)\right]^2$ $= 6.901\,25$	0.675	0.98

Table 2.12 Summary of sensitivity coefficients for Model 6.

Factors	PEAR (Y, X_j)	PCC (Y, X_j)	$\frac{V_{X_j}[E(Y \mid X_j)]}{V(Y)}$	$\widehat{\text{SPEA}}$ (Y, X_j)	$\widehat{\text{PRCC}}$ (Y, X_j)
Configuration (a)					
X_1	− 0.45	− 0.98	0.2023	− 0.42	− 0.97
X_2	0.89	0.99	0.7690	0.90	0.99
Configuration (b)					
X_1	− 0.47	− 0.64	0.261 909	− 0.43	− 0.95
X_2	0.67	0.76	0.510 979	0.89	0.99

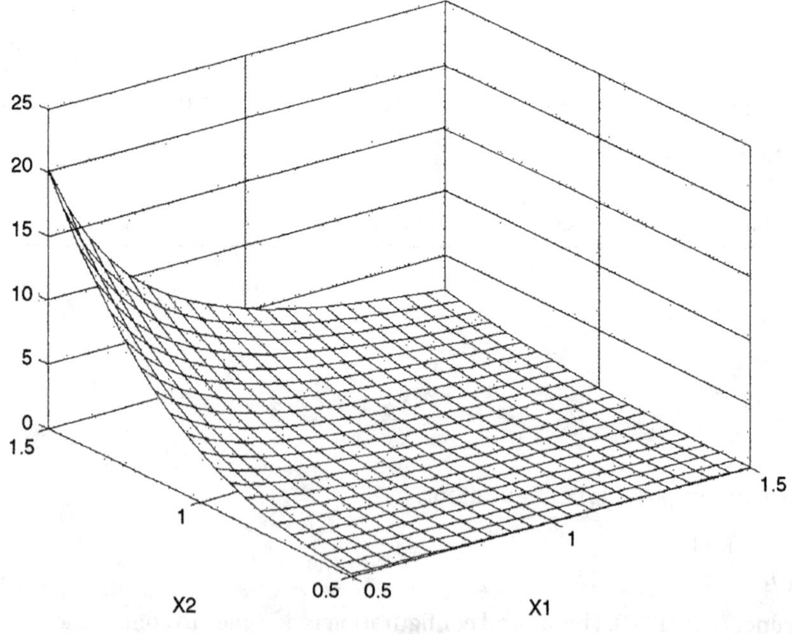

Figure 2.2 3D graphical representation of $Y = X_2^4/X_1^2$ for Case (b).

2.9.3 Non-Monotonic Test Problems

Model 7: The non-monotonic Sobol' g-function

$$Y = \prod_{j=1}^{k} g_j(X_j),$$

where

$$g_j(X_j) = \frac{|4X_j - 2| + a_j}{1 + a_j},$$

with $a_j \geqslant 0$, $k = 8$, $a_j = \{0, 1, 4.5, 9, 99, 99, 99, 99\}$, and $X_j \sim U(0, 1)$ (source: Saltelli and Sobol', 1995). The respective summary statistics for the output and sensitivity coefficients are shown in Tables 2.13 and 2.14, and scatterplots are shown in Figure 2.3.

$\hat{R}_y^{2*} = 0$, rank based statistics are not computed, since the model is non-monotonic.

Table 2.13 Summary statistics of the output.

$E(Y)$	$V(Y)$	$\hat{R}_y^2$
1	0.4652	0

Table 2.14 Summary of sensitivity coefficients.

Factor	$\frac{V_{X_j}[E(Y \mid X_j)]}{V(Y)}$
X_1	0.7165
X_2	0.1791
X_3	0.0237
X_4	0.0072
$X_{5,6,7,8}$	0.0001

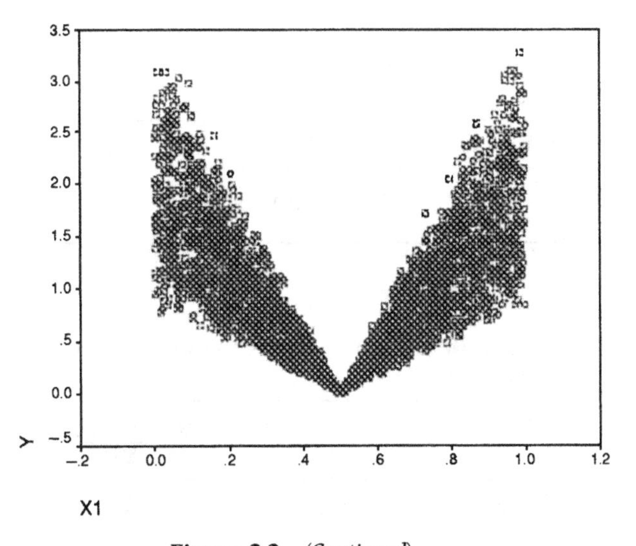

Figure 2.3 *(Continued)*

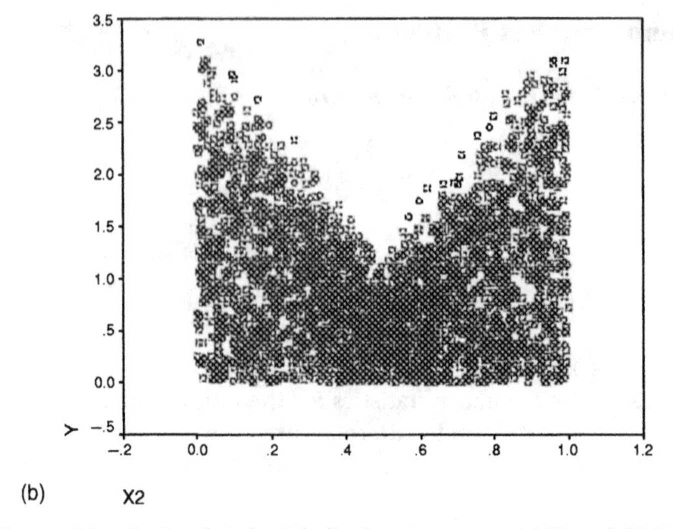

Figure 2.3 Scatterplots for Sobol' g-function versus (a) X_1 and (b) X_2.

Model 8: The non-monotonic Legendre polynomials in X_1 of degree X_2

$$Y = L(X_1, X_2) = h(X_2) \sum_{m=0}^{N} c_m(X_2) g_m(X_1)$$

where $N = [X_2/2]$ is the largest integer $\leqslant X_2/2$,

$$h(X_2) = \frac{1}{2^{X_2}}, \qquad c_m = (-1)^m \binom{X_2}{m} \binom{2X_2 - 2m}{X_2}, \qquad g_m(X_1) = X_1^{X_2 - 2m},$$

and $X_1 \sim U(-1, 1)$, $X_2 \sim DU(5)$ (source: McKay, 1996). The respective summary statistics for the output and sensitivity coefficients are shown in Tables 2.15 and 2.16, and a plot of the Legendre polynomials is shown in Figure 2.4.

Table 2.15 Summary statistics of the output for Model 8.

$E(Y)$	$V(Y)$	$\hat{R}_y^2$
0	$\frac{3043}{17\,325} \approx 0.1756$	0

Table 2.16 Summary of sensitivity coefficients for Model 8.

Factor	$\frac{V_{X_j}[E(Y \mid X_j)]}{V(Y)}$
X_1	1/5
X_2	0

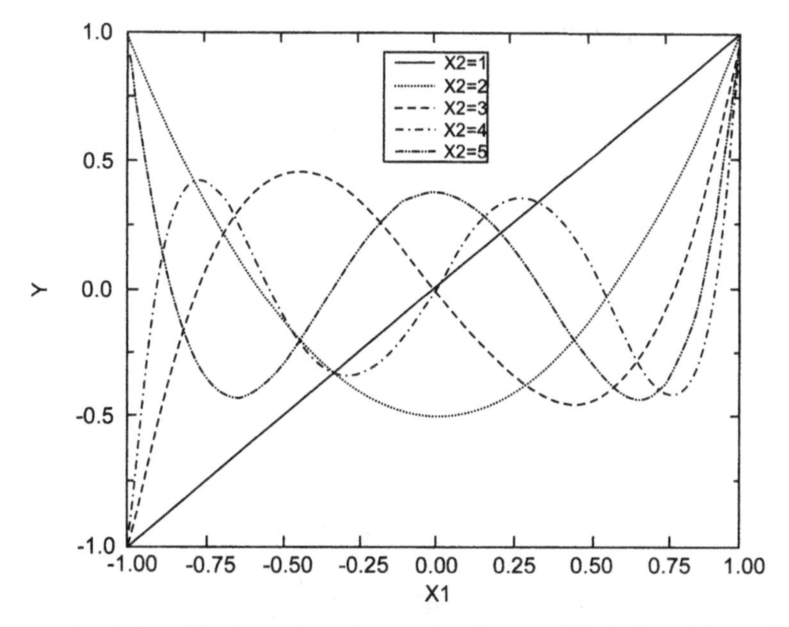

Figure 2.4 A plot of the Legendre polynomials with $X_1 \sim U(-1, 1)$ and $X_2 \sim DU(5)$.

As $R_y^2 = 0$, none of the correlation/regression-based statistics are meaningful in this test case. As $R_y^{2^*} = 0$, rank-based statistics are not computed, since the model is non-monotone.

Model 9: The non-monotonic **Ishigami function**

$$Y = \sin X_1 + A \sin^2 X_2 + BX_3^4 \sin X_1,$$

where $X_j \sim U(-\pi, \pi)$ and $A = 7$, $B = 0.1$ (source: Ishigami and Homma, 1990). The main peculiarity of this model is the dependence on X_3: there is no additive effect on Y and there is interaction with X_1. The respective summary statistics for the output and sensitivity coefficients are shown in Tables 2.17 and 2.18, a comparison between regression coefficients is given in Table 2.19, and scatterplots are shown in Figure 2.5.

Table 2.17 Summary statistics of the output for Model 9.

$E(Y)$	$V(Y)$	$\hat{R}_y^2$	$\hat{R}_y^{2*}$
3.5	$\frac{0.1\pi^4}{5} + \frac{0.1^2\pi^8}{18} + \frac{1}{2} + \frac{49}{8} = 13.8445$	0.19	0.19

Table 2.18 Summary of sensitivity coefficients for Model 9.

Factor	$\frac{V_{X_j}[E(Y \mid X_j)]}{V(Y)}$
X_1	0.3139
X_2	0.4424
X_3	0

Table 2.19 Comparison between regression coefficients for Model 9.

Factor	$\widehat{SRC}$	$\widehat{SRRC}$
X_1	0.435	0.436
X_2, X_3	0	0

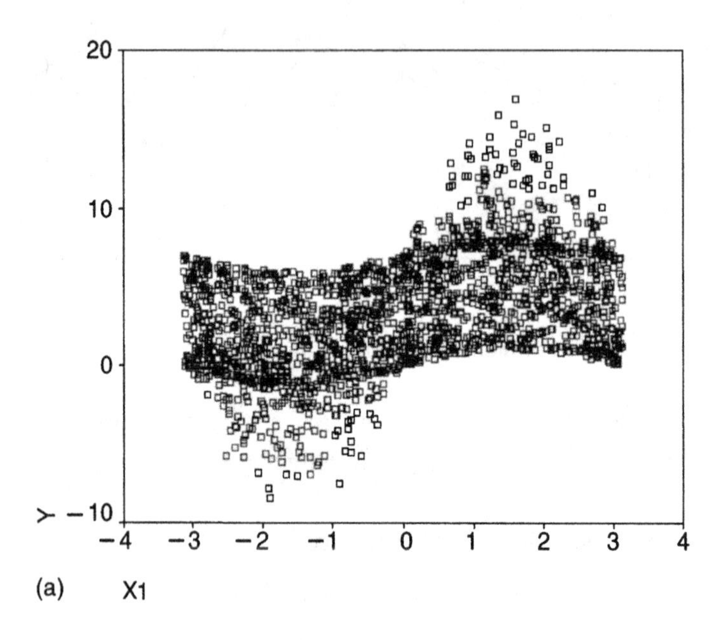

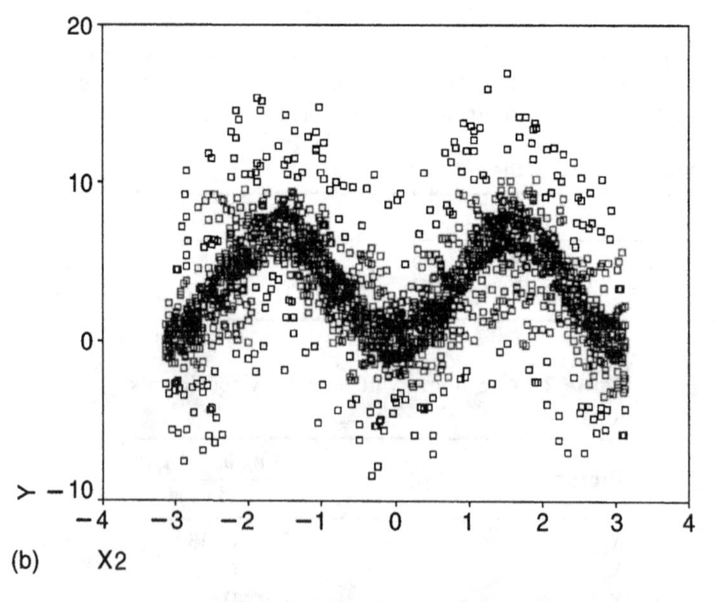

Figure 2.5 *(Continued)*

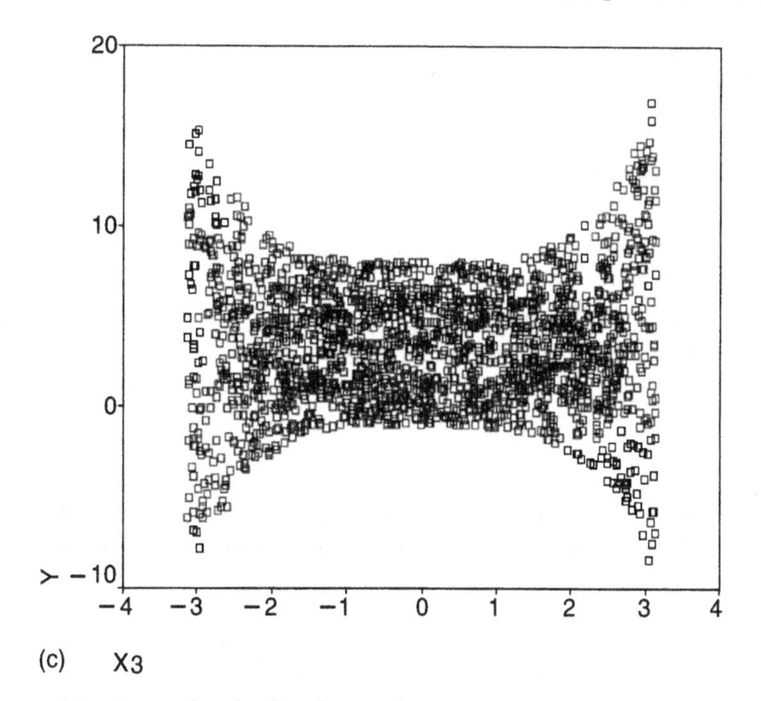

Figure 2.5 Scatterplots for the Ishigami function versus (a) X_1, (b) X_2, and (c) X_3.

Model 10: The non-monotonic function of Morris

$$Y = \beta_0 + \sum_{i=1}^{20} \beta_i w_i + \sum_{i<j}^{20} \beta_{ij} w_i w_j + \sum_{i<j<l}^{20} \beta_{ijl} w_i w_j w_l + \sum_{i<j<l<s}^{20} \beta_{ijls} w_i w_j w_l w_s$$

where

$$w_i = \begin{cases} 2(1.1X_i/(X_i + 0.1) - 0.5) & \text{for } i = 3, 5, 7, \\ 2(X_i - 0.5) & \text{otherwise,} \end{cases}$$

and $X_i \sim U(0, 1)$. Note that $\sum_{i<j<l<s}^{20}$ stands for $\sum_{i=1}^{20} \sum_{j=i+1}^{20} \sum_{l=j+1}^{20} \sum_{s=l+1}^{20}$, and similarly for the other summands. The β_i are assigned as

$$\begin{aligned} \beta_i &= 20 && \text{for } i = 1, \ldots, 10; \\ \beta_{ij} &= -15 && \text{for } i, j = 1, \ldots, 6; \\ \beta_{ijl} &= -10 && \text{for } i, j, l = 1, \ldots, 5; \\ \beta_{ijls} &= +5 && \text{for } i, j, l, s = 1, \ldots, 4. \end{aligned}$$

The remaining third- and fourth-order coefficients are set to zero. The remaining first- and second-order coefficients are generated independently from a normal distribution with zero mean and unit standard deviation. These values can be generated by Monte Carlo simulations, and should not affect the outcome of the analysis. Alternatively, a deterministic rule is given to assign values to the remaing β_i and β_j, so that the same model can be evaluated. They are defined as $\beta_i = (-1)^i$ and $\beta_{ij} = (-1)^{i+j}$. (Source: Morris, 1991.)

Model 11: A nonlinear test model: the chemical reaction system

$$A + A \xrightarrow{k} \text{products},$$

with rate constant $k = X_1 \exp(-X_2/T)$. The absolute temperature T is set to 300 K. The equation governing the concentration f of species A is

$$\frac{dY}{dt} = -2kY^2,$$

with an initial condition X_3, where $X_1 \sim U(8.97 \times 10^6, 3.59 \times 10^7)$, $X_2 \sim U(0, 1000)$, as suggested in the reference paper, and $X_3 \sim U(1.0, 1.2)$. (Source: Tilden *et al.*, 1980). The time-dependent analytical solution

$$Y(t) = \left[2x_1 \exp\left(-\frac{x_2}{T} t\right) + \frac{1}{x_3}\right]^{-1}$$

shows a rapid evolution of the chemical system. It is recommended that one explores the time interval at least in the range $(10^{-10}, 10^{-3})$ s where the transient features can be well exploited. This study is aimed to show how different techniques perform (including the capability to illustrate results) when dealing with time-dependent output. Table 2.20 represents estimated statistics from a sample of size 16 000 for each of the 14 time points. Figure 2.6 shows plots of $\hat{E}(Y)$ and the standard deviation of Y, while Figure 2.7 shows $\hat{R}_y^{2*}$ and $\hat{R}_y^2$. The sample explores the space (X_1, X_2, X_3).

Table 2.20 Summary statistics of the output for Model 11.

t	$\hat{E}(Y)$	$\hat{V}(Y)$	$\hat{R}_y^2$	$\hat{R}_y^{2*}$
1.00×10^{-10}	1.10	3.3×10^{-3}	1	1
3.30×10^{-10}	1.09	3.3×10^{-3}	1	1
1.00×10^{-9}	1.08	3.4×10^{-3}	0.99	0.99
3.30×10^{-9}	1.05	4.8×10^{-3}	0.92	0.92
1.00×10^{-8}	0.98	1.3×10^{-2}	0.89	0.92
3.30×10^{-8}	0.81	3.9×10^{-2}	0.95	0.97
1.00×10^{-7}	0.58	6.0×10^{-2}	0.99	0.99
3.30×10^{-7}	0.33	4.4×10^{-2}	0.96	0.99
1.00×10^{-6}	0.15	1.6×10^{-2}	0.88	0.99
3.30×10^{-6}	0.06	3.1×10^{-3}	0.80	0.99
1.00×10^{-5}	0.02	4×10^{-4}	0.78	0.99
3.30×10^{-5}	0.00	4×10^{-5}	0.76	0.99
1.00×10^{-4}	0.00	5×10^{-6}	0.76	0.99
3.30×04^{-4}	0.00	4×10^{-7}	0.76	0.99

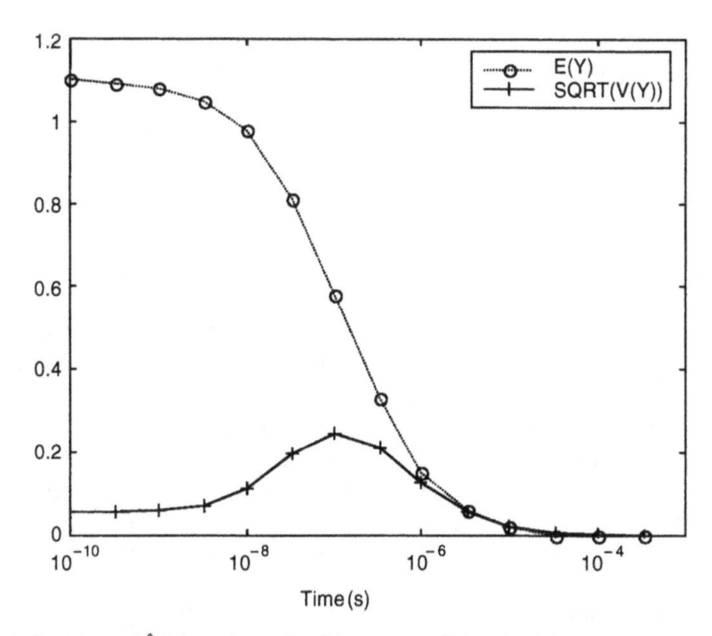

Figure 2.6 Plots of $\hat{E}(Y)$ and standard deviation of Y at the 14 time points considered.

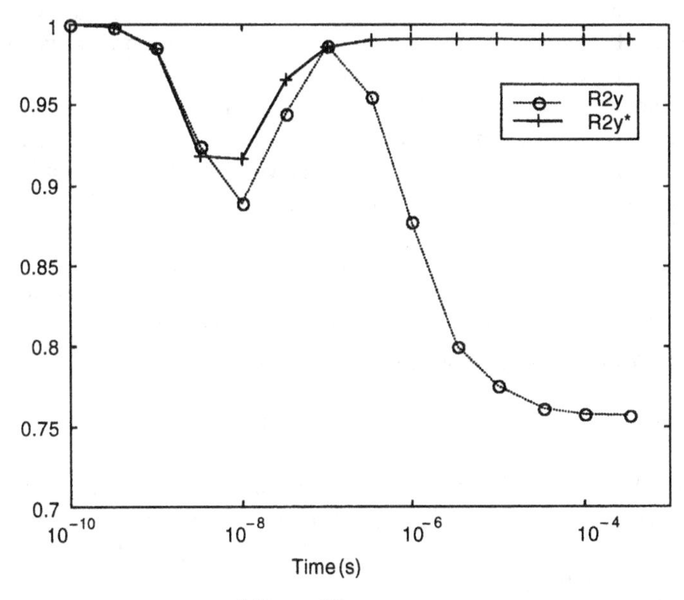

Figure 2.7 Plots of $\hat{R}_y^{2*}$ and $\hat{R}_y^2$ at the 14 time points considered.

2.10 WHEN TO USE WHAT

The performance of a sensitivity analysis method is model-dependent. The difficulty of the analysis increases with the model pathology, i.e. with:

- nonlinearity;
- non-monotonicity;
- non-additivity.

The modeller usually does not know *a priori* how his/her model behaves with respect to these properties, and hence the preferred technique would be the one that copes with all these difficulties.

Further, the analysis also depends on the number and nature of the factors, e.g.:

- with the dimensionality of the factor space: the larger the number of factors to be explored, the more expensive and difficult the analysis;
- with the range of variation of each factor.

In our experience, independently of the model, any increase in either the number of factors or their range (or both) increases the likelihood of non-negligible interactions among factors. The existence of points of singularity within the space of the inputs is also a possible concern. Finally, different sensitivity analysis estimators point to different types of variables, and the objective of the analysis may often force the selection of a method.

All the above reasons make the choice of the SA method to adopt a very delicate step. Although a 'universal recipe' does not exist, in this chapter we have presented a number of considerations that should help an investigator in making such a choice.

(a) Local methods are well established and available in application-oriented software; their use in chemical kinetics is the best example. They can be used for mechanism identification in complex systems, especially when used in conjunction with descriptive statistics, and for the solution of inverse problems. Their use for predictive models should be limited to the case of linear models, or of models with modest factor variations.
(b) The computational execution time of the model is a major concern. Models with several factors call for a screening method that is computationally compatible with the problem.
(c) Within the class of screening methods, Morris is well established, global and computationally efficient. IFFD is also appealing, and to be preferred, when the number of factors is huge (hundreds).
(d) Estimators such as the SRC are easy to implement, relatively inexpensive, and intuitive. They have the limit of being as good as the regression on which they are based. This can be ascertained by computing the model coefficient of determination R_y^2. Such a coefficient should always be determined, since it gives an indication of the model behavior (linearity, monotonicity) and comes at no extra cost.
(e) The ranked version, SRRC, although improving the R_y^2 value, has the drawback of altering the model under analysis.
(f) Variance-based techniques have the advantage of being intuitive and quantitative. Within this class of methods, we strongly favor those capable of computing the total effect, especially since the analyst cannot know in advance whether his/her model will be additive in all its factors.
(g) The Sobol' theory offers the ideal theoretical background to the problem. The Sobol' method allows the computation of the total effect.
(h) FAST (conventional) is shown to be computationally more efficient than the Sobol' method in computing the main effects. Yet the upgraded algorithms discussed in Chapter 8 should be used, even if one decided not to use the extended FAST.
(i) The extended FAST seems at the moment the most efficient variance-based measure available. Like Sobol', it can cope with nonlinear and non-monotonic models, but its efficiency seems to be superior to that of Sobol' (in the sense of more rapid convergence). Like Sobol', it is capable of computing the total sensitivity indices, but, unlike Sobol', the same sample set can be used to compute the first-order indices. (Note that Winding

Stairs in conjunction with the Sobol' method could perhaps offer a computationally viable alternative to FAST—see Chapter 8.)

(j) In general, for a moderate number of factors (tens) and a model whose execution on the given machine does not exceed a minute, the variance-based methods are ideal, either in the Sobol' or FAST implementations.

(k) For the reliability user community, FORM and SORM offer the advantage of measuring the quantity of most direct interest, e.g. the distance from the failure surface.

The tendency of model developers to invent their own method for their application, on the basis that it will give them an indication of the model behavior, should be resisted. Worked analysis of sensitivities often reveal more complexity than one would imagine. Further, the use of objective quantitative methods should be preferred to combat the unavoidable bias introduced by the model developer in the analysis. The rules of the game should be established beforehand, to avoid the analysis working toward a pre-established end.

ACKNOWLEDGMENTS

The authors are grateful to Nicla Giglioli for her assistance in computing test case functions and acknowledge financial support from the European Commission (DG XIII) within the project 'Integration Demonstration and Marketing of the Uncertainty & Sensitivity Analysis Software PREP/SPOP', Ref. CSA98124.

II

Methods

3
Design of Experiments

Francesca Campolongo and Andrea Saltelli

European Commission, Joint Research Centre, Ispra, Italy

3.1 CONTENT

Why is there a chapter on DOE in a sensitivity analysis book? DOE (design of experiments) is an important element of the planning of a physical experiment. With time, DOE has been extended to computer experiments, and, in our view, it can be considered, as one of the fore-runners of sensitivity analysis. Although there are several differences between physical and simulation experiments, sensitivity analysis is based on the same principles as those under-lying DOE. The selection of inputs at which to run a computer code is still an experimental design problem, and statistical ideas for design are helpful (Sacks *et al.*, 1989a). Further, much of the terminology used in SA has originated in a DOE setting.

The purposes of the present chapter are:

(i) to illustrate those DOE principles of most relevance to SA experimentation by referring to classic textbook examples;
(ii) to highlight the links between DOE and SA, as well as the differences between physical and computational experiments.

Point (ii) is mainly based on the work of investigators who first attempted the customization of DOE methods to simulation studies.

3.2 INTRODUCTION

In physical experimentation, experiments followed by analysis are frequently performed to measure the 'effects' of one or more factors on a response. For this purpose, it is extremely important to design *a priori* an experiment that can provide information at the right cost. The use of a 'brute force' approach, which evaluates the impact of changing the values of each factor in turn (also known as *ceteris paribus*), without following a predetermined design, should be avoided. A good experimental design is crucial. Conclusions are easily drawn from

Sensitivity Analysis. Edited by A. Saltelli *et al.*

a well-designed experiment, even when elementary methods of analysis are employed, but even the most sophisticated analyses fail when an experiment is badly designed.

The basic problem of experimental design is deciding what pattern of factor combinations (design points) will best reveal the properties of the response and how it is affected by the factors. Unfortunately, the question of where to place the points is often a circular one: if we knew the response function, we might easily decide where the points should be placed—but the response function is the very object of the investigation!

Design of experiments (DOE)—first introduced by Fisher (1935)—can be defined as selecting the combinations of factor values to be employed that will provide the most information on the input–output relationship in the presence of variation. Many classical designs are presented in numerous publications, to which we address the reader for further reading on DOE. Two authoritative textbooks are Box and Draper (1987) and Box *et al.* (1978b).

In this chapter, we aim to describe the spirit of DOE, in order to make a comparison with SA and to introduce fundamental concepts such as effect, interaction, and aliasing. For this purpose, we focus on a class of the DOE techniques, namely factorial designs, and we briefly summarize their main features.

The methods described in this chapter are based on elementary statistical principles and have very wide applicability. When only quantitative variables are involved in the analysis, more sophisticated methods, such as least squares (regression analysis) or response surface methods, can be employed to analyze the results of a designed experiment. These methods mostly address objectives that are different from sensitivity analysis, such as model validation, model optimization, and construction of an efficient predictor, and are hence not treated here.

We start by discussing a class of designs of great practical importance: the class of *factorial designs (FDs)*. Such designs are easy to implement, and the interpretation of the observations produced can proceed largely by using common sense and elementary arithmetic. FDs are created to measure the effects of the input factors on the response: not only additive effects for each factor but also the effects of interactions between factors can be estimated. If k is large, the basic FD requires an extremely high number of runs. In such cases, *fractional* factorial designs (Box *et al.*, 1978b) are usually employed. These fractional designs employ carefully selected subsets of all possible combinations of factor values, and provide the desired information economically, as we shall see below.

One important requisite of the data, almost always assumed true for SA in numerical experiments, is that factors are quantitative, defined over a cardinal (0.5, 1.0, 2.3, . . .) or at least ordinal (low, medium, high, . . .) scale. In DOE one can *also* use qualitative factors, defined over nominal scales (white, yellow, red, . . .).

Whether a factor's scale is weak (e.g. nominal) or strong (e.g. cardinal), in DOE, one assumes a variation among two (or more) possible values, called *levels*, for each factor, i.e. one would use level = 1, 2, 3 for each and all of the three examples above. In SA, factors are most often defined on cardinal scales and sampled from continuous distributions (an important exception is, for instance, when SA includes among its factors a trigger to activate different (sub)models; see the discussion on structural uncertainty in Chapter 1).

Often a DOE and the associated outcome are fed into a regression analysis algorithm, in order, for instance, to build a predictor for the system under analysis. This problem setting is substantially different from that of DOE discussed here (*identification of effects and their magnitude*), and is hence not treated. An exception is our Section 3.7, where a DOE coupled with regression links to the use of a class of sensitivity methods described in Chapter 8.

Most of Sections 3.3 and 3.4 are taken from Box *et al.* (1978b), and describe FDs. Other methods in the context of DOE for physical experiments (still aiming at the identification of effects) are mentioned in Section 3.5. Section 3.6 contains a discussion on the use of DOE for simulation models: similarities and dissimilarities between physical and computer experiments are outlined. Section 3.7 describes designs appositely created for computer experiments, and addresses the problem of building an efficient predictor of the response for untried data.

3.3 FACTORIAL DESIGNS

We start with the classical Box *et al.* (1978b) example. An experiment has to be performed that involves two quantitative variables, temperature T and concentration C, and a single qualitative variable, catalyst K. The response is a chemical yield. An FD can be performed with the goal of estimating the effects of each factor on the response. Each factor is assumed to take two possible values, called 'levels', which are denoted as '+' (on) or '−' (off). The design simulates all possible combinations of factor values. Hence, the computational cost is $2^3 = 8$ runs. In general, the computational cost of an FD is l^k, where k is the number of factors and l the (constant) number of levels. More generally, if the number of selected levels varies from one factor to another, the cost is $l_1 l_2 \ldots l_k$.

The combinations of levels to be simulated in a design are displayed in a matrix called the *design matrix*. In our example, the design matrix, with elements '+' and '−', has the form shown in Table 3.1. This is an example of a 2^k factorial design in standard order: the jth column consists of 2^{j-1} minus signs followed by 2^{j-1} plus signs, etc.

The design matrix determines the n experimental runs (factor combinations). For example, Table 3.1 may be interpreted as follows: in run (say) 2, T is high, while C and K are low. We denote the outcome of experiment i by y_i, with $i = 1, \ldots, n$; Table 3.1 has $n = 2^3$.

The design matrix extended with the outputs $y_1, \ldots, y_n$ is used to estimate main effects and interactions, as follows. In Table 3.1, the difference between the output values y_1 and y_2 is due to the variation in the level of variable 1 (T) only, since the other two variables are kept the same. This difference is an individual measure of the effect of the temperature on the chemical yield. In the example, there are four of these individual measures for the temperature. The *main* effect of a variable is defined as the average effect of that variable over all conditions of other factors. For example, the main effect of the temperature is estimated by averaging four individual measures:

$$\frac{(y_2 - y_1) + (y_4 - y_3) + (y_6 - y_5) + (y_8 - y_7)}{4}.$$

Table 3.1 A 2^3 design.

Run	T	C	K
1	−	−	−
2	+	−	−
3	−	+	−
4	+	+	−
5	−	−	+
6	+	−	+
7	−	+	+
8	+	+	+

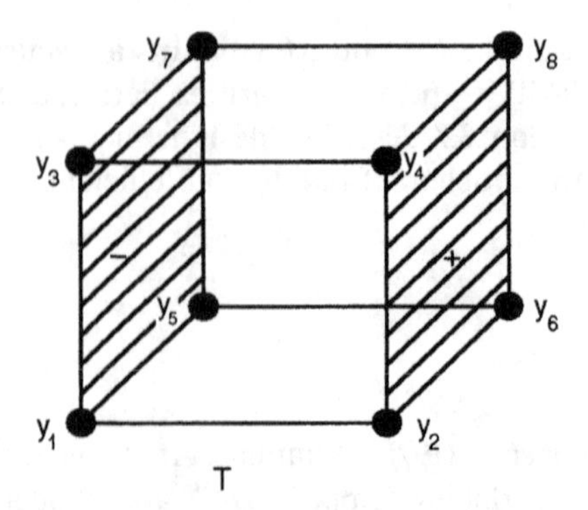

Figure 3.1 Geometric representation of contrasts corresponding to main effects.

This can also be rewritten as

$$\frac{(y_2 + y_4 + y_6 + y_8)}{4} - \frac{(y_1 + y_3 + y_5 + y_7)}{4}.$$

which is the difference between the average response for the plus level and the minus level of temperature. In matrix notation, this can also be written as $\mathbf{x}_1'\mathbf{y}$ where $\mathbf{x}_1$ is the column vector for temperature in Table 3.1. (In general, the 2^k design has k orthogonal columns $\mathbf{x}_j$, with $j = 1, \ldots, k$). The geometric representation is shown in Figure 3.1.

In general, the main effect of each variable is estimated by the difference between two averages (say) $\bar{y}_+$ and $\bar{y}_-$ where $\bar{y}_+$ and $\bar{y}_-$ are the average responses for the plus and minus levels of the variable respectively.

This idea can be extended to interactions: an interaction between two variables is the difference between two averages, half of the outputs being included in one average and half in the other. For instance, as the main effect was viewed as a *contrast* between responses on parallel faces of the cube (Figure 3.1), the interaction between temperature and concentration is the contrast between results on two diagonal planes: see Figure 3.2. Similarly, the three-factor interaction is the difference between the two tetrahedra in Figure 3.3.

To compute all effects by using their definitions would be very tedious. A more rapid way is that of using the *table of contrast coefficients*. This table is obtained by extending the design

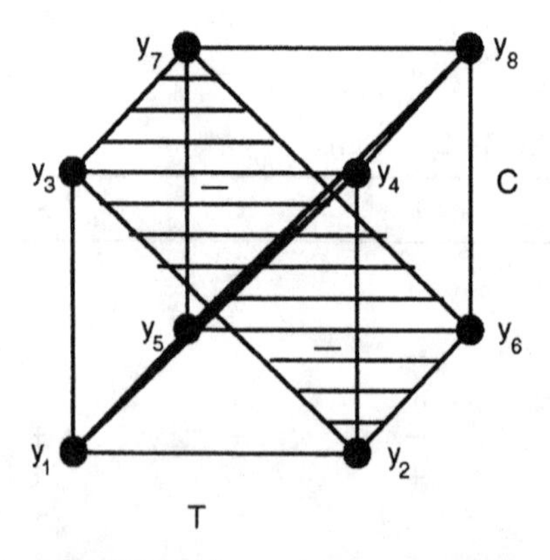

Figure 3.2 Geometric representations of contrasts corresponding to two-factor interactions.

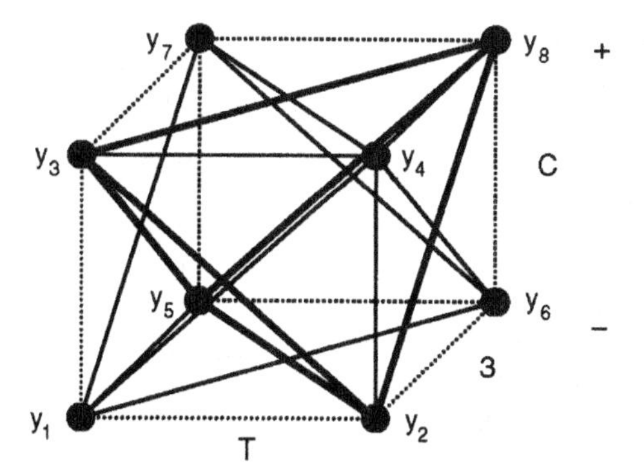

Figure 3.3 Geometric representations of contrasts corresponding to three-factor interactions.

matrix to add columns that represent the interactions among variables, plus a column containing the mean (an example is given in Table 3.2). The signs of the interactions are obtained by simply multiplying the signs of their respective variables. For example, in the table of Contrast Coefficients for the three-variable example (Table 3.2), the column of signs for the $T \times C$ interaction is obtained by multiplying together the signs for T and C. A 2^k design gives a table with all $n = 2^k$-columns being *orthogonal* and can be considered as a matrix **S**.

Because this matrix $\mathbf{S} = (S_{ij})$ is orthogonal, an estimator of the jth effect is

$$E_j = \frac{\sum_{i=1}^{n}(S_{ij}y_j)}{n/2}. \tag{3.1}$$

For example, the $T \times C$ interaction effect is given by

$$E_4 = E_{TC} = \frac{+y_1 - y_2 - y_3 + y_4 + y_5 - y_6 - y_7 + y_8}{4}$$

More generally, an estimate for the quantity E_j represented by the jth column of the table of contrasts can be obtained as

$$E_j = \frac{\sum_{i=1}^{n}(S_{ij}y_j)}{P}, \tag{3.2}$$

Table 3.2 Table of contrast coefficients for 2^3 design.

Run	*T*	*C*	*K*	*TC*	*TK*	*CK*	*TCK*	Mean	*y*
1	−	−	−	+	+	+	−	+	y_1
2	+	−	−	−	−	+	+	+	y_2
3	−	+	−	−	+	−	+	+	y_3
4	+	+	−	+	−	−	−	+	y_4
5	−	−	+	+	−	−	+	+	y_5
6	+	−	+	−	+	−	−	+	y_6
7	−	+	+	−	−	+	−	+	y_7
8	+	+	+	+	+	+	+	+	y_8

Table 3.3 The calculated effects.

Effect	**Estimate**
Average	64.25
Main effect T	23.0
Main effect C	−5.0
Main effect K	1.5
Interaction TC	1.5
Interaction TK	10.0
Interaction CK	0.0
Interaction TCK	0.5

where P is the number of plus signs in the jth column. Thus, an estimate of the mean is obtained as

$$M = \sum_{i=1}^{8} y_i/8.$$

Factorial designs were developed in the precomputer age, so the analysis of the results was cumbersome and 'tricks' were developed, such as Yates's algorithm (Box *et al.*, 1978b). Nowadays there is ample software (SPSS, SAS, NCSS, Statistica) to analyze the results of a designed experiment.

The analysis of the yield values is used to calculate the effects of the three factors (displayed in Table 3.3).

To evaluate whether or not the effects are significant, an estimate of the standard error is necessary. In our example, the estimated standard error of an effect (based on two replicate runs) is $\sqrt{2} = 1.4$ (Box *et al.*, 1978b, p. 320). Comparison of the estimates given in Table 3.3 with their standard errors suggests that only the effects of T, C, and $T \times K$ are significant. Also, it has to be kept in mind that the total effect of a variable results from its main effect plus interactions involving it. For instance, the main effect of K in the example seems non-significant, but its interaction with T makes it an important factor.

In the example we have shown how, in physical experimentation, design of experiments is a *systematic* way of separating important effects from background noise. *Replication* has also been employed to reduce the effect of random error. Alternative ways to reduce the effect of random error are *blocking* and *randomization*.

To illustrate the idea of *blocking*, we illustrate an example again taken from Box *et al.* (1978b, p. 336). Assume that a 2^3 factorial design has to be performed and that, to make the 2^3 runs under conditions as homogeneous as possible, it is desirable that batches of raw material sufficient to complete the experiment be blended together. Also assume that, the blender is not large enough to contain the material for 8 runs, but only that for half of them. Thus, two different blends have to be used. The experimenter now has to decide which is the 'best' way to arrange the 8 runs of the design in two blocks of 4, to neutralize the effects of possible blend difference. Figure 3.4 shows the two blocks: runs 1, 4, 6, 7 are assigned to block I, while runs 2,3,5, 8 are assigned to block II.

The main effect of each variable is estimated by the difference between two average responses on opposite faces of the cube (see Figure 3.1). Since each face contains two runs of blend I and two runs of blend II, any additive effect due to the blend difference is neutralized. For example, say each run of block I is higher by a certain amount h than run in block II. In estimating the main effects, the value of h will cancel out. A similar argument

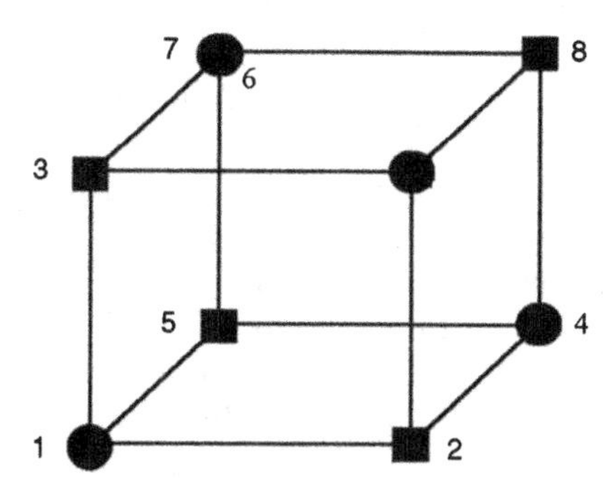

Figure 3.4 A 2^3 factorial design arranged in two blocks.

holds for the two-factor interactions (see Figure 3.2). In contrast, when estimating the three-factor interaction effect (see Figure 3.3), the effect due to the blend difference is not eliminated, but is mixed up and confused with the three-factor interaction. Using the classical terminology of DOE, we shall say that the three-factor interaction effect and the blend difference effect are *confounded* or *aliased* with each other.

Blocking has led to two main consequences:

- the three-factor interactions cannot be estimated, because they are deliberately confounded with the effect due to blocking;
- on the other hand, the estimates of first- and second-order effects are more precise than would have been otherwise.

Blocking can be used in physical experiments to eliminate 'known' sources of variation during an experiment. To deal with 'unknown' sources of variation, *randomization* is employed. Fisher (1935) pointed out that randomization is a precaution to be taken to ensure the validity of comparative experiments. The goal of randomization is that of forcing the unknown sources of variation to contribute homogeneously to different runs (experiments).

3.4 FRACTIONAL FACTORIAL DESIGNS

The number of runs required by a full 2^k FD increases geometrically as k increases. When the cost of such a design becomes too high, a useful alternative is the *fractional factorial design* (Box *et al.*, 1978b). Fractional factorial designs are based on the idea that a full FD often makes possible the estimation of more effects than are detectably different from noise. At some point, higher-order interactions tend to become negligible. Thus, if k is not small, in a full factorial there is *redundancy* in terms of the number of effects that are estimated.

A fractional design assumes some of those higher-order interactions to be unimportant. For an example, we return to Table 3.2, which was given for three factors. Now, we assume *a priori* that the interaction among all three factors—see the column **123**—is unimportant. Then we can examine four factors in only eight runs: identify **4** with the interaction **123**; that is, to the design for three factors shown in Table 3.1 we add a column for a fourth factor with elements identical to the column **123** in Table 3.2.

Note that this idea of using aliasing in the design (due to Fisher) is the same used when blocking. Imagine that, in the example above, the effect due to the blend difference was treated as if it was the effect due to a fourth variable, **4**. In building our design, we have deliberately aliased the three-factor interaction effect with the effect of this fourth variable. This corresponds to assigning in our design matrix the same elements to columns **4** and **123**. The 'blocking' may be said to be *generated* by the relationship **4** = **123**.

We give one more example. Consider a two-level full factorial for $k = 5$ factors, which requires $2^5 = 32$ runs. Assuming that some effects are non-significant, we perform a fractional FD with only 16 instead of 32 runs. This design is called a *half factorial*, denoted as 2^{5-1}. A possible procedure to select these 16 runs is to first write a full 2^4 design for the four variables **1**, **2**, **3**, and **4**, and then to define the levels of variable **5** as those of the column for the **1234** interaction. This design is *generated* by the generator **5** = **1234**. The generator **5** = **1234** can be rewritten by multiplying both sides by **5**, so $\mathbf{5}^2 = \mathbf{12345}$; more conveniently, defining **I** as a column with ones' only, we write **I** = **12345**, and call this the *defining relation* of the design.

If we wish a smaller fractional design, we need more generators, which results in more members of the defining relation. For example, instead of 16 runs we may want to use only the eight runs of a 2^{5-2} design. We than choose two generators; for example, **4** = **12** and **5** = **23**. Then the defining relation is **I** = **124** = **235** = **1345**, where the last member follows from multiplying the first two members and $\mathbf{2}^2 = \mathbf{I}$.

The half factorial generated by **5** = **1234** is successful in estimating first- and second-order effects. In fact, the estimates of the main and of the two-factor interaction effects obtained by using the 16 runs are not significantly different from those obtained when performing the full factorial. However, when we consider the column **123** of the design matrix, to estimate the three-factor interaction, we notice that this is exactly the same as column **45**. The interaction effects due to **123** and to **45** are confounded, or, in the DOE terminology, they are *aliases* of each other. Thus, the use of such a design is acceptable if third- and higher-order effects are assumed to be negligible.

The smaller design generated by **I** = **124** = **235** = **1345** assumes unimportant two-factor interactions. In fact, for such a design, the columns of the design matrix are such that **1** = **24**, ..., **5** = **23**. Obviously, higher-order interactions are also assumed unimportant: **5** = **134** (the previous 2^{5-1} design assumed only that the **1234** interaction was unimportant).

In general, the confounding pattern (aliases) of a fractional FD can be derived from its defining relation: this relation implies particular identities among the columns of the matrix of signs. For example, from the relation **I** = **1345**, by multiplying both sides by **5** we derive **5** = **134**. This leads us to the concept of *resolution*.

A *resolution*-3 (R-3) design does not confound main effects with one another, but it does confound main effects with two-factor interactions. An example was presented above, namely the 2^{5-2} design (with **4** = **12** and **5** = **23**). A *resolution*-4 (R-4) design does not confound main effects with two-factor interactions, but does confound two-factor interactions with other two-factor interactions. A resolution-4 design can be easily constructed from a resolution-3 design: double the resolution-3 design by adding its 'mirror' image (+ becomes −; − becomes +): *foldover principle* (Box *et al.*, 1978b). A resolution-5 design does not confound main effects with two-factor interactions, nor does it confound two-factor interactions with each other. An example was presented above, namely the 2^{5-1} design (with **5** = **1234**).

In general, the *resolution* of a two-level fractional design is *the length of the shortest word in the defining relation*. For example, a design whose defining relation is **I** = **1234** has resolution 4, whereas a design with defining relation **I** = **12345** has resolution 5. So, a design of resolution R has no p-factor effect confounded with any other effect containing less than $R - p$ factors.

A fractional FD of resolution R contains complete factorials (possibly replicated) in every set of $R-1$ factors (Box *et al.*, 1978b). Hence, when modellers correctly assume that no more than $R-1$ factors have detectable effects but they do not know which ones, then a resolution R design will be a complete factorial in the effective factors.

In general, when constructing a fractional FD, the goal is to obtain the highest possible resolution. To construct a half fractional design (2^{k-1}) of highest possible resolution, the following steps should be followed:

1. Write a full factorial design for the first $k-1$ factors.
2. Assign to the kth factor the signs obtained by multiplying the signs of the first $k-1$ factors.

The effects estimable in a fractional design are calculated in exactly the same way as in the full factorial.

When small fractions are considered, the design may become *saturated*. A design is *saturated* when every available contrast is associated with a variable. For example, assume that seven variables are involved and a 2^{7-4} design is constructed by (1) writing a full factorial design for the first three factors; (2) associating additional variables **4**, **5**, **6**, **7** with all the interaction columns **12**, **13**, **23**, **123**. Such a design is saturated. It is possible also to construct supersaturated designs, i.e. designs where the number of runs required, n, is smaller than the number of factors under study, k. For further reading on saturated/supersaturated designs, see Andres and Hajas (1993), Morris (1987), Srivastava (1975), Watson (1961), and Chapter 4.

Just as for full FDs, fractional designs may also be run in blocks, with suitable contrasts used as 'block factors'. A design in 2^q blocks is defined by q independent contrasts. All effects associated with these basic contrasts and all their interactions are confounded with blocks (Box *et al.*, 1978b).

3.5 OTHER DESIGNS

The FDs illustrated above are classical and very well-known designs in the context of DOE for physical experiments. However, the literature covers many other designs, aimed at the identification of factor effects in physical experiments, and based on different kinds of assumptions. The Plackett–Burman designs (Plackett and Burman, 1946) are among the oldest designs used in physical experimentation to detect the subset of most influential input factors on the response function. A more recent approach is the Bayesian analysis suggested by Box and Meyer (1986), which aims to identify active factors assuming factor 'sparsity' (only a few important factors). Other alternatives can be found in the supersaturated design referenced above, which tries to determine which of the k input factors are important using $n < k$ runs.

Another well known approach is the one-factor-at-a-time (OAT) approach, which, assuming absence of interactions, varies experimental factors one at a time, with the remaining factors held constant. This method provides unbiased estimators of the effects of an individual factor, provided that these effects are the same at different settings of the other factors; that is, the factors act additively over the range of interest. The OAT approach has been often criticized in the literature because it relies on the above assumption, which, although advantageous to simplify the problem, can be rarely accepted. In fact, interactions are usually relevant, and need to be estimated by varying factors simultaneously.

In the next section, we shall discuss how the use of DOE methodologies can be extended to computer experiments, and we shall mention some designs that were specifically introduced for computer experimentation. Indeed, the whole class of sensitivity analysis methods that are developed and used in the context of numerical (computer) experimentation may be seen as the natural evolution of DOE. The guiding philosophies of DOE (when aimed at estimating effects) and SA are in fact identical. In both cases, the issue is that of choosing the inputs at which to run the experiment (either physical or numerical) and the statistical tools for an efficient data analysis. These similarities between DOE and SA are particularly evident for the class of screening designs (see Chapter 4). These designs tend to be conceptually simple and computationally light, aiming to identify a subset of few important input factors among a large number contained in a computer model.

3.6 DOE FOR COMPUTER EXPERIMENTS

In numerical experiments, one may be faced with a *deterministic* or *stochastic* simulation model. Unless otherwise specified, the simulation models addressed in this chapter are deterministic: running the model with the same inputs will give identical observations.

Although created for physical experiment, DOE can be extended in the context of numerical experimentation (i.e. computer simulations). When experimenting with a simulation model, DOE can be defined as selecting the combinations of factor values that will be actually simulated. Next the simulation program is run for these combinations. Then the resulting input/output (I/O) data of the experiment are analyzed to derive conclusions about the importance of the factors (outputs are also called responses or criteria).

However, it has to be kept in mind that the lack of random (or replication) error that characterizes deterministic simulation makes computer experiment different from physical experiments, which have substantial random error due to variability in the experimental units and the environment.

The lack of random error leads to a number of differences (Sacks *et al.*, 1989a). In deterministic computer experiments:

- The absence of random error allows the complexity of the computer code to emerge.
- In the absence of random error, the rationale for least-squares fitting of a response surface (often coupled to a DOE) is not clear, unless least squares are interpreted as simple curve fitting and do not rely on the assumption that the departures (differences between the response and regression model) behave like white noise. Taking this point of view, the adequacy of a response-surface model fitted to the observed computer data is determined solely by systematic bias.
- In physical experiments, the application of blocking is an important technique to contrast one source of variation in the results. For example, tire wear differs among the four positions on the car: left–front through right–rear. In simulation, however, complete control over the experiment eliminates the need for blocking.
- As for blocking, the concepts of experimental unit, replicate and randomization in deterministic simulations are irrelevant.

Although the most important, the lack of random error is not the only cause of difference between computer and physical experiments. In the simulation literature, there are a number of codes, called stochastic, that incorporate substantial random error through random or pseudo-random number generators. These stochastic codes differ from deterministic codes because they model the random noise in physical experiments. Therefore, for stochastic codes, the use of standard techniques for physical experiments is appropriate. However, even in this case, there are important differences (Kleijnen, 1998):

- The pseudo-random numbers give the analysts much more control over the noise in their experiments than the investigators have in physical experiments. 'Seeds' are used to initialize the pseudo-random number sequence, and a given 'seed' may be used twice, to replicate exactly the numerical experiment; a different seed will lead to a new sequence, and so on.
- Randomization is of major concern in physical experiments: assign the 'experimental units' (e.g. patients) to the treatments (e.g. types of medication) in a random, non-systematic way so as to avoid bias (healthy patients receive medication of type 1 only). In stochastic simulation, this randomization problem disappears: pseudo-random numbers take over.

Moreover, computer and physical experiments (either deterministic or stochastic) have the following important distinctions of a practical nature:

- The number of factors varied in a computer experiment can be much higher than the number of factors in physical experiments.
- The range of variation for each of those factors can be much higher in numerical experiments than in physical experiments.

The above imply that the methodologies employed for the design and analysis of physical experiments, may not be ideal for complex computer models. McKay *et al.* (1979) were the first to explicitly consider experimental design specifically for deterministic computer codes. These authors introduced Latin hypercube sampling (LHS), an extension of stratified sampling that ensures that each of the input factors has all portions of its range represented. LHS is described in detail in Chapter 6. Other designs of experiment for computer codes can be found in Chapter 4.

3.7 MORE ON DOE FOR COMPUTER EXPERIMENTS: THE PREDICTION PROBLEM

The DOE techniques mentioned above (LHS and screening) are of particular interest to SA practitioners because they were specifically created to investigate the sensitivity of the output of a computer code to its inputs. In general, when designing a computer experiment, the goals of an investigator may vary. Sacks *et al.* (1989a) presented three primary objectives (none of them specific to sensitivity analysis):

- predict the response at untried inputs;
- optimize a functional of the response;
- tune the internal variables of the computer code to physical data.

The design problem concerns the input 'sites' $\{s_1, \ldots, s_n\}$ at which the data $Y(s_1), \ldots, Y(s_n)$ should be collected, while the analysis problem concerns the way in which this data should be used to meet the objective. Both problems have solutions that depend on the objective addressed.

Sacks *et al.* (1989b) focused on the first objective (the prediction problem), and proposed an interesting design for computer experiments. In the context of prediction, the design problem is to select the inputs to make efficient predictions. Sacks *et al.* (1989b) assumed that the response is modelled by response = linear model + systematic departures, i.e.

$$Y(x) = \sum_{j=1}^{k} \beta_j f_j(x) + Z(x), \tag{3.3}$$

and treated the systematic departures Z as a realization of a stochastic process. The covariance structure of Z relates to the smoothness of the response. Once the correlation structure of Z has been specified (the choice of the correlation structure is discussed in Sacks *et al.*, 1989b), such a correlation can be used in (3.3) to provide predictions of $Y(x)$ from data $Y(s_1), \ldots, Y(s_n)$ drawn from a design $S = \{s_1, \ldots, s_n\}$, a set of eligible inputs. Then the best linear predictor (BLP) of $Y(x)$, namely $\hat{Y}(x)$, and its mean-square error, MSE, are computed. Of the many possible design criteria, Sacks *et al.* choose the one minimizing the integrated mean-square error (IMSE) of a prediction. Thus, the design problem is to choose a design S minimizing IMSE. Other alternative criteria are maximum mean-squared error (MMSE) or entropy (see Sacks *et al.*, 1989a). A limitation of this procedure is that the choice of the design S is conditional upon the choice of the correlation structure of Z, which cannot be known before taking observations from an experiment. The effect of the correlation function on the efficiency and robustness of the design and predictor is discussed in Sacks *et al.* (1989a, b). The problem of the computational cost is also tackled and a solution involving a sequential design algorithm is proposed in Sacks *et al.* (1989a).

Although the source of intense debates (see the discussion at the end of Sacks *et al.* 1989a), the originality of the contribution given by Sacks and co-authors has to be appreciated. Before their work, the majority of the existing designs for fitting predictors were those developed for physical experiments, not clearly appropriate for computer experiments, and design of deterministic computer experiments had only partly been addressed (Sacks and Ylvisaker 1984, 1985; Welch, 1983). Furthermore, these authors explicitly considered the problem of the sensitivity of the response function to its inputs, and introduced in DOE the concept of sensitivity measures. For insight into the relative effects of the inputs, Sacks *et al.* (1989a) used a decomposition of the response function $y(x)$ into an average, main effect for each input, two-input interactions, and higher-order interactions. The average of $y(x)$ over the experimental region is defined to be

$$\mu_0 = \int y(x) \prod_{h=1,\ldots,k} dx_h, \tag{3.4}$$

where k is the number of inputs and has the dimension of the vector x. The main effect of input x_i, averaged over the other inputs, is the function

$$\mu_i(x_i) = \int y(x) \prod_{h \neq i} dx_h - \mu_0; \tag{3.5}$$

the interaction effect of x_i and x_j is

$$\mu_{ij}(x_i, x_j) = \int y(x) \prod_{h \neq i,j} dx_h - \mu_i(x_i) - \mu_j(x_j) - \mu_0; \tag{3.6}$$

and so on for higher-order effects. The true effects of the inputs are estimated by replacing $y(x)$ with $\hat{y}(x)$, thus obtaining measures of the sensitivity of the response to its inputs. It is worthwhile to note that the sensitivity measures illustrated by Sacks *et al.* (1989a) are based on the same idea used by Sobol' in the 1990s to introduce the Sobol' sensitivity indices (see Chapter 8), i.e. the decomposition of the function of interest into terms of increasing dimensionality. Note that the variance-based measures described in Chapter 8 are based on the decomposition of the variance of the function rather than on the function itself.

The methods described by Sacks *et al.* (1989a,b) are computationally feasible for a small ($\leqslant 10$) number of inputs. An extended version that is able to cope with up to 30–40 inputs can be found in Welch *et al.* (1992). Given the increased number of inputs that it considers, the extended method can be used not only to build an accurate predictor but also for screening purposes (for a definition of 'screening', see Chapter 4). The idea of screening in DOE is not new. In 1946, Plackett and Burman introduced the two-level Plackett–Burman designs (Plackett and Burman, 1946), a method used for screening at least in physical experiments with random error. However, Welch *et al.* (1992) showed an example where their method performs better than the Plackett–Burman design, and other screening designs such as Latin hypercube. A relevant feature of the method is that it is data-adaptive: the nonparametric model defined by Welch *et al.* (1992) automatically adapts to nonlinear and interaction effects in the data. In this sense, the method is very flexible, because it allows those effects to emerge without explicitly modelling them.

3.8 CONCLUSIONS

Full factorial designs (Section 3.3) enable estimation of all high-order interactions. A drawback is that they have relatively high computational costs. Fractional factorial designs (Section 3.4) of resolution 5 and higher do permit the estimation of two-factor interaction effects, but they still require many simulation runs. OAT designs (Section 3.5) provide unbiased estimators of the effects of an individual factor, provided that factors act additively over the range of interest. Plackett–Burman designs and the Bayesian analysis suggested by Box and Meyer, 1986 (Section 3.5) may be used to detect the subset of input factor most influential on the response function. Other designs of experiment specifically created for computer codes (such as LHS) can be found in Chapters 8 and 4.

As mentioned in the introduction, DOE can be considered as one of the forerunners of sensitivity analysis. In this chapter, we have introduced many concepts that will be useful elsewhere in this volume, such as effect, interaction, and aliasing, and several specific experimental designs useful in computer experiments.

In Section 3.6, we highlighted the strong conceptual similarity between DOE in physical experimentation and SA in simulation. Nevertheless, we also noted that the different contexts of applications gave rise to a number of important distinctions. Thus, all methods that are valuable in DOE may not be appropriate for simulation experiments.

On the other hand, we do not want to discount the possibility that some among our readers might find it useful to straightforwardly apply a DOE method provided by a standard statistical package to a numerical experiment.

The statistical DOE techniques surveyed in this chapter are only a small subset of those available in the literature. They were chosen since, in our opinion, they are the most representatives of the philosophy underlying DOE as well as the most established in the literature.

4

Screening Methods

Francesca Campolongo

European Commission, Joint Research Centre, Ispra, Italy

Jack Kleijnen

Tilburg University, The Netherlands

Terry Andres

AECL Whiteshell Labs, Pinawa, Manitoba, Canada

4.1 INTRODUCTION

The central question of screening in the context of modelling and computer simulation is: Which factors—among the many potentially important factors—are really important? One of the aims in modelling is to come up with a *short list* of important factors (this is sometimes called the principle of parsimony or Occam's razor), and to do this, the choice of a well-designed experiment is essential.

This chapter gives a survey of statistical designs and analysis for screening, focusing on computer experiments. It is often assumed as a working hypothesis that the number of important factors is small compared with the total number of factors in a model (see e.g. Morris, 1987). This assumption is based on the idea that the influence of factors in models is distributed as income in nations, i.e. it follows Pareto's law, with a few, very influential factors and a majority of non-influential ones (see also Saltelli *et al.*, 1999a).

Screening designs are organized to deal with models containing hundreds of input factors. For this reason, they should be economical. As a drawback, these economical methods tend to provide qualitative sensitivity measures, i.e. they rank the input factors in order of importance, but do not quantify how much a given factor is more important than another. There is clearly a trade-off between computational cost and information.

Several approaches to the problem of screening have been proposed in the literature (see for instance, references given in Section 4.2 about screening techniques developed in the

Sensitivity Analysis. Edited by A. Saltelli *et al.*

context of physical experimentation). In this chapter, we shall describe a few of them through numerical examples and case studies. The methods described here include the one-factor-at-a-time (OAT) experiment proposed by Morris (1991), the design of Cotter (1979), the iterated fractional factorial designs (IFFDs) introduced by Andres (Andres and Hajas, 1993), and sequential bifurcation proposed by Bettonvil (Bettonvil, 1990; Bettonvil and Kleijnen, 1997). The examples used in this chapter demonstrate that these techniques can be simple, efficient, and effective.

The main conclusions of this chapter are as follows:

(i) Screening may involve several techniques, which are quite simple and yet efficient and effective.
(ii) These screening techniques have already been applied to several practical simulation studies, in different domains.

The remainder of this chapter is organized as follows. Section 4.2 gives notation, definitions, and key assumptions. Section 4.3 evaluates OAT designs. Section 4.4 describes the OAT design proposed by Morris. Section 4.5 describes the design type proposed by Cotter. Section 4.6 describes Andres' IFFD. Section 4.7 describes Bettonvil's sequential bifurcation. A summary and conclusions are given in Section 4.8.

4.2 DEFINITIONS

We use the term *factor* to denote any input included in the sensitivity study. Thus, a factor is any quantity that can be changed in the model prior to its running. This quantity can be a parameter, an input variable, or a module of the model. *Input variables* are directly observable in the corresponding real system, whereas *parameters* are not (they may be estimated). For example, parameters in simulation models of queuing problems in supermarkets and telecommunication systems are customer arrival and service rates; an input variable may be the number of parallel servers; a module may be the submodel for the priority rules (first-in-first-out or FIFO, shortest-processing-time or SPT, and so on).

By definition, factors are not kept constant during the whole experiment: a factor takes at least two *levels* or 'values'. A factor may also be *qualitative*. A brief discussion on quantitative and qualitative factors can be found in Chapter 3.

The term *metamodel* is used to denote the model of the underlying simulation model, i.e. the approximation of the simulation program's input/output (I/O) transformation.

Since the 1980s, screening design has become a common term. Some authors (e.g. Klejnen, 1998) restrict the term *screening designs* only to *supersaturated* designs, i.e. designs with fewer runs than factors. On the other hand, some others speak of screening designs when the number of runs is larger than the number of factors. For example, Myers and Montgomery (1995) refer to fractional factorials as being used as screening designs, with no hint of using supersaturated designs. Similarly, Rahni *et al.* (1997) refer to parameter screening designs as being equivalent to two-level experimental designs (but go on to discuss more efficient group screening methods). In this chapter, we shall use the term *screening design* to indicate any preliminary activity that, independently of the number of experimental runs it uses, aims to discover which of the input factors involved in a model are *important*, i.e. control most of the output variability.

Note that such a general definition does not necessarily imply the existence of a computer code, and can be applied in the case of a physical experiment. In the context of physical experimentation, a screening design indicates an experimental design that aims to discover which of a collection of experimentally controlled factors have an important influence upon an observable response. Screening methods aimed at physical experiments with random error have been proposed by Plackett and Burman (1946), whose designs are among the oldest used in physical experimentation for screening purposes, and by Satterthwaite (1959), who used supersaturated random-balance designs for screening the factors in linear models. Box and Meyer (1986) suggested a Bayesian analysis based on the assumption of few important factors. Other designs are the supersaturated designs proposed by Watson (1961), Srivastava (1975), and Morris (1987), who, following the approach introduced by Watson (1961), proposed a generalization of the group screening technique for finding non-negligible factors in a first-order model. Similar works, which are still developments of Watson's idea, are those of Mauro and Smith (1982), Patel (1962), and Patel and Ottieno (1984). However, in the present chapter we shall not illustrate these methods. We shall focus instead on screening methods developed in the context of computer experiments. For screening designs in the context of physical experimentation, see also Chapter 3.

Also excluded from the present chapter is the design proposed by Welch *et al.* (1992). Such a design in fact focuses on an objective different from that treated here: it aims not only at identifying the subset of most important factors in the model, but also wants to determine the way in which those factors jointly affect the model's predictions. In other words, Welch *et al.* (1992) do not separate the screening and prediction objectives so strongly, and propose a design whose results can be used to both screen and to build an accurate predictor.

A highly desirable property of a screening method is its low computational cost. The *computational cost* of the experiment is defined as the number of model evaluations (computer runs) required. This cost is usually a function of the number of factors involved in the analysis and of the complexity of the input/output behavior. A screening exercise that requires a high number of model evaluations would be inappropriate, especially when the evaluation of the model requires much time. Thus, the main goal in developing a screening method is to provide adequate information about the sensitivity of the model to its inputs, while keeping the computational cost of the whole experiment low.

4.3 ONE-AT-A-TIME (OAT) DESIGNS

The simplest class of screening designs is that of the one-at-a-time (OAT) experiments. In these designs, the impact of changing the values of each factor is evaluated in turn (Daniel, 1973). This approach is also known sometimes as *ceteris paribus*.

The standard OAT designs use the 'nominal' or 'standard' value per factor; often this value is taken from the literature. The combination of nominal values for the k factors is called the 'control' scenario. Two extreme values are usually proposed to represent the range of likely values for each factor; normally the 'standard' value of a factor is 'midway' between the two extremes. The magnitudes of the differences between the outputs for the extreme inputs and the 'control' are then compared to find those factors that significantly affect the model.

Although most commonly used, the standard strategy is not the only one followed when implementing an OAT experiment. According to Daniel (1973), OAT designs can be classified into five categories:

- standard OAT designs, which vary one factor from a standard condition;
- strict OAT designs, which vary one factor from the condition of the last preceding experimental run;

- paired OAT designs, which produce two observations and hence one simple comparison at a time;
- free OAT designs, which make each new run under new conditions;
- curved OAT designs, which produce a subset of results by varying only one easy-to-vary factor.

In physical experiments, the practice of an OAT experiment is acceptable if the random error is small compared with expected main effects. Nevertheless, such a design may give biased estimates, the biases originating from the effect of interactions. Daniel (1973) proposed methods for refinement of estimates provided by an OAT design; the improvement comes in by removing two-factor interaction biases from main effect estimates. Daniel's methods have been shown to work best for strict OAT; they do not work so well for standard designs that have the limitation of their conservatism (Daniel, 1973).

In general, the number of model evaluations required of an OAT design is of the order of k (often, $2k+1$), k being the number of factors examined. The low computational cost is one of the main advantages of the OAT design. However, Kleijnen (1998, p. 195) argues that a resolution-3 design (see Chapter 3), requires only roughly $k+1$ runs, and provides more accurate estimators *of the main effects*.

One limitation of OAT designs is that they do not enable estimation of interactions among factors. (Likewise, resolution-3 designs allow only the estimation of the main effects.)

Furthermore, many OAT experiments in the literature are *local* experiments; that is, factors are changed over small intervals around their nominal values. These nominal values represent a specific point of the input space (the 'control' scenario). Results of such a local experiment are thus dependent on the choice of this point, and the model behavior is identified only locally in the input space (namely, around the selected point). This is acceptable only if the input–output relationship can be adequately approximated through a first-order polynomial. If the model shows strong nonlinearity then a change in the selected nominal values provides totally different sensitivity results. The limitation of such local experiments (sometimes also called 'elementary OAT', or EOAT) is highlighted in an application example illustrated in Chapter 18.

An OAT design that is not dependent on the choice of the specific point in the input space is that proposed by Morris (1991).

4.4 MORRIS'S (1991) OAT DESIGNS

We call the OAT design proposed by Morris (1991) a *global* sensitivity experiment, because his experiment covers the entire space over which the factors may vary (in a local experiment, the factors vary only around their nominal values, and the results depend on the choice of these values). Morris estimates the main effect of a factor by computing a number (say) r of local measures, at different points $\mathbf{x}_1, \ldots, \mathbf{x}_r$ in the input space, and then taking their average (this reduces the dependence on the specific point that a local experiment has). These r values are selected such that each factor is varied over its interval of experimentation.

Morris assumes an expensive (in computer power/time) computational model, or a model with a large number of factors; the number of computer runs needed by his design is proportional to k (number of factors). His design does not need simplifying assumptions about the input/output behavior. Morris wishes to determine which factors have (a) negligible effects, (b) linear and additive effects, or (c) nonlinear or interaction effects. His design is composed of individual randomized OAT designs, in which the impact of changing the value of each of the chosen factors is evaluated in turn. In the terminology proposed by Daniel (1973), the Morris design is a strict OAT.

The k-dimensional factor vector $\mathbf{x}$ for the simulation model has components x_i that have p values in the set $\{0, 1/(p-1), 2/(p-1), \ldots, 1\}$. The region of experimentation Ω is then a k-dimensional p-level *grid*. In practical applications, the values sampled in Ω are subsequently rescaled to generate the actual (non-standardized) values of the simulation factors. Let Δ be a predetermined multiple of $1/(p-1)$. Then Morris defines the *elementary effect* of the ith factor at a given point $\mathbf{x}$ as

$$d_i(\mathbf{x}) = \frac{[y(x_1, \ldots, x_{i-1}, x_i + \Delta, x_{i+1}, \ldots, x_k) - y(\mathbf{x})]}{\Delta}, \tag{4.1}$$

where $\mathbf{x}$ is any value in Ω selected such that the perturbed point $\mathbf{x} + \Delta$ is still in Ω. A finite distribution (say) F_i of elementary effects for the ith input factor is obtained by sampling $\mathbf{x}$ from Ω. The number of elements of each F_i is $p^{k-1}[p - \Delta(p-1)]$.

The characterization of the distribution F_i through its mean μ and standard deviation σ gives useful information about the influence of the ith factor on the output. A high mean indicates a factor with an important overall influence on the output; a high standard deviation indicates either a factor interacting with other factors or a factor whose effect is nonlinear.

In its simplest form, the total computational effort required for a random sample of r values from each distribution F_i is $n = 2rk$ runs (k is the number of factors): each elementary effect requires the evaluation of y twice. Morris defines the *economy* of a design as the number of elementary effects estimated by the design, divided by the number of runs. The larger the value of the *economy* for a particular design, the better it is in terms of providing information for sensitivity and uncertainty analysis. The simplest form of Morris' design has an *economy* of $rk/2rk = 1/2$.

Morris proposed a more economical design than the simple design discussed so far. This design is based on the construction of a matrix $\mathbf{B}^*$ with rows that represent input vectors $\mathbf{x}$, for which the corresponding experiment provides k elementary effects (one for each input factor) from $k+1$ runs. This increases the economy of the design to $k/(k+1)$. In the development of such a design, it is convenient to assume that p is even and $\Delta = p/[2(p-1)]$. With such assumptions, each of the $p^{k-1}[p - \Delta(p-1)] = p^k/2$ elementary effects for the ith input factor has equal probability of being selected (see Morris, 1991). The key idea of the Morris design is the following:

1. A 'base' value $\mathbf{x}^*$ is randomly chosen for the vector $\mathbf{x}$, each component x_i being sampled from the set $\{0, 1/(p-1), \ldots, 1-\Delta\}$.
2. One or more of the k components of $\mathbf{x}^*$ are increased by Δ such that a vector (say) $\mathbf{x}^{(1)}$ results that is still in Ω.
3. The estimated elementary effect of the ith component of $\mathbf{x}^{(1)}$ (if the ith component of $\mathbf{x}^{(1)}$ has been changed by Δ) is (see Equation (4.1))

$$d_i(\mathbf{x}^{(1)}) = \frac{y(x_1^{(1)}, \ldots, x_{i-1}^{(1)}, x_i^{(1)} + \Delta, x_{i+1}^{(1)}, \ldots, x_k^{(1)}) - y(\mathbf{x}^{(1)})}{\Delta} \tag{4.2a}$$

if $\mathbf{x}^{(1)}$ has been increased by Δ, or

$$d_i(\mathbf{x}^{(1)}) = \frac{y(\mathbf{x}^{(1)}) - y(x_1^{(1)}, \ldots, x_{i-1}^{(1)}, x_i^{(1)} - \Delta, x_{i+1}^{(1)}, \ldots, x_k^{(1)})}{\Delta} \tag{4.2b}$$

if $\mathbf{x}^{(1)}$ has been decreased by Δ.

4. Let $\mathbf{x}^{(2)}$ be the new vector $(x_1^{(1)}, \ldots, x_{i-1}^{(1)}, x_i^{(1)} \pm \Delta, x_{i+1}^{(1)}, \ldots, x_k^{(1)})$ defined in the above step. Select a third vector $\mathbf{x}^{(3)}$ such that $\mathbf{x}^{(3)}$ differs from $\mathbf{x}^{(2)}$ for *only one component j*: either $x_j^{(3)} = x_j^{(2)} + \Delta$ or $x_j^{(3)} = x_j^{(2)} - \Delta$, with $j \neq i$. The estimated elementary effect of factor j is then

$$d_j(\mathbf{x}^{(2)}) = \frac{y(\mathbf{x}^{(3)}) - y(\mathbf{x}^{(2)})}{\Delta} \tag{4.3a}$$

if $\Delta > 0$, or

$$d_j(\mathbf{x}^{(2)}) = \frac{y(\mathbf{x}^{2}) - y(\mathbf{x}^{(3)})}{\Delta} \tag{4.3b}$$

otherwise.

Step 4 is repeated such that a succession of $k+1$ input vectors $\mathbf{x}^{(1)}, \mathbf{x}^{(2)}, \ldots, \mathbf{x}^{(k+1)}$ is produced with two consecutive vectors differing in only one component. Furthermore, any component i of the 'base vector' $\mathbf{x}^*$ is selected at least once to be increased by Δ to estimate one elementary effect for each factor. The successive vectors $\mathbf{x}^{(1)}, \mathbf{x}^{(2)}, \ldots, \mathbf{x}^{(k+1)}$ define a *trajectory* in the parameter space; an example is given in Figure 4.1 for $k = 3$ and $p = 4$.

Each component x_i of the 'base' vector $\mathbf{x}^*$ can only be increased (not decreased) by Δ (see the example below). Thus, any point of the trajectory in the parameter space will have Euclidean distance from the origin (the k-dimensional $\mathbf{0}$ vector) greater than the distance of the 'base' vector. This does not imply that each point has a greater distance than the previous one: a component x_i of $\mathbf{x}^*$ that was increased at a certain stage can be decreased in a successive step (maintaining a greater value than the 'base' one).

The rows of a $\mathbf{B}^*$ matrix are the vectors $\mathbf{x}^{(1)}, \mathbf{x}^{(2)}, \ldots, \mathbf{x}^{(k+1)}$ described above. This matrix is called the *orientation* matrix. It corresponds to one trajectory of k steps in the parameter space, with starting point $\mathbf{x}^{(1)}$. This provides a single elementary effect per factor.

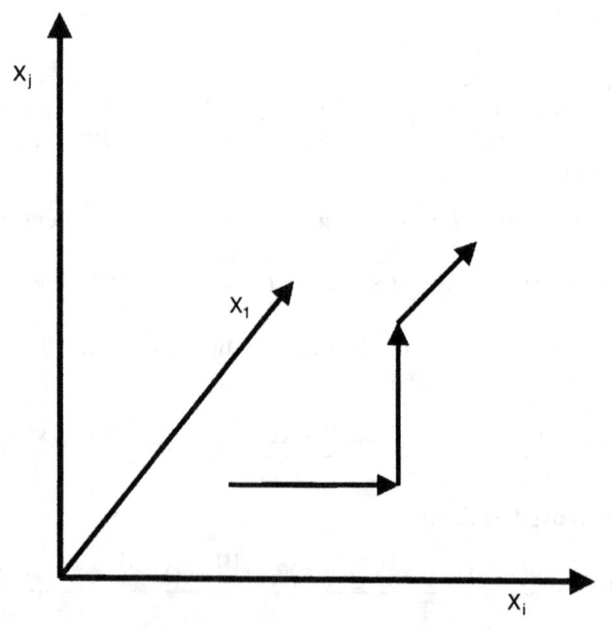

Figure 4.1 An example trajectory in the input factor space for $k = 3$ and $p = 4$.

To build a $\mathbf{B}^*$, the first step is the selection of a $(k+1) \times k$ matrix $\mathbf{B}$ with elements that are 0s and 1s such that for every column there are two rows of $\mathbf{B}$ that differ in only one element. In particular, $\mathbf{B}$ may be chosen to be a strictly lower triangular matrix of 1s. Consider the transposed matrix $\mathbf{B}'$ given by

$$\mathbf{B}' = \mathbf{J}_{k+1,1}\,\mathbf{x}^* + \Delta\mathbf{B}, \tag{4.4}$$

where $\mathbf{J}_{k+1,k}$ is a $(k+1) \times k$ matrix of 1s, and $\mathbf{x}^*$ is a randomly chosen 'base value' of $\mathbf{x}$. This $\mathbf{B}'$ could be used as a design matrix, since it would provide k elementary effects, one for each input factor, with a computational cost of $k+1$ runs. However, the problem is that the k elementary effects $\mathbf{B}'$ produces would not be randomly selected. A randomized version of the design matrix is given by

$$\mathbf{B}^* = (\mathbf{J}_{k+1,1}\mathbf{x}^* + (\Delta/2)[(2\,\mathbf{B} - \mathbf{J}_{k+1,k})\mathbf{D}^* + \mathbf{J}_{k+1,k}])\mathbf{P}^*, \tag{4.5}$$

where $\mathbf{D}^*$ is a k-dimensional diagonal matrix with elements either $+1$ or -1 with equal probability, and $\mathbf{P}^*$ is a $k \times k$ random permutation matrix, in which each column contains one element equal to 1 and all the others equal to 0, and no two columns have 1s in the same position. $\mathbf{B}^*$ provides one elementary effect per factor that is randomly selected.

4.4.1 Example

We suppose that $p = 4$, $k = 2$, and $\Delta = \frac{2}{3}$; that is, we examine two factors that may have values in the set $\{0, \frac{1}{3}, \frac{2}{3}, 1\}$. Then $\mathbf{B}$ is given by

$$\mathbf{B} = \begin{bmatrix} 0 & 0 \\ 1 & 0 \\ 1 & 1 \end{bmatrix},$$

and the randomly generated $\mathbf{x}^*$ $\mathbf{D}^*$ and $\mathbf{P}^*$ happen to be

$$\mathbf{x}^* = (0, \tfrac{1}{3}), \qquad \mathbf{D}^* = \begin{bmatrix} 1 & 0 \\ 0 & -1 \end{bmatrix}, \qquad \mathbf{P}^* = \mathbf{I}.$$

This gives

$$(\Delta/2)[(2\mathbf{B} - \mathbf{J}_{k+1,k})\mathbf{D}^* + \mathbf{J}_{k+1,k}] = \begin{bmatrix} 0 & \Delta \\ \Delta & \Delta \\ \Delta & 0 \end{bmatrix} = \begin{bmatrix} 0 & \frac{2}{3} \\ \frac{2}{3} & \frac{2}{3} \\ \frac{2}{3} & 0 \end{bmatrix},$$

and, from Equation (4.5),

$$\mathbf{B}^* = \begin{bmatrix} 0 & 1 \\ \frac{2}{3} & 1 \\ \frac{2}{3} & \frac{1}{3} \end{bmatrix},$$

or

$$\mathbf{x}^{(1)} = (0, 1), \qquad \mathbf{x}^{(2)} = (\tfrac{2}{3}, 1), \qquad \mathbf{x}^{(3)} = (\tfrac{2}{3}, \tfrac{1}{3}).$$

To estimate the mean and variance of the distribution $F_i(i = 1, \ldots, k)$, Morris takes a random sample of r elements; that is, he samples r mutually independent orientation matrices (corresponding to r different trajectories, each with a different starting point). Since each orientation matrix provides one elementary effect per factor, the r matrices together provide *rk-dimensional* samples, one for each $F_i (i = 1, \ldots, k)$.

This design gives k correlated estimators per trajectory (orientation matrix), whereas the r independent trajectories give r independent estimators. Therefore, the mean μ and standard deviation σ for each of the k factors can be estimated through the classic estimators for an independent random sample.

The main advantage of Morris' design is its relatively low computational cost. The design requires about one model evaluation per computed elementary effect, and a number r of elementary effects is computed for each factor. Thus, the design requires a total number of runs that is a linear function of the number of examined factors, k. The economy of the design is $rk/r(k+1) = k/(k+1)$.

The main disadvantage of the method is that individual interactions among factors can not be estimated. The method can provide an 'overall' measure of the interactions of a factor with the rest of the model, but it does not give specific information about the identity of the interactions.

An analytical test case, proposed by Morris (1991, see Chapter 2), is presented below.

4.4.2 Analytical Test Case

The computational model constructed by Morris (1991, Chapter 2) contains 20 input factors and has the following form:

$$y = \beta_0 + \sum_{i=1}^{20} \beta_i w_i + \sum_{i<j}^{20} \beta_{i,j} w_i w_j + \sum_{i<j<l}^{20} \beta_{i,j,l} w_i w_j w_l + \sum_{i<j<l<s}^{20} \beta_{i,j,l,s} w_i w_j w_l w_s,$$

where $w_i = 2 \times (x_i - \frac{1}{2})$ except for $i = 3, 5$, and 7, where $w_i = 2 \times (1.1x_i/(x_i + 0.1) - \frac{1}{2})$. Coefficients with relatively large values are assigned as

$$\beta_i = +20 \quad (i = 1, \ldots, 10), \qquad \beta_{i,j} = -15 \quad (i, j = 1, \ldots, 6),$$
$$\beta_{i,j,l} = -10 \quad (i, j, l = 1, \ldots, 5), \qquad \beta_{i,j,l,s} = +5 \quad (i, j, l, s = 1, \ldots, 4).$$

The remaining first- and second-order coefficients are independently generated from a normal distribution with zero mean and unit standard deviation; the remaining third- and fourth-order coefficients are set to zero.

Parameters of the experiment were set respectively to $l = 4$, $\Delta = \frac{2}{3}$, and $r = 4$. Using the same representation as in Morris (1991), the values obtained for the sensitivity measures μ and σ are displayed in Figure 4.2. The pattern described in Figure 4.2 almost reproduces the one shown in Figure 4.1 of Morris (1991). Input variables 1, ..., 10, which are supposed to have a significant effect on the output, are well separated from the others. In particular, as shown in Morris (1991), variables 8, 9, and 10 are separated from the others because of their high value of the mean (abscissa). Hence, considering both means and standard deviations together, one can conclude that the first ten factors are important; of these, the first seven have significant effects that involve either interactions or curvatures; the other three are important mainly because of their first-order effect.

The present volume also contains two examples of application of the Morris method to environmental models: a chemical kinetics model of the tropospheric oxidation pathways of dimethylsulfide (DMS), a sulfur-bearing compound of interest in climatic studies (see Chapter 18), and a model of fish population dynamics (Chapter 19).

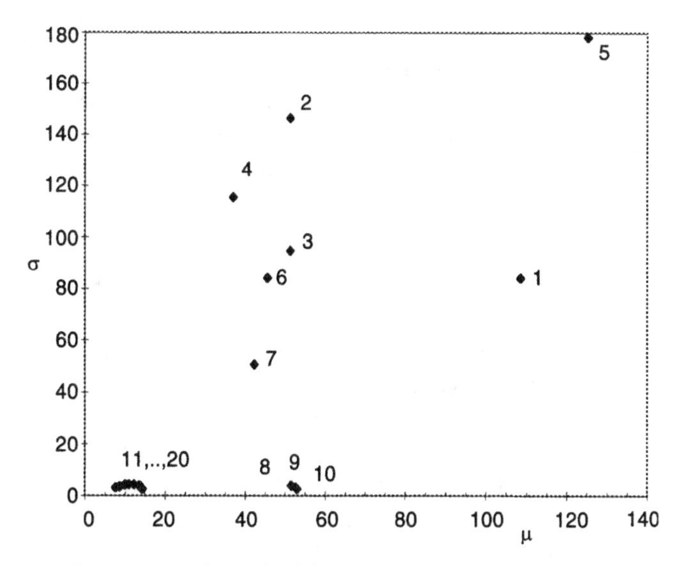

Figure 4.2 Estimated means μ and standard deviations σ of the distribution of elementary effects in the analytical example of Section 4.4.

A software package containing the Morris method, namely PREP-SPOP (SIMLAB), is described in the software Appendix.

4.5 COTTER'S DESIGN

Cotter (1979) proposed a type of design that does not require any prior assumptions about interactions, so his design results in less ambiguity of interpretation. His design is a *systematic fractional replicate design*. It requires the following $2k + 2$ runs for k factors:

- one initial run with all factors at their low levels;
- k runs with each factor in turn at its upper level, while all other $k - 1$ factors remain at their low levels;
- k runs with each factor in turn at its low level, while all other factors remain at their upper levels;
- one run with all factors at their upper levels.

Denote the resulting outputs by $y_0, y_1, \ldots, y_k, y_{k+1}, \ldots, y_{2k}, y_{2k+1}$. Let $S_o(j)$ and $S_e(j)$ denote the sum of odd-order and even-order effects involving factor j respectively, with expected values estimated by

$$C_o(j) = \tfrac{1}{4}[(y_{2k+1} - y_{k+j}) + (y_j - y_o)], \tag{4.6}$$

$$C_e(j) = \tfrac{1}{4}[(y_{2k+1} - y_{k+j}) - (y_j - y_o)]. \tag{4.7}$$

Then the measures $M(j) = |C_e(j)| + |C_o(j)|$ can be used to estimate the order of importance for the factors.

A major problem in the method proposed by Cotter is that an important factor may remain undetected. In fact, when a factor has effects that cancel each other out, the measures defined in (4.6) and (4.7) may fail. This occurrence, although unlikely, is not impossible, and an investigator has no means to protect him/herself against it.

Moreover, this design has the disadvantage of lack of precision: Equations (4.6) and (4.7) imply that, for one replicate, $\operatorname{var} C_o(j) = \operatorname{var} C_e(j) = \frac{1}{4}\sigma^2$, whereas a fractional replicate with n runs has its effects estimated with variance σ^2/n. The latter design, however, requires strict assumptions on the input/output behavior to be valid.

4.6 ANDRES' ITERATED FRACTIONAL FACTORIAL DESIGN (IFFD)

Andres developed the *iterated fractional factorial design* (IFFD), which requires fewer runs n than there are factors k: $n < k$ (Andres and Hajas, 1993). IFFD estimates the main effects, quadratic effects, and two-factor interactions of influential factors.

IFFD belongs to *group screening* designs; that is, initially individual factors are aggregated into clusters. Next, to identify an influential factor, IFFD groups factors into smaller subsets of the original set. Factors are randomly assigned to groups. These groups are investigated through a fractional factorial design. An influential group must contain an influential parameter. The procedure is then repeated with a different random grouping; influential factors lie in the intersection of influential groups in the two iterations.

IFFD samples three levels per factor, designated L (low), M (middle), and H (high). IFFD ensures that this sampling is balanced: different combinations of values for two or three factors appear with equal frequency. Therefore sampling follows an orthogonal fractional factorial design (FFD). More specifically, IFFD is a composite design consisting of multiple iterations of a basic FFD.

The first step of IFFD constructs a basic design with two levels only (L and H). It is a resolution-4 design, ensuring that main and two-factor interaction effects are not mutually confounded. To construct this design, IFFD follows Raghavarao (1971):

1. By definition, an order-n Hadamard matrix (say) $\mathbf{H}_n$ is a square $n \times n$ orthogonal matrix containing only 1s and -1s such that $\mathbf{H}_n'\mathbf{H}_n = n\mathbf{I}_n$, where $\mathbf{I}_n$ is the $n \times n$ identity matrix. An example is

$$\mathbf{H}_2 = \begin{pmatrix} 1 & -1 \\ 1 & 1 \end{pmatrix}.$$

By convention, the first column of a Hadamard matrix is shown with all 1s. (Any Hadamard matrix can be transformed to this form by inverting the signs in all rows having a -1 in the first column. The matrix remains orthogonal because inverting all the signs in any single row negates all cross-products with that row, leaving their values as 0). To construct a Hadamard matrix for any n that is a power of 2, perform a sequence of Kronecker products with $\mathbf{H}_2$. For example, to obtain $\mathbf{H}_4$, multiply each entry in $\mathbf{H}_2$ by the entire matrix $\mathbf{H}_2$ (the resulting matrix has four 2×2 submatrices rather than four scalar entries, and so it is a 4×4 matrix); it can be easily verified that it has the required properties. Since the number of input factors in a design can always be increased by adding dummy factors, one can assume without loss of generality that k is a power of 2. A Hadamard matrix $\mathbf{H}_k$ describes a balanced (i.e. having equal numbers of 1s and -1s) two-level FFD for $k-1$ factors in which each factor can take only the values 1 and -1 (which represent H and L respectively). Each column in the matrix except the first represents the values for a particular factor, and each row represents the combination of factor values for a simulation to be conducted. $\mathbf{H}_k[i,j]$ represents the value in the (i,j)th position of $\mathbf{H}_k$, the initial design matrix. Conversely, a complete set of fractional factorial designs of resolution 3 (see Chapter 3) can be used to generate a Hadamard matrix.

2. To obtain a resolution-4 design for k factors, double the Hadamard matrix $\mathbf{H}_k$ through the foldover principle (see Chapter 3), and use all the columns including the first:

$$\mathbf{J}_k[i,j]\begin{cases}\mathbf{H}_k[i,j] & (1 \leqslant i \leqslant k; 1 \leqslant j \leqslant k), \\ -\mathbf{H}_k[i-k,j] & (k+1 \leqslant i \leqslant 2k; 1 \leqslant j \leqslant k).\end{cases} \tag{4.8}$$

3. Randomly assign original (non-standardized) factors to columns of the folded matrix. Each column of the design matrix corresponds to a model factor. By randomly assigning factors to columns, the same design matrix $\mathbf{J}_k$ can be reused to generate a large number of different designs. The factor A gets its value $\mathbf{X}_A$ from the randomly chosen column $\mathbf{C}_A$; that is, $\mathbf{X}_A[i] = \mathbf{J}_k[i, \mathbf{C}_A]$.

Note that, if k is a power of 2 (as assumed before), a two-level fractional factorial design described above requires $2k$ simulations. This design is clearly not supersaturated, but iteration of the design allows it to acquire new features.

The following procedure describes the construction of a group-screening, supersaturated IFFD.

1. Generate a fractional factorial design, following the three steps above, to select a design matrix $\mathbf{J}_k$ for k groups of factors, where k is a power of 2 much smaller than the number of individual factors, K. Typically k is chosen as the power of 2 just larger than the number of influential factors to be identified. Commonly used values are $k = 8$ or $k = 16$ (Andres and Hajas, 1993).
2. Group the K individual factors: randomly assign each factor to one of the k groups (design variables or columns of the design matrix). As before, the column assigned to factor A is given by $\mathbf{C}_A$; now, however, many factors have the same column assignment.
3. To each factor, randomly assign a sign: to factor A randomly assign the sign s_A such that either $s_A = +1$ or $s_A = -1$ with equal probability $\frac{1}{2}$. Factors assigned to the same column of the design matrix may be given different signs:

$$\mathbf{X}_A[i] = s_A \mathbf{J}_k[i, \mathbf{C}_A]. \tag{4.9}$$

4. Iterate the preceding three steps M times, with M selected such that the total cost of the experiment (namely, $2kM$) is as close as possible to the budgeted number of simulation runs. In each of the iterations, the factors are randomly assigned to groups, and signs are randomly assigned to factors. The design matrix $\mathbf{J}_k$ remains the same. The notation given in Step 3 is modified to indicate the mth iteration:

$$\mathbf{X}_A^m[i] = s_A^m \mathbf{J}_k[i, \mathbf{C}_A^m]. \tag{4.10}$$

5. Create a three-level design: the sign s_A^m is set to zero in a specified proportion of the iterations, usually $\frac{1}{4}$. The positions of these zeros are randomly selected for each factor. Those iterations have $\mathbf{X}_A^m[i] = 0$ for all i, which means that the middle value (M) is used for factor A in the mth iteration. This step converts the entire design from a two-level into a three-level design, even though each of the iterations has either a single middle level M or two extreme levels (L, H) per individual factor.

First, each IFFD iteration m can be analysed separately; then results can be combined for an analysis of the entire composite design. Let $\mathbf{Y}^m[i]$ denote the output in the ith simulation of the mth iteration. Let $Z_j, j = 1, \ldots, k$, represent a factor that takes its values from the jth column of the basic design matrix $\mathbf{J}_k$. The main effect of Z_j on $\mathbf{Y}^m[i]$ is then

$$\mathrm{ME}_m(Z_j, \mathbf{Y}^m) = \frac{1}{k}\sum_{p=1}^{2k} \mathbf{J}_k[p, j]\mathbf{Y}^m[p]. \tag{4.11}$$

The main effect of factor A through the entire design is given by

$$\begin{aligned}\mathrm{ME}(A, \mathbf{Y}) &= \mathrm{avg}_m(s_A^m \,\mathrm{ME}(Z_{C_A^m}, \mathbf{Y}^m) | s_A^m \neq 0) \\ &= \frac{\sum_{m=1}^{M} s_A^m \,\mathrm{ME}(Z_{C_A^m}, \mathbf{Y}^m)}{\sum_{m=1}^{M} |s_A^m|}.\end{aligned} \tag{4.12}$$

Note that the denominator equals the number of iterations that have non-zero s_A^m.

Quadratic effects are estimated by

$$\begin{aligned}\mathrm{QE}(A, Y) &= \mathrm{avg}(Y|s_A = 0) - \mathrm{avg}(Y|s_A \neq 0) \\ &= \frac{\sum_{m=1}^{M}(1 - |s_A^m|)\sum_{i=1}^{2k}\mathbf{Y}^m[i]}{2k\sum_{m=1}^{M}(1 - |s_A^m|)} - \frac{\sum_{m=1}^{M}|s_A^m|\sum_{i=1}^{2k}\mathbf{Y}^m[i]}{2k\sum_{m=1}^{M}|s_A^m|}.\end{aligned} \tag{4.13}$$

The computation of the main and quadratic effects can be simplified by noting those terms that are reused in evaluating the effects of different factors.

The analysis of IFFD data sets is best accomplished through *stepwise regression*. Each important factor will give rise to 'copycats', which are unimportant factors that by chance share a column of the design matrix with an important factor, several times. Copycats' linear effects will be correlated with that of the important factor. Stepwise regression removes the contribution of each important factor, as that factor is identified. Stepwise regression can be simplified by taking advantage of the inherent structure of the IFFD (Saltelli *et al.*, 1995).

In practice, IFFD allows identification of the most influential factors among several thousand factors, in fewer than 100 simulation runs. In fact, the number of runs is relatively insensitive to the total number of factors in the simulation model, but it is very sensitive to the distribution of squared main effects (see Andres and Hajas, 1993). For best results, the output should be dominated by a few highly influential factors.

IFFD's performance was investigated with respect to two test cases by Saltelli *et al.* (1993, 1995). Hence, it is not necessary to perform any additional tests on the analytical functions proposed in Chapter 2. Saltelli *et al.* (1993) found that IFFD showed the best reproducibility among the methods tested; Saltelli *et al.* (1995) showed that IFFD predictions are very reproducible; IFFD performs as well as rank-based importance measures (see Chapter 8) and better than SRRC (Chapter 6). Furthermore, IFFD was shown to be more robust than SRRC in that it detected quadratic effects, and less robust than rank-based importance measures since it could not cope with higher-order effects (Saltelli *et al.*, 1995).

Andres' IFFD, with the improved procedure described below, is available in the software package SAMPLE2 described in the software Appendix.

A recent paper by Andres (1997) suggests two improvements to the IFFD procedure described here:

1. Combine the FFD with a Latin hypercube design (LHD) instead of setting some sign coefficients s_a^m to 0, so that levels are $0, 1, \ldots, k-1$ instead of $-1, 0$ and 1;
2. Use the design entries to designate intervals rather than fixed values, and randomly sample factor values from these intervals.

These two changes may be implemented independently, but are most effective when implemented together.

The first change (combining with a LHD) preserves the ability to compute main effects, for the lower half of the levels (i.e. 0 to $k/2-1$) can be considered 'low', and the higher half (i.e. $k/2$ to $k-1$) can be considered 'high'. A main effect for factor A now becomes the difference between average function values for A taking any high value versus A taking any low value. With this change, every simulation contributes to the calculation of the main effect, whereas in the original method, runs where A would take a middle value did not contribute. The fractional factorial Latin hypercube designs (FFLHD) constructed by Andres ensured that main effects sampled this way would have the same properties as in a simple FFD. Not only were the original advantages of the IFFD method maintained and improved, but also new features were added. With a balanced design on k levels, rather than two or three levels, one could now perform quadratic, cubic, and higher-order polynomial regression fits to the data for a single factor.

The second change (sampling from intervals) meant that the same simulations used for sensitivity analysis could also be used for unbiased estimation of mean values. The most obvious mean value is that of the objective function, obtained by averaging all of the simulation values. With the original IFFD approach, this average bore no clear connection to the theoretical mean of the objective function, no matter how many simulations were performed, because only three discrete values from each factor were used in the simulations. Once factor values were selected randomly from a designated level, any combination of factor values became a possible simulation that could be sampled by this design. Hence, the average value for all simulations became an unbiased estimator of the mean. Any mean value computable from the data could be treated this way. For example, to get an unbiased estimate of the cumulative distribution function of the function at a value x, one needs only to average an indicator variable that takes the value 1 for every function value less than or equal to x, and 0 for function values greater than x. Not only were the averages obtained this way unbiased estimators but they were very efficient estimators, because of the combined properties of the FFLH designs (Andres, 1997).

4.7 BETTONVIL'S SEQUENTIAL BIFURCATION

Like Andres' IFFD, sequential bifurcation (SB) is a *group-screening* technique. This technique was originally proposed in Bettonvil (1990); see also Bettonvil and Kleijnen (1997). At the start of the simulation experiment, this technique groups the individual factors into clusters. To know with certainty that individual factor effects within a group do not cancel out, SB must assume that the analyst knows whether a specific individual factor has a positive or negative effect on the simulation response; that is, the factor effects must have *known signs*. In practice, this assumption may not be too restrictive. For example, an ecological study performed in the Netherlands (discussed in Kleijnen, 1998) had 281 factors influencing the output, namely the future carbon dioxide concentration (greenhouse effect); the ecological experts could specify in which direction a specific factor affected the response. Another example is a queueing simulation of a supermarket or a telecommunications system: it may be known that increasing the service rates—while keeping all other factors constant (*ceteris paribus* assumption)—decreases waiting time. (But it is unknown how big this decrease is; therefore the analysts use a simulation model.) Moreover, if a few individual factors have unknown signs then these factors can be investigated separately, outside SB!

A specific group factor is said to be at its 'high' level (denoted as +), if each of its components or individual factors is at a level that gives a higher or equal response. Analogously, a group is at its 'low' level (−), if each component gives a lower response (see also IFFD).

A *sequential* design is one where factor combinations to be simulated are selected as the experimental results become available. So, as simulation runs are executed, insight into factor effects is accumulated and used to select the next run. It is well known that in general, sequential designs require fewer observations; the price is more cumbersome analysis and data handling (Kleijnen, 1998).

As the experiment proceeds, SB eliminates groups as soon as the procedure concludes that these clusters contain no important factors. Also, as the experiment proceeds, the groups become smaller. More specifically, each group that seems to include one or more important factors is split into two subgroups of the same size: *bifurcation*. Obviously, at the end of bifurcation, individual factor effects are estimated.

To illustrate the way in which SB works, we consider an academic example taken from Jacoby and Harrison (1962). Suppose the simulation has 128 factors, and we know—but SB does not—that three factors are important, namely factors #68, #113, and #120. We use the symbol $y_{(h)}$ to denote the simulation output when the factors $1, \ldots, h$ are switched on (are at high value) and the remaining factors $(h+1, \ldots, k)$ are off. We saw above that the assumption of known signs implies that the method can define the low level of an individual factor as the level that gives a low response value. Consequently, the response sequence $\{y_{(h)}\}$ is non-decreasing in the index h (as individual factors are switched on, the response may increase). The main effect of factor h is usually denoted by β_h. We introduce the symbol $\beta_{h-h'}$ to denote the sum of individual effects β_h through $\beta_{h'}$ with $h' > h$. For example, β_{1-128} denotes the sum of β_j through β_{128}; see Figure 4.3, line 1.

At the start (stage #0) of the procedure, SB always observes the two 'extreme' factor combinations, namely $y_{(0)}$ (no factor high) and $y_{(k)}$ (all factors high). The presence of (three) important factors gives $y_{(0)} < y_{(128)}$. Hence, SB infers that the sum of all individual main effects is important: $\beta_{1-128} > 0$. SB works such that any important sum of effects leads to a new observation that splits that sum into two subsums: see the symbol $\downarrow$ in Figure 4.3. Because stage #0 gives $\beta_{1-128} > 0$, SB proceeds to the next stage.

Stage #1 gives $y_{(64)}$. SB first compares $y_{(64)}$ with $y_{(0)}$, and notices that these two outputs are equal (remember that only factors #68, #113, and #120 are important). Hence, SB concludes that the first 64 individual factors are unimportant! So after only three simulation runs and based on the comparison of two runs, SB eliminates all factors in the first half of the total group of 128 factors.

Next, SB compares $y_{(64)}$ with $y_{(128)}$, and notices that these two outputs are not equal. Hence, the procedure concludes that the second subgroup of 64 factors is important; that is, there is at least one important factor in the second half of the group of 128 factors.

In stage #2, SB concentrates on the remaining factors (#65 through #128). That subgroup is again bifurcated. And so the procedure goes on; see Figure 4.3.

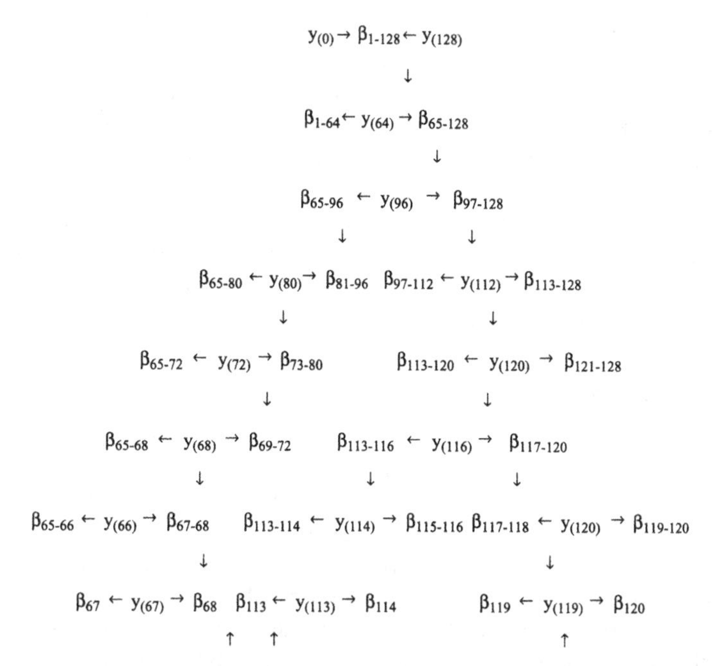

Figure 4.3 Finding $k = 3$ important factors among $K = 128$ factors in Jacoby and Harrison's (1962) example. Reproduced by permission of John Wiley & Sons, Ltd from Kleijnen (1998), Figure 6.1, p. 179.

At the end, SB finds the three important factors (#68, #113, #120) in this example. In total, SB requires only 16 observations to search for the three important factors among the 128 factors. The procedure also determines the individual main effects of the important factors: see the symbol in the last line of Figure 4.3.

Next we discuss some SB details. It can be proved that the individual factors with the largest main effects are identified first. For example, in an academic example with 24 factors, the two most important factors were identified after 13 runs. After four more runs, six other factors were identified; all remaining factors were known to have main effects smaller than 140 units. So the analysts could decide to stop screening at stage 17, if they felt it was not worthwhile to spend more time on the identification of factors with effects smaller than 140. So, it is not necessary to quantify a critical threshold (namely, 140) *a priori*. See Bettonvil and Kleijnen (1997, p. 185).

If the analysts assume that there are *interactions* between individual factors then the number of runs in SB is *double* the number required in the 'main effects only' case: foldover principle (see Chapter 3). This doubling gives estimators of main effects that are not biased by interactions between factors. In general, we recommend such a 'conservative' design strategy.

The analysts may wish to allow for individual factors with *quadratic effects*; that is, the *metamodel* may show curvature, implying a response maximum for the optimal combination of factors. We propose to apply SB, assuming main effects and two-factor interactions only. After this screening, individual factors may be further investigated, now allowing for more complicated metamodels (so more than two values per individual factor must be simulated; see also alternative screening designs). Obviously, such an approach does not identify a factor that has such a quadratic effect, because simulating only the two extreme values gives (roughly) the same response values. However, we conjecture that such a situation occurs rarely.

Finally, we discuss some case studies that use SB. The ecological case study mentioned at the beginning of this section took 154 simulation runs to identify and estimate the 15 most important factors among the original 281 factors. Some of these 15 factors surprised the ecological experts, so SB may be a powerful statistical (black box) technique. Notice that on hindsight it turned out that there are no important interactions between factors, so only 154/2 = 77 runs would have sufficed (no foldover).

Another case study is the building thermal deterministic simulation in De Wit (1997). In his simulation, SB gave the 16 most important inputs among the 82 factors, after only 50 runs. He checked these results by applying Morris' screening technique described in Section 4.4; the latter technique took 328 runs.

4.8 CONCLUSIONS

In the initial phase of a simulation, it is often necessary to find out which factors amongst the multitude of potential factors are really important. The goal is then to reduce the number of factors to be further explored in the next phase.

Some designs (called supersaturated designs) require fewer runs than factors! In this chapter, we have surveyed several types of design: one-at-a-time (OAT) designs (including Morris's design), the systematic fractional replicate design proposed by Cotter, the iterated fractional factorial design (IFFD), and Bettonvil's sequential bifurcation (SB). Each type has its own advantages and disadvantages.

OAT designs have as a major limitation the neglect of factor interactions. The advantage is that OAT does not make simplifying assumptions such as that only a few factors have important effects or that the input/output (I/O) function is monotonic. Moreover, the computational cost of OAT designs is linear in the number of factors. However, OAT methods provide unbiased estimators of the effects of each individual factor, provided that these effects are the same at different settings of the other factors; that is, the factors act additively, over the range of interest. Such an assumption, although advantageous to simplify the problem, can be rarely accepted. In fact, interactions are usually relevant, and need to be estimated by varying factors simultaneously. On the other hand, when the model is expensive to run, and there are many factors, Morris' method is both efficient and easy to implement. The Morris' method is available in the software package PREP-SPOP (SIMLAB) (see the software Appendix).

The systematic fractional replicate design of Cotter (1979) is computationally efficient and does not require any prior assumption about interactions or few important factors. However, the design lacks precision and cannot detect factors having effects that cancel each other out.

IFFD estimates the main effects, quadratic effects, and two-factor interactions of the most influential factors, with a number of runs that is small compared with the total number of factors. However, for good results, the model output should be determined by only a few highly influential factors. Andres' IFFD is available in the package SAMPLE2 (see the software Appendix).

Sequential bifurcation is simple, efficient, and effective (as several case studies have illustrated). Its major limitations are that the signs of main effects must be known, and that metamodels with only main effects and two-factor interactions must be adequate I/O approximations.

The screening designs described in this chapter are only a small subset of the total number available in the literature. We have presented only those designs that focus on the problem of the identification of the few important factors in a model.

5
Local Methods

Tamas Turányi

Eötvös University (ELTE), Budapest, Hungary, and Chemical Research Center, Budapest, Hungary

Herschel Rabitz

Princeton University, USA

5.1 INTRODUCTION

Mathematical models are widely used in various disciplines, and most of these models are based on systems of algebraic and differential equations. A growth in the number of variables and parameters of mathematical models has been observed over recent years. The basic reason for this is that, in the course of refinement of physical insight, models become more sophisticated. In addition, since the capacity of computers has grown, models that are more complex can be handled more easily. A common problem is that, in large models, the role of various parameters is not obvious. Usually it is not clear which are the important parameters, what is the effect of changing parameters, what is the uncertainty of the model results, originating from the uncertainty of parameters, and so on.

Local sensitivities provide the slope of the calculated model output in the parameter space at a given set of values. In many applications, this is exactly the information needed. In other areas, such as uncertainty analysis, local sensitivity analysis is a computationally efficient technique that allows a rapid preliminary exploration of the model.

There have been a number of reviews of local methods: A comprehensive review of sensitivity analysis was given by Rabitz *et al.* (1983). This review dealt mainly with local methods, and concentrated on distributed-parameter systems. Applications, mainly in chemical kinetics and molecular dynamics, were presented there. The review of Turányi (1990a) described both global and local methods, and provided an almost complete list of applications in chemical kinetics up to 1989. The review by Radhakrishnan (1990) dealt with the numerical aspects of local sensitivity methods, with an emphasis on combustion chemical modeling. The review by Tomlin *et al.* (1997) discussed the applications of several mathematical methods, including sensitivity analysis, to combustion kinetics.

Sensitivity Analysis. Edited by A. Saltelli *et al.*

5.2 FEATURES OF LOCAL SENSITIVITIES

Time-independent (stationary) systems can be characterized by the following system of algebraic equations:

$$0 = \mathbf{f}(\mathbf{y}, \mathbf{k}) \tag{5.1}$$

where $\mathbf{y}$ is the n-vector of variables and $\mathbf{k}$ is the m-vector of parameters. The solution of the implicit algebraic Equation (5.1) is denoted by $\mathbf{y}^s$. The solution changes when the values of parameters $\mathbf{k}$ are changed, and the new solution can be obtained from the following equation:

$$\mathbf{y}^s(\mathbf{k} + \Delta\mathbf{k}) = \mathbf{y}^s(\mathbf{k}) + \sum_{j=1}^{m} \frac{\partial y_i}{\partial k_j} \Delta k_j + \frac{1}{2} \sum_{l=1}^{m} \sum_{j=1}^{m} \frac{\partial^2 y_i}{\partial k_l \, \partial k_j} \Delta k_l \, \Delta k_j + \cdots. \tag{5.2}$$

A chemical example of such a system is the concentration in a well-stirred (i.e. spatially homogeneous) stationary reactor.

Non-stationary systems can be described by differential or differential–algebraic systems of equations. Consider the following initial-value problem:

$$\frac{d\mathbf{y}}{dt} = f(\mathbf{y}, \mathbf{k}), \qquad \mathbf{y}(0) = \mathbf{y}^0. \tag{5.3}$$

Here again, $\mathbf{y}$ is the n-vector of variables and $\mathbf{k}$ is the m-vector of system parameters, and $\mathbf{y}^0$ is the array of initial values. Solution of the initial-value problem (5.1) provides the time evolution of the system variables.

The effect of parameter change on the solution can be expressed through a Taylor series expansion:

$$\mathbf{y}^s(t, \mathbf{k} + \Delta\mathbf{k}) = \mathbf{y}^s(t, \mathbf{k}) + \sum_{j=1}^{m} \frac{\partial y_i}{\partial k_j} \Delta k_j + \frac{1}{2} \sum_{l=1}^{m} \sum_{j=1}^{m} \frac{\partial^2 y_i}{\partial k_l \, \partial k_j} \Delta k_l \, \Delta k_j + \cdots. \tag{5.4}$$

In both the time-dependent and time-independent cases, the partial derivatives $\partial y_i / \partial k_j$ are called first-order local sensitivities, $\partial^2 y_i / \partial k_l \, \partial k_j$ are called second-order local sensitivities, and so on. The first-order local sensitivities form the sensitivity matrix $\mathbf{S} = \{s_{ij}\} = \{\partial y_i / \partial k_j\}$.

Global sensitivity coefficients depend on the assumed probability density function of the parameters, and usually also on the method of calculation chosen. In contrast, local sensitivity coefficients are defined exactly by Equations (5.2) and (5.4). There are several numerical methods for the calculation of local sensitivities, but the calculated values should be identical within the numerical accuracy of the method used. Also, calculation of local sensitivities is much faster than that of global sensitivities. However, local sensitivities have some special limiting features that have to be kept in mind.

For all models of real systems, the values of the parameters are subject to some uncertainty. In most cases, such uncertainties can be very high, and sometimes when the parameters are changed within the range of uncertainty, a qualitatively different model is obtained. Unlike global sensitivities, local sensitivities are totally incapable of providing information on the effect of significant parameter changes. Local sensitivities are really local, and the information provided is related to a single point in the space of parameters. The point investigated is usually the point of best parametric estimate, also called the nominal value of parameters. Small variations in parameter values usually do not change the local sensitivities dramatically, but a significantly different parameter set may result in a completely different sensitivity pattern.

Sensitivity analysis of time-dependent systems has another characteristic feature. In most cases, sensitivity analysis can be considered as probing the model using another set of parameters. However, sensitivity analysis can also be used for the analysis of a model via perturbation of the parameters. In the former case, the parameters are changed at simulation time zero, and therefore the initial time of sensitivity calculation is equal to the initial time of simulation. In the general case, however, the initial times of the model and of the sensitivities are different. Let the simulation be started at time 0, let the parameters be perturbed at time t_1, and let the effect of the perturbation be studied at time t_2. The perturbed solution $\mathbf{y}'$ can be approximated from the original solution $\mathbf{y}$ and sensitivity matrix $\mathbf{S}$:

$$\mathbf{y}'(t_2) \approx \mathbf{y}(t_1) + \mathbf{S}(t_2, t_1)\Delta\mathbf{k}_{t_1}. \tag{5.5}$$

This means that the sensitivity matrix $\mathbf{S}$ has double time dependence in the general case, and the time limits t_1 and t_2 provide a degree of freedom in the analysis of models.

5.3 NUMERICAL METHODS FOR THE CALCULATION OF LOCAL SENSITIVITIES

5.3.1 Finite-Difference Approximation

The simplest way to calculate local sensitivities is based on slightly changing one parameter at a time and rerunning the model. Using the *finite-difference approximation*, elements of the sensitivity matrix can be approximated by

$$\frac{\partial \mathbf{y}}{\partial k_j} \approx \frac{\mathbf{y}(k_j + \Delta k_j) - \mathbf{y}(k_j)}{\Delta k_j}, \qquad j = 1, \ldots, m. \tag{5.6}$$

This procedure is also called the *brute force method* or the *indirect method*. The main advantage of this method is that no modification to the original model or extensive extra coding is needed. However, the brute force method is slower and less accurate than more sophisticated methods.

Calculation of local sensitivities in this way requires $m + 1$ simulations of the original model. If central differences are used, $2m$ simulations are required. The accuracies of the sensitivities calculated depend on the parameter change Δk_j. In the case of nonlinear models, parameter changes that are too large (e.g. $>5\%$) would damage the assumption of local linearity. If the parameter change is too small, the difference between the original and perturbed solutions is too small and the round-off error is too high. In most cases, a 1% perturbation is a good practical choice, but finding the best (or acceptable) value is a trial-and-error process.

5.3.2 Direct Method

Differentiation of Equation (5.3) with respect to k_j gives the following system of *sensitivity differential equations*:

$$\frac{d}{dt}\frac{\partial \mathbf{y}}{\partial k_j} = \mathbf{J}\frac{\partial \mathbf{y}}{\partial k_j} + \frac{\partial \mathbf{f}}{\partial k_j}, \tag{5.7}$$

or, in matrix form,

$$\dot{\mathbf{S}} = \mathbf{JS} + \mathbf{F}. \qquad (5.8)$$

Here $\mathbf{J} = \{\partial f_i/\partial y_l\}$ is the derivative of the right-hand side of the differential equation with respect to the system variables (called the Jacobian matrix) and $\mathbf{F} = \{\partial f_i/\partial k_j\}$ is the derivative with respect to the parameters, sometimes called the parametric Jacobian. The initial condition of the differential equation (5.7) is a zero vector.

Direct methods are based on the solution of the ODE (5.7). Numerical solution of Equation (5.7) requires knowledge of the values of the matrices $\mathbf{J}$ and $\mathbf{F}$ at each step of the ODE solver. To evaluate these matrices, the actual values of the system variables have to be known, and therefore a simultaneous or preceding solution of the ODE (5.3) is needed. In the first realizations of the direct method, Equations (5.3) and (5.7) were solved independently but simultaneously, and the solution of Equation (5.3) was used for setting up Equation (5.7). All variants of this algorithm were relatively slow.

Dunker (1981, 1984) was the first to show that a special relation between Equation (5.3) and Equation (5.7) allows a numerical shortcut, and called this algorithm the *decoupled direct method* or *DDM*. Equations (5.3) and (5.7) have the same Jacobian, and therefore a stiff ODE solver selects the same step size and order of approximation for the solution of both equations. In Dunker's method, the ODE solver decomposes the Jacobian only once, and then takes a timestep solving Equation (5.3) and then solving Equation (5.7) with all parameters one after the other. Since the triangularization of the Jacobian is the most time-consuming part of a stiff ODE solution, using the decoupled direct method, sensitivities can be calculated with relatively little extra cost.

Several implementations of the DDM exist, and the DDM has proved to be the best general method for the numerical calculation of local sensitivities.

In the case of stationary systems, if the stationary point is asymptotically stable, the stationary sensitivity coefficients are limits in time of the dynamic ones, and their time derivatives tend to zero. Therefore, Equation (5.8) can be transformed to

$$\mathbf{S}^s = -\mathbf{J}^{-1}\mathbf{F}. \qquad (5.9)$$

The matrix $\mathbf{S}^s$ is the *stationary sensitivity matrix* and the matrices $\mathbf{J}$ and $\mathbf{F}$ are evaluated at the variable values of the stationary point. Equation (5.9) can also be applied when the original model is defined as a system of algebraic equations (5.1).

5.3.3 The Green Function Method

Differentiating Equation (5.3) with respect to the initial values $\mathbf{y}^0$, the following equation is obtained:

$$\frac{d}{dt}\mathbf{K}(t, t_1) = \mathbf{J}(t)\mathbf{K}(t, t_1), \qquad (5.10)$$

where t_1 and t are the time of perturbation and the time of observation, respectively, and $\mathbf{K}$ is the initial value sensitivity matrix, that is

$$\mathbf{K}(t, t_1) = \left\{\frac{\partial c_i(t)}{\partial c_j^0(t_1)}\right\}, \qquad \mathbf{K}(t_1, t_1) = \mathbf{I}, \qquad t \geqslant t_1.$$

Equation (5.7) is a linear inhomogeneous system of differential equations, and therefore it can be solved by first determining the homogeneous part (5.10) and then calculating the particular solution:

$$\mathbf{S}(t_1, t_2) = \int_{t_1}^{t_2} \mathbf{K}(t_2, s)\mathbf{F}(s)\, ds. \tag{5.11}$$

In this equation, $\mathbf{K}$ is known as the *Green function* or *kernel*, and the numerical method based on the solution of Equation (5.11) is called the *Green function method*.

The Green function method also has several variants. The most developed of these is called the analytically integrated Magnus version of the Green function method (GFM/AIM) (Kramer *et al.*, 1981). In this version, the matrix $\mathbf{K}$ is approximated by a matrix exponential:

$$\mathbf{K}(t + \Delta t, t) = \exp\left[\int_{t}^{t+\Delta t} \mathbf{J}(s)\, ds\right]. \tag{5.12}$$

The GFM/AIM method is several times faster than other versions of the Green function method.

Applying the direct method, the numerical effort increases linearly with the number of parameters. In the case of Green function methods, the numerical effort is proportional to the number of variables. In practice, however, the GFM is faster than the DDM only at a very high ratio of the number of parameters to the number of variables, and the numerical error is less easily controllable than in the case of the much simpler DDM algorithm.

5.3.4 Other Methods

Other methods have also been described in the literature, but they are much less widespread. It is frequently useful (Miller and Frenklach, 1983) to approximate the integrated solution of the original model by an array of simpler empirical equations as a function of parameters in a parameter region. Preparation of such an empirical model is very time-consuming, and cannot be justified by only the sensitivity calculations. However, if such an empirical function is available, differentiating it provides an estimate of the local sensitivities as a by-product.

According to the polynomial approximation method (Hwang, 1983), the solution of the sensitivity differential equations (5.7) is approximated by Lagrange interpolation polynomials. Although high computational speed and good numerical stability were demonstrated, this method was never applied to real problems.

5.4 DERIVED SENSITIVITIES

In the case of models defined by the differential equations (5.3), not only the actual values of variables are interesting but also their rates of change at a given time. The rates of change of variables are given by the left-hand side of Equation (5.3). Since

$$\frac{d}{dt}\left(\frac{\partial y_i}{\partial k_j}\right) = \frac{\partial (dy_i/dt)}{\partial k_j},$$

the sensitivities of the rates of change of variables can be calculated by Equation (5.7), knowing the local sensitivity coefficients.

The *rate sensitivity matrix* $\dot{\mathbf{S}}(t_1, t_2)$ also has double time dependence. If the two times coincide ($t_1 = t_2$), the instantaneous effect of parameter change is obtained. It is clear from Equation (5.8) that the matrix $\mathbf{F} = \{\partial f_i/\partial k_j\}$ can be considered as an instantaneous rate sensitivity matrix. Knowing the values of variables at a given time, $\mathbf{F}$ can be calculated analytically and therefore the solution of the sensitivity ODE (5.8) is not needed. It has been shown (Turányi *et al.*, 1989) that $\mathbf{F}$ can provide valuable information on the structures of models.

Mathematical models may provide qualitative information. Such information can be whether a model oscillates, if a given variable reaches a threshold value during the time interval inspected, and so on. Sensitivity analysis cannot be used for the study of such information. On the other hand, frequently the information desired is quantitative, but may not be among the primary outputs of the model, although it can be deduced from the time histories of variables. Such information might be the maximum value of a variable, the time needed for a variable to reach a threshold value, or, in the case of periodic solutions, the period time. Such quantitative information can be called a feature, and its sensitivity is named *feature sensitivity*.

The brute force method offers a direct way to calculate feature sensitivities (Frenklach, 1984). A particular feature is evaluated from the original and perturbed solutions, and the feature sensitivity is calculated using finite differences.

In many cases, the feature sensitivities can also be calculated from the local sensitivities of variables. As an example, assume that variable i has a maximum (or minimum) at time t^*. This implies that the time derivative of the variable is zero:

$$\dot{y}_i(\mathbf{k}, t)\big|_{t=t^*} = 0. \tag{5.13}$$

Differentiating Equation (5.13) with respect to the parameter k_j, the following equation is obtained (Rabitz *et al.*, 1983) for the calculation of the sensitivity of the location of the maximum:

$$\frac{\partial t^*}{\partial k_j} = -\frac{\dfrac{\partial^2 y_i(t^*)}{\partial t\, \partial k_j}}{\dfrac{\partial^2 y_i(t^*)}{\partial^2 t}}. \tag{5.14}$$

The numerator contains the appropriate rate sensitivity coefficient $\dot{s}_{ij}(0, t^*)$, while the denominator can be calculated from the Jacobian and the right-hand side of the original ODE:

$$\frac{\partial^2 \mathbf{y}}{\partial t^2} = \mathbf{J}\mathbf{f}(\mathbf{y}). \tag{5.15}$$

Another frequently applied feature sensitivity is the sensitivity of the period time of periodic (oscillating) models. Period time sensitivities can also be calculated (Edelson and Thomas, 1981) approximately from the local variable sensitivities:

$$\frac{\partial \tau}{\partial k_j} = \frac{\dfrac{\partial y_i(t_2)}{\partial k_j(t_1)} - \dfrac{\partial y_i(t_2 + \tau)}{\partial k_j(t_1)}}{\dfrac{dy_i(t_2)}{dt}}. \tag{5.16}$$

5.5 INTERPRETATION OF SENSITIVITY INFORMATION

5.5.1 Effect of Changing One Parameter on a Single Variable

The sensitivity coefficient $\partial y_i/\partial k_j$ is a linear estimate of the number of units change in the variable y_i as a result of a unit change in the parameter j. This also means that the sensitivity result depends on the physical units of variables and parameters, and is meaningful only when the units of the model are known. In the general case, the variables and the parameters each have different physical units, and therefore the sensitivity coefficients cannot be compared with each other.

To make the sensitivity results independent of the units of the model, usually normalized sensitivity coefficients are applied. The normalized local sensitivity matrix is denoted by $\tilde{\mathbf{S}}$ and is defined as

$$\tilde{\mathbf{S}} = \left\{ \frac{k_j}{y_i} \frac{\partial y_i}{\partial k_j} \right\}. \tag{5.17}$$

These coefficients represent a linear estimate of the percentage change in the variable y_i caused by a one percent change in the parameter k_j. The normalized sensitivity coefficients are independent of the original units of the model, and are comparable with each other.

A practical difficulty in handling sensitivity matrices comes from their size. A reasonably sized model may consist of 50 variables and 100 parameters. This results in a sensitivity matrix of 5000 elements. In addition, if the sensitivities are studied at 20 time points then 10^5 numbers have to be compared and analyzed. It is inevitable that some methods have to be used for summarizing the sensitivity information.

5.5.2 Effect of Changing one Parameter on Several Variables

In model optimization, the improvement of the fit is expressed by the change in a single number. This is achieved by introducing an objective function, which converts the multivariate output of the model to a single value. As an example, such an objective function can be:

$$e(t_1, t_2) = \sum_{i=1}^{n} w_i \left[\frac{y_i^*(t_1, t_2) - y_i(t_2)}{y_i(t_2)} \right]^2, \tag{5.18}$$

where $y_i(t_2)$ is the solution of the model at time t_2 at the nominal parameter set and $y_i^*(t_1, t_2)$ is the solution of the model at time t_2 using a parameter set perturbed at time t_1. The weights w_i allow the expression of the relative importance of the model variables according to the modeller. For some variables, this weight can be zero, showing that the variable has to be present in the model as an auxiliary variable, but its value is not interesting at all.

The sensitivity of the objective function above can be calculated from the local variable sensitivities:

$$\frac{\partial e}{\partial k_j}(t_1, t_2) = \sum_{i=1}^{n} w_i \left[\frac{1}{y_i} \frac{\partial y_i}{\partial k_j}(t_2, t_1) \right]^2. \tag{5.19}$$

Investigation of the sensitivity of objective functions significantly decreases the number of sensitivities to be inspected. However, for a fixed time of perturbation t_1, the sensitivities still have to be studied at several time points t_2 to get an impression of the change of sensitivities in time.

The next stage of information compression is the application of time-integrated sensitivities. The corresponding objective function is

$$e = \int_{t_1}^{t_2} \sum_{i=1}^{n} w_i \left[\frac{y_i^*(t_1, t_2) - y_i(t_2)}{y_i(t_2)} \right]^2 dt \tag{5.20}$$

The sensitivity of this objective function can be approximately calculated by

$$\frac{\partial e}{\partial k_j} = \sum_{h=2}^{l} \sum_{i=1}^{n} w_i \left[\frac{1}{y_i} \frac{\partial y_i}{\partial k_j}(t_h, t_1) \right]^2 . \tag{5.21}$$

Sensitivities of objective functions, calculated from normalized sensitivities are called *overall sensitivities* (Vajda *et al.*, 1985). Selecting proper weights w_i, the overall sensitivities provide information on the importance of model parameters.

5.5.3 Effect of Simultaneously Changing Several Parameters on Several Variables

The overall sensitivities give information on the change of single parameters only. However, changing several parameters simultaneously can strengthen or weaken the effect of single parameter changes. First-order local sensitivities always correspond to 'changing one parameter at a time', and do not show the effect of simultaneous parameter changes. *Principal component analysis* (Vajda *et al.*, 1985; Vajda and Turányi, 1986) can, however, be used to estimate the effect of simultaneous parameter changes on several variables, based on local sensitivities only.

Use the time-integrated objective function (5.20) to assess the effect of parameter changes and replace the integral by a summation:

$$e(\boldsymbol{\alpha}) = \sum_{h=2}^{l} \sum_{i=1}^{n} \left[\frac{y_i^*(t_h) - y_i(t_h)}{y_i(t_h)} \right]^2 . \tag{5.22}$$

Assuming that all parameters are positive, normalized parameters, $\boldsymbol{\alpha}$, defined as $\boldsymbol{\alpha} = \ln \mathbf{k}$, can be used. If some of the parameters are negative, a simple modification of the model can lead to all-positive parameters. For simplicity, let the weights now be either 1 or 0. Weight 0 deletes the corresponding row from the sensitivity matrix. The local change of the objective function above around the nominal values of parameters $\boldsymbol{\alpha}^0$ can be approximated by the local sensitivity matrix:

$$e(\boldsymbol{\alpha}) \approx (\Delta\boldsymbol{\alpha})^{\mathrm{T}} \tilde{\mathbf{S}}^{\mathrm{T}} \tilde{\mathbf{S}} (\Delta\boldsymbol{\alpha}), \tag{5.23}$$

where $\Delta\boldsymbol{\alpha} = \boldsymbol{\alpha} - \boldsymbol{\alpha}^0$, and the matrix $\tilde{\mathbf{S}}$ has been composed from a series of local sensitivity matrices, belonging to times $(t_1, t_2), \ldots, (t_1, t_h)$:

$$\tilde{\mathbf{S}} = \begin{bmatrix} \tilde{\mathbf{S}}_2 \\ \tilde{\mathbf{S}}_3 \\ \vdots \\ \tilde{\mathbf{S}}_h \\ \vdots \\ \tilde{\mathbf{S}}_l \end{bmatrix} . \tag{5.24}$$

Equation (5.23) is a quadratic approximation to the real shape of the objective function. Any cross-section of this approximate objective function is a hyperellipsoid, defined by the matrix $\tilde{\mathbf{S}}^{\mathrm{T}}\tilde{\mathbf{S}}$. The orientation of the ellipsoid with respect to the parameter axes is defined by the eigenvectors of the matrix $\tilde{\mathbf{S}}^{\mathrm{T}}\tilde{\mathbf{S}}$, while the relative lengths of the axes of the ellipsoid are revealed by the eigenvalues of this matrix.

If the axes of the ellipsoid are parallel to the axes of the parameter space, there is no synergistic effect among the parameters, and the relative lengths of the axes define the relative importance of parameters. However, if, say, the direction of the longest axis of the ellipsoid is at 45° on the plane of two of the parameter axes, this means that the effect on all variables by changing one parameter can be well corrected by also changing another parameter.

A similar interpretation can be given using the term *principal component*. A principal component is a new parameter, obtained via a linear combination of the original parameters. Let matrix $\mathbf{U}$ denote the matrix of normalized eigenvectors of $\tilde{\mathbf{S}}^{\mathrm{T}}\tilde{\mathbf{S}}$. Principal components are defined as

$$\boldsymbol{\Psi} = \mathbf{U}^{\mathrm{T}}\boldsymbol{\alpha}, \tag{5.25}$$

and, using principal components, the objective function (5.23) can be given in a simpler form:

$$e = \sum_{i=1}^{n} \lambda_i (\Delta \Psi_i)^2, \tag{5.26}$$

where $\Delta\boldsymbol{\Psi} = \mathbf{U}^{\mathrm{T}}\Delta\boldsymbol{\alpha}$ and $\boldsymbol{\lambda}$ is the vector of eigenvalues. Equation (5.26) provides another explanation of why the eigenvectors of matrix $\tilde{\mathbf{S}}^{\mathrm{T}}\tilde{\mathbf{S}}$ reveal the related parameters and why the corresponding eigenvalues show the relative weights of these parameter groups.

From a practical point of view, principal component analysis is an inexpensive post-processing technique that extracts otherwise-unavailable information from the local sensitivity matrices.

5.6 INITIAL SENSITIVITIES

The solution of the initial-value problem (5.3) depends on the values of the parameters, but also on the initial values of the variables. Calculation of the *initial-value sensitivity matrix* has been introduced as a first step in the calculation of local sensitivities, according to the Green function method. It has been shown in Section 5.3.3 that the initial-value sensitivity matrix $\mathbf{K}(t, t_1) = \{\partial c_i(t)/\partial c_j^0(t_1)\}$ can be obtained as the solution of the following initial-value problem (Equation (5.10)):

$$\frac{d}{dt}\mathbf{K}(t, t_1) = \mathbf{J}(t)\mathbf{K}(t, t_1).$$

The initial value of $\mathbf{K}$ is a unit matrix. The initial-value sensitivities can be considered as if a unit perturbation were applied to the initial values, one-by-one, and the fate of this perturbation were monitored.

Initial-value sensitivities are interesting because they are related to time scales of models. If the time scales are well separated, variables can be categorized as fast or slow. The slow variables respond very slowly to a perturbation, since the perturbation puts them on a trajectory almost parallel to their original one, and therefore the initial value sensitivity of a slow variable (i.e. the diagonal element belonging to a slow variable) remains close to unity for a long time. Fast variables quickly return to their original trajectory after the perturbation, and therefore their initial-value sensitivities decay to zero quickly. If the initial-value sensitivity of a variable exceeds the unit value instead of remaining close to unity or decaying this indicates that a slight increase in the variable increases its production rate. Such behavior is called *autocatalysis* in chemical kinetics.

The point of the quasi-steady-state approximation (QSSA) is that the values of slow variables determine the values of fast variables (Turányi *et al.*, 1993). This means that it is enough to solve a system of differential equations for the slow variables, and the values of fast variables can be calculated from the values of the slow ones using algebraic equations. The critical step in the application of the quasi-steady-state approximation is appropriate division of variables into fast and slow ones. Initial-variable sensitivities can do the job, but there are other approximate techniques, which provide similar information in a computationally less expensive way.

During the solution of initial-value problem (5.10), the values of variables change, and therefore the elements of the matrix $\mathbf{J}$ are continuously changing. On fixing the elements of $\mathbf{J}$ at the starting time, Equation (5.10) becomes a homogeneous linear system of differential equations with constant parameters. The solution of such a system is

$$\mathbf{K}' = \exp[\mathbf{J}(t_1)t]. \tag{5.27}$$

It has been shown that the Jacobians of chemical kinetic differential equations can frequently be rearranged to approximately lower triangular form (Turányi *et al.*, 1993). It is possible that a similar observation holds for many models in other disciplines. Consequently, for most chemical kinetic systems, the eigenvalues of the Jacobian are close to the diagonal elements of the Jacobian, $\lambda_i \approx j_{ii}$, where j_{ii} is the ith diagonal element of the Jacobian. Since the lifetime can be defined as $\tau_i = -1/j_{ii}$, this relation supports the traditional observation that short-lifetime variables decay rapidly after perturbation and behave as fast variables. This also means that the time history of the diagonal of the initial-value sensitivity matrix can be approximated as

$$k_{ii}(t) \approx \exp(j_{ii}t). \tag{5.28}$$

A more sophisticated handling of timescales takes into account that eigenvectors of the Jacobian define variable groups. The time scale separation is better if variable groups, not single variables, are considered and therefore a more accurate quasi-steady-state approximation with fewer variables can be applied. The corresponding numerical techniques (Lam and Goussis, 1988; Maas and Pope, 1992) represent a further development of the classical QSSA.

So far, only the interpretation of the diagonal elements of the initial-value sensitivity matrix has been discussed. The off-diagonal elements of the matrix $\mathbf{K}$ also contain important dynamic information, but their interpretation depends on the actual physical model. In general, the off-diagonal elements show the displacement of the trajectory of all other variables, in response to perturbing a given variable slightly. As an example, a large off-diagonal element indicates strong coupling between a fast and a slow variable, introducing large error into the QSSA calculation (Turányi *et al.*, 1993).

The whole initial-value sensitivity matrix can also be approximated based on an eigenvector–eigenvalue analysis of $\mathbf{J}(t_1)$ (Maas and Pope, 1994). Let t_2 be the time of observation of the initial-value sensitivity calculation and let $\mathbf{V}$ and $\tilde{\mathbf{V}}$ denote the matrices of right- and left-eigenvectors of $\mathbf{J}(t_1)$, respectively. The matrices $\mathbf{V}_f$ and $\tilde{\mathbf{V}}_f$ are truncated arrays, obtained by deleting the columns and rows, respectively, belonging to the eigenvalues of the Jacobian larger than $-1/(t_2 - t_1)$. This means that the matrices $\mathbf{V}_f$ and $\tilde{\mathbf{V}}_f$ have dimensions $n \times n_f$ and $n_f \times n$, respectively, and belong to the n_f fast eigenvectors of the Jacobian. The initial-value sensitivity matrix can now be approximated by

$$\mathbf{K}(t_1, t_2) \approx \mathbf{P} = \mathbf{I} - \mathbf{V}_f \tilde{\mathbf{V}}_f. \tag{5.29}$$

5.7 FUNCTIONAL SENSITIVITIES

Many physical models contain input functions $k_i = k_i(\mathbf{r}, t), i = 1, 2, \ldots,$ that depend on spatial coordinates $\mathbf{r}$ and/or time t. All of the same general questions about parametric sensitivity carry over to this function case, where the system output y is a *functional* of the inputs. Thus, Equations (5.2) and (5.4) have functional analogs at any order. For example, to first order, we have

$$\delta y = \sum_i \int \frac{\delta y}{\delta k_i(\mathbf{r}, t)} \delta k_i(\mathbf{r}, t)\, d\mathbf{r}\, dt, \tag{5.30}$$

and the functional sensitivity density is given by $S_i(\mathbf{r}, t) = \delta y / \delta k_i(\mathbf{r}, t)$. Keeping in mind that the model output y can also have position and/or time dependence, it is evident that the functional sensitivities provide a detailed input–output map. The analogy with parametric sensitivities extends beyond those defined in Equation (5.30), to include the full family of derived sensitivities for various applications.

Input functions $k_i = k_i(\mathbf{r}, t), i = 1, 2, \ldots,$ can arise in many physical circumstances, but the most common case occurs in atomic and molecular physics, where the input involves fundamental intermolecular interactions between the atoms and molecules, and the goal is to reveal how these input functions influence the observable chemical and physical properties. In turn, the sensitivity of these properties to the input functions provides a basis for attempting to extract these functions from suitable observed laboratory output data. The basis for such inversions is rooted in Equation (5.30), where δy is the deviation between the observed value and that of the current theoretical model, with $\delta k_i(\mathbf{r}, t)$ being the deviation of the input function from its true value. Such inverse problems are typically ill-posed, calling for suitable regularization, and a number of inversions along these lines have been carried out (Ho and Rabitz, 1993).

5.8 SCALING AND SELF-SIMILARITY RELATIONS

Substantial effort can be involved in calculating sensitivity coefficients. The recognition of any patterns of behavior amongst these coefficients would be of considerable significance, not only for simplifying the sensitivity information, but also for the fundamental insight gained about the intimate workings of the system. There is certainly no *a priori* reason to expect the existence of particular patterns or relationships amongst the numerous sensitivity coefficients in a system, since this would imply the presence of hidden dynamical couplings between the system dependent and independent variables. However, such relationships amongst sensitivity coefficients have been identified through patterns of similar behavior in a variety of sensitivity calculations arising from problems in chemical kinetics, especially of a combustion nature (Rabitz and Smooke, 1988). Such connections have been referred to as scaling and self-similarity relations, and the possibility of their existence has potentially important implications for model analysis, as well as system simplification. To be specific, the discussion here will be confined to the treatment of one-dimensional steady problems described by reaction–diffusion equations, often arising, for example, in combustion problems.

In typical case studies, the system differential equations are strongly coupled, and, even more importantly, it is generally possible to identify a distinct and *dominant* member of the dependent variable set, denoted without loss of generality as y_1. The assumed role of y_1 is to provide the strong coupling linkage between all of the N differential equations or dependent variables. A typical example of this behavior in combustion might be the identification of y_1 as the temperature or the concentration of some particularly important chemical species. This dominance is asserted to imply total coordinate and parametric entrainment such that

$$y_n(x, \boldsymbol{\alpha}) \approx F_n(y_1(x, \boldsymbol{\alpha})), \tag{5.31}$$

where F_n is an appropriate non-determined function. Clearly, this relation is an approximation, and we take it as a working anstatz to explore its consequences. Simple differentiation of Equation (5.31) with respect to the system parameter α_j, as well as to x, will lead to

$$\frac{\partial y_n(x)}{\partial \alpha_j} \approx \frac{\partial y_1(x)}{\partial \alpha_j} \frac{\partial y_n}{\partial x} \left(\frac{\partial y_1}{\partial x} \right)^{-1}. \tag{5.32}$$

Equation (5.32) is referred to as a scaling relation in that the sensitivity of the nth dependent variable is prescribed in terms of the sensitivity of the first member and relevant slope information. Also, note that these relations are independent of the unknown function F_n.

Although the result in Equation (5.32) is based on the hypothesis in Equation (5.31) that $y_1(x, \boldsymbol{\alpha})$ is dominant, it is a simple matter to show that the scaling relations are in fact fully symmetrical with regard to all of the dependent variables. Consideration of Equation (5.32), along with the same equation for the n'th dependent variable immediately leads to the following result:

$$\frac{\partial y_n(x)}{\partial \alpha_j} \approx \frac{\partial y_{n'}(x)}{\partial \alpha_j} \frac{\partial y_n}{\partial x} \left(\frac{\partial y_{n'}}{\partial x} \right)^{-1} \tag{5.33}$$

for all n and n' strongly coupled dependent variables. This implies that Equation (5.31) may be used as reciprocal relations such that the special role provided by y_1 may be inverted and replaced by any member of the strongly coupled dependent variable set. In cases of non-monotonic coordinate dependence, this inversion has to be done on a piecewise basis. Similarly, singular points where $\partial y_l / \partial x = 0, l = 1, \ldots,$ indicate changes in the monotonicity of the dependent variables, and it is clear from Equation (5.32) that the scaling relations can exhibit singularities at these points (corrections to the scaling relations may be especially significant near these points).

The scaling relations have been shown to be remarkably accurate in a number of numerical calculations. The actual presence of scaling was only identified subsequent to finding evidence for the more powerful self-similarity conditions. The arguments leading to self-similarity involve a number of operations with the system dynamical equations and the use of the Green function analog of Equation (5.32). The net result is the identification of the approximate similarity relationship

$$\frac{\partial y_n(x)}{\partial \alpha_j} \approx \frac{\partial y_n}{\partial x}\left(\frac{\partial y_1}{\partial x}\right)^{-1}\lambda(x)\sigma_j. \tag{5.34}$$

The term $\lambda(x)$ is a function and σ_j is a constant, with both being characteristic of the particular dynamic system. The self-similarity condition in Equation (5.34) has a surprisingly simple structure that states that, under its conditions of validity, all system sensitivities reduce to knowledge of a scalar function $\lambda(x)$, the dependent variable spatial slopes, and a vector of characteristic constants $\boldsymbol{\sigma}$. The vector $\boldsymbol{\sigma}$ has the same length as the parameter vector; however, its components are generally complicated functions of all the system parameters.

The simple form of Equation (5.34), upon substitution into Equation (5.31), leads to the prediction

$$\frac{\partial y_n(x)/\partial \alpha_j}{\partial y_n(x)/\partial \alpha_{j'}} \approx \sigma_j/\sigma_{j'}. \tag{5.35}$$

This equation states that the sensitivity of a given dependent variable, with respect to a sequence of parameters, may be approximately described by a self-similar set of curves in (coordinate) space, all related by constants in the vector $\boldsymbol{\sigma}$. The scaling behavior suggested by Equation (5.35) is often seen to be valid (Rabitz and Smooke, 1988) for at least a subset (i.e. the strongly coupled subset of dependent variables).

The essential assumption underlying the self-similarity and scaling results in Equations (5.32) and (5.35) is the basic entrainment conditions in Equation (5.31). A growing body of numerical results has justified these relations at least qualitatively, and even quantitatively in some cases. The consequences of scaling and self-similarity behavior go beyond mere simplification of the sensitivity coefficients. The existence of this behavior suggests that the physical system itself may be simplified.

The basic implication behind the existence of dominant variable dependence is that strongly coupled systems, in fact, may behave in a simpler fashion than was at first believed. It is curious that this behavior appears likely to be more valid in problems that are inherently nonlinear and normally thought of as having more complex behavior than arising in linear problems. In a sense, the strong mixing often found in nonlinear systems can lead to an unusual level of parametric simplicity under appropriate conditions.

5.9 APPLICATIONS OF LOCAL SENSITIVITIES

5.9.1 Uncertainty Analysis Based on Local Sensitivities

In some cases, many measurements are available for model parameters, and therefore the probability density functions or at least the variances of the parameters are known. The task of uncertainty analysis is to determine the probability density function (pdf) of the model output at a given time, if the pdfs of the parameters are known. A less ambitious task is the calculation of the variance of the model output, knowing the variance of parameters.

Capability for uncertainty analysis is one of the major features of global sensitivity analysis methods. However, a first estimate can also be made, based on local sensitivities (Atherton *et al.*, 1975).

Using the equations for the propagation of error, a linear estimate can be given for the variance of model output $\sigma^2(y_i)$:

$$\sigma_j^2(y_i) = \left(\frac{\partial y_i}{\partial k_j}\right)^2 \partial^2(k_j), \tag{5.36}$$

$$\sigma^2(y_i) = \sum_j \sigma_j^2(y_i). \tag{5.37}$$

$\sigma^2(y_i)$ is the sum of the contributions of the uncertainties of each parameter k_j to model output y_i, denoted by $\sigma_j^2(y_i)$. The partial variances $S\%_{ij}$ give the percentage contribution of the uncertainty of parameter j to the total uncertainty of model output y_l:

$$S\%_{ij} = \frac{\sigma_j^2(y_i)}{\sigma^2(y_i)} 100. \tag{5.38}$$

Uncertainty analysis using local sensitivities is not a substitute for the better-based global methods, like FAST, but may provide an order-of-magnitude estimation.

One of the applications of uncertainty analysis is the determination of strategies for the improvement of a model. The most uncertain parameters should be studied in more detail for the most effective improvement of model reliability.

5.9.2 Global Parametric Mapping

The predictions of local sensitivity analysis are best in the neighborhood of the reference operating point in parameter space. Nevertheless, there is interest in extracting as much information from the analysis as possible, particularly regarding parameter behavior over larger domains. Short of employing techniques attempting to fully explore this issue, local gradient analysis has some special contributions to make. First, if a sufficient number of derivatives are available in Equations (5.2) or (5.4) then the results may often be extended by Padé approximates. In addition, power-law or other types of scaling relations may also be postulated to exist over the parameter space.

Feature sensitivity analysis (Kramer *et al.*, 1984) provides a systematic means of non-linearly probing a region of parameter space. As an explicit illustration of this procedure, consider $y(\mathbf{r}, t, \alpha)$ as the objective of interest, where the parameter dependence is explicitly indicated. By an examination of the $\mathbf{r}$ and t dependence of this observation, it is assumed that meaningful characteristic features may be identified and an explicit functional form $\tilde{y}(\mathbf{r}, t, \beta)$ chosen that contains the feature parameters $\beta_1, \beta_2, \ldots$. By implication, the two forms of the observation are equivalent:

$$y(\mathbf{r}, t, \alpha) \equiv \tilde{y}(\mathbf{r}, t, \beta(\alpha)). \tag{5.39}$$

Equation (5.39) implies a relationship between β and α. In practice, we shall only know a solution of the model equations and the sensitivities at a reference point α^0 in parameter space. This information will not be sufficient to determine the functional relation $\beta = \beta(\alpha)$; however, we may determine $\beta^0 = \beta(\alpha^0)$ at the system reference point and the corresponding sensitivity coefficients $(\partial\beta_i/\partial\alpha_j)_{\alpha^0}$. In order to achieve this goal, the feature parameters in Equation (5.39) must be adjusted consistently with that relation. One technique is to employ minimization of the least-squares functional

$$R = \iint d\mathbf{r}\, dt [y(\mathbf{r}, t, \alpha) - \tilde{y}(\mathbf{r}, t, \beta)]^2. \tag{5.40}$$

Minimization of R with respect to the feature parameters will yield the equation

$$\frac{\partial R}{\partial \beta_i} = \int\int d\mathbf{r}\, dt [y(\mathbf{r}, t, \alpha) - \tilde{y}(\mathbf{r}, t, \beta)] \frac{\partial \tilde{y}}{\partial \beta_i}(\mathbf{r}, t, \beta) = 0. \tag{5.41}$$

The derivative in the integrand of Equation (5.41) may be explicitly evaluated by recalling that $\tilde{y}$ has a known functional form with respect to its variables. Equation (5.41) implies the existence of the relationship $\beta = \beta(\alpha^0)$, but again it must be recalled that $y(\mathbf{r}, t, \alpha^0)$ is assumed known only at the parameter reference point. Therefore, differentiation of Equation (5.41) with respect to one of the input parameters will yield an equation that may be solved for the desired feature sensitivity coefficients $(\partial \beta_i / \partial \alpha_j)_{\alpha^0}$. In carrying out this last differentiation, it is evident that the system sensitivity coefficients $\partial y_i(\mathbf{r}, t, \alpha)/\partial \alpha_j$ (or, if appropriate, their functional analog) will enter. The implementation of this overall procedure of feature sensitivity analysis is quite straightforward, and, in practice, it is only limited by one's ingenuity in choosing simple but flexible functional forms $\tilde{y}(\mathbf{r}, t, \beta)$.

The technique of feature sensitivity analysis embodied by the relation in Equation (5.39) has an immediate spin-off application to global parameter mapping. Equation (5.39), for the present purposes, may be recast into the following form:

$$y(\mathbf{r}, t, \alpha + \Delta\alpha) \equiv \tilde{y}(\mathbf{r}, t, \beta(\alpha + \Delta\alpha)). \tag{5.42}$$

This equivalence cannot be directly applied, since we do not have full knowledge about the relation between β and α. However, the feature sensitivity analysis based on Equation (5.39) leads to knowledge of β and the sensitivities of β about the nominal operating point. Therefore, we may consider the expansion

$$\beta(\alpha + \Delta\alpha) \approx \beta(\alpha) + \frac{\partial \beta}{\partial \alpha}\Delta\alpha. \tag{5.43}$$

Substitution of Equation (5.43) into Equation (5.42) will yield a *nonlinear* scaling expression with respect to the parameters $\Delta\alpha$. This feature parameter scaling approach is both computationally practical as well as likely to give acceptable results over an extended neighborhood around the system operating point. A clear example of this situation arises in the singular perturbation problem of parameter dependence in oscillating flames. In those cases where the parameters influence the system frequency (Kramer *et al.*, 1984), a local sensitivity analysis will produce secular growth. In contrast, a feature analysis on the system frequency should be stable.

5.9.3 Parameter Estimation

Some parameter estimation methods, such as the simplex method, do not use local sensitivities. However, in most cases, calculation of the slope of the objective function in the space of parameters is a part of the parameter estimation algorithm. Strangely, while much work was devoted to finding better algorithms for the calculation of local sensitivities, this knowledge was not recycled to the parameter estimation programs. Most parameter estimation programs, even nowadays, use the brute force method for calculation of the slope of the objective function. The inaccurate calculation of the slope usually does not spoil the final result of parameter estimation, but may slow the procedure. Application of the decoupled direct method (Section 5.3.2) and the conversion of variable sensitivities to the sensitivity of the objective function (Equation (5.21)) should be used to improve the numerical efficiency of most programs for the estimation of parameters of ordinary differential equations.

All parameter estimation procedures fail if an ill-conditioned problem is encountered. An ill-conditioned problem means that the data do not carry enough information to provide an estimate of all parameters fitted. Usually the only sign of this is that the parameter estimation algorithm fails to converge. Local sensitivities and principal component analysis (Turányi, 1990a) may help to avoid this problem in the following way.

In the case of parameter estimation, the normalized sensitivity matrix, to be investigated, is defined as

$$\tilde{\mathbf{R}} = \frac{\mathbf{k}}{\mathbf{h}} \frac{\partial \mathbf{h}}{\partial \mathbf{y}} \frac{\partial \mathbf{y}}{\partial \mathbf{k}}, \tag{5.44}$$

where $\mathbf{h}(\mathbf{y})$ is the instrumental function, which converts the calculated variables to the calculated observable quantities (e.g. signals of the experimental apparatus). The matrix $\tilde{\mathbf{P}}$ corresponds to the matrix $\tilde{\mathbf{S}}^{\mathrm{T}}\tilde{\mathbf{S}}$ of Equation (5.23):

$$\tilde{\mathbf{P}} = \sum_{l=1}^{L} \tilde{\mathbf{R}}^{\mathrm{T}}(t_l)\mathbf{W}_l(t_l)\tilde{\mathbf{R}}(t_l) \tag{5.45}$$

The matrix $\mathbf{W}$ is the weight matrix. In general, it is the inverse of the covariance matrix. If the covariances are not known or assumed to be zero, $\mathbf{W}$ is diagonal, where the diagonal elements are the inverses of the variances. In the unweighted case, $\mathbf{W}$ is the identity matrix.

Eigenvector–eigenvalue analysis of the matrix $\tilde{\mathbf{P}}$ reveals which parameters can be determined from a given experiment. Parameters that are not related to large eigenvector elements of large eigenvalues cannot be determined. Parameters that are not coupled to other parameters and are linked to large eigenvalues can be fitted easily. A typical situation is when several parameters are strongly coupled, for example when only the ratio of two parameters has an influence on the objective function. In this case, the corresponding eigenvector has the form $(\sqrt{2}, \sqrt{2}, 0, 0, \ldots, 0)$. If both parameters are fitted simultaneously, the result is a deep-valley-like objective function, and the fitting procedure fails. To avoid this problem, one of the parameters should be fixed at a nominal value, and only the other parameter has to be fitted. However, the result of fitting is always the ratio of the two parameters and not the individual values of parameters. Individual values can be obtained from independent experiments or other sources.

The matrix $\tilde{\mathbf{P}}$ has been calculated as the first guess of the parameter values. During the parameter estimation procedure, improved estimates of the parameters become available. A substantially different parameter set may provide a qualitatively different picture, and therefore the analysis should be repeated at every stage of the parameter estimation. Carrying out a principal component analysis at the beginning and during the parameter estimation helps to avoid many problems, and should always be encouraged.

5.9.4 Experimental Design

Experimental design is a branch of mathematical statistics where the aim is to find experimental conditions that provide the most information for the determination of some parameters in a model. Most experimental design algorithms are applicable only for linear models. The above procedure, based on eigenvector–eigenvalue analysis of the matrix $\tilde{\mathbf{P}}$, can be used also for experimental design in the case of any nonlinear model. By selecting the method of measurement (the function h), and the times of measurement t_l, the information content of the experiment for the determination of a given parameter can be optimized. Note, however, that this optimization is based on an *a priori* assumption of the parameters and of the model structure. The experimental design, experiment, and parameter estimation cycle has to be repeated several times until a satisfactory result is obtained.

5.9.5 Stability Analysis

A common concern is the stability of dynamical systems to disturbances, either in the operating parameters or the state of the system during its evolution. An analysis of the system Green function in Equation (5.10) addresses both of these issues. This is evident, since $K_{ij}(t, t_1) = \partial C_i(t)/\partial C_j^0(t_1)$. The Green function dictates the response to all parameter disturbances, as is evident in Equation (5.11). Since Equation (5.10) for the Green function is a linear differential equation driven by the Jacobian, its eigenvalue analysis can reveal the stability of the dynamics. Any eigenvalues with positive real parts indicate growth behavior with respect to time, and, hence, instability with regard to disturbances. These eigenvalues may also be expressed in terms of Lyapunov stability numbers. A complete analysis would also include the general case where the Jacobian is time-dependent. Cases of this type have been explored (Hedges and Rabitz, 1985) for explosive chemical kinetics, limit-cycle oscillations, and classical dynamics.

5.9.6 Investigation of Models

Local sensitivity analysis can be considered as a perturbation study of models. In the case of time-dependent models, the change of a parameter value influences first the values of those variables that contain that parameter in their rate expression. This effect spreads further to other variables. By inspecting this spread, much new information can be gathered on the structure and behavior of the model. As has been shown in Section 5.2, the local sensitivity matrix $\mathbf{S}(t_1, t_2)$ depend on both the time of perturbation, t_1, and the time of observation of the effect, t_2. Selection of these times provides a wide range of opportunities (Hwang, 1988) for the study of models.

5.9.7 Reduction of Models

Reduction of models means that the same phenomenon is described by a smaller, simpler model, derived from the larger model. The derived model can be entirely different from the original one, e.g. when a dynamical system is modeled by a system of difference equations instead of a system of differential equations (see e.g. Turányi, 1994). Another way of reduction is variable lumping (see e.g. Tomlin *et al.*, 1994), when the array of variables is replaced by a smaller set of variables and the new and old set of variables are related to each other by linear or nonlinear functions. Also, effective model reduction can be based on the time scale separation of models (see e.g. Maas and Pope, 1992; Turányi *et al.*, 1993; Tomlin *et al.*, 1997).

In this section, a more restrictive meaning of model reduction is used. The reduced model is obtained from the original model by setting some of its parameters to zero. This might mean that some of the variables are also cut out from the model.

Detection of redundant variables should be the first step in model reduction. In the case of most models, the user is interested in only some of the variables and their effect on the model output. These variables can be called *important variables*. In most models, there are also auxiliary variables. They should be there for making the model work, but their actual value is not interesting for the modeler. Such variables are termed here *necessary variables*. Also, in most models, there are *redundant variables* as well, which can be deleted without any change of the model output.

Two methods have been proposed (Turányi, 1990b) for the detection of redundant variables in reaction kinetic models. The first method is based on the preparation and simulation of a series of reduced models. If all parameters related to a given variable are set to zero and the calculated model output for the important variables is practically identical to that of the original model, then this variable and the corresponding parameters can be eliminated from the model. However, in many cases, the elimination of a smaller number of parameters provides a more accurate reduced model. An algorithm was given to find the minimal number of parameters to be eliminated in this step, but it exploits the special structure of the kinetic differential equations, and therefore cannot be applied to any model.

The second method is based on the investigation of the Jacobian, and is more general, although it provides a suggestion only, which has to be checked by the preparation of the appropriate reduced model. The Jacobian $\mathbf{J} = \{\partial f_i / \partial y_j\}$ can also be considered as a sensitivity matrix. It indicates the sensitivity of the calculated variable rates to perturbing the values of variables. According to the ideas described in Sections 5.5.1 and 5.5.2, further processing of this matrix makes the information more readily available. Application of the normalized Jacobian $\tilde{\mathbf{J}} = \{(y_j/f_i)\, \partial f_i / \partial y_j\}$ makes the information independent of the units of variables, and the corresponding overall sensitivity shows the effect of variable perturbation on a group of N variables:

$$B_i = \sum_{j=1}^{N} \left(\frac{y_i}{f_j} \frac{\partial f_j}{\partial y_i} \right)^2 . \tag{5.46}$$

B_i shows the instantaneous or direct effect of changing variable i on the values of N other variables. Variable i influences the rate of variable j directly, if variable i is present in the rate term of variable j on the right-hand side of the ODE. Of course, an indirect effect is also possible, when variable i influences the rate of variable k, while k controls the rate of j. The Jacobian shows the direct effects only.

Redundant variables can be detected by using the following algorithm: Consider first the N important variables only, and calculate B_i, which expresses the strengths of direct effect of each variable on the important variables. The variables most closely connected directly to the important variables are added to the group of N variables to be investigated, and the procedure is repeated. The algorithm usually converges in a few steps, and the group now contains the important variables and all variables that have a strong influence on their rate, directly or indirectly. The variables left out are the redundant variables.

This algorithm is local in time, and therefore has to be repeated at several time points. Also, it is based on a local linear approximation, and therefore its findings have to be confirmed by trial calculation of appropriate reduced models. Having eliminated the redundant variables from the model, it contains important and necessary variables only. The next step is the identification of redundant parameters.

It is generally assumed, wrongly, that if the sensitivity of a parameter is small for all important variables, then this parameter can be eliminated from the model. However, the local sensitivities show only the effect of small changes of parameters (which may be called 'parameter tuning'). The order of importance of parameters deduced from the sensitivity of important variables can be called *tuning importance*. Considering a group of important variables, the order of tuning importance can be deduced from the overall sensitivity values (5.21).

Setting a parameter to zero is a drastic effect, and such a change may alter significantly the calculated value of an important variable even if the corresponding local sensitivity is small. The reason is the indirect effect again: setting a parameter to zero may significantly influence the value of a necessary variable, and this effect extends to the important variable. However, strong influence on a necessary variable can usually be detected at the nominal point of parameters, and such a parameter has high sensitivity for some of the necessary variables. A rule of thumb is that a parameter can be eliminated (i.e. can be set to zero) in a model, if the sensitivity of all important and necessary variables of the corresponding parameter is small at any time during the interval considered (Hwang, 1982). It is important to scan the whole time interval, because if a parameter is influential only at the beginning and its value does not effect the location of the stationary point, then the calculated sensitivity goes to zero as time advances. Overall sensitivities, as defined in Equation (5.21), provide a good guess for *reduction importance*, if the summation has been extended to all important and necessary variables. Using principal component analysis, parameters of high reduction importance all appear with large eigenvector elements of large eigenvalues, if again all important and necessary variables are considered.

An alternative method for model reduction is based on the study of the normalized rate sensitivity matrix $\tilde{\mathbf{F}} = \{(k_j/f_i)\,\partial f_i/\partial k_j\}$. This matrix shows the instantaneous effect of changing a parameter on the rate of variables. Principal component analysis of $\tilde{\mathbf{F}}$, considering the important and necessary variables, reveals all parameters that can be eliminated from the model without significant changes in the values of important parameters (Turányi *et al.*, 1989). Since $\tilde{\mathbf{F}}$ can be calculated easily in an algebraic way, this technique is fast and simple. However, studying the effect of parameters on the rates instead of the model output is a less direct approach, and the reduced model that is found has to be validated by comparing its solution with that of the full model.

5.10 CONCLUSIONS

All modelling work includes the following steps: collection of information on the parameters and on the model structure, setting up the model, and validation of the model against experimental or observation data. The next step should be an analysis of the model, which includes the assessment of the importance of parameters and reduction of the model by eliminating the redundant parts.

Global sensitivity analysis methods have been designed to study the model in a wide range of parameters. However, in the case of large models, calculation of global sensitivities is computationally prohibitive, but local sensitivities can provide useful information on the behavior of the model near the nominal values of parameters.

A general purpose program package, called KINAL (Turányi, 1990c) is available for manipulating and processing the local sensitivity matrix in many ways discussed in this chapter. KINAL and a specific program, called KINALC, for the analysis of combustion and gas kinetic problems are available through the World Wide Web. The functionality of KINAL and KINALC are briefly described in the software Appendix.

ACKNOWLEDGMENTS

H. R. acknowledges support from the National Science Foundation and T. T. wishes to thank the support of the Hungarian Science Foundation OTKA via Contract T025875.

6

Sampling-Based Methods

Jon C. Helton

Arizona State University, Tempe, USA

Freddie J. Davis

Sandia National Laboratories, Albuquerque, USA

6.1 INTRODUCTION

Sampling-based methods for uncertainty and sensitivity analysis involve the generation and exploration of a mapping from uncertain analysis inputs to analysis results (Iman *et al.*, 1981a,b; Iman, 1992; Helton, 1993). Conceptually, the analysis or model under consideration can be represented by a vector function

$$\mathbf{y} = [y_1, y_2, \ldots, y_{n_Y}], \tag{6.1.1}$$

and the associated input can be represented by a vector

$$\mathbf{x} = [x_1, x_2, \ldots, x_{nX}], \tag{6.1.2}$$

where n_X and n_Y are the dimensions of **x** and **y**, respectively, and each value of **x** produces a corresponding value $\mathbf{y}(\mathbf{x})$. Most real analyses are quite complex, with the result that the dimensions of **x** and **y** can be large.

If the value for **x** was known unambiguously, then $\mathbf{y}(\mathbf{x})$ could be determined and presented as the unique outcome of the analysis. However, there is uncertainty with respect to the appropriate value to use for **x** in most analyses, with the result that there is also uncertainty with respect to the value of $\mathbf{y}(\mathbf{x})$. The uncertainty in **x** and its associated effect on $\mathbf{y}(\mathbf{x})$ lead to two closely related questions:

(i) What is the uncertainty in $\mathbf{y}(\mathbf{x})$ given the uncertainty in **x**?
(ii) How important are the individual elements of **x** with respect to the uncertainty in $\mathbf{y}(\mathbf{x})$?

Attempts to answer these two questions are typically referred to as uncertainty analysis and sensitivity analysis, respectively.

Sensitivity Analysis. Edited by A. Saltelli *et al.*

An assessment of the uncertainty in $\mathbf{y}$ derives from a corresponding assessment of the uncertainty in $\mathbf{x}$. In particular, $\mathbf{y}$ is assumed to have been developed so that appropriate analysis results are obtained if the appropriate value for $\mathbf{x}$ is used in the evaluation of $\mathbf{y}$. Unfortunately, it is impossible to unambiguously specify the appropriate value of $\mathbf{x}$ in most analyses; rather, there are many possible values for $\mathbf{x}$ of varying levels of plausibility. Such uncertainty is often given the designation subjective or epistemic (Helton, 1994, 1997), and is characterized by assigning a distribution

$$D_1, D_2, \ldots, D_{nX} \tag{6.1.3}$$

to each element x_j of $\mathbf{x}$. Correlations and other restrictions involving the x_j are also possible. These distributions and any associated conditions characterize a degree of belief as to where the appropriate value of each variable x_j is located for use in evaluation of $\mathbf{y}$, and in turn lead to distributions for the individual elements of $\mathbf{y}$. Given that the distributions in (6.1.3) characterize a degree of belief with respect to where the appropriate input to use in the analysis is located, the resultant distributions for the elements of $\mathbf{y}$ characterize a corresponding degree of belief with respect to where the appropriate values of the outcomes of the analysis are located.

Sampling-based methods for uncertainty and sensitivity analysis are based on a sample

$$\mathbf{x}_k = [x_{k1}, x_{k2}, \ldots, x_{knX}], \qquad k = 1, 2, \ldots, nS, \tag{6.1.4}$$

of size nS from the possible values for $\mathbf{x}$ as characterized by the distributions in (6.1.3) and on the corresponding evaluations

$$\mathbf{y}(\mathbf{x}_k) = [y_1(\mathbf{x}_k), y_2(\mathbf{x}_k), \ldots, y_{nY}(\mathbf{x}_k)], \qquad k = 1, 2, \ldots, nS, \tag{6.1.5}$$

of $\mathbf{y}$. The pairs

$$[\mathbf{x}_k, \mathbf{y}(\mathbf{x}_k)], \quad k = 1, 2, \ldots, nS, \tag{6.1.6}$$

form a mapping from the uncertain analysis inputs (i.e., the $\mathbf{x}_k$) to the corresponding uncertain analysis results (i.e., the $\mathbf{y}(\mathbf{x}_k)$). When an appropriate probabilistic procedure has been used to generate the sample in (6.1.4) from the distributions in (6.1.3), the resultant distributions for the elements of $\mathbf{y}$ characterize the uncertainty in the results of the analysis (i.e., constitute the outcomes of an uncertainty analysis). Further, examination of scatterplots, regression analysis, partial correlation analysis and other procedures for investigating the mapping in (6.1.6) provide a way to determine the effects of the elements of $\mathbf{x}$ on the elements of $\mathbf{y}$ (i.e., constitute procedures for sensitivity analysis).

When viewed at a high level, performance of a sampling-based uncertainty and sensitivity analysis involves five components:

(i) definition of the distributions in (6.1.3) that characterize uncertainty;
(ii) generation of the sample in (6.1.4) from the distributions in (6.1.3);
(iii) evaluation of $\mathbf{y}$ for the individual elements of the sample in (6.1.4) to produce the model evaluations in (6.1.5);
(iv) generation of displays of the uncertainty in $\mathbf{y}$ from the analysis outcomes in (6.1.5);
(v) exploration of the mapping in (6.1.6) to determine the effects of the elements of $\mathbf{x}$ on the elements of $\mathbf{y}$.

The preceding components of a sampling-based uncertainty and sensitivity analysis are discussed in Sections 6.2–6.6.

Analysis procedures are easier to understand and assess when they are illustrated by real examples. For this reason, a nontrivial example from a performance assessment (PA) carried out in support of the 1996 Compliance Certification Application for the Waste Isolation Pilot Plant (WIPP) will be used to illustrate the procedures under consideration (US DOE, 1996; Helton *et al.*, 1998). The WIPP is under development near Carlsbad, NM, by the U.S. Department of Energy (DOE) for the geologic (i.e., deep underground) disposal of transuranic waste (Rechard, 1999; NRC, 1996). A number of mathematical models are involved in assessing the potential behavior of the WIPP, its surrounding environment, and the radionuclides emplaced there. Most of these models involve the numerical solution of systems of partial differential equations used to represent material deformation, fluid flow, and radionuclide transport. The model used to represent two-phase (i.e., gas and brine) fluid flow in the vicinity of the repository will be used for illustration, with this model implemented by the BRAGFLO program (Helton *et al.*, 1998). The chapter ends with a summary discussion (Section 6.7).

6.2 DEFINITION OF DISTRIBUTIONS FOR SUBJECTIVE UNCERTAINTY

The definition of the distributions in (6.1.3) used to characterize subjective uncertainty is, in many ways, the most important single part of a sampling-based uncertainty and sensitivity analysis, because these distributions determine both the uncertainty in $\mathbf{y}$ and the relative importance of the individual elements of $\mathbf{x}$ that give rise to this uncertainty. However, the determination of such distributions is not the primary focus of this presentation, and thus will be treated rather briefly.

It is important for everyone involved in the definition of the distributions in (6.1.3) to understand the type of information that is being quantified. In particular, the purpose of these distributions is to characterize a degree of belief with respect to where the appropriate value of each element of $\mathbf{x}$ for use in the analysis is located. In concept, the analysis structure has been developed to the point that a single value for each element of $\mathbf{x}$ is required, but the precise values for these elements, and hence for $\mathbf{x}$, are not known.

A common error is to define the D_j so that they characterize spatial, temporal, or experimental variability. If the analysis uses a quantity that is held constant over an extended period of time or over an extended area, then the corresponding distribution D_j should not be defined to characterize temporal or spatial variability. Rather, given that the model uses a spatially or temporally averaged input, the distribution D_j should characterize the uncertainty in this averaged quantity rather than the variability that is averaged over. Similarly, experimental variability is not the same as the uncertainty in an analysis input derived from variable experimental outcomes.

Owing to its importance and pervasiveness, the characterization of subjective uncertainty has been widely studied (see, e.g., Berger, 1985; Cook and Unwin, 1986; Mosleh *et al.*, 1988; Hora and Iman, 1989; Keeney and von Winterfeldt, 1991; Bonano *et al.*, 1990; Bonano and Apostolakis, 1991; Cooke, 1991; Meyer and Booker, 1991; Ortiz *et al.*, 1991; NRC, 1992; Thorne, 1993). Perhaps the largest example of an analysis to use a formal expert review process to assess the uncertainty in its inputs is the US Nuclear Regulatory Commission's reassessment of the risks from commercial nuclear power stations (US NRC, 1990–91; Harper *et al.*, 1990, 1991, 1992; Breeding *et al.*, 1992a,b). Another large example is an assessment of seismic risks in the eastern USA (EPRI, 1989).

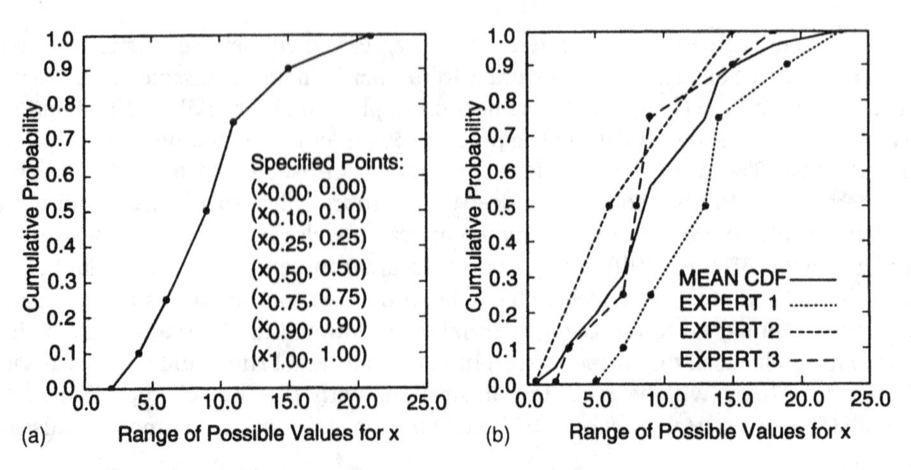

Figure 6.1 Construction of CDF to characterize subjective uncertainty: (a) from specified quantile values for single expert, and (b) by averaging of CDFs defined by individual experts with equal weight (i.e., $1/n_E = \frac{1}{3}$, where $n_E = 3$ is the number of experts) given to each expert.

Although formal statistical procedures can be useful in the construction of the distributions in (6.1.3) in some situations, in most cases such distributions are probably best developed by specifying selected quantile values without making an attempt to specify a particular distribution type and its associated parameters (e.g., normal, lognormal, beta,...) (Helton, 1993, Section 3.1). For example, the construction procedure might start by specifying minimum, median, and maximum values for the variable under consideration (i.e., the points $(x_{0.00}, 0.00)$, $(x_{0.5}, 0.5)$, and $(x_{1.00}, 1.00)$ on the cumulative distribution function (CDF) in Figure 6.1a). Then, resolution could be added by specifying additional quantile values (e.g., the points $(x_{0.10}, 0.10)$, $(x_{0.25}, 0.25)$, $(x_{0.75}, 0.75)$ and $(x_{0.90}, 0.90)$ in Figure 6.1a). The process can be continued until it is felt that the distribution is providing an adequate characterization of the uncertainty in the variable under consideration. Hopefully the expert, or experts, whose knowledge is being quantified by this distribution should be able to provide a documentable rationale for the selection of specific quantile values. The expert is more likely to be able to justify the selection of specific quantile values than the choice of specific parameters to define a beta distribution or some other formal distribution.

When several experts are used to develop a distribution for a variable, one possibility is first to have each expert independently develop a distribution as indicated in Figure 6.1 (a). Then, these distributions can be vertically averaged to produce a new distribution based on the distributions supplied by the individual experts (Figure 6.1b). This is easiest to do if each expert's distribution is assigned equal weight (i.e., the divisor in the averaging process is n_E, where n_E is the number of experts). In practice, the assigning of different weights to different experts is very difficult. A review of procedures for combining distributions from multiple experts is given by Clemen and Winkler (1999).

As indicated in Section 6.1, this presentation uses an example from the 1996 WIPP PA. The variables that comprise the elements of **x** in this example are indicated in Table 6.1. Specifically, **x** is of the form

$$\mathbf{x} = [ANHBCEXP, ANHBCVGP, \ldots, WRGSSAT], \tag{6.2.1}$$

and corresponds to the vector **x** in (6.1.2) with $n_X = 31$. The distributions assigned to the elements of **x** were defined by appropriate staff members of the experimental programs that

Table 6.1 Examples of the $n_X = 31$ uncertain variables used as input to BRAGFLO in the 1996 WIPP PA (see Helton *et al.*, 1998, Table 5.2.1, and US DOE, 1996, Appendix PAR, for additional information).

ANHBCEXP	: Brooks-Corey pore distribution parameter for anhydrite (dimensionless).
ANHBCVGP	: Pointer variable for selection of relative permeability model for use in anhydrite.
ANHCOMP	: Bulk compressibility of anhydrite (Pa^{-1}). Correlation: −0.99 rank correlation (Sections 6.3.5 and 6.6.6) with *ANHPRM*.
ANHPRM	: Logarithm of anhydrite permeability (m^2). Correlation: −0.99 rank correlation with *ANHCOMP*.
BHPRM	: Logarithm of borehole permeability (m^2).
HALCOMP	: Bulk compressibility of halite (Pa^{-1}). Correlation: −0.99 rank correlation with *HALPRM*.
HALPOR	: Halite porosity (dimensionless).
HALPRM	: Logarithm of halite permeability (m^2). Correlation: −0.99 rank correlation with *HALCOMP*.
WMICDFLG	: Pointer variable for microbial degradation of cellulose.
WRBRNSAT	: Residual brine saturation in waste (dimensionless).
WRGSSAT	: Residual gas saturation in waste (dimensionless).

Names of variables not listed: *ANRBRSAT, ANRGSSAT, BPCOMP, BPINTPRS, BPPRM, BPVOL, SALPRES, SHBCEXP, SHPRMASP, SHPRMCLY, SHPRMCON, SHPRMDRZ, SHPRMHAL, SHRBRSAT, SHRGSSAT, WASTWICK, WFBETCEL, WGRCOR, WGRMICH*, and *WGRMICI*.

were being carried out at Sandia National Laboratories to support the development of the WIPP, with these distributions intended to characterize a degree of belief with respect to where the appropriate values of these variables are located for use in the 1996 WIPP PA. The distributions assigned to *WMICDFLG* and *WSOLAM3C* are illustrated in Figure 6.2, with *WMICDFLG* having a discrete distribution and *WSOLAM3C* having a piecewise uniform distribution (i.e., the type of distribution that results when quantiles are defined as indicated in Figure 6.1a and then connected by straight lines).

The care and effort used in the definition of the distributions in (6.1.3) are dependent on both the purpose of an analysis and the amount of time and resources available for its

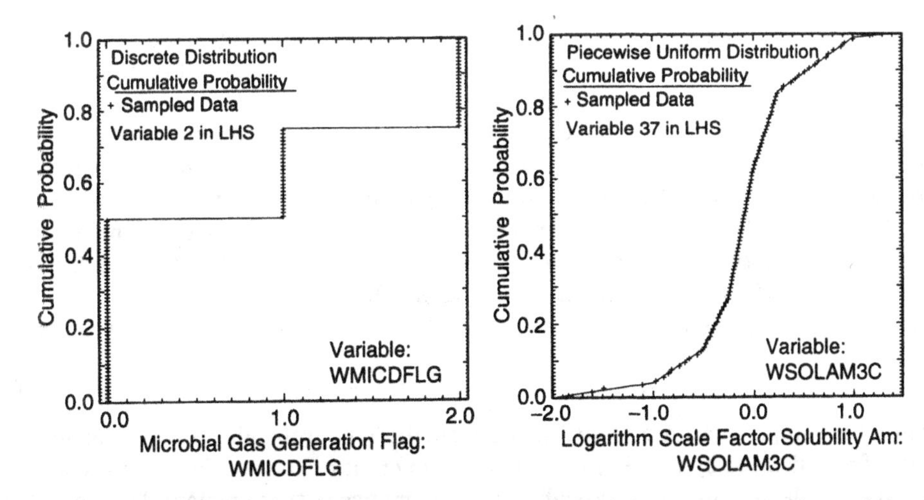

Figure 6.2 Examples of uncertain variables, their associated distributions, and sampled values obtained with a Latin hypercube sample (LHS, see Section 6.5.6) of size 100 (see US DOE, 1996 Appendix PAR, and Helton *et al.*, 1998, Appendix, for distributions of the $n_X = 31$ variables in $\mathbf{x}$).

implementation. If the analysis is primarily exploratory in nature or if limited time and resources are available, then rather crude specifications for these distributions might be used (e.g., uniform and loguniform for variables with uncertainty ranges less than and greater than one order of magnitude, respectively). As long as the ranges are not unreasonably small or large, such an approach can lead to considerable insights into the behavior of a system and the variables that influence this behavior. However, more robust insights would require greater effort in the definition of the distributions. An efficient approach is to carry out an initial screening analysis with uniform and loguniform distributions to identify the most important variables and then to characterize more carefully the uncertainty in these variables for use in a second analysis. This iterative approach allows resources to be concentrated on characterizing the uncertainty in the most important variables. If a variable has little effect on the outcome of an analysis, then the accuracy with which its uncertainty is characterized is not very important to the outcome of the analysis.

6.3 SAMPLING PROCEDURES

Sampling-based methods for uncertainty and sensitivity analysis obviously require sampling procedures. Three sampling procedures are discussed in this section: Random sampling (Section 6.3.1), importance sampling (Section 6.3.2), and Latin hypercube sampling (Section 6.3.3). Random and Latin hypercube sampling are compared (Section 6.3.4), and a correlation control procedure for use in conjunction with random and Latin hypercube sampling is discussed (Section 6.3.5). Finally, the use of Latin hypercube sampling to generate the sample used in the example analysis is described (Section 6.3.6).

6.3.1 Random Sampling

For notational convenience, assume that the variable under consideration is represented by

$$\mathbf{x} = [x_1, x_2, \ldots, x_{nX}] \tag{6.3.1}$$

and that the corresponding probability space is $(\mathcal{S}, \mathfrak{S}, p)$. As a reminder, a probability space $(\mathcal{S}, \mathfrak{S}, p)$ is the formal structure on which the mathematical development of probability is based, and consists of three components: (i) a set $\mathcal{S}$ that contains everything that could occur in the particular universe under consideration, (ii) a suitably restricted collection $\mathfrak{S}$ of subsets of $\mathcal{S}$ for which probabilities are defined, and (iii) a function p that defines the probabilities of the elements of $\mathfrak{S}$ (Feller, 1971, p. 116). In the usual terminology of probability theory, $\mathcal{S}$ is the sample space, the elements of $\mathcal{S}$ are elementary events, the elements of $\mathfrak{S}$ are events, and p is a probability measure. In practice, $(\mathcal{S}, \mathfrak{S}, p)$ is defined by specifying a distribution D_j for each element x_j of $\mathbf{x}$ as indicated in (6.1.3).

In random sampling, sometimes also called simple random sampling, the observations

$$\mathbf{x}_k = [x_{k1}, x_{k2}, \ldots, x_{knX}], \quad k = 1, 2, \ldots, nR, \tag{6.3.2}$$

where nR is the sample size, are selected according to the joint probability distribution for the elements of $\mathbf{x}$ as defined by $(\mathcal{S}, \mathfrak{S}, p)$. Points from different regions of the sample space $\mathcal{S}$ occur in direct relationship to the probability of occurrence of these regions. Further, each sample element is selected independently of all other sample elements. As illustrated in Figure 6.3 for $x_1 = U$, $x_2 = V$, $nX = 2$, and $nR = 5$, the numbers $RU(1), RU(2), \ldots, RU(5)$

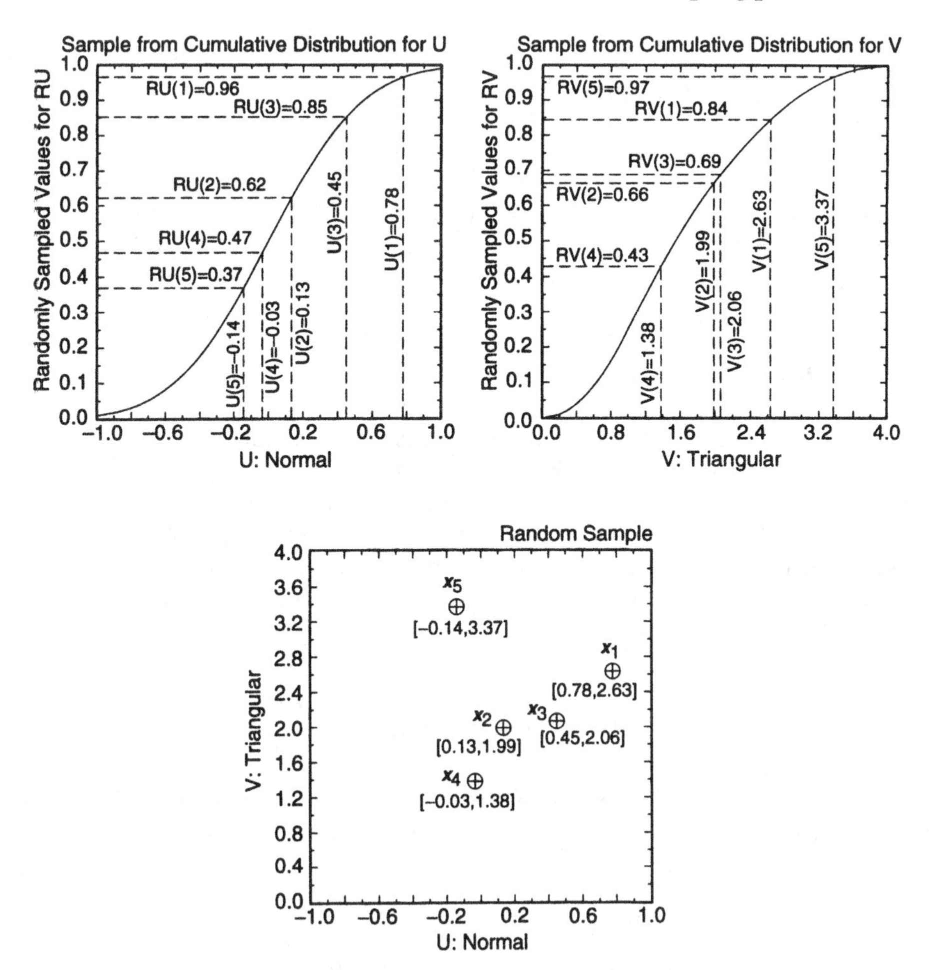

Figure 6.3 Example of random sampling to generate a sample of size $n_R = 5$ from $\mathbf{x} = [U, V]$, with U normal on $[-1, 1]$ (mean $= 0, 0.01$ quantile $= -1, 0.99$ quantile $= 1$) and V triangular on [0, 4] (mode $= 1$).

are sampled from a uniform distribution on [0, 1] and in turn lead to a sample $U(1)$, $U(2), \ldots, U(5)$ from U based on the CDF for U. Similarly, the numbers $RV(1), RV(2), \ldots$, $RV(5)$ lead to a sample $V(1), V(2), \ldots, V(5)$ from V. The pairs

$$\mathbf{x}_k = [U(k), V(k)], \qquad k = 1, 2, \ldots, n_R = 5, \tag{6.3.3}$$

then constitute a random sample from $\mathbf{x} = [U, V]$, where U has a normal distribution on $[-1, 1]$ and V has a triangular distribution on [0, 4] in this example.

Random samples are generated in an analogous manner when $\mathbf{x}$ has a dimension greater than 2 (e.g., $n_X = 100$). Specifically, if the elements of $\mathbf{x}$ are represented by $U, V, \ldots, W$ and a random sample of size n_R is to be generated, then random numbers $RU(1)$, $RU(2), \ldots$, $RU(n_R)$ are sampled uniformly from [0,1] and used to obtain corresponding values $U(1), U(2), \ldots, U(n_R)$ for U; random numbers $RV(1), RV(2), \ldots, RV(n_R)$ are sampled uniformly from [0,1] and used to obtain corresponding values $V(1), V(2), \ldots, V(n_R)$ for V, and so on, with the process continuing through all elements of $\mathbf{x}$ and ending with the selection of random numbers $RW(1), RW(2), \ldots, RW(n_R)$ from [0,1] and the generation of the corresponding values $W(1), W(2), \ldots, W(n_R)$ for W. The vectors

$$\mathbf{x}_k = [U(k), V(k), \ldots, W(k)], \qquad k = 1, 2, \ldots, n_R, \tag{6.3.4}$$

then constitute a random sample from $\mathbf{x} = [U, V, \ldots, W]$.

The preceding sampling procedure depends on the generation of random samples from a uniform distribution on [0, 1] (i.e., uniform random variates). The generation of such samples is widely discussed (see, e.g., Press *et al.*, 1992; Barry, 1996; Fishman, 1996; L'Ecuyer, 1998), and the capability to do so is taken for granted in this chapter.

6.3.2 Importance Sampling

In random sampling, there is no assurance that points will be sampled from any given subregion of the sample space $\mathcal{S}$. Also, it is possible for an inefficient sampling of $\mathcal{S}$ to occur owing to several sampled values falling very close together. The preceding problems can be partially ameliorated by using importance sampling. With this technique, $\mathcal{S}$ is exhaustively divided into a number of nonoverlapping subregions (i.e., strata) $\mathcal{S}_i, i = 1, 2, \ldots, n_S$. Then, n_{S_i} values for $\mathbf{x}$ are randomly sampled from $\mathcal{S}_i$, with the random sampling carried out in consistency with the definition of $(\mathcal{S}, \mathfrak{S}, p)$ and the restriction of $\mathbf{x}$ to $\mathcal{S}_i$. The resultant vectors

$$\mathbf{x}_k = [x_{k1}, x_{k2}, \ldots, x_{kn_X}], \qquad k = 1, 2, \ldots, \sum_{i=1}^{n_S} n_{S_i}, \tag{6.3.5}$$

then constitute an importance-based sample from $\mathcal{S}$ (i.e., a sample obtained by importance sampling). Typically, only one value is sampled from each $\mathcal{S}_i$, with the result that the sample has the form

$$\mathbf{x}_k = [x_{k1}, x_{k2}, \ldots, x_{kn_X}], \qquad k = 1, 2, \ldots, n_S. \tag{6.3.6}$$

The name 'importance sampling' derives from the fact that the $\mathcal{S}_i$ are in part defined on the basis of how important the **x**s contained in each set are to the final outcome of the analysis. Often, importance sampling is used to ensure the inclusion in an analysis of **x**s that have high consequences but low probabilities (i.e., the probabilities $p(\mathcal{S}_i)$ are small for the $\mathcal{S}_i$ that contain such **x**s). When importance sampling is used, the probabilities $p(\mathcal{S}_i)$ and number of observations $n(S_i)$ taken from each $\mathcal{S}_i$ must be folded back into the analysis before results can be meaningfully presented.

Several examples of importance sampling for $\mathbf{x} = [U, V]$ are given in Figure 6.4. The two top frames are for strata of equal probability (i.e., all $p(\mathcal{S}_i)$ are equal). For two uniform distributions, this results in all strata having the same area (upper left frame). For two nonuniform distributions, different strata can have different areas even though they have the same probability (upper right frame). The two lower frames are for strata of unequal probability. In this case, the variable distributions and the strata probabilities interact to determine the area of the strata. However, it is important to recognize that specifying variable distributions, number of strata, and strata probabilities does not uniquely define an importance sampling procedure; rather, there are many ways in which the strata $\mathcal{S}_i$ can be defined that are consistent for the preceding constraints. In particular, appropriate definition of strata will depend on specific properties of individual analyses. Similar ideas also hold for more than two variables, in which case the strata become volumes in a space with the same dimension as **x**.

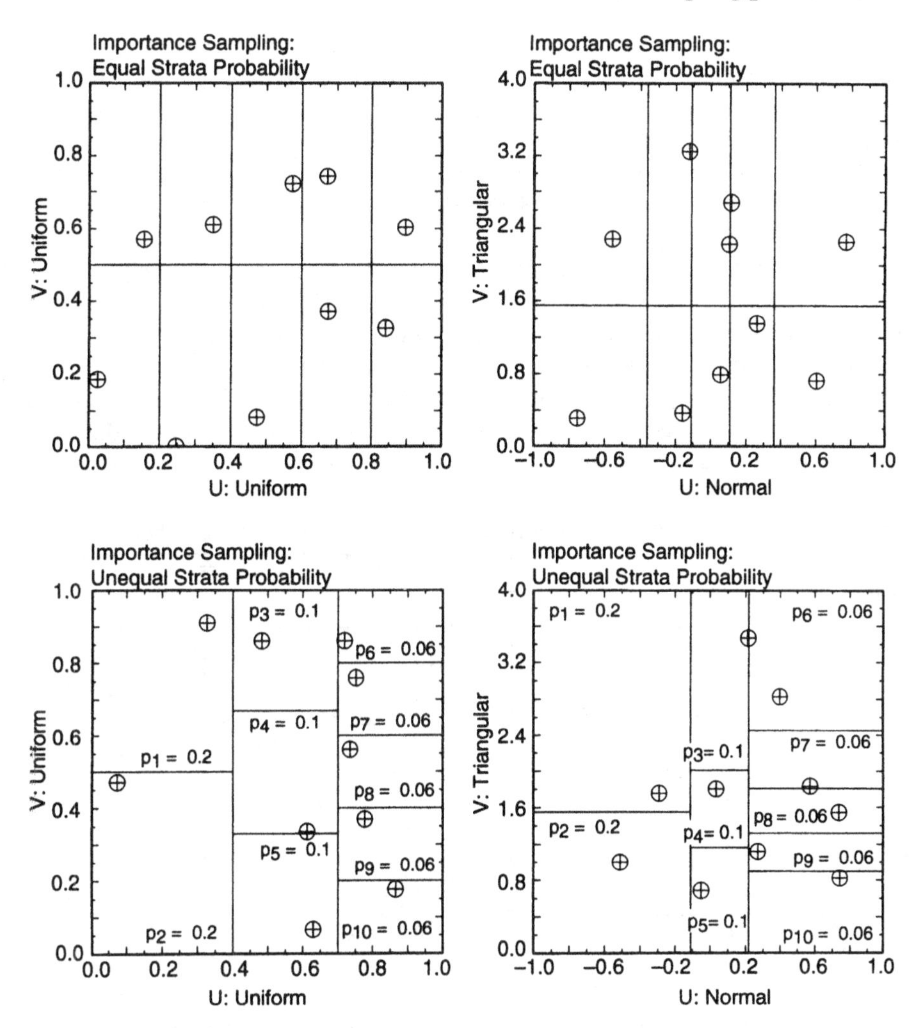

Figure 6.4 Examples of importance sampling with 10 strata (i.e., $n_S = 10$), one random sample per strata (i.e., $n_{S_i} = 1$), equal strata probability (i.e., $p(S_i) = \frac{1}{10}$, upper frames), unequal strata probability (i.e., $p(S_i) = 0.2, 0.2, 0.1, 0.1, 0.1, 0.06, 0.06, 0.06, 0.06, 0.06$, lower frames), U and V uniform on $[0, 1]$ (left frames), and U normal on $[-1, 1]$ (mean $= 0, 0.01$ quantile $= -1, 0.99$ quantile $= 1$) and V triangular on $[0, 4]$ (mode $= 1$) (right frames).

6.3.3 Latin Hypercube Sampling

Importance sampling operates to ensure the full coverage of specified regions in the sample space. This idea is carried farther in Latin hypercube sampling (McKay *et al.*, 1979) to ensure the full coverage of the range of each variable. Specifically, the range of each variable (i.e., the x_j) is divided into n_{LHS} intervals of equal probability and one value is selected at random from each interval. The n_{LHS} values thus obtained for x_1 are paired at random and without replacement with the n_{LHS} values obtained for x_2. These n_{LHS} pairs are combined in a random manner without replacement with the n_{LHS} values of x_3 to form n_{LHS} triples. This process is continued until a set of n_{LHS} n_X-tuples is formed. These n_X-tuples are of the form

$$\mathbf{x}_k = [x_{k1}, x_{k2}, \ldots, x_{kn_X}], \qquad k = 1, \ldots, n_{LHS}, \tag{6.3.7}$$

and constitute the Latin hypercube sample (LHS). The individual x_j must be independent for the preceding construction procedure to work; a method for generating Latin hypercube and random samples from correlated variables has been developed by Iman and Conover (1982), and is discussed in Section 6.3.5. Latin hypercube sampling is an extension of quota sampling (Steinberg, 1963), and can be viewed as an nX-dimensional randomized generalization of Latin square sampling (Raj, 1968, pp. 206–209).

The generation of an LHS of size $n_{LHS} = 5$ from $\mathbf{x} = [U, V]$ is illustrated in Figure 6.5. Initially, the ranges of U and V are subdivided into five intervals of equal probability, with this subdivision represented by the lines that originate at 0.2, 0.4, 0.6, and 0.8 on the ordinates of the two upper frames in Figure 6.5, extend horizontally to the CDFs, and then drop vertically to the abscissas to produce the five indicated intervals. Random values $U(1), U(2), \ldots, U(5)$ and $V(1), V(2), \ldots, V(5)$ are then sampled from these intervals. The sampling of these random values is implemented by (i) sampling $RU(1)$ and $RV(1)$ from a uniform distribution on [0, 0.2], $RU(2)$ and $RV(2)$ from a uniform distribution on [0.2, 0.4], and so on, and (ii) then

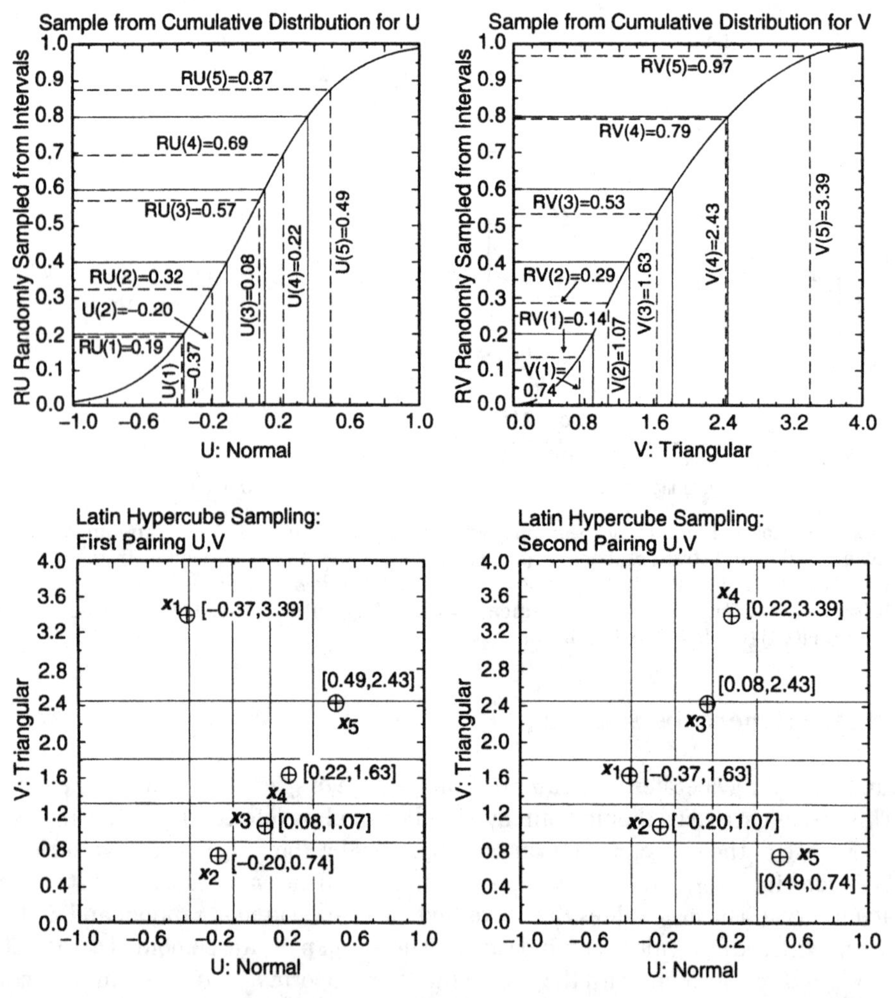

Figure 6.5 Example of Latin hypercube sampling to generate a sample of size $n_{LHS} = 5$ from $\mathbf{x} = [U, V]$, with U normal on $[-1, 1]$ (mean $= 0, 0.01$ quantile $= -1, 0.99$ quantile $= 1$) and V triangular on [1, 4] (mode $= 1$).

using the CDFs to identify (i.e., sample) the corresponding U and V values, with this identification represented by the dashed lines that originate on the ordinates of the two upper frames in Figure 6.5, extend horizontally to the CDFs, and then drop vertically to the abscissas to produce $U(1), U(2), \ldots, U(5)$ and $V(1), V(2), \ldots, V(5)$. The generation of the LHS is then completed by randomly pairing (without replacement) the resulting values for U and V. As this pairing is not unique, many possible LHSs can result. Two such LHSs are shown in the lower two frames in Figure 6.5, with one LHS resulting from the pairings $[U(1), V(5)]$, $[U(2), V(1)]$, $[U(3), V(2)]$, $[U(4), V(3)]$, $[U(5), V(4)]$ (lower left frame) and the other LHS resulting from the pairings $[U(1), V(3)]$, $[U(2), V(2)]$, $[U(3), V(4)]$, $[U(4), V(5)]$, $[U(5), V(1)]$ (lower right frame).

The generation of an LHS for $n_X > 2$ proceeds in a manner similar to that shown in Figure 6.5 for $n_X = 2$. The sampling of the individual variables for $n_X > 2$ takes place in the same manner as shown in Figure 6.5. However, the n_X variables define an n_X-dimensional solid rather than a two-dimensional rectangle in the plane. Thus, the two lower frames in Figure 6.5 would involve a partitioning of an n_X-dimensional solid rather than a rectangle.

6.3.4 Comparison of Random and Latin Hypercube Sampling

Random sampling is the preferred technique when sufficiently large samples are possible because it is easy to implement, easy to explain, and provides unbiased estimates for means, variances, and distribution functions. The possible problems with random sampling derive from the rather vague phrase 'sufficiently large' in the preceding sentence. When the underlying models are expensive to evaluate (e.g., many hours of CPU time per evaluation) or estimates of extreme quantiles are needed (e.g., the 0.999 999 quantile), the required sample size to achieve a specific purpose may be too large to be computationally practicable. In the 1996 WIPP PA, random sampling was used for the estimation of complementary cumulative distribution functions (CCDFs) for radionuclide releases to the accessible environment because it was possible to develop a computational strategy that allowed the use of a sample of size $n_S = 10\,000$ to estimate an exceedance probability of 0.001 (Helton *et al.*, 1998).

When random sampling is not computationally feasible for the estimation of extreme quantiles, importance sampling is often employed. However, the use of importance sampling on nontrivial problems is not easy due to the difficulty of defining the necessary strata and also of calculating the probabilities of these strata. For example, the fault and event tree techniques used in probabilistic risk assessments for nuclear power stations and other complex engineered facilities can be viewed as algorithms for defining importance sampling procedures. The bottom line is that the definition and implementation of an importance sampling procedure can be difficult. Further, without extensive *a priori* knowledge, the strata may end up being defined more finely than is necessary, with the result that the importance sampling procedure ends up requiring more calculations than the use of random sampling to calculate the same outcomes. For example, the number of strata in the importance sampling procedure used to estimate CCDFs in the 1991 and 1992 WIPP PAs (Helton and Iuzzolino, 1993) greatly exceeds the size of the random samples used in the 1996 WIPP PA to estimate CCDFs. The unequal strata probabilities also make the outcomes of analyses based on importance sampling inconvenient for use in sensitivity analyses (e.g., how does one interpret a scatterplot or a regression analysis derived from results obtained from an importance sampling procedure?).

Latin hypercube sampling is used when large samples are not computationally practicable and the estimation of very high quantiles is not required. The preceding is typically the case in uncertainty and sensitivity studies to assess the effects of subjective uncertainty. First, the models under consideration are often computationally demanding, with the result that the number of calculations that can be performed to support the analysis is necessarily limited. Second, the estimation of very high quantiles is generally not required in an analysis to assess the effects of subjective uncertainty. Typically, a 0.90 or 0.95 quantile is adequate to establish where the available information indicates a particular analysis outcome is likely to be located; in most analyses, a 0.99, 0.999, or 0.9999 quantile is not needed in assessing the effects of subjective uncertainty.

Desirable features of Latin hypercube sampling include unbiased estimates for means and distribution functions and dense stratification across the range of each sampled variable (McKay *et al.*, 1979). In particular, uncertainty and sensitivity analysis results obtained with Latin hypercube sampling have been observed to be quite robust even when relatively small samples (i.e., $n_{LHS} = 50\text{–}200$) are used (Iman and Helton, 1988, 1991; Helton *et al.*, 1995a; Kleijnen and Helton, 1999b).

For perspective, Latin hypercube and random sampling are illustrated in Figure 6.6 for two different distribution pairs. To facilitate comparisons, the grid that underlies the LHSs is also shown for the random samples, although it plays no role in the generation of these samples. The desirability of Latin hypercube sampling derives from the full coverage of the range of the sampled variables; specifically, each equal probability interval for U and also each equal probability interval for V has exactly one value sampled from it. In contrast, random sampling makes less efficient use of the sampled points, with the possibility existing that significant parts of a variable's range will be omitted (e.g., only one value below the 0.5 quantile for U in the lower left frame and no values for U below the 0.19 quantile nor above the 0.85 quantile in the lower right frame) and that other parts will be overemphasized (e.g., 5 out of 10 values for U fall between the 0.5 and 0.7 quantiles for U in the lower left frame, and two pairs of sampled points fall close together in the lower right frame). The enforced stratification in Latin hypercube sampling prevents such inefficient samplings while still providing unbiased estimates for means and distribution functions.

The outcome of the enforced stratification associated with Latin hypercube sampling is that estimates of means and distribution functions tend to be more stable when generated by Latin hypercube sampling than by random sampling. Here, stability refers to the amount of variation between results obtained with different samples generated by the particular sampling technique under consideration. This stability has been illustrated in several comparisons of Latin hypercube and random sampling (McKay *et al.*, 1979; Iman, 1992; Helton *et al.*, 1998, Section 6.1).

From the perspective of uncertainty and sensitivity analysis, the full stratification over the range of each sampled variable is a particularly desirable property of Latin hypercube sampling. In a large study, there are potentially hundreds of predicted variables that will be examined at some point in associated uncertainty and sensitivity analyses. Further, it is likely that almost every sampled variable will be important with respect to at least one of these predicted variables. With Latin hypercube sampling, every variable gets equal treatment (i.e., full stratification) within the sample; should a variable be important with respect to a particular output variable, it has been sampled in a way that will permit this importance to be identified. In contrast, it is very difficult to design an importance sampling procedure that provides acceptable results for a large number of sampled and predicted variables. In some sense, Latin hypercube sampling can be viewed as a compromise importance sampling procedure when *a priori* knowledge of the relationships between the sampled and predicted variables is not available. When random sampling is used with a small sample size in an analysis that involves a large number of sampled and predicted variables, the possibility

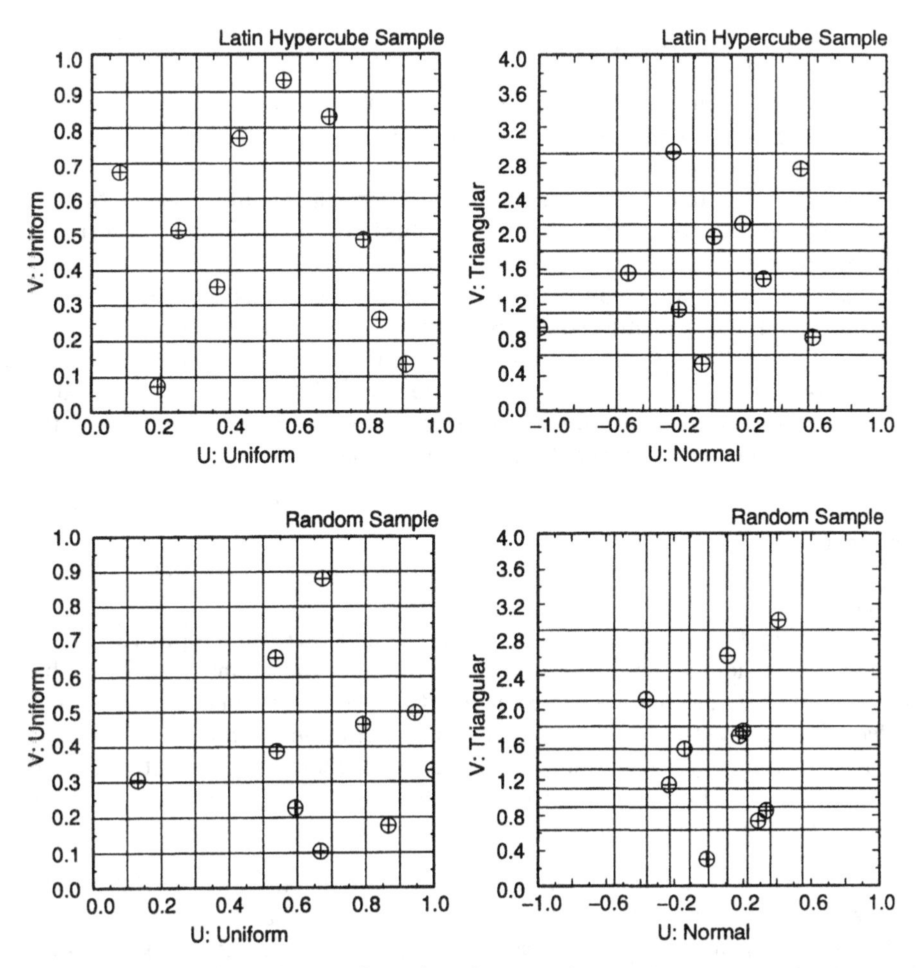

Figure 6.6 Examples of Latin hypercube and random sampling to generate a sample of size 10 from variables U and V with (1) U and V uniform on $[0, 1]$ (left frames), and (2) U normal on $[-1, 1]$ (mean $= 0, 0.01$ quantile $= -1, 0.99$ quantile $= 1$) and V triangular on $[0, 4]$ (mode $= 1$) (right frames).

exists that the chance structure of the sample will result in a poor representation of the relationships between some of the sampled and predicted variables. Such poor representations can also occur for Latin hypercube sampling when several sampled variables affect a given predicted variable, but are less likely to occur than is the case with random sampling.

Formal results involving Latin hypercube sampling and other sampling procedures are available in a number of publications (e.g., Owen, 1992; Stein, 1987; Iman and Conover, 1980; McKay *et al.*, 1979).

6.3.5 Correlation Control

Control of correlation within a sample can be very important. If two or more variables are correlated, then it is necessary that the appropriate correlation structure be incorporated into the sample if meaningful results are to be obtained in subsequent uncertainty/

sensitivity studies. On the other hand, it is equally important that variables do not appear to be correlated when they are really independent.

It is often difficult to induce a desired correlation structure on a sample. Indeed, multivariate distributions can be incompatible with correlation patterns that are proposed for them. Thus, it is possible to encounter analysis situations where the proposed variable distributions and the suggested correlations between the variables are inconsistent; that is, it is not possible to have both the desired variable distributions and the requested correlations between the variables.

In response to this situation, Iman and Conover (1982) proposed a restricted pairing technique for controlling the correlation structure in random and Latin hypercube samples that is based on rank correlation (i.e., on rank-transformed variables; see Section 6.6.6) rather than sample correlation (i.e., on the original untransformed data; see Section 6.6.4). With their technique, it is possible to induce any desired rank-correlation structure onto the sample. This technique has a number of desirable properties:

(i) It is distribution–free. That is, it may be used with equal facility on all types of distribution functions.
(ii) It is simple. No unusual mathematical techniques are required to implement the method.
(iii) It can be applied to any sampling scheme for which correlated input variables can logically be considered, while preserving the intent of the sampling scheme. That is, the same numbers originally selected as input values are retained; only their pairing is affected to achieve the desired rank correlations. This means that in Latin hypercube sampling the integrity of the intervals is maintained. If some other structure is used for selection of values, that same structure is retained.
(iv) The marginal distributions remain intact.

For many, if not most, uncertainty/sensitivity analysis problems, rank correlation is probably a more natural measure of congruent variable behavior than is the more traditional sample correlation. What is present in most situations is some idea of the extent to which variables tend to move up or down together; more detailed assessments of variable linkage are usually not available. This is precisely the level of knowledge that a rank correlation captures.

The restricted pairing technique begins with a sample of size m from the n input variables under consideration. This sample can be represented by the $m \times n$ matrix

$$\mathbf{x} = \begin{bmatrix} x_{11} & x_{12} & \dots & x_{1n} \\ x_{21} & x_{22} & \dots & x_{2n} \\ \vdots & \vdots & & \vdots \\ x_{ml} & x_{m2} & \dots & x_{mn} \end{bmatrix}, \qquad (6.3.8)$$

where x_{ij} is the value for variable j in sample element i. Thus, the rows of $\mathbf{x}$ correspond to sample elements, and the columns of $\mathbf{x}$ contain the sampled values for individual variables. The technique is based on rearranging the values in the individual columns of $\mathbf{x}$ so that a desired rank-correlation structure results between the individual variables.

The numerical details of the Iman/Conover restricted pairing technique to induce a desired rank-correlation structure are given in the original article (Iman and Conover, 1982). Further, the technique is implemented in the widely used LHS program (Iman and Shortencarier, 1984). The results of various rank-correlation assumptions are illustrated in

Table 6.2 Example rank correlations in replicate 1.

	WGRCOR	WMICDFLG	HALCOMP	HALPRM	ANHCOMP	ANHPRM	BPCOMP	BPPRM
WGRCOR	1.0000							
WMICDFLG	0.0198	1.0000						
HALCOMP	0.0011	0.0235	1.0000					
HALPRM	– 0.0068	– 0.0212	– 0.9879	1.0000				
ANHCOMP	0.0080	0.0336	– 0.0123	– 0.0025	1.0000			
ANHPRM	0.0049	– 0.0183	0.0037	0.0113	– 0.9827	1.0000		
BPCOMP	0.0242	0.1071	– 0.0121	0.0057	– 0.0184	0.0078	1.0000	
BPPRM	– 0.0514	– 0.0342	0.0035	0.0097	0.0283	– 0.0202	– 0.7401	1.0000
	WGRCOR	*WMICDFLG*	*HALCOMP*	*HALPRM*	*ANHCOMP*	*ANHPRM*	*BPCOMP*	*BPPRM*

Iman and Davenport (1980, 1982). As illustrated in Section 6.3.6, this technique was used to control correlations in the example analysis.

6.3.6 Latin Hypercube Sampling in the 1996 WIPP PA

This chapter uses an example from the 1996 WIPP PA. In this analysis, the LHS program (Iman and Shortencarier, 1984) was used to produce three independently generated LHSs of size $n_{LHS} = 100$ each from the 31 variables indicated in Table 6.1, for a total of 300 sample elements. Each individual replicate is an LHS of the form

$$\mathbf{x}_k = [x_{k1}, x_{k2}, \ldots, x_{kn_X}], \qquad k = 1, 2, \ldots, n_{LHS} = 100, \tag{6.3.9}$$

with $n_X = 31$. The three replicated samples were generated to provide a way to observe the stability of results obtained with Latin hypercube sampling. For notational convenience, the replicates are designated by R1, R2, and R3 for replicates 1, 2, and 3, respectively.

The restricted pairing technique indicated in Section 6.3.5 was used to induce requested correlations and also to ensure that uncorrelated variables had correlations close to zero. The variable pairs (*ANHCOMP, ANHPRM*), (*HALCOMP, HALPRM*) and (*BPCOMP, BPPRM*) were assigned rank correlations of -0.99, -0.99, and -0.75, respectively (Table 6.1). Further, all other variable pairs were assigned rank correlations of zero. The restricted pairing technique was quite successful in producing these correlations (Table 6.2). Specifically, the correlated variables have correlations that are close to their specified values, and uncorrelated variables have correlations that are close to zero.

6.4 EVALUATION OF MODEL

Once the sample in (6.1.4) has been generated, the corresponding model evaluations in (6.1.5) must be carried out. The nature of these evaluations is model-specific and outside the scope of this presentation. However, this brief section is included to emphasize that these evaluations are something that must be done as part of a sampling-based uncertainty and sensitivity analysis. If the model under consideration is expensive to evaluate, then this will probably be the most computationally demanding part of the analysis, and may significantly influence the sample size selected for use and possibly other aspects of the analysis. For example, the model for two-phase fluid flow used as an example in this presentation requires approximately 4–5 hours of CPU time on a VAX Alpha per evaluation (i.e., for each sample element) and produces a large quantity of temporally and spatially variable results. Thus, for this example, carrying out and then saving the necessary model evaluations involved a significant expenditure of human and computational resources. In contrast, this part of the analysis can be relatively undemanding for models that are less complex and computationally intensive.

6.5 UNCERTAINTY ANALYSIS

After the sample in (6.1.4) has been generated and the corresponding model evaluations in (6.1.5) have been carried out, the primary computational portions of the uncertainty analysis component of a sampling-based analysis have been completed. What remains to be done is to display the uncertainty information contained in the mapping between analysis inputs and analysis results in (6.1.6). Two cases are considered: results represented by single numbers (Section 6.5.1), and results represented by functions (Section 6.5.2). Finally, example analysis outcomes illustrating the stability of uncertainty analysis results obtained with Latin hypercube sampling are presented (Section 6.5.3).

6.5.1 Scalar Results

When a single real-valued result is under consideration, the vector-valued function $\mathbf{y}(\mathbf{x}_k)$ in (6.1.5) and (6.1.6) becomes the scalar-valued function

$$y_k = y(\mathbf{x}_k), \qquad k = 1, 2, \ldots, n_S. \tag{6.5.1}$$

One possibility is to summarize the uncertainty in y with a mean and a variance. If random or Latin hypercube sampling was used to generate the results in (6.5.1), then estimates $\hat{E}(y)$ and $\hat{V}(y)$ for the expected value and variance of y are given by

$$\hat{E}(y) = \sum_{k=1}^{n_S} \frac{y_k}{n_S}, \qquad \hat{V}(y) = \sum_{k=1}^{n_S} \frac{[y_k - \hat{E}(y)]^2}{n_S - 1}. \tag{6.5.2}$$

If importance sampling was used in the generation of the results in (6.5.1), then the probabilities of the individual strata in the importance sampling procedure would have to be used in the determination of $\hat{E}(y)$ and $\hat{V}(y)$.

Although the estimation of means and variances is a possibility for summarizing the uncertainty in scalar-valued results, these quantities do not provide very good summaries of subjective uncertainty for at least two reasons. First, information is always lost in the calculation of means and variances. Specifically, there is more information in the n_S numbers in (6.5.1) and their associated weights (i.e., the reciprocal of the sample size for random and Latin hypercube sampling and the strata probabilities for importance sampling) than there is in the two numbers in (6.5.2). Second, means and variances are not very natural quantities for summarizing subjective uncertainty. Specifically, the quantiles associated with a distribution summarizing subjective uncertainty convey more meaningful information about where the quantity under consideration is believed to be located.

Distribution functions provide a more effective summary of the information associated with the mapping in (6.5.1) than means and variances. In particular, this mapping can be summarized with either a CDF or a CCDF, with the CCDF simply being one minus the CDF (Figure 6.7). The presence of the included and excluded points in Figure 6.7 results from the

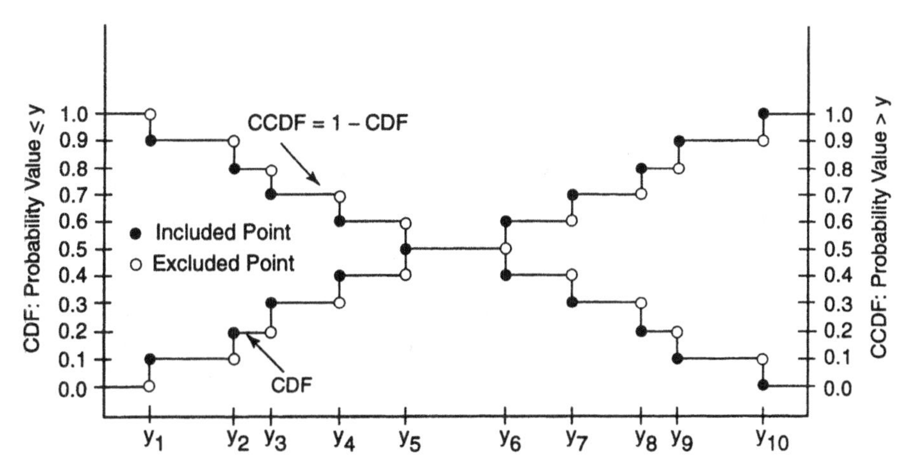

Figure 6.7 Example of construction of CDFs and CCDFs for a sample of size $n_S = 10$ (i.e., $y_k = y(\mathbf{x}_k), k = 1, 2, \ldots, n_S = 10$ in (6.5.1)).

use of a finite number of y values and the inequalities in the definitions of CDFs and CCDFs. Technically, the vertical lines should not be present in the CDF and CCDF in Figure 6.7 but these lines are typically included to make plots of CDFs and CCDFs easier to read. For the same reason, the distinction between included and excluded points is typically omitted. When random or Latin hypercube sampling is used, the step heights in the definitions of CDFs and CCDFs are the reciprocal of the sample size n_S (i.e., $1/n_S$ and thus $\frac{1}{10}$ in Figure 6.7); when importance sampling is used, the step heights correspond to the strata probabilities. An example with real data is given in Figure 6.8.

The value of CDFs and CCDFs is that they provide a display of all the information associated with the mapping in (6.5.1). In particular, they allow an easy extraction of the probabilities of having values in different subsets of the range of y. Although CDFs and CCDFs are equivalent in their information content, CCDFs are often used for display purposes when large samples are in use and it is important to display the effects of low-probability but high-consequence analysis outcomes (i.e., unlikely but large y values); further, CCDFs answer the question 'How likely is it to be this bad or worse?', which is typically the question of interest in risk assessments. Given that the distributions assigned to the elements of $\mathbf{x}$ are characterizing subjective uncertainty, the resultant probabilities extracted from CDFs and CCDFs are also characterizing subjective uncertainty and are thus providing quantitative measures of where the value of y is believed to be located.

Many individuals prefer density functions rather than CDFs or CCDFs for the display of distributions. Density functions have the advantage that they make it easy to identify the mode of a distribution, but they do not allow an easy extraction of the probabilities associated with various subranges of the dependent variable. Further, unless smoothing procedures are used, the best that can be obtained from the results in (6.5.1) is a histogram that approximates the shape of the density function, with the potential that the shape of this histogram will be significantly influenced by the resolution at which the y_k are binned (Silverman, 1986). As recommended by Ibrekk and Morgan (1987), an alternative display is to plot the CDF, the mean, and the associated density function on the same plot frame (Figure 6.9).

One disadvantage associated with CDFs, CCDFs, and density functions is that displays using these distributional summaries can become quite cluttered when results for a number

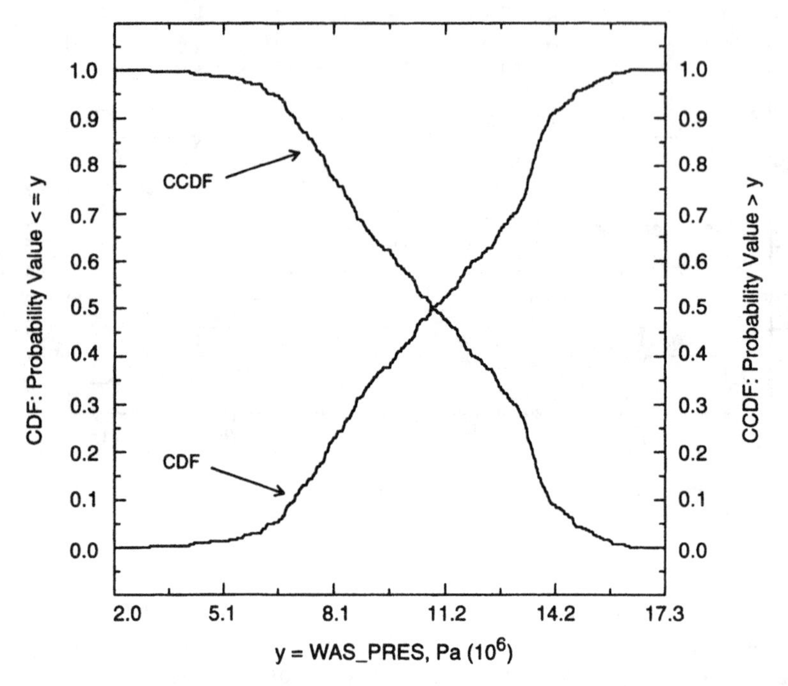

Figure 6.8 Example of estimated CDF and CCDF for repository pressure at 10 000 years under undisturbed conditions (i.e., $y = WAS\_PRES$) obtained from the 300 LHS elements that result from pooling replicates R1, R2 and R3 (see Section 6.3.6).

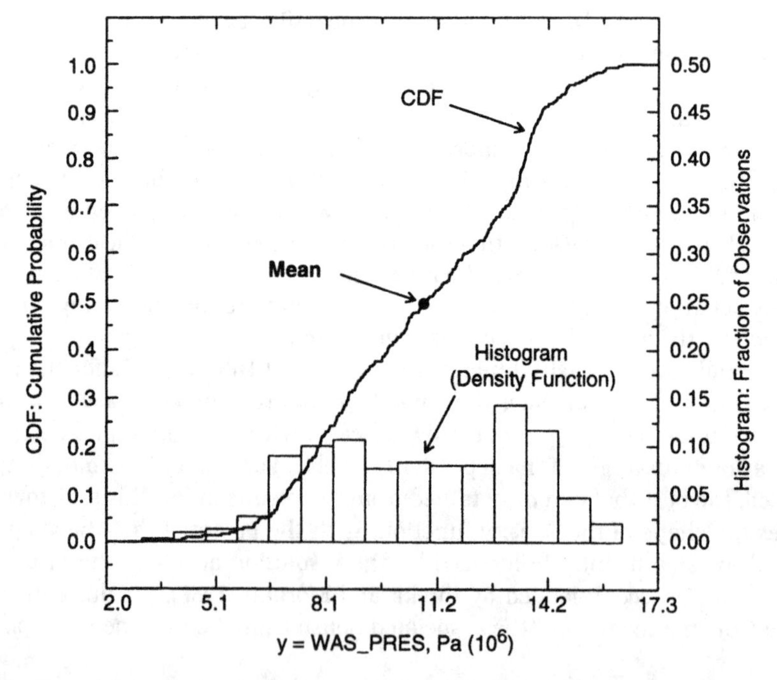

Figure 6.9 Uncertainty display including estimated distribution function, density function, and mean for repository pressure at 10 000 years under undisturbed conditions (i.e., $y = WAS\_PRES$).

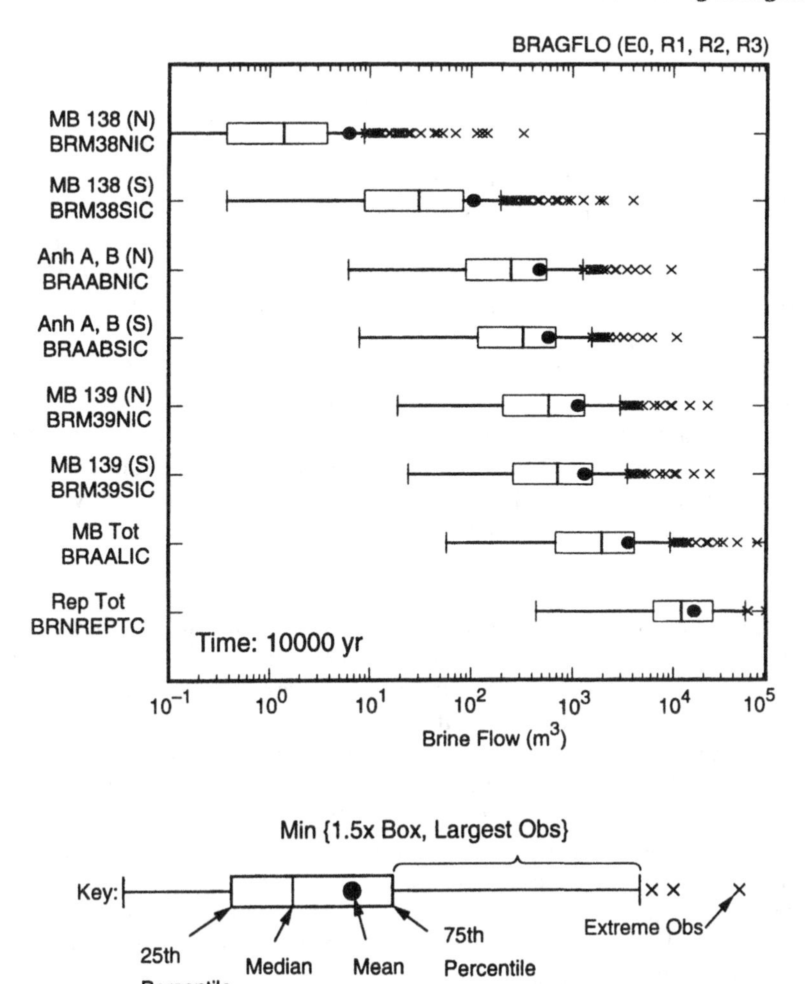

Figure 6.10 Examples of box plots for cumulative brine flow over 10 000 years into various regions in disturbed rock zone surrounding repository and into repository (*BRNREPTC*) under undisturbed conditions in the 1996 WIPP PA (Helton *et al.*, 1998, Figure 7.2.2).

of different analysis outcomes are presented in a single-plot frame (e.g., a plot involving CDFs, CCDFs, or density functions for 10 different analysis outcomes can be hard to read owing to the tendency of the individual curves to repeatedly cross each other). Box plots provide an alternative, less congested display of multiple distributions (Figure 6.10). In such plots, the endpoints of the boxes are formed by the lower and upper quartiles of the data, namely, $x_{0.25}$ and $x_{0.75}$. The vertical line within the box represents the median, $x_{0.50}$. The mean is identified by the large dot. The bar on the right of the box extends to the minimum of $x_{0.75} + 1.5(x_{0.75} - x_{0.25})$ and the maximum value. In a similar manner, the bar on the left of the box extends to the maximum of $x_{0.25} - 1.5(x_{0.75} - x_{0.25})$ and the minimum value. The observations falling outside of these bars are shown with crosses. Box plots contain the same information as a distribution function, but in a somewhat reduced form. Further, their flattened shape makes it convenient to place many distributions on a single plot and also to compare different distributions.

6.5.2 Functions

In many analyses, outcomes of interest are functions of one or more variables. In the example used in this presentation, many results are functions of time. Thus, time is the independent variable (i.e., function argument). However, there is also subjective uncertainty in the variables required in the estimation of these functions, with this uncertainty leading to multiple possible functions as illustrated in Figure 6.11 (a). The estimated distribution presented in Figure. 6.11 (a) was obtained from the LHS in (6.3.9) associated with replicate R1 (i.e., each curve in the figure was calculated conditional on the occurrence of one of the sample elements $\mathbf{x}_k$ in (6.3.9) for replicate R1).

The curves in Figure 6.11 (a) are an approximation obtained with an LHS of size 100 to the actual distribution that results from subjective uncertainty. As presented, these curves provide an impression of the shape of the associated distribution, but they do not directly provide probabilistic information. In concept, this distribution can be summarized by presenting density functions for the values on the ordinate for a sequence of values on the abscissa. In practice, the construction and informative presentation of such density functions is not easy.

An alternative and often effective representation is to determine mean values and percentiles conditional on individual values on the abscissa and then to plot these means and percentiles above the values for which they were determined. Conceptually, a vertical is drawn through the curves above a given value on the abscissa. The locations where this line passes through the individual curves identify the corresponding pressure values, with the number of pressure values equal to the sample size in use. These values can be used to produce a mean value and also selected percentile values. If desired, the definition of the mean and percentile values can be represented formally by integrals over the possible values for $\mathbf{x}$ (Helton, 1996), with the sampling procedure being used to provide approximations to these integrals. Once the mean and percentile values have been determined, they can be plotted above the corresponding values on the abscissa and then connected to form continuous curves (Figure 6.11b). With this summary procedure, the percentile values are defined conditional on individual times on the abscissa; as a result, the percentile curves (Figure 6.11b) should not be viewed as being percentiles for the distribution of curves (i.e., it is

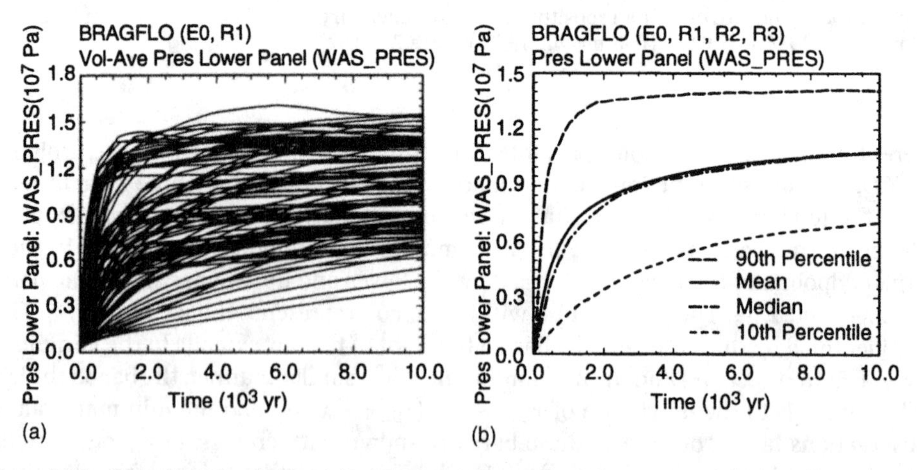

Figure 6.11 Repository pressure under undisturbed conditions (i.e., $y = WAS\_PRES$): (a) individual pressure curves for 100 LHS elements in replicate R1, and (b) mean and percentile curves obtained from the 300 observations that result from pooling replicates R1, R2, and R3.

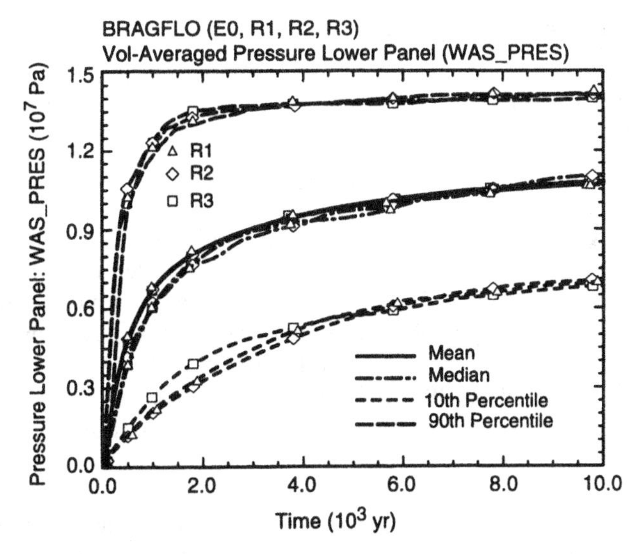

Figure 6.12 Mean and percentile curves for three replicated LHSs for repository pressure under undisturbed conditions (i.e., $y = WAS\_PRES$ in Figure 6.11a).

inappropriate to assume that there is a probability of 0.9 that a randomly selected value for $\mathbf{x}$ will produce a curve that falls below the 0.9 percentile curve in Figure 6.11b). Results such as those given in Figure 6.11(b) provide a more quantitative summary of the distribution of curves in Figure 6.11(a) than the intuitive impression that is obtained by visually examining the curves themselves.

6.5.3 Stability of Results

The LHS in (6.3.9) was replicated three times (i.e., $n_R = 3$) to provide a measure of the stability of the results obtained in the 1996 WIPP PA. For the pressure results in Figure 6.11, the results were quite stable from sample to sample (Figure 6.12). Indeed, the results obtained with the individual replicates were quite stable across the large number of predicted outcomes examined in the analysis, with no instance occurring where different replicates would have lead to different conclusions with respect to system behavior.

6.6 SENSITIVITY ANALYSIS

Sensitivity analysis involves an exploration of the mapping in (6.1.6) to determine the effects of individual components of $\mathbf{x}$ on the analysis outcomes contained in $\mathbf{y}(\mathbf{x})$. A number of procedures and topics associated with the exploration of this mapping are discussed, including examination of scatterplots (Section 6.6.1), regression analysis (Sections 6.6.2 and 6.6.3), correlation and partial correlation analysis (Section 6.6.4), stepwise regression analysis (Section 6.6.5), the rank transformation to facilitate regression and correlation analysis (Section 6.6.6), effects of correlations on sensitivity analysis (Section 6.6.7), identification of nonmonotonic patterns (Section 6.6.8), and identification of non-random patterns (Section 6.6.9).

6.6.1 Examination of Scatterplots

The generation of scatterplots is undoubtedly the simplest sensitivity analysis technique and only involves plotting the points

$$(x_{kj}, y_k), \qquad k = 1, 2, \ldots, n_S, \tag{6.6.1}$$

for each element x_j of $\mathbf{x}$ for $j = 1, 2, \ldots, n_X$ (see (6.1.2) and (6.1.4)). This produces n_X scatterplots that can be examined for relationships between y and the elements of $\mathbf{x}$ (i.e., the x_j). As an example, the scatterplot in Figure 6.13(a) shows a nonlinear but monotonic relationship between borehole permeability (*BHPRM*) and cumulative brine flow down an intruding borehole, with no brine flow taking place for small values of *BHPRM* and brine flow increasing rapidly for larger values of *BHPRM* (for additional discussion, see Helton *et al.*, 1998, Section 8.2). As another example, the scatterplot in Figure 6.13(b) shows a complex relationship between *BHPRM* and repository pressure that is both nonlinear and nonmonotonic, with repository pressure decreasing as *BHPRM* increases and then undergoing a sudden jump at $BHPRM \doteq -11.7$ (i.e., at a permeability of $10^{-11.7}\,\text{m}^2 \doteq 2 \times 10^{-12}\,\text{m}^2$) (for additional discussion, see Helton *et al.*, 1998, Section 8.4). In contrast to the well-defined patterns in Figure 6.13, the individual points will be randomly spread over the plot when there is no relationship between y and a particular x_j.

Sometimes scatterplots alone will completely reveal the relationships between model input (i.e., elements of $\mathbf{x}$) and model predictions (i.e., y). This is often the case when only one or two inputs dominate the outcome of the analysis. Further, scatterplots often reveal nonlinear relationships, thresholds, and variable interactions that facilitate the understanding of model behavior and the planning of more sophisticated sensitivity studies. Iman and Helton (1988) provide an example where the examination of scatterplots revealed a rather complex pattern of variable interactions. The examination of scatterplots is always a good starting point in a sensitivity study. The examination of such plots when Latin hypercube

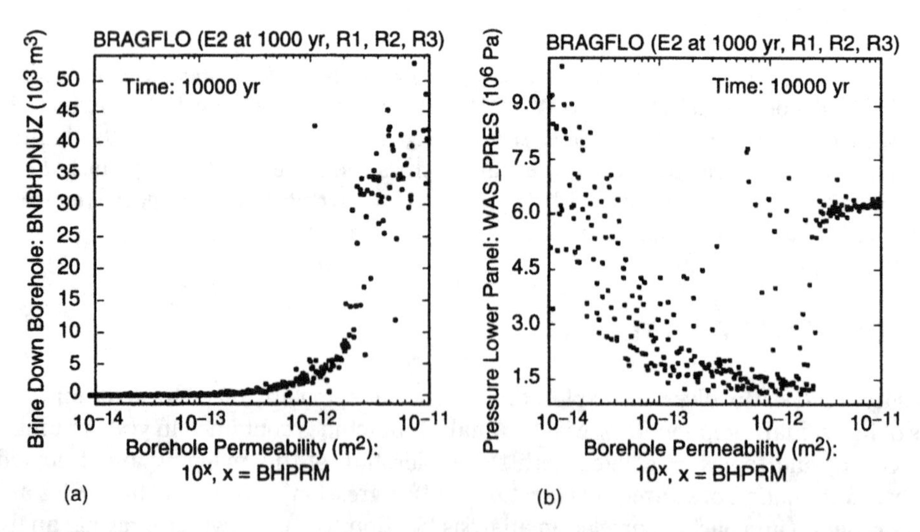

Figure 6.13 Example scatterplots: (a) cumulative brine flow through borehole into upper disturbed rock zone (DRZ) over 10 000 years for intrusion at 1000 years into lower waste panel (i.e., $y = BNBHDNUZ$) versus borehole permeability (*BHPRM*), and (b) repository pressure (Pa) at 10 000 years for intrusion at 1000 years into lower waste panel (i.e., $y = WAS\_PRES$ at 10 000 years) versus borehole permeability (*BHPRM*).

sampling is used can be particularly revealing due to the full stratification over the range of each input variable.

6.6.2 Regression Analysis

A more formal investigation of the mapping

$$[\mathbf{x}_k, y(\mathbf{x}_k)], \qquad k = 1, 2, \ldots, n_S, \tag{6.6.2}$$

can be based on regression analysis. In this approach, a model of the form

$$\hat{y} = b_0 + \sum_{j=1}^{n_X} b_j x_j \tag{6.6.3}$$

is developed from the mapping between analysis inputs and analysis results, where the x_j are the input variables under consideration and the b_j are coefficients that must be determined. The coefficients b_j and other aspects of the construction of the regression model in (6.6.3) can be used to indicate the importance of the individual x_j with respect to the uncertainty in y.

The construction of the regression model in (6.6.3) is considered first. To keep the notation from becoming unwieldy, n will be used to denote the number of independent variables under consideration (i.e., $n = n_X$ as used in (6.1.2) and (6.6.3)) and m will be used to denote the number of observations under consideration (i.e., $m = n_S$ as used in (6.1.4) and (6.6.2)). As shown in (6.6.2), there exists a sequence $y_k = y(\mathbf{x}_k), k = 1, 2, \ldots, m$, of values for the output variable. When expressed in the form of the model in (6.6.3), each y_k becomes

$$y_k = b_0 + \sum_{j=1}^{n} b_j x_{kj} + \varepsilon_k, \qquad k = 1, \ldots, m, \tag{6.6.4}$$

where the error terms ε_k, $k = 1, \ldots, m$, are defined by $\varepsilon_k = y_k - \hat{y}_k$ and thus equal the difference between the observed value y_k and the corresponding predicted value $\hat{y}_k$ defined by (6.6.3). At this point, the b_j are still unknown. What is desired is to determine the b_j in some suitable manner. The method of least squares is widely used and will be employed here (Eisenhart, 1964; Harter, 1983). As a result of its extensive use, there exist a number of excellent textbooks on least squares regression analysis (e.g., Neter and Wasserman, 1974; Seber, 1977; Daniel and Wood, 1980; Draper and Smith, 1981; Weisberg, 1985; Myers, 1990). The following discussion presents just enough information to be able to describe some of the applications of regression-based techniques in sensitivity analysis. The indicated textbooks, as well as many others, provide far more information on regression analysis than can be presented here.

To determine the b_j, it is convenient to use the following matrix representation for the equalities in (6.6.4):

$$\mathbf{y} = \mathbf{x}\mathbf{b} + \boldsymbol{\varepsilon}, \tag{6.6.5}$$

where

$$\mathbf{y} = \begin{bmatrix} y_1 \\ \vdots \\ y_m \end{bmatrix}, \quad \mathbf{x} = \begin{bmatrix} 1 & x_{11} & \ldots & x_{1n} \\ \vdots & \vdots & & \vdots \\ 1 & x_{m1} & \ldots & x_{mn} \end{bmatrix}, \quad \mathbf{b} = \begin{bmatrix} b_0 \\ \vdots \\ b_n \end{bmatrix}, \quad \boldsymbol{\varepsilon} = \begin{bmatrix} \varepsilon_1 \\ \vdots \\ \varepsilon_m \end{bmatrix}.$$

In the least-squares approach, the b_j are determined such that the sum

$$S(\mathbf{b}) = \sum_{k=1}^{m}\left(y_k - b_0 - \sum_{j=1}^{n} b_j x_{kj}\right)^2 = (\mathbf{y} - \mathbf{xb})^{\mathrm{T}}(\mathbf{y} - \mathbf{xb}) \tag{6.6.6}$$

is a minimum. Put another way, the b_j are determined such that the sum of the squared error terms, $\sum_k \varepsilon_k^2$, is a minimum. The determination of the b_j in the least-squares approach is a calculus rather than a statistics problem, and is based on consideration of the first derivatives of $S(\mathbf{b})$ with respect to the individual b_j (Draper and Smith, 1981, Section 2.6).

This determination leads to the following matrix equation that defines the coefficient vector $\mathbf{b}$ for which the sum $S(\mathbf{b})$ in (6.6.6) is a minimum:

$$\mathbf{x}^{\mathrm{T}}\mathbf{xb} = \mathbf{x}^{\mathrm{T}}\mathbf{y}. \tag{6.6.7}$$

For the analysis to produce a unique value for $\mathbf{b}$, the matrix $\mathbf{x}^{\mathrm{T}}\mathbf{x}$ must be invertible. Then, $\mathbf{b}$ is given by

$$\mathbf{b} = (\mathbf{x}^{\mathrm{T}}\mathbf{x})^{-1}\mathbf{x}^{\mathrm{T}}\mathbf{y}. \tag{6.6.8}$$

The matrix $\mathbf{x}^{\mathrm{T}}\mathbf{x}$ will always be invertible when the columns of $\mathbf{x}$ are linearly independent. This is usually the case in a sampling-based study in which the number of sample elements (i.e., m) exceeds the number of independent variables (i.e., n).

The following identity holds for the least-squares regression model, and plays an important role in assessing the adequacy of such models:

$$\sum_{k=1}^{m}(y_k - \bar{y})^2 = \sum_{k=1}^{m}(\hat{y}_k - \bar{y})^2 + \sum_{k=1}^{m}(\hat{y}_k - y_k)^2, \tag{6.6.9}$$

where $\hat{y}_k$ denotes the estimate of y_k obtained from the regression model and $\bar{y}$ is the mean of the y_k (Myers, 1990, Section 3.4). For notational convenience, the preceding equality is often written as

$$\mathrm{SS}_{\mathrm{tot}} = \mathrm{SS}_{\mathrm{reg}} + \mathrm{SS}_{\mathrm{res}}, \tag{6.6.10}$$

where

$$\mathrm{SS}_{\mathrm{tot}} = \sum_{k=1}^{m}(y_k - \bar{y})^2, \quad \mathrm{SS}_{\mathrm{reg}} = \sum_{k=1}^{m}(\hat{y}_k - \bar{y})^2, \quad \mathrm{SS}_{\mathrm{res}} = \sum_{k=1}^{m}(\hat{y}_k - y_k)^2.$$

The three preceding summations are called the total sum of squares, regression sum of squares, and residual sum of squares, respectively.

Since $\mathrm{SS}_{\mathrm{res}}$ provides a measure of variability about the regression model, the ratio

$$R^2 = \mathrm{SS}_{\mathrm{reg}}/\mathrm{SS}_{\mathrm{tot}} \tag{6.6.11}$$

provides a measure of the extent to which the regression model can match the observed data. Specifically, when the variation about the regression model is small (i.e., when $\mathrm{SS}_{\mathrm{res}}$ is small relative to $\mathrm{SS}_{\mathrm{reg}}$), the corresponding R^2 value is close to 1, which indicates that the regression model is accounting for most of the uncertainty in the y_k. Conversely, an R^2 value close to zero indicates that the regression model is not very successful in accounting for the uncertainty in the y_k. Another name for R^2 is the coefficient of multiple determination.

An important situation occurs when the rows of the matrix $\mathbf{x}$ (i.e., the variable values at which the model is evaluated) are selected so that $\mathbf{x}^{\mathrm{T}}\mathbf{x}$ is a diagonal matrix. In this case, the columns of $\mathbf{x}$ are said to be orthogonal, and the estimated regression coefficients are given by

$$\mathbf{b} = (\mathbf{x}^{\mathrm{T}}\mathbf{x})^{-1}\mathbf{x}^{\mathrm{T}}\mathbf{y}$$
$$= \begin{bmatrix} d_0 & 0 & \ldots & 0 \\ 0 & d_1 & & 0 \\ \vdots & \vdots & & \vdots \\ 0 & 0 & \ldots & d_n \end{bmatrix}^{-1} \begin{bmatrix} 1 & 1 & \ldots & 1 \\ x_{11} & x_{21} & \ldots & x_{m1} \\ \vdots & \vdots & & \vdots \\ x_{1n} & x_{2n} & \ldots & x_{mn} \end{bmatrix} \begin{bmatrix} y_1 \\ y_2 \\ \vdots \\ y_m \end{bmatrix}, \tag{6.6.12}$$

and so each element b_j of $\mathbf{b}$ is given by

$$b_j = \sum_{k=1}^{m} \frac{x_{kj} y_k}{d_j} = \sum_{k=1}^{m} x_{kj} y_k \Bigg/ \sum_{k=1}^{m} x_{kj}^2 . \tag{6.6.13}$$

Thus, the estimate of the regression coefficient b_j for the variable x_j depends only on the values for x_j in the design matrix $\mathbf{x}$ (i.e., $x_{1j}, \ldots, x_{mj}$). This is true regardless of the number of variables included in the regression. As long as the design is orthogonal, the addition or deletion of variables from the model will not change the regression coefficients for the remaining variables. Further, when the design matrix $\mathbf{x}$ is orthogonal, the R^2 value for the regression can be expressed as

$$R^2 = \frac{SS_{\mathrm{reg}}}{SS_{\mathrm{tot}}} = R_1^2 + R_2^2 + \cdots + R_n^2, \tag{6.6.14}$$

where R_j^2 is the R^2 value that results from regressing y on only x_j (Draper and Smith, 1981, p. 99). Thus, R_j^2 is equal to the contribution of x_j to R^2 when the design matrix $\mathbf{x}$ is orthogonal.

The regression model in (6.6.3) can be algebraically reformulated as

$$\frac{\hat{y} - \bar{y}}{\hat{s}} = \sum_{j=1}^{n} \frac{b_j \hat{s}_j}{\hat{s}} \frac{x_j - \bar{x}_j}{\hat{s}_j}, \tag{6.6.15}$$

where

$$\bar{y} = \sum_{k=1}^{m} \frac{y_k}{m}, \qquad \hat{s} = \left[\sum_{k=1}^{m} \frac{(y_k - \bar{y})^2}{m-1} \right]^{1/2},$$
$$\bar{x}_j = \sum_{k=1}^{m} \frac{x_{kj}}{m}, \qquad \hat{s}_j = \left[\sum_{k=1}^{m} \frac{(x_{kj} - \bar{x}_j)^2}{m-1} \right]^{1/2}.$$

The coefficients $b_j \hat{s}_j / \hat{s}$ appearing in (6.6.15) are called standardized regression coefficients (SRCs). When the x_j are independent, the absolute value of the SRCs can be used to provide a measure of variable importance. Specifically, the coefficients provide a measure of importance based on the effect of moving each variable away from its expected value by a fixed fraction of its standard deviation while retaining all other variables at their expected values. Calculating SRCs is equivalent to performing the regression analysis with the input and output variables normalized to mean zero and standard deviation one.

An example regression analysis is now given. The output variable (i.e., y) is pressure in the repository at 10 000 years under undisturbed conditions (i.e., the pressure values above 10 000 years in Figure 6.11(a)). To keep the example at a convenient size, three independent variables (i.e., x_j) will be considered: pointer variable for microbial degradation of cellulose (*WMICDFLG*), halite porosity (*HALPOR*), and corrosion rate for steel (*WGRCOR*). The following regression model is obtained using the preceding three variables and the pooled LHS indicated in conjunction with (6.3.9) (i.e., $n = 3$ and $m = 300$):

$$y = 5.73 \times 10^6 + 2.46 \times 10^6 \; WMICDFLG + 1.55 \times 10^8 \; HALPOR + 1.52 \times 10^{20} \; WGRCOR. \tag{6.6.16}$$

The coefficients in the preceding model show the effect of a one unit change in an input variable (i.e., an x_j) on the output variable (i.e., y). The sign of a regression coefficient indicates whether y tends to increase (a positive regression coefficient) or tends to decrease (a negative regression coefficient) as the corresponding input variable increases. Thus, y tends to increase as each of *WMICDFLG*, *HALPOR* and *WGRCOR* increases.

It is hard to assess variable importance from the regression coefficients in (6.6.16) because of the effects of units and distribution assumptions. In particular, the regression coefficient for *WGRCOR* is much larger than the regression coefficients for *WMICDFLG* and *HALPOR*, which does not necessarily imply that *WGRCOR* has greater influence on the uncertainty in y than *WMICDFLG* or *HALPOR*. Variable importance is more clearly shown by the following reformulation of (6.6.16) with SRCs:

$$y = 0.722 \; WMICDFLG + 0.468 \; HALPOR + 0.246 \; WGRCOR, \tag{6.6.17}$$

where y, *WMICDFLG*, *HALPOR*, and *WGRCOR* have been standardized to mean zero and standard deviation one as indicated in (6.6.15). The SRCs in (6.6.17) provide a better characterization of variable importance than the unstandardized coefficients in (6.6.16). For perturbations equal to a fixed fraction of their standard deviation, the impact of *WMICDFLG* is approximately 50% larger than the impact of *HALPOR* (i.e., $(0.722-0.468)/0.468 = 0.54$) and almost 200% larger than the impact of *WGRCOR* (i.e., $(0.722 - 0.246)/0.246 = 1.96$). Both regression models have an R^2 value of 0.79, and thus can account for approximately 79% of the uncertainty in y. Standardized regression coefficients are a popular way of ranking variable importance in sampling-based sensitivity analysis, and many examples of their use exist (e.g., Ma *et al.*, 1993; Ma and Ackerman, 1993; Whiting *et al.*, 1993; Hamby, 1995; Chan, 1996; Helton *et al.*, 1996).

6.6.3 Statistical Tests in Regression Analysis

Determination of the regression coefficients $b_0, b_1, b_2, \ldots, b_n$ that constitute the elements of the vector $\mathbf{b}$ in (6.6.8) involves no statistics. Rather, as already indicated, this determination is based entirely on procedures involving minimization of functions and algebraic manipulations. If desired, formal statistical procedures can be used to indicate if these coefficients appear to be different from zero. However, such procedures are based on assumptions that are not satisfied in sampling-based sensitivity studies of deterministic models (i.e., models for which a given input always produces the same result), and thus the outcome of using formal statistical procedures to make assessments about the significance of individual coefficients or other entities in sampling-based sensitivity studies should be regarded simply as one form of guidance as to whether or not a model prediction appears to be affected by a particular model input.

In the usual construction of tests for the significance of regression coefficients, the relationship between the dependent and independent variables is assumed to be of the form

$$y = \beta_0 + \sum_{j=1}^{n} \beta_j x_j + \varepsilon, \tag{6.6.18}$$

where ε is normally distributed with mean 0 and standard deviation σ and characterizes the variation in y that is observed when y is repeatedly evaluated for $\mathbf{x} = [x_1, x_2, \ldots, x_n]$. Further, σ is assumed to be the same for all values of $\mathbf{x}$. It is the distributional assumptions involving ε that allows the construction of statistical tests for the coefficients $\beta_0, \beta_1, \beta_2, \ldots, \beta_n$. These assumptions are not satisfied in sampling-based sensitivity studies with deterministic models because a given $\mathbf{x}$ always produces the same value for y.

Given the preceding assumptions involving ε, the relationship in (6.6.9) can be used in the development of tests to indicate if various of the β_j in (6.6.18) appear to be different from zero. For notational convenience, let

$$\mathrm{SS}_{\mathrm{reg}}(\beta_1, \beta_2 \ldots, \beta_n \,|\, \beta_0) = \sum_{k=1}^{m} (\hat{y}_k - \bar{y})^2 \tag{6.6.19}$$

when the vector $\mathbf{b}$ in (6.6.8), and hence the associated regression model, contains estimates for $\beta_0, \beta_1, \beta_2, \ldots, \beta_n$. The preceding quantity is called the regression sum of squares and constitutes the part of the total sum of squares (i.e., the left-hand side of (6.6.9)) that can be explained by the regression model. More generally, if $\beta_1, \beta_2, \ldots, \beta_n$ are partitioned into vectors $\boldsymbol{\beta}_1$ and $\boldsymbol{\beta}_2$, where $\boldsymbol{\beta}_1$ contains p_1 of the coefficients $\beta_1, \beta_2, \ldots, \beta_n$ and $\boldsymbol{\beta}_2$ contains the remaining $p_2 = n - p_1$ coefficients, then

$$\mathrm{SS}_{\mathrm{reg}}(\beta_1, \beta_2, \ldots, \beta_n \,|\, \beta_0) = \mathrm{SS}_{\mathrm{reg}}(\boldsymbol{\beta}_1 \,|\, \boldsymbol{\beta}_2, \beta_0) + \mathrm{SS}_{\mathrm{reg}}(\boldsymbol{\beta}_2 \,|\, \beta_0), \tag{6.6.20}$$

where $\mathrm{SS}_{\mathrm{reg}}(\boldsymbol{\beta}_1 \,|\, \boldsymbol{\beta}_2, \beta_0)$ is the increase in the regression sum of squares that results from extending a regression model involving estimates for β_0 and the β_j in $\boldsymbol{\beta}_2$ to a regression model involving estimates for β_0 and the coefficients in $\boldsymbol{\beta}_1$ and $\boldsymbol{\beta}_2$.

Given the assumptions involving ε indicated in conjunction with (6.6.18), $\mathrm{SS}_{\mathrm{reg}}(\boldsymbol{\beta}_1 \,|\, \boldsymbol{\beta}_2, \beta_0)$ can be used to test the hypothesis that $\boldsymbol{\beta}_1 = \mathbf{0}$. In particular, if $\boldsymbol{\beta}_1 = \mathbf{0}$ and the assumptions involving ε are satisfied, then

$$F = \frac{\mathrm{SS}_{\mathrm{reg}}(\boldsymbol{\beta}_1 \,|\, \boldsymbol{\beta}_2, \beta_0)/p_1}{\hat{s}^2} \tag{6.6.21}$$

can be regarded as a randomly sampled value from an F-distribution with $(p_1, m - n - 1) = (p_1, m - p_1 - p_2 - 1)$ degrees of freedom, where

$$\hat{s}^2 = \sum_{k=1}^{m} \frac{(y_k - \hat{y}_k)^2}{m - n - 1} \tag{6.6.22}$$

is an approximation to σ^2 (see Myers, 1990, Section 3.4, or any other standard text on regression analysis). The probability $\mathrm{prob}_F(\tilde{F} > F|\eta_1, \eta_2)$ of exceeding an F-statistic value of F calculated with (η_1, η_2) degrees of freedom can be estimated by

$$\mathrm{prob}_F(\tilde{F} > F|\eta_1, \eta_2) = I_\nu(\tfrac{1}{2}\eta_2, \tfrac{1}{2}\eta_1), \qquad \nu = \frac{\eta_2}{\eta_2 + \eta_1 F}, \tag{6.6.23}$$

where $I_\nu(a, b)$ denotes the incomplete beta function (Press *et al.*, 1992, p. 222). Thus, under the assumption that $\boldsymbol{\beta}_1 = \mathbf{0}$, the probability that a larger value for $SS_{reg}(\boldsymbol{\beta}_1 | \boldsymbol{\beta}_2, \beta_0)$ would result from chance alone can be calculated and used to make an assessment as to whether or not it appears to be reasonable to reject the assumption that $\boldsymbol{\beta}_1 = \mathbf{0}$, with this probability typically called the *p*-value or α-value for F and the corresponding vector $\boldsymbol{\beta}_1$. Small *p*-values indicate that the observed value for F is unlikely to have occurred owing to chance, and thus suggest that $\boldsymbol{\beta}_1 \neq \mathbf{0}$.

The statistic F in (6.6.21) can be used to test the hypothesis that

$$\boldsymbol{\beta} = [\beta_1, \beta_2, \ldots, \beta_n] = \mathbf{0}. \tag{6.6.24}$$

In this case,

$$F = \frac{SS_{reg}(\boldsymbol{\beta} | \beta_0)/n}{\hat{s}^2} \tag{6.6.25}$$

can be regarded as a randomly sampled value from an *F*-distribution with $(n, m - n - 1)$ degrees of freedom. A small *p*-value for F suggests that $\boldsymbol{\beta} \neq \mathbf{0}$.

Another important special case occurs when a single regression coefficient (i.e., β_j) is under consideration, with the result that $p_1 = 1$ and $p_2 = n - 1$ in (6.6.21). Then,

$$F = \frac{SS_{reg}(\beta_j | \boldsymbol{\beta}_2, \beta_0)/1}{\hat{s}^2} \tag{6.6.26}$$

can be used to indicate if β_j appears to differ from zero given that estimates for β_0 and the coefficients in $\boldsymbol{\beta}_2$ are included in the regression model. Specifically, under the assumption that $\beta_j = 0$, F can be regarded as a randomly sampled value from an *F*-distribution with $(1, m - n - 1)$ degrees of freedom. An equivalent test involving β_j can be based on the statistic

$$t = \frac{b_j}{\hat{s}\sqrt{c_{jj}}}, \tag{6.6.27}$$

where b_j is the estimated value for β_j, $\hat{s}$ is defined in (6.6.22), and c_{jj} is the jth diagonal element of the matrix $(\mathbf{x}^T\mathbf{x})^{-1}$ in (6.6.8) (Myers, 1990, p. 98). Under the assumption that $\beta_j = 0$, t can be regarded as a randomly sampled value from a *t*-distribution with $m - n - 1$ degrees of freedom. The probability $\text{prob}_t(|\tilde{t}| > |t| \,|\, m - n - 1)$ of obtaining a value $\tilde{t}$ from the preceding distribution for which $|\tilde{t}|$ exceeds $|t|$ is given by

$$\text{prob}_t(|\tilde{t}| > |t| \,|\, m - n - 1) = 1 - I_x[\tfrac{1}{2}(m - n - 1), \tfrac{1}{2}], \qquad x = \frac{m - n - 1}{m - n - 1 + t^2}, \tag{6.6.28}$$

where $I_x(a, b)$ denotes the incomplete beta function (Press *et al.*, 1992, p. 222). Thus, t as defined in (6.6.27) can also be used to test if an individual regression coefficient appears to be different from zero. The equality $F = t^2$ holds for F and t as defined in (6.6.26) and (6.6.27). Further, identical significance results (i.e., *p*- or α-values) are produced by the use of F in conjunction with the relationship in (6.6.23) and the use of t in conjunction with the relationship in (6.6.28)

As already indicated, the distributional assumptions that lead to the *p*-values defined by (6.6.23) and (6.6.28) are not satisfied in sampling-based sensitivity studies. However, these *p*-values still provide a useful criterion for assessing variable importance, because they provide an indication of how viable the relationships between input and output variables would appear to be in a study in which the underlying distributional assumptions were satisfied.

As an illustration, results of a formal statistical analysis of the regression models in (6.6.16) and (6.6.17) are presented in Table 6.3, with the coefficients in these models appearing in the columns labeled 'Regression coefficient' and 'Standardized regr Coeff., respectively. The *p*-value for the regression model containing all three variables (Footnote *e* in the table) is less than 10^{-4}, as are the *p*-values for adding individual variables to the regression model (Footnote *n* in the Table). Thus, in a study in which the necessary distributional assumptions were satisfied (see (6.6.18)), the implication would be that *WMICDFLG, HALPOR*, and *WGRCOR* have significant influences (i.e., nonzero regression coefficients) on y. The *p*-values for the individual variables (Footnote *n* in the table) are more useful from a sensitivity analysis perspective than the *p*-value for all three variables (Footnote *e* in the table), since they indicate whether or not individual variables appear to affect y. In contrast, the *p*-value for the variables collectively only indicates that at least one of the variables appears to affect y.

The regression analysis summarized in (6.6.16), (6.6.17) and Table 6.3 only involves the variables *WMICDFLG, HALPOR*, and *WGRCOR*, with these variables selected for illustrative purposes on the basis of *a priori* knowledge that they had identifiable effects on y. As a result, these variables produce regression models with small *p*-values. In a sensitivity analysis with no *a priori* knowledge, all of the variables indicated in Table 6.1 would have to be investigated

Table 6.3 Summary of regression analysis for repository pressure at 10 000 years under undisturbed conditions (i.e., $y = WAS\_PRES$ at 10 000 years in Figure 6.11a), $x_1 = WMICDFLG$, $x_2 = HALPOR$, and $x_3 = WGRCOR$.

Source	DofF[a]	SS[b]	MS[c]	F[d]	SIGNIF[e]
Regression	3	1.9009×10^{15}	6.3365×10^{14}	3.7643×10^{2}	0.0000
Residual	296	4.9827×10^{14}	1.6833×10^{12}		
Total	299	2.3992×10^{15}			

R-Square[f] = 0.792 32   Intercept[g] = 5.7274×10^{6}

Variable[h]	Regression[i] coefficient	Standardized[j] regr. coeff.	Partial[k] SSQ	t-test[l] values	R-square[m] deletes	Alpha[n] hats
WMICDFLG	2.4625×10^{6}	0.722 01	1.2482×10^{15}	27.231	0.272 06	0.0000
HALPOR	1.5529×10^{8}	0.468 09	5.2479×10^{14}	17.657	0.573 59	-4.4409×10^{-16}[o]
WGRCOR	1.5210×10^{20}	0.246 49	1.4559×10^{14}	9.3000	0.731 64	-4.4409×10^{-16}

[a] Degrees of freedom associated with regression (SS_{reg}), residual (SS_{res}), and total (SS_{tot}) sums of squares; see (6.6.9) and (6.6.10).
[b] Regression (SS_{reg}), residual (SS_{res}), and total (SS_{tot}) sums of squares.
[c] Mean sums of squares (SS_{reg}/n, $SS_{res}/(m-n-1)$, where estimates for $\beta_1, \beta_2, \ldots, \beta_n$ are obtained from m observations).
[d] F-statistic ($[SS_{reg}/n]/[SS_{res}/(m-n-1)]$); see (6.6.25).
[e] *p*- or α-value for *F*; see (6.6.23).
[f] R^2 value for regression model with estimates for $\beta_0, \beta_1, \beta_2, \ldots, \beta_n$; see (6.6.11).
[g] Estimate for β_0.
[h] Variables in regression model $(x_1, x_2, \ldots, x_n)$.
[i] Regression coefficients $(b_1, b_2, \ldots, b_n)$; see (6.6.8).
[j] Standardized regression coefficients; see (6.6.15).
[k] Partial sum of squares for variable (i.e., x_j) in row ($SS_{reg}(\beta_j \mid \boldsymbol{\beta}_2, \beta_0)$); see (6.6.20) and (6.6.26).
[l] *t*-statistic for variable in row; see (6.6.27).
[m] For variable (i.e. x_j) in row, R^2 value for regression model constructed with x_i, $i = 1, 2, \ldots, n$ and $i \neq j$.
[n] For variable (i.e. x_j) in row, *p*- or α-value for addition of x_j to regression model containing x_i, $i = 1, 2, \ldots, n$ and $i \neq j$; use of *F*-statistic or *t*-statistic produces the same value; see (6.6.23) and (6.6.26) for *F*-statistic and (6.6.27) and (6.6.28) for *t*-statistic.
[o] Negative values result from numerical errors in the calculation of very small *p*-values with the STEPWISE program (Iman *et al.*, 1980).

for their effects on y, which implies the construction of a regression model with 31 variables. The result of such an analysis would be a table similar to Table 6.3 but containing 31 variables. Such a table would be quite unwieldy, with much of the table involving variables that have no effect on y. Stepwise regression analysis provides a more informative and less cumbersome procedure for constructing and presenting regression models, and will be described in Section 6.6.5.

6.6.4 Correlation and Partial Correlation

Correlation and partial correlation are useful concepts that often appear in sampling-based uncertainty/sensitivity studies. For a sequence of observations $(x_k, y_k), k = 1, \ldots, m$, the (sample or Pearson) correlation r_{xy} between x and y is defined by

$$r_{xy} = \frac{\sum_{k=1}^{m}(x_k - \bar{x})(y_k - \bar{y})}{\left[\sum_{k=1}^{m}(x_k - \bar{x})^2\right]^{1/2}\left[\sum_{k=1}^{m}(y_k - \bar{y})^2\right]^{1/2}}, \tag{6.6.29}$$

where $\bar{x}$ and $\bar{y}$ are defined in conjunction with (6.6.15). The correlation coefficient (CC) r_{xy} provides a measure of the linear relationship between x and y. For the regression model defined by (6.6.8), the R^2 value in (6.6.11) is equal to the square of the correlation between y and $\hat{y}$ (i.e., $R^2 = r_{y\hat{y}}^2$) (Draper and Smith, 1981, p. 91).

The nature of r_{xy} is perhaps most readily understood by considering the regression

$$\hat{y} = b_0 + b_1 x. \tag{6.6.30}$$

The definition of r_{xy} in (6.6.29) is equivalent to the definition

$$r_{xy} = \text{sign}(b_1)(R^2)^{1/2}, \tag{6.6.31}$$

where $\text{sign}(b_1) = 1$ if $b_1 \geqslant 0, \text{sign}(b_1) = -1$ if $b_1 < 0$, and R^2 is the coefficient of determination that results from regressing y on x. With respect to interpretation, r_{xy} provides a measure of the linear relationship between x and y, and the regression coefficient b_1 characterizes the effect that a unit change in x will have on y. The definition of r_{xy} in (6.6.29) is also equivalent to the definition

$$r_{xy} = b_1 \hat{s}_1 / \hat{s}, \tag{6.6.32}$$

where $\hat{s}_1$ and $\hat{s}$ are defined in conjunction with (6.6.15) with x assumed to correspond to x_1. Thus, r_{xy} is also equal to the SRC that results from regressing y on x. Hence, r_{xy} can be viewed as characterizing the effect that changing x by a fixed fraction of its standard deviation will have on y, with this effect being measured relative to the standard deviation of y. The CC can also be viewed as a parameter in a joint normal distribution involving x and y (Myers, 1990, Section 2.13); however, this interpretation is not as intuitively appealing as the two preceding interpretations involving the regression model in (6.6.30). Further, x and y typically do not have normal distributions in sampling-based sensitivity analyses.

When more than one input variable is under consideration, partial correlation coefficients (PCCs) can be used to provide a measure of the linear relationships between the output variable y and the individual input variables. The PCC between an individual variable x_j and y is obtained from the use of a sequence of regression models. First, the following two regression models are constructed:

$$\hat{x}_j = c_0 + \sum_{\substack{p=1 \\ p \neq j}}^{n} c_p x_p, \qquad \hat{y} = b_0 + \sum_{\substack{p=1 \\ p \neq j}}^{n} b_p x_p. \tag{6.6.33}$$

Then, the results of the two preceding regressions are used to define the new variables $x_j - \hat{x}_j$ and $y - \hat{y}$. The PCC $p_{x_j y}$ between x_j and y is the CC between $x_j - \hat{x}_j$ and $y - \hat{y}$. Thus, the PCC provides a measure of the linear relationship between x_j and y with the linear effects of the other variables removed. The preceding provides a rather intuitive development of what a PCC is. A formal development of PCCs is provided by Iman *et al.* (1985).

The PCC characterizes the strength of the linear relationship between two variables after a correction has been made for the linear effects of the other variables in the analysis, and the SRC characterizes the effect on the output variable that results from perturbing an input variable by a fixed fraction of its standard deviation. Thus, PCCs and SRCs provide related, but not identical, measures of variable importance. In particular, the PCC provides a measure of variable importance that tends to exclude the effects of other variables, the assumed distribution for the particular input variable under consideration, and the magnitude of the impact of an input variable on an output variable. In contrast, the value for an SRC is more influenced by the distribution assigned to an input variable and the impact that this variable has on an output variable.

The following relationship exists between $p_{x_j y}$ and the SRC $c_j = b_j \hat{s}_j / \hat{s}$ in (6.6.15):

$$p_{x_j y} = c_j [(1 - R_j^2)/(1 - R_y^2)]^{1/2}, \tag{6.6.34}$$

where R_j^2 is the R^2 value that results from regressing x_j on y and the $x_i, i = 1, 2, \ldots, n$ with $i \neq j$, and R_y^2 is the R^2 value that results from regressing y on the $x_i, i = 1, 2, \ldots, n$ (Iman *et al.*, 1985, Equation (1)). If the x_i are orthogonal, then

$$R_y^2 = \sum_{i=1}^{n} R_i^2 = \sum_{i=1}^{n} r_{x_i y}^2 = \sum_{i=1}^{n} c_i^2, \tag{6.6.35}$$

with the first equality following from (6.6.14), and the second and third equalities following from (6.6.31) and (6.6.32). Thus,

$$\begin{aligned} p_{x_j y} &= c_j \left[(1 - c_j^2) \bigg/ \left(1 - \sum_{i=1}^{n} c_i^2 \right) \right]^{1/2} \\ &= r_{x_j y} \left[(1 - r_{x_j y}^2) \bigg/ \left(1 - \sum_{i=1}^{n} r_{x_i y}^2 \right) \right]^{1/2} \end{aligned} \tag{6.6.36}$$

Because of the inequality

$$b(1 - b^2)^{1/2} > a(1 - a^2)^{1/2} \tag{6.6.37}$$

for $a^2 + b^2 < 1$ and $0 \leqslant a < b$ (see Kleijnen and Helton, 1999a, Figure 7), an ordering of variable importance based on $|p_{x_j y}|$, $|c_j|$, or $|r_{x_j y}|$ produces the same results when the x_i are orthogonal; further, the values for c_j and $r_{x_j y}$ will be the same and generally different from $p_{x_j y}$.

Many output variables are functions of time or location. A useful way to present sensitivity results for such variables is with plots of PCCs or SRCs. An example of such a presentation for the pressure curves in Figure 6.11(a) is given in Figure 6.14, which displays two sets of curves. The left set contains SRCs plotted as a function of time; the right set contains PCCs plotted in a similar manner. For both sets of curves, the dependent variables are pressures at fixed times, and each curve displays the values of SRCs or PCCs relating these pressures to a single input variable as a function of time. The SRCs and PCCs in Figure 6.14 were calculated for 24 of the 31 variables indicated in Table 6.1, with two variables omitted from consideration because of assumed -0.99 rank correlations with other variables and five variables omitted from consideration because they were not involved in the calculation of the dependent variable under analysis. Many additional examples of the use of PCCs in sampling-based sensitivity analysis also exist (see, e.g., Breshears *et al.*, 1992; Whiting *et al.*, 1993; Hamby, 1995; Helton *et al.*, 1996).

Determination of CCs and PCCs involves no statistical assumptions. However, as previously discussed for regression coefficients in Section 6.6.3, statistical tests can be performed conditional on suitable assumptions. For example,

$$t = \frac{r_{xy}(m-2)^{1/2}}{(1-r_{xy}^2)^{1/2}} \tag{6.6.38}$$

can be regarded as a random sample from a t-distribution with $m-2$ degrees of freedom when (i) r_{xy} is calculated from the observations (x_k, y_k), $k = 1, 2, \ldots, m$, and (ii) x and y are uncorrelated and have a bivariate normal distribution (Press *et al.*, 1992, p. 631). Then, the probability of observing a stronger correlation due to chance variation is given by the relationship in (6.6.28). The preceding test is identical to the test involving the t-statistic described in Section 6.6.3 for the significance of b_1 in (6.6.30) (Myers, 1990, p. 70). Further,

$$z = r_{xy}\sqrt{m} \tag{6.6.39}$$

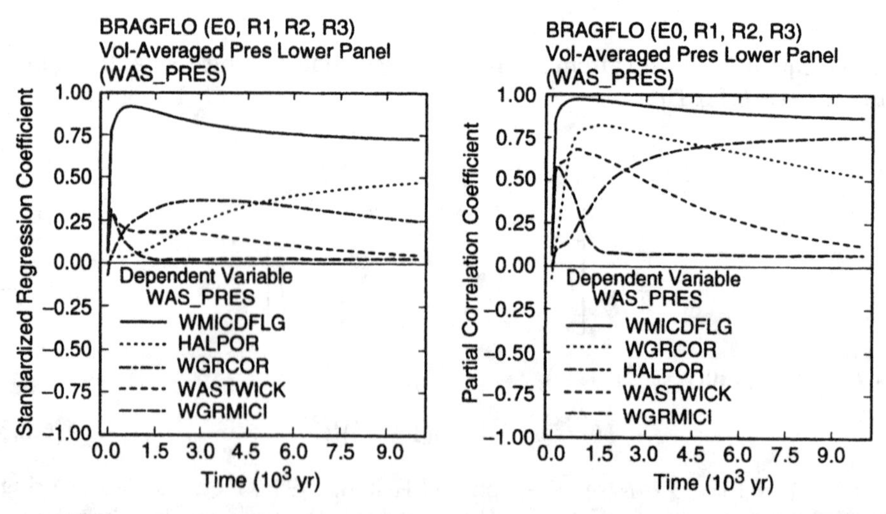

Figure 6.14 Standardized regression coefficients (SRCs) and partial correlation coefficients (PCCs) for five variables having the largest PCCs, in absolute value, with repository pressure (Pa) under undisturbed conditions (i.e., $y = WAS\_PRES$ in Figure 6.11a).

is distributed approximately normally with mean 0 and standard deviation 1 when x and y are uncorrelated, x and y have enough convergent moments (i.e., the tails of their distributions die off sufficiently rapidly), and m is sufficiently large (Press *et al.*, 1992, p. 631). Given the preceding assumptions, the probability $\text{prob}_n(|\tilde{r}_{xy}| > |r_{xy}|)$ of obtaining a value $\tilde{r}_{xy}$ for which $|\tilde{r}_{xy}|$ exceeds $|r_{xy}|$ is given by

$$\text{prob}_n(|\tilde{r}_{xy}| > |r_{xy}|) = \text{erfc}(|r_{xy}|\sqrt{m}/\sqrt{2}), \tag{6.6.40}$$

where erfc is the complementary error function (i.e., $\text{erfc}(x) = (2/\sqrt{\pi}) \int_x^\infty \exp(-t^2)\, dt$) (Press *et al.*, 1992, p. 631).

Significance results obtained with the statistics in (6.6.38) and (6.6.39) converge as m increases. Related significance results can also be defined for PCCs (Quade, 1989).

As an example, CCs, SRCs, and PCCs for repository pressure at 10 000 years are shown in Table 6.4. Of the variables under consideration, five have CCs with p-values less than 0.1. The CCs and SRCs have similar values, with equality failing to exist because of small correlations between the variables in the sample (see Table 6.2). The PCCs tend to be larger than the CCs and SRCs. Because PCCs provide a measure of the strength of linear relationships after corrections have been made for the effects of other variables, large PCCs have the potential to produce misleading impressions of variable importance; therefore, care should be exercised in the use and interpretation of PCCs. In particular, a large PCC does not necessarily imply that the corresponding input variable makes a large contribution to the uncertainty in the output variable under consideration. However, when the sampled variable values are independent (i.e., orthogonal), use of CCs, SRCs, and PCCs will produce identical rankings of variable importance as previously noted. The effect of correlations within the sample can be seen in Table 6.4, with *SALPRES* ranked 5 with CCs and 6 with SRCs and PCCs.

6.6.5 Stepwise Regression Analysis

When many input variables are involved, the direct construction of a regression model containing all input variables as shown in (6.6.3) may not be the best approach for several reasons. First, the large number of variables makes the regression model tedious to examine

Table 6.4 Correlation coefficients (CCs), standardized regression coefficients (SRCs), and partial correlation coefficients (PCCs) for repository pressure under undisturbed conditions at 10 000 years (i.e., $y = WAS\_PRES$ at 10 000 years in Figure 6.11a).

Variable name[a]	CC[b]			SRC[c]		PCC[d]	
	p-value	Rank	Value	Rank	Value	Rank	Value
WMICDFLG	0.0000	1.0	0.7124	1.0	0.7234	1.0	0.8642
HALPOR	0.0000	2.0	0.4483	2.0	0.4651	2.0	0.7469
WGRCOR	0.0000	3.0	0.2762	3.0	0.2460	3.0	0.5113
ANHPRM	0.0241	4.0	0.1302	4.0	0.1277	4.0	0.2953
SALPRES	0.0855	5.0	0.0993	6.0	0.0639	6.0	0.1526

[a]Variables for which CC with y has a p-value less than 0.1; variables ordered by p-values for CCs.
[b]p-value for CC, variable rank based on p-value for CC, and value of CC.
[c]Variable rank based on SRC and value for SRC for regression model containing 24 independent variables used in Figure 6.14.
[d]Variable rank based on PCC and value for PCC calculated for 24 independent variables used in Figure 6.14.

and unwieldy to display. Second, only a relatively small number of input variables typically has an impact on the output variable. As a result, there is no reason to include the remaining variables in the regression model. Third, correlated variables result in unstable regression coefficients (i.e., coefficients whose values are sensitive to the specific variables included in the regression model; see Section 6.6.7). When this occurs, the regression coefficients in a model containing all the input variables can give a misleading representation of variable importance. As a side point, if several input variables are highly correlated, consideration should be given to either removing all but one of the correlated variables or transforming the variables to correct for (i.e., remove) the correlations between them. Fourth, an overfitting of the data can result when variables are arbitrarily forced into the regression model. This phenomenon occurs when the regression model attempts to match the predictions associated with individual sample elements rather than match the trends shown by the sample elements collectively.

Stepwise regression analysis provides an alternative to constructing a regression model containing all the input variables. With this approach, a sequence of regression models is constructed. The first regression model contains the single input variable that has the largest impact on the uncertainty in the output variable (i.e., the input variable that has the largest correlation with the output variable y). The second regression model contains the two input variables that have the largest impact on the output variable: the input variable from the first step plus whichever of the remaining variables has the largest impact on the uncertainty not accounted for by the first variable (i.e., the input variable that has the largest correlation with the uncertainty in y that cannot be accounted for by the first variable). The third regression model contains the three input variables that have the largest impact on the output variable: the two input variables from the second step plus whichever of the remaining variables has the largest impact on the uncertainty not accounted for by the first two variables (i.e., the input variable that has the largest correlation with the uncertainty in y that cannot be accounted for by the first two variables). Additional models in the sequence are defined in the same manner until a point is reached at which further models are unable to meaningfully increase the amount of the uncertainty in the output variable that can be accounted for. Further, at each step of the process, the possibility exists for an already-selected variable to be dropped out if this variable no longer has a significant impact on the amount of uncertainty in the output variable that can be accounted for by the regression model; this only occurs when correlations exist between the input variables.

Several aspects of stepwise regression analysis provide insights on the importance of the individual variables. First, the order in which the variables are selected in the stepwise procedure provides an indication of their importance, with the most important variable being selected first, the next most important variable being selected second, and so on. Second, the R^2 values (see (6.6.11)) at successive steps of the analysis also provide a measure of variable importance by indicating how much of the uncertainty in the dependent variable can be accounted for by all variables selected through each step. When the input variables are uncorrelated, the differences in the R^2 values for the regression models constructed at successive steps equals the fractions of the total uncertainty in the output variable that can be accounted for by the individual input variables being added at each step (see (6.6.14)). Third, the absolute values of the SRCs (see (6.6.15)) in the individual regression models provide an indication of variable importance. Further, the sign of an SRC indicates whether the input and output variable tend to increase and decrease together (a positive coefficient) or tend to move in opposite directions (a negative coefficient).

An important situation occurs when the input variables are uncorrelated. In this case, orderings of variable importance based on order of entry into the regression model, size of the R^2 values attributable to the individual variables, the absolute values of the SRCs, the absolute values of correlation coefficients, and the absolute values of the PCCs are the same. In situations where the input variables are believed to be uncorrelated, one of the important applications of the previously discussed restricted pairing technique of Iman and Conover (Section 6.3.5) is to ensure that the correlations between variables within a Latin hypercube or random sample are indeed close to zero. When variables are correlated, care must be used in the interpretation of the results of a regression analysis, since the regression coefficients can change in ways that are basically unrelated to the importance of the individual variables as correlated variables are added to and deleted from the regression model (see Section 6.6.7 for an example of the effects of correlated variables on the outcomes of a regression analysis).

When the stepwise technique is used to construct a regression model, it is necessary to have some criterion to stop the construction process. When there are many independent variables, there is usually no reason to let the construction process continue until all the variables have been used. It is also necessary to have some criterion to determine when a variable is no longer needed and thus can be dropped from the regression model. As indicated earlier, this latter situation only occurs when the input variables are correlated. The usual criterion for making the preceding decisions is based on whether or not the regression coefficient associated with an input variable appears to be significantly different from zero. Specifically, an F-test or t-test is used to determine the probability that a regression coefficient with absolute value as large as or larger than the one constructed in the analysis would be obtained if, in reality, there was no relationship between the input and output variable, and, as a result, the apparent relationship that led to the constructed regression coefficient was due entirely to chance (see (6.6.26), (6.6.27), and associated text). Sensitivity studies often use an α-value of 0.01 or 0.02 to add a variable to a regression model and a somewhat larger value to drop a variable from the model.

As models involving more variables are developed in a stepwise regression analysis, the possibility exists of overfitting the data. Overfitting occurs when the regression model in essence 'chases' the individual observations rather than following an overall pattern in the data. For example, it is possible to obtain a good fit to a set of points by using a polynomial of high degree. However, in doing so, it is possible to overfit the data and produce a spurious model that makes poor predictions.

To protect against overfit, the predicted error sum of squares (PRESS) criterion can be used to determine the adequacy of a regression model (Allen, 1971). For a regression model containing q variables and constructed from m observations, PRESS is computed in the following manner. For $k = 1, 2, \ldots, m$, the kth observation is deleted from the original set of m observations and then a regression model containing the original q variables is constructed from the remaining $m - 1$ observations. With this new regression model, the value $\hat{y}_q(k)$ is estimated for the deleted observation y_k. Then, PRESS is defined from the preceding predictions and the m original observations by

$$\text{PRESS}_q = \sum_{k=1}^{m} [y_k - \hat{y}_q(k)]^2. \tag{6.6.41}$$

The regression model having the smallest PRESS value is preferred when choosing between two competing models, since this is an indication of how well the basic pattern of the data has been characterized versus an overfit or an underfit. In particular, PRESS values will decrease in size as additional variables are added to the regression model without an overfitting of the data (i.e., $\text{PRESS}_q > \text{PRESS}_{q+1}$), with an increase in the PRESS values (i.e.,

$PRESS_q < \text{PRESS}_{q+1}$) indicating an overfitting of the data. In addition to PRESS, there are also a number of other diagnostic tools that can be used to investigate the adequacy of regression models (Belsley *et al.*, 1980; Cook and Weisberg, 1982).

It is important to use scatterplots, PRESS values, and other procedures to examine the reasonableness of regression models. This is especially true when regression models are used for sensitivity analysis. Such analyses often involve many input variables and large uncertainties in these variables. The appearance of spurious patterns is a possibility that must be checked for.

An example stepwise regression analysis follows for the variable y and the 24 independent variables considered in the calculation of SRCs and PCCs in Figure 6.14. The first step selects the input variable x_j that has the largest impact on the output variable y. Specifically, this is defined to be the variable that has the largest correlation, in absolute value, with y (see (6.6.29) and (6.6.31)). Thus, it is necessary to calculate the correlations between y and each of the 24 input variables under consideration. For illustration, Table 6.5 shows the 7×7 correlation matrix for y and the six input variables ultimately selected in the stepwise regression, although the full correlation matrix would actually be $(24 + 1) \times (24 + 1)$. Each element in the correlation matrix is the correlation between the variables in the corresponding row and column. As examination of the correlation matrix in Table 6.5 shows, *WMICDFLG* has the highest correlation with $y = WAS\_PRES$. Thus, the first step in the analysis selects the variable *WMICDFLG*. Here and elsewhere in the stepwise procedure, the selection of variables to enter the regression model could equivalently be made on the basis of *F*-test or *t*-test values as defined in (6.6.26) and (6.6.27). A regression model relating y to *WMICDFLG* is then developed as shown in (6.6.8) with $n = 1$ and $m = 300$. The resultant regression model is

$$\hat{y} = 6.94 \times 10^6 + 2.43 \times 10^6 \quad WMICDFLG, \tag{6.6.42}$$

which has an R^2 value of 0.508, an α-value of 0.0000, an SRC of 0.712, and a PRESS value of 1.20×10^{15}. This model is summarized as Step 1 in Table 6.6.

The second step selects the input variable x_j that has the largest impact on the uncertainty in the output variable y that cannot be accounted by *WMICDFLG*, the variable selected in the first step. This selection is made by defining a new variable

$$\tilde{y} = y - \hat{y} = y - (6.94 \times 10^6 + 2.43 \times 10^6 \quad WMICDFLG), \tag{6.6.43}$$

where $\hat{y}$ is defined in (6.6.42), and then calculating the correlations between $\tilde{y}$ and the remaining variables. The variable with the largest correlation, in absolute value, with $\tilde{y}$ is selected as the second variable for inclusion in the model. In this example, the selected

Table 6.5 Correlation matrix for variables selected in stepwise regression analysis for pressure in the repository at 10 000 years under undisturbed conditions (i.e., $y = WAS\_PRES$ at 10 000 years in Figure 6.11a).

WMICDFLG	1.0000						
HALPOR	– 0.0348	1.0000					
WGRCOR	0.0272	0.0216	1.0000				
ANHPRM	0.0008	– 0.0039	0.0130	1.0000			
SHRGSSAT	– 0.0026	0.0395	– 0.0171	– 0.0042	1.0000		
SALPRES	0.0560	– 0.0072	0.0010	– 0.0117	0.0061	1.0000	
WAS_PRES	0.7124	0.4483	0.2762	0.1303	0.0820	0.0993	1.0000
	WMICDFLG	*HALPOR*	*WGRCOR*	*ANHPRM*	*SHRGSSAT*	*SALPRES*	*WAS_PRES*

Table 6.6 Results of stepwise regression analysis for pressure in the repository at 10 000 years under undisturbed conditions (i.e., $y = WAS\_PRES$ at 10 000 years in Figure 6.11a).

Step [a]	Variables [b]	SRC [c]	α-values [d]	R^2 values [e]	PRESS [f]
1	*WMICDFLG*	0.712	0.0000	0.508	1.20×10^{15}
2	*WMICDFLG*	0.729	0.0000	0.732	6.59×10^{14}
	HALPOR	0.474	0.0000		
3	*WMICDFLG*	0.722	0.0000	0.792	5.14×10^{14}
	HALPOR	0.468	0.0000		
	WGRCOR	0.246	0.0000		
4	*WMICDFLG*	0.722	0.0000	0.809	4.79×10^{14}
	HALPOR	0.469	0.0000		
	WGRCOR	0.245	0.0000		
	ANHPRM	0.128	0.0000		
5	*WMICDFLG*	0.722	0.0000	0.814	4.70×10^{14}
	HALPOR	0.466	0.0000		
	WGRCOR	0.246	0.0000		
	ANHPRM	0.129	0.0000		
	SHRGSSAT	0.070	0.0056		
6	*WMICDFLG*	0.718	0.0000	0.818	4.63×10^{14}
	HALPOR	0.466	0.0000		
	WGRCOR	0.246	0.0000		
	ANHPRM	0.129	0.0000		
	SHRGSSAT	0.070	0.0055		
	SALPRES	0.063	0.0012		

[a] Steps in the analysis.
[b] Variables selected at each step.
[c] Standardized regression coefficients (SRCs) for variables in the regression model at each step; see (6.6.15).
[d] p- or α-values for variables in the regression model at each step; see (6.6.26) and (6.6.27).
[e] R^2 value for the regression model at each step; see (6.6.11).
[f] Predicted error sum of squares (PRESS) value for the regression model at each step; see (6.6.41).

variable is *HALPOR*. The regression model at this step will thus involve the two variables *WMICDFLG* and *HALPOR*, and is constructed as shown in (6.6.8) with $n = 2$ and $m = 300$. The resultant regression model is

$$\hat{y} = 6.89 \times 10^6 + 2.49 \times 10^6 \quad WMICDFLG + 1.57 \times 10^8 \quad HALPOR. \tag{6.6.44}$$

This model is summarized as Step 2 in Table 6.6.

The third step selects the input variable x_j that has the largest impact on the uncertainty in the output variable y that cannot be accounted for by *WMICDFLG* and *HALPOR*, the two variables from the second step. This selection is made by defining a new variable

$$\tilde{y} = y - \hat{y} = y - (6.89 \times 10^6 + 2.49 \times 10^6 \quad WMICDFLG + 1.57 \times 10^8 \quad HALPOR), \tag{6.6.45}$$

where $\hat{y}$ is defined in (6.6.44). The variable with the largest correlation, in absolute value, with $\tilde{y}$ is selected as the third variable for inclusion in the model. In this example, the selected variable is *WGRCOR*. The regression model for this step will thus involve the three variables *WMICDFLG*, *HALPOR*, and *WGRCOR*. The resultant regression model is summarized as Step 3 in Table 6.6.

As shown in Table 6.6, the stepwise procedure then continues in the same manner through a total of six steps, until no more variables can be found with an α-value less than 0.02. At this point, the stepwise procedure stops.

At each step, the stepwise procedure also checks to see if any variable selected at a prior step now has an α-value that exceeds a specified level, which is 0.05 in this analysis. If such a situation occurs, the variable will be dropped from the analysis, with the possibility that it may be reselected at a later step as other variables are added and deleted from the model. This type of behavior only occurs when there are correlations between the input variables. As shown in the correlation matrix in Table 6.5, the restricted pairing technique has been successful in keeping the correlations between the input variables close to zero. Thus, no variables meet the criterion to be dropped from the regression model once they have been selected at a prior step.

Another result of this lack of correlation is that the regression coefficients do not change significantly as additional variables are added to the regression model. As examination of Table 6.6 shows, the regression coefficients for a specific variable are essentially the same in all regression models containing that variable. Further, as indicated in (6.6.14), the R^2 values obtained for successive models can be subtracted to obtain the contribution to the uncertainty in y due to the newly added variable. Thus, for example, *WMICDFLG* accounts for approximately 51% of the uncertainty in y (i.e., $R^2 = 0.508$), while *WMICDFLG* and *HALPOR* together account for approximately 73% of the uncertainty (i.e., $R^2 = 0.732$). As a result, *HALPOR* by itself accounts for approximately $73\% - 51\% = 22\%$ of the uncertainty in y. Similar results hold for the other variables selected in the analysis.

Table 6.6 also shows the PRESS values for the regression models obtained at the individual steps in the analysis. A decreasing sequence of PRESS values indicates that the regression models are not overfitting the data on which they are based. An increase in the PRESS values suggests that a model is overfitting the data, and thus that the stepwise procedure should probably be stopped at the preceding step. As shown by the decreasing PRESS values in Table 6.6, the regression models in this analysis are probably not overfitting the data from which they were constructed.

Typically, a certain amount of discretion is involved in selecting the exact point at which to stop a stepwise regression analysis. Certainly, α-values and the behavior of PRESS values provide two criteria to consider in selecting a stopping point. Other criteria include the changes in the R^2 values that take place as additional variables are added to the regression models and whether or not spurious variables are starting to enter the regression models. When only very small changes in R^2 values are taking place (e.g., $\leqslant 0.01$), there is often little reason to continue the stepwise process. When α-values approach or exceed 0.01 and a large number of input variables are being considered, it is fairly common to start getting spurious variables in the regression (see Kleijnen and Helton, 1999b, Figure 1). Such variables appear to have a small effect on the output variable, which, in fact, is due to chance variation. In such situations, a natural stopping point may be just before spurious variables start being selected. Another possibility is to delete spurious variables from the regression model.

When the input variables are uncorrelated, a display of the results of a stepwise regression analysis as shown in Table 6.6 contains a large amount of redundant information. A more compact display can be obtained by listing the variables in the order that they entered in the regression model, the R^2 values obtained with the entry of successive variables into the

Table 6.7 Compact summary of stepwise regression analyses for pressure in the repository at 10 000 years under undisturbed conditions (i.e., $y = WAS\_PRES$ at 10 000 years in Figure 6.11a).

Step [a]	Variable [b]	SRC [c]	R^{2} [d]
1	*WMICDFLG*	0.718	0.508
2	*HALPOR*	0.466	0.732
3	*WGRCOR*	0.246	0.792
4	*ANHPRM*	0.129	0.809
5	*SHRGSSAT*	0.070	0.814
6	*SALPRES*	0.063	0.818

[a] Steps in stepwise analysis.
[b] Variables listed in the order of selection in regression analysis.
[c] Standardized regression coefficients (SRCs) for variables in final regression model.
[d] Cumulative R^2 value with entry of each variable into regression model.

regression model, and the SRCs for the variables contained in the final model. Table 6.7 shows what this summary looks like for the stepwise regression analysis in Table 6.6.

Numerous examples of the use of stepwise regression analysis in sampling-based sensitivity analyses are available in various articles by Helton *et al.* (1989, 1995b, 1996).

6.6.6 The Rank Transformation

Regression and correlation analyses often perform poorly when the relationships between the input and output variables are nonlinear. This is not surprising, since such analyses are based on developing linear relationships between variables. The problems associated with poor linear fits to nonlinear data can often be mitigated by use of the rank transformation (Iman and Conover, 1979; Conover and Iman, 1981; Saltelli and Sobol', 1995). The rank transformation is a simple concept: data are replaced with their corresponding ranks, and then the usual regression and correlation procedures are performed on these ranks. Specifically, the smallest value of each variable is assigned the rank 1, the next largest value is assigned the rank 2, and so on up to the largest value, which is assigned the rank m, where m denotes the number of observations. Further, averaged ranks are assigned to equal values of a variable. The analysis is then performed with these ranks being used as the values for the input and output variables. In essence, the use of rank-transformed data results in an analysis based on the strength of monotonic relationships rather than on the strength of linear relationships.

As an example, the strength of the monotonic relationship between x and y can be measured with Spearman's rank CC for x and y, R_{xy}, which is simply Pearson's CC in (6.6.29) calculated on ranks. The test for zero rank correlation uses a table of quantiles for $|R_{xy}|$ (e.g., Conover, 1980, Table A10). For a sample size of $m \geqslant 30$,

$$z = R_{xy}\sqrt{m-1} \tag{6.6.46}$$

approximately follows a standard normal distribution if the rank correlation between x and y is zero (Conover, 1980, p. 456). Thus, similarly to (6.6.40) for r_{xy},

$$\text{prob}_n(|\tilde{R}_{xy}| > |R_{xy}|) = \text{erfc}(|R_{xy}|\sqrt{m-1}/\sqrt{2}), \tag{6.6.47}$$

where $\text{prob}_n(|\tilde{R}_{xy}| > |R_{xy}|)$ is the probability that random variation would produce a value $\tilde{R}_{xy}$ larger in absolute value than the observed value R_{xy}. Further, standardized rank regression coefficients (SRRCs) and partial rank-correlation coefficients (PRCCs) can be calculated analogously to the corresponding coefficients for raw data.

For perspective, analyses for cumulative brine flow over 10 000 years (i.e., the value for y at 10 000 years in Figure 6.15) with CCs, SRCs, and PCCs calculated with both raw and rank-transformed data are presented in Table 6.8. The general patterns exhibited by the analyses with raw data and by the analyses with rank-transformed data are similar to those discussed in conjunction with Table 6.4. However, the two analyses differ in the importance assigned to individual variables. In particular, the analysis with rank-transformed data identifies *WMICDFLG* as the most important variable, with an RCC of -0.6521; in contrast, the analysis with raw data identifies *WMICDFLG* as the second most important variable, with a CC of -0.3210. The preceding is a nontrivial difference, because an RCC of -0.6521 implies that *WMICDFLG* can account for 42.5% of the uncertainty in y in rank-transformed space (i.e., $0.6521^2 \doteq 0.425$; see (6.6.31)), while a CC of -0.3210 implies that *WMICDFLG* can account for only 10.3% of the uncertainty in y in the original untransformed space (i.e., $0.3210^2 \doteq 0.103$). Numerous other differences also exist.

Additional perspective on the use of raw and rank-transformed data in the analysis of y can be obtained from examination of the results of stepwise regression analyses (Table 6.9). In particular, the use of rank-transformed data leads to a regression model with seven variables and an R^2 value of 0.869. In contrast, the use of raw data leads to a regression model with six variables and an R^2 value of only 0.496. Thus, the use of rank-transformed data is resulting in an analysis that can account for more of the uncertainty in y than can be accounted for in an analysis with raw data. As a result, the coefficients in Table 6.8 obtained with rank-transformed data (i.e., RCCs, SRRCs, and PRCCs) are more informative with respect to the sources of the uncertainty in y than are the coefficients obtained with raw data.

When the relationship between the dependent and independent variables is linear, the use of raw and rank-transformed data tends to produce similar results. When rank-transformed

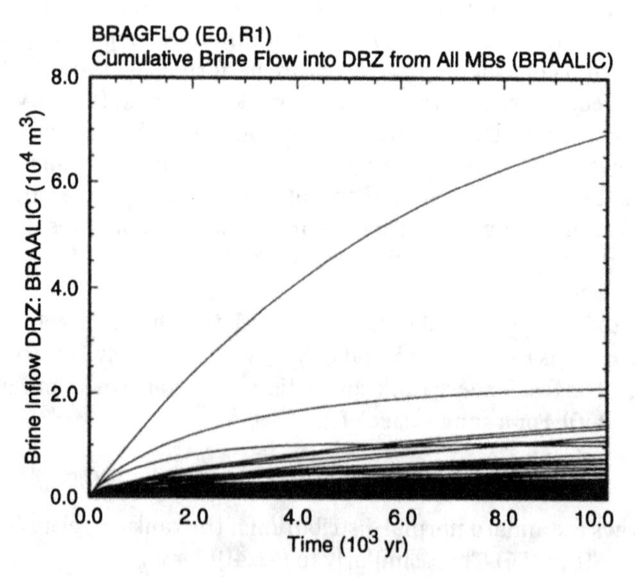

Figure 6.15 Cumulative brine flow (m^3) into disturbed rock zone (DRZ) from all anhydrite marker beds (MBs) under undisturbed conditions (i.e., $y = BRAALIC$).

Table 6.8 Correlation coefficients (CCs, RCCs), standardized regression coefficients (SRCs, SRRCs) and partial correlation coefficients (PCCs, PRCCs) with raw and rank-transformed data for cumulative brine flow over 10 000 years under undisturbed conditions from the anhydrite marker beds to the disturbed rock zone that surrounds the repository (i.e., $y = BRAALIC$ at 10 000 years in Figure 6.15).[a]

Raw data							
Variable name	**CC**			**SRC**		**PCC**	
	***p*-value**	**Rank**	**Value**	**Rank**	**Value**	**Rank**	**Value**
ANHPRM	0.0000	1.0	0.5655	1.0	0.5568	1.0	0.6317
WMICDFLG	0.0000	2.0	− 0.3210	2.0	− 0.2931	2.0	− 0.3878
WASTWICK	0.0045	3.0	− 0.1639	4.0	− 0.1451	4.0	− 0.2075
WGRCOR	0.0048	4.0	− 0.1628	3.0	− 0.1669	3.0	− 0.2370
ANHBCEXP	0.0095	5.0	− 0.1497	5.0	− 0.1155	5.0	− 0.1663
WFBETCEL	0.0555	6.0	− 0.1105	8.0	− 0.0757	8.0	− 0.1098
WRBRNSAT	0.0615	7.0	− 0.1080	9.0	− 0.0733	9.0	− 0.1065
HALPOR	0.0934	8.0	− 0.0969	6.0	− 0.0993	6.0	− 0.1435

Rank-transformed data							
Variable name	**RCC**			**SRRC**		**PRCC**	
	***p*-value**	**Rank**	**Value**	**Rank**	**Value**	**Rank**	**Value**
WMICDFLG	0.0000	1.0	− 0.6521	1.0	− 0.6533	1.0	− 0.8787
ANHPRM	0.0000	2.0	0.5804	2.0	0.5937	2.0	0.8619
HALPRM	0.0014	3.0	0.1850	5.0	0.1443	5.0	0.3817
WGRCOR	0.0057	4.0	− 0.1598	4.0	− 0.1509	4.0	− 0.3963
HALPOR	0.0087	5.0	− 0.1518	3.0	− 0.1539	3.0	− 0.4031
WASTWICK	0.0405	6.0	− 0.1185	7.0	− 0.0948	7.0	− 0.2617

[a]Table structure analogous to Table 6.4.

Table 6.9 Comparison of stepwise regression analyses with raw and rank-transformed data for cumulative brine flow over 10 000 years under undisturbed conditions from the anhydrite marker beds to the disturbed rock zone that surrounds the repository (i.e., $y = BRAALIC$ at 10 000 years in Figure 6.15).

	Raw data			**Rank-transformed data**		
Step[a]	**Variable**[b]	**SRC**[c]	R^{2}[d]	**Variable**[b]	**SRRC**[e]	R^{2}[d]
1	*ANHPRM*	0.562	0.320	*WMICDFLG*	− 0.656	0.425
2	*WMICDFLG*	− 0.309	0.423	*ANHPRM*	0.593	0.766
3	*WGRCOR*	− 0.164	0.449	*HALPOR*	− 0.155	0.802
4	*WASTWICK*	− 0.145	0.471	*WGRCOR*	− 0.152	0.824
5	*ANHBCEXP*	− 0.120	0.486	*HALPRM*	0.143	0.845
6	*HALPOR*	− 0.101	0.496	*SALPRES*	0.120	0.860
7				*WASTWICK*	− 0.010	0.869

[a] Steps in stepwise regression analysis.
[b]Variables listed in order of selection in regression analysis with *ANHCOMP* and *HALCOMP* excluded from entry into regression model.
[c] Standardized regression coefficient (SRCs) in final regression model.
[d] Cumulative R^2 value with entry of each variable into regression model.
[e] Standardized rank regression coefficients (SRRCs) in final regression model.

data are used and there are no ties in the data, the resulting values for regression coefficients and SRRCs are equal; thus, the rank transform results in an automatic standardization of the data in this case.

The analysis with rank-transformed data is more effective than the analysis with raw data because the rank transformation tends to linearize the relationships between the independent variables (i.e., the x_j) and the dependent variable (i.e., y). In particular, both *WMICDFLG* and *ANHPRM* show a stronger linear relationship with y after the rank transformation (Figure 6.16). The rank transformation improves the analysis when nonlinear but monotonic relationships exist between the independent variables and the dependent variable. When more complex relationships exist, the rank transformation may do little to improve the quality of an analysis. In such cases, more sophisticated procedures are required. For example, various tests can be used to check for deviations from randomness in scatterplots

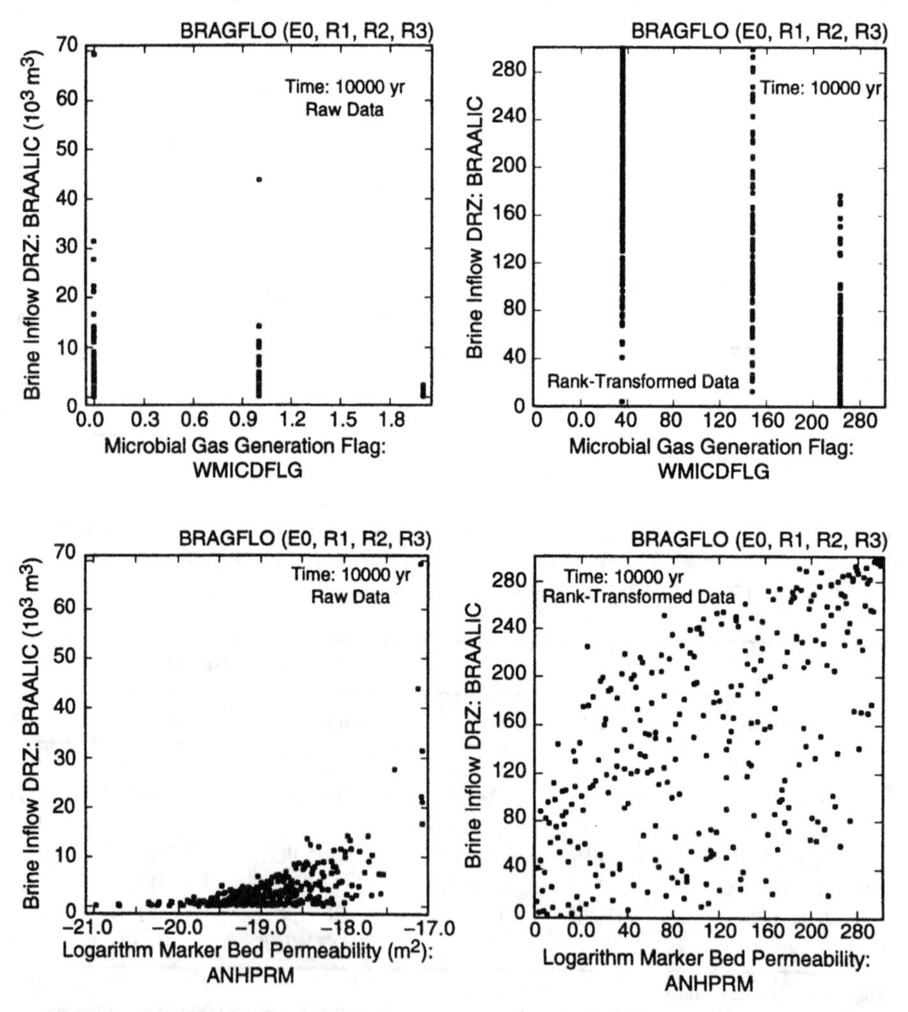

Figure 6.16 Scatterplots for cumulative brine flow (m^3) from the marker beds over 10 000 years under undisturbed conditions (i.e., $y = BRAALIC$ at 10 000 years in Figure 6.15) versus microbial gas generation flag (*WMICDFLG*) and marker bed permeability (*ANHPRM*) with raw (i.e., untransformed) and rank-transformed data.

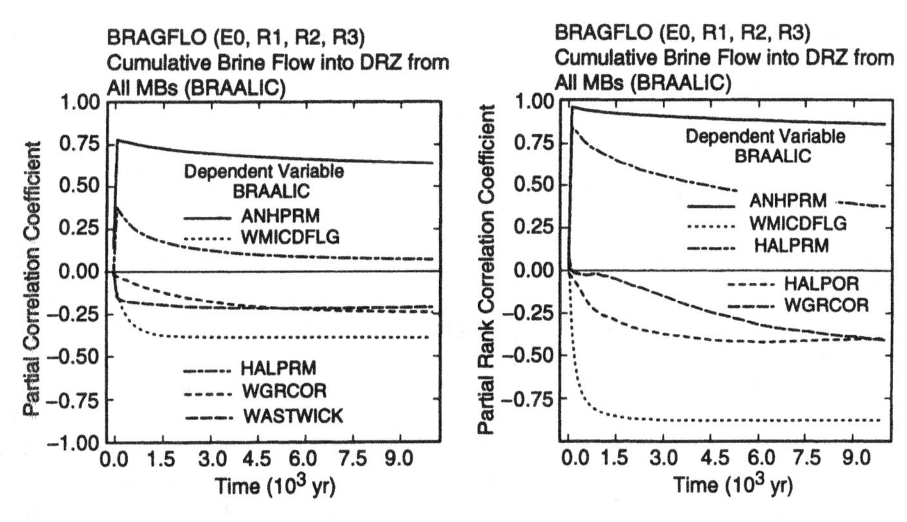

Figure 6.17 Partial correlation coefficients calculated with raw and rank-transformed data for cumulative brine flow from anhydrite marker beds to disturbed rock zone (DRZ) under undisturbed conditions (i.e., $y = BRAALIC$ in Figure 6.15), with *ANHCOMP* and *HALCOMP* excluded from the calculation.

(Sections 6.6.8 and 6.6.9; see also Saltelli and Marivoet, 1990; Hamby, 1994; Saltelli and Sobol', 1995; Kleijnen and Helton, 1999a).

As for stepwise regression analyses, analyses with SRCs and PCCs of the type presented in Figure 6.14 can often be improved with the use of rank-transformed data. When the rank transform is used, the resultant plots will contain SRRCs and PRCCs. As an example, the results of analyzing the cumulative brine inflows in Figure 6.15 with PCCs and PRCCs are presented in Figure 6.17, with each plot frame showing the five variables with the largest, in absolute value, PCCs or PRCCs as appropriate. As in the comparisons of stepwise regression analyses with raw and rank-transformed data (Table 6.9), the analyses with rank-transformed data in Figure 6.17 produce outcomes that indicate stronger effects for individual variables than is the case for the analyses with raw data.

The rank transformation has become quite popular in sampling-based sensitivity analyses and many additional examples of its use exist (see, e.g., MacDonald and Campbell, 1986; Helton *et al.*, 1989, 1996; Whiting *et al.*, 1993; Blower and Dowlatabadi, 1994; Hamby, 1995; Gwo *et al.*, 1996; Sanchez and Blower, 1997).

6.6.7 Effects of Correlations on Sensitivity Analyses

The presence of correlations between uncertain (i.e., sampled) variables can greatly complicate the interpretation of sensitivity analysis results. When no correlations exist between the sampled variables in a stepwise regression model, the regression coefficients will decrease monotonically in absolute value. In this case, an ordering of the variables by the absolute value of their regression coefficients provides a way to rank variable importance. However, when correlated variables are included in a regression model, the sizes and even the signs of the associated regression coefficients may not properly indicate the effects of these variables. In the example analysis, such complications result from the rank correlations of −0.99 that are assigned to the variable pairs (*ANHPRM*, *ANHCOMP*) and (*HALPRM*, *HALCOMP*) (Table 6.1). Regression-based sensitivity analyses for the variable in Figure 6.15 at 10 000 years (i.e., $y = BRAALIC$) will be used as an example (Table 6.10).

Table 6.10 Stepwise regression analyses with rank-transformed data for cumulative brine flow from all marker beds over 10 000 years under undisturbed conditions (i.e., $y = BRAALIC$ at 10 000 years in Figure 6.15).

	All variables included		*ANHCOMP HALCOMP* excluded		*ANHPRM HALPRM* excluded	
Step [a]	**Variable** [b]	**SRRC** [c]	**Variable**	**SRRC**	**Variable**	**SRRC**
1	*WMICDFLG*	−0.65	*WMICDFLG*	−0.66	*WMICDFLG*	−0.66
2	*ANHPRM*	0.59	*ANHPRM*	0.59	*ANHCOMP*	−0.58
3	*HALPOR*	−0.16	*HALPOR*	−0.16	*HALPOR*	−0.16
4	*WGRCOR*	−0.15	*WGRCOR*	−0.15	*WGRCOR*	−0.15
5	*HALPRM*	0.51	*HALPRM*	0.14	*HALCOMP*	−0.14
6	*SALPRES*	0.12	*SALPRES*	0.12	*SALPRES*	0.13
7	*WASTWICK*	−0.10	*WASTWICK*	−0.10	*WASTWICK*	−0.09
8	*HALCOMP*	0.37	R^2	0.87	R^2	0.85
	R^2 [d]	0.87				

[a] Steps in stepwise regression analysis.
[b] Variables in regression model.
[c] Standardized rank regression coefficients (SRRCs) for variables in final regression model.
[d] R^2 value for final regression model.

The following three regression analyses for y are shown in Table 6.10: (i) all 31 sampled variables allowed as candidates for inclusion in the regression model, (ii) *ANHCOMP* and *HALCOMP* excluded as candidates for inclusion in the regression model, and (iii) *ANHPRM* and *HALPRM* excluded as candidates for inclusion in the regression model. When all sampled variables are included as candidates, the regression coefficients decrease monotonically until Step 8, when *HALCOMP* enters the regression model. With entry of *HALCOMP*, the regression coefficient for *HALPRM* jumps from a value of 0.14 at Step 5 (not shown) to a value of 0.51; further, *HALCOMP* has a regression coefficient of 0.37 even though it has essentially no effect on the R^2 value for the regression model (i.e., $R^2 = 0.868\,89$ at Step 7 and $R^2 = 0.872\,03$ at Step 8). When *ANHCOMP* and *HALCOMP* are excluded as candidates for entry into the regression model, a sequence of seven regression models is produced that is identical to the first seven regression models is produced when all variables are allowed as candidates for inclusion. However, a different sequence of regression models is constructed when *ANHPRM* and *HALPRM* are excluded. In this case, *ANHPRM* and *HALPRM* are replaced in the regression models with *ANHCOMP* and *HALCOMP*, and the signs of the regression coefficients are reversed. Thus, *ANHCOMP* and *HALCOMP* appear with negative regression coefficients where *ANHPRM* and *HALPRM* appear with positive regression coefficients. In contrast, *HALPRM* and *HALCOMP* both have positive regression coefficients when they appear together in the regression model constructed at Step 8 when all variables are included as candidates for entry into the analysis. Thus, care must be used in interpreting regression analyses that involve highly correlated variables.

6.6.8 Identification of Nonmonotonic Patterns

Sometimes regression-based sensitivity analyses perform poorly. The rank transformation has been introduced as a possible analysis procedure for such situations (Section 6.6.5).

However, when viewed broadly, the rank transformation provides only a variant on linear regression analysis, with a model that seeks to identify linear relationships being replaced by a model that seeks to identify monotonic relationships. A more general approach is to attempt to determine if the scatterplots of a dependent (i.e., predicted) variable versus individual independent (i.e., sampled) variables appear to display nonmonotonic patterns (Kleijnen and Helton, 1999a; Shortencarier and Helton, 1999).

As an example, time-dependent pressure in the repository subsequent to a drilling intrusion and an associated sensitivity analysis based on PRCCs are shown in Figure 6.18, with the PRCCs having small values after the occurrence of the drilling intrusion at 1000 years. Further, as indicated by the regression analyses in Table 6.11 for pressure at 10 000 years, the use of neither raw nor rank-transformed data produces a successful regression model (i.e., R^2 values of 0.22 and 0.20 result for raw and rank-transformed data, respectively). Yet, unless there is an error in the calculations, the uncertainty in the sampled variables must be giving rise to the variations in the pressure curves in Figure 6.18(a).

As discussed in Section 6.6.1, examination of scatterplots often provides an effective way to identify influential variables. In particular, examination of scatterplots shows that *BHPRM* is the dominant variable with respect to the uncertainty in repository pressure subsequent to a drilling intrusion (Figure 6.13b). This is rather disconcerting, since *BHPRM* was not identified in either the PRCC analysis in Figure 6.18(b) or the regression analyses in Table 6.11. Thus, the clearly dominant variable has been completely missed in the formal analyses in Figure 6.18(b) and Table 6.11, and was only identified by an exhaustive examination of the scatterplots for the individual variables. Clearly, some type of formal procedure for identifying patterns in scatterplots is desirable; otherwise, the analyst is confronted with the requirement to manually examine large numbers of scatterplots and also to subjectively assess the relative strengths of the individual patterns appearing in these plots.

In this section, three procedures for identifying nonmonotonic patterns are introduced. Each of these procedures is based on determining if some measure of central tendency for the dependent variable is a function of individual independent variables. In particular, the F-test for equal means, the χ^2 test for equal medians, and the Kruskal–Wallis test are introduced as means of determining if measures of central tendency for a dependent variable change as a function of the values of individual independent variables (Kleijnen and

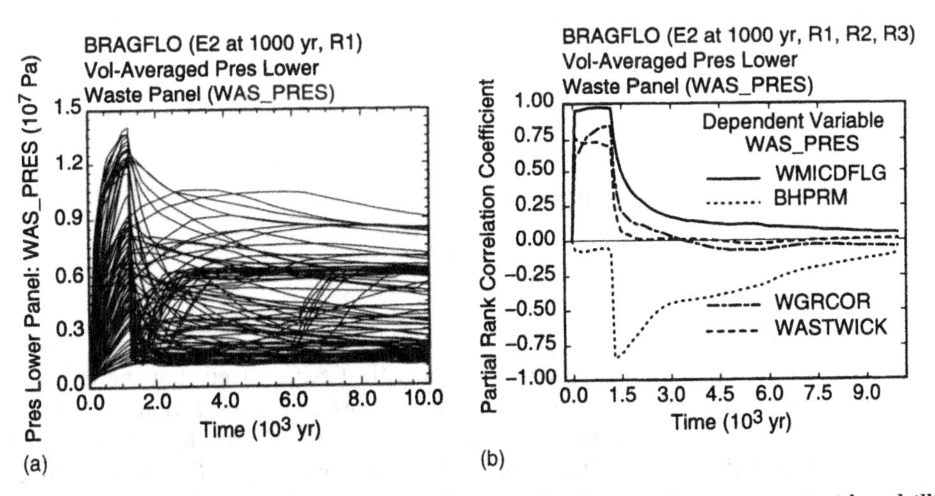

Figure 6.18 Uncertainty and sensitivity analysis results for repository pressure with a drilling intrusion into lower waste panel at 1000 years (i.e., $y = WAS\_PRES$).

Table 6.11 Stepwise regression analyses with raw and rank-transformed data with pooled results from replicates R1, R2, and R3 (i.e., for a total of 300 observations) for repository pressure at 10 000 years after a drilling intrusion at 1000 years (i.e., $y = WAS\_PRES$ at 10 000 years in Figure 6.18a).

	Raw data, *E2:WAS_PRES*			Rank-transformed data, *WAS_PRES*		
Step [a]	**Variable** [b]	**SRC** [c]	$R^{2\,d}$	**Variable** [b]	**SRRC** [e]	$R^{2\,d}$
1	*HALPRM*	0.37	0.14	*HALPRM*	0.36	0.13
2	*ANHPRM*	0.24	0.20	*ANHPRM*	0.24	0.19
3	*HALPOR*	0.14	0.22	*HALPOR*	0.14	0.20

[a] Steps in stepwise regression analysis.
[b] Variables listed in order of selection in regression analysis with *ANHCOMP* and *HALCOMP* excluded from entry into regression model because of −0.99 rank correlation within the pairs (*ANHPRM, ANHCOMP*) and (*HALPRM, HALCOMP*).
[c] Standardized regression coefficients (SRCs) in final regression model.
[d] Cumulative R^2 value with entry of each variable into regression model.
[e] Standardized rank regression coefficients (SRRCs) in final regression model.

Helton, 1999a, Section 5). For convenience, the preceding tests will be designated as tests for common means (CMNs), common medians (CMDs), and common locations (CLs).

The procedures discussed in this section involve an assessment of the relationship between a dependent and an independent variable. For notational convenience, these variables will be represented by y and x, respectively. This assessment is based on dividing the values of x (i.e., $x_k, k = 1, 2, \ldots, m$) into n_X classes and then testing to determine if y has a common measure of central tendency across these classes. The required classes are obtained by dividing the range of x into a sequence of mutually exclusive and exhaustive subintervals containing equal numbers of sampled values (Figure 6.19). When an x is discrete (see e.g., *WMICDFLG* in Figure 6.16), individual classes are defined for each of the distinct values. For notational convenience, let $q, q = 1, 2, \ldots, n_X$, designate the individual classes into which the values of x have been divided; let $\mathcal{X}_q$ designate the set such that $k \in \mathcal{X}_q$ only if x_k belongs to class q; and let n_{X_q} equal the number of elements contained in $\mathcal{X}_q$ (i.e., the number of x_k associated with class q).

The F-test can be used to test for the equality of the mean values of y for the classes into which the values of x have been divided (e.g., the intervals defined on the abscissas of the scatterplots in Figure 6.19). Specifically, if the y values conditional on each class of x values are normally distributed with equal expected values, then

$$F = \frac{\left(\sum_{q=1}^{n_X} n_{X_q} \bar{y}_q^2 - m\bar{y}^2\right) \Big/ (n_X - 1)}{\left(\sum_{k=1}^{m} y_k^2 - \sum_{q=1}^{n_X} n_{X_q} \bar{y}_q^2\right) \Big/ (m - n_X)} \qquad (6.6.48)$$

follows an F-distribution with $(n_X - 1, m - n_X)$ degrees of freedom, where $\bar{y}_q = \sum_{k \in \mathcal{X}_q} y_k / n_{X_k}$ and $\bar{y}$ is defined in conjunction with (6.6.15). Given that the indicated assumptions hold, the probability of obtaining an F-statistic of value $\tilde{F}$ that exceeds the value of F in (6.6.48) can be estimated by $\text{prob}_F(\tilde{F} > F \mid n_X - 1, m - n_X)$ as defined in (6.6.23). A low probability (i.e., p-value) of obtaining a larger value for F suggests that the observed pattern involving x and y did not arise by chance and hence that x has an effect on the behavior of y.

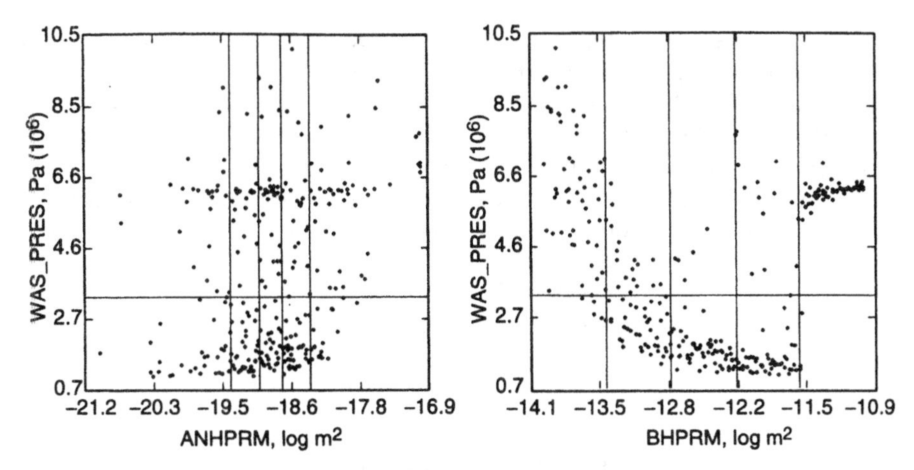

Figure 6.19 Partitioning of the ranges of *ANHPRM* and *BHPRM* into $n_X = 5$ classes in an analysis for repository pressure at 10 000 years after a drilling intrusion at 1000 years (i.e., $y = WAS\_PRES$ at 10 000 years in Figure 6.18a); horizontal lines correspond to the median, $y_{0.5}$, of y.

The χ^2 test for contingency tables can be used to test for the equality of the median values of y for the classes into which the values of x have been divided (Conover, 1980, pp. 143–178). First, the median, $y_{0.5}$, is estimated for all m observations. Specifically,

$$y_{0.5} = \begin{cases} y_{[0.5m]} & \text{if } 0.5\,\text{m is an integer,} \\ \frac{1}{2}\left[y_{([0.5m])} + y_{([0.5m]+1)}\right] & \text{otherwise,} \end{cases} \tag{6.6.49}$$

where $y_{(k)}, k = 1, 2, \ldots, m$, denotes the ordering of the y-values such that $y_{(k)} \leqslant y_{(k+1)}$ and $[\sim]$ designates the greatest integer function (David, 1970, p. 14). The individual classes of x-values are then further subdivided on the basis of whether y-values fall above or below $y_{0.5}$ (Figure 6.19). For class q, let $n_{X_{1q}}$ equal the number of y-values that exceed $y_{0.5}$, and let $n_{X_{2q}}$ equal the number of y-values that are less than or equal to $y_{0.5}$.

The result of this partitioning is a $2 \times n_X$ contingency table with $n_{X_{rq}}$ observations in each cell. The following statistic can now be defined:

$$T = \sum_{q=1}^{n_X} \sum_{r=1}^{2} \frac{\left(n_{X_{rq}} - n_{E_{rq}}\right)^2}{n_{E_{rq}}}, \tag{6.6.50}$$

where

$$n_{E_{rq}} = \left(\sum_{r=1}^{2} n_{X_{rq}}\right)\left(\sum_{q=1}^{n_X} n_{X_{rq}}\right) \Big/ n_X$$

and corresponds to the expected number of observations in cell (r, q). If the individual classes of x-values, $q = 1, 2, \ldots, n_X$, have equal medians, then T approximately follows a χ^2 distribution with $(n_X - 1)(2 - 1) = n_X - 1$ degrees of freedom (Conover, 1980, p. 156). The probability $\text{prob}_{\chi^2}(\tilde{T} > T \,|\, n_X - 1)$ of obtaining a value $\tilde{T}$ that exceeds T in the presence of equal medians is given by

$$\text{prob}_{\chi^2}(\tilde{T} > T | n_X - 1) = Q[\tfrac{1}{2}(n_X - 1), \tfrac{1}{2}T], \tag{6.6.51}$$

where $Q(a, b)$ denotes the complement of the incomplete gamma function (Press *et al.*, 1992, p. 215). A small value (i.e., *p*-value) for $\text{prob}_{\chi^2}(\tilde{T} > T|n_X - 1)$ indicates that the *ys* conditional on individual classes have different medians and hence that *x* has an influence on *y*. To maintain the validity of the χ^2 test in the analysis of contingency tables, Conover (1980, p. 156) suggests using a partition in which $n_{E_{rq}} \geqslant 1$.

The Kruskal–Wallis test statistic *T* is based on rank-transformed data, and uses the same classes of *x*-values as the *F*-statistic in (6.6.48) (Conover, 1980, pp. 229–230). Specifically,

$$T = \left[\sum_{q=1}^{n_X} \frac{R_q^2}{n_{X_q}} - \frac{1}{4}m(m+1)^2\right] \bigg/ s^2, \tag{6.6.52}$$

where

$$R_q = \sum_{k \in \mathscr{X}_q} r(y_k), \qquad s^2 = \left[\sum_{k=1}^{m} r(y_k)^2 - \tfrac{1}{4}m(m+1)^2\right] \bigg/ (m-1),$$

and $r(y_k)$ denotes the rank of y_k. If the *y*-values conditional on each class of *x*-values have the same distribution, then the statistic *T* in (6.6.52) approximately follows a χ^2 distribution with $n_X - 1$ degrees of freedom (Conover, 1980, pp. 230–231). Given this approximation, the probability $\text{prob}_{\chi^2}(\tilde{T} > T|n_X - 1)$of obtaining a value $\tilde{T}$ that exceeds *T* in the presence of identical *y*-distributions for the individual classes is given by (6.6.51). A small value for $\text{prob}_{\chi^2}(\tilde{T} > T|n_X - 1)$ (i.e., a *p*-value) indicates that the *ys* conditional on individual classes have different distributions and thus, most likely, different means and medians. Hence, a small *p*-value indicates that *x* has an effect on *y*.

For waste pressure subsequent to a drilling intrusion (Figure 6.18a), the three tests for non-monotonic relationships introduced in this section (i.e., CMNs, CMDs, and CLs) all identify *BHPRM* as the most influential variable (Table 6.12). In contrast, the effect of *BHPRM* was missed in analyses based on correlation coefficients with raw and rank-transformed data (Table 6.12). Further, the three tests assign identical rankings to all variables with *p*-values below 0.1. After *BHPRM*, the next two most important variables as indicated by *p*-values are *HALPRM* and *ANHPRM*. These variables were also indicated as having effects with correlation coefficients with raw and rank-transformed data; however, as indicated by the

Table 6.12 Comparison of variable rankings with different analysis procedures for repository pressure at 10 000 years after a drilling intrusion at 1000 years (i.e., y = *WAS_PRES* at 10 000 years in Figure 6.18a) and a maximum of five classes of values for each variable (i.e., $n_X = 5$).

Variable name[a]	CC[b]		RCC[b]		CMN: 1 × 5[b]		CMD: 2 × 5[b]		CL: 1 × 5[b]	
	Rank	***p*-value**	**Rank**	***p*-value**	**Rank**	***p*-value**	**Rank**	***p*-value**	**Rank**	***p*-value**
HALPRM	1.0	0.0000	1.0	0.0000	2.0	0.0000	2.0	0.0000	2.0	0.0000
ANHPRM	2.0	0.0000	2.0	0.0000	3.0	0.0002	3.0	0.0007	3.0	0.0000
HALPOR	3.0	0.0090	3.0	0.0184	5.0	0.0415	5.0	0.0700	5.0	0.0940
ANHBCEXP	7.0	0.1786	8.0	0.2373	4.0	0.0405	4.0	0.0595	4.0	0.0602
BHPRM	10.0	0.3651	6.0	0.1704	1.0	0.0000	1.0	0.0000	1.0	0.0000
ANRBRSAT	19.0	0.7133	14.0	0.4378	7.0	0.1513	6.0	0.0823	7.0	0.1304

[a] Variables for which at least one of the tests (i.e., CC, RCC, CMN: 1 × 5, CMD: 2 × 5, CL: 1 × 5) has a *p*- or α-value less than 0.1; variables ordered by *p*-values for CCs.

[b] Ranks and *p*-values for CCs, RCCs, CMNs test with 1 × 5 grid, CMDs test with 2 × 5 grid, and CLs (Kruskal–Wallis) test with 1 × 5 grid as indicated.

low R^2 values in the associated regression models (i.e., 0.20 and 0.19 in Table 6.11), these variables by themselves are not very effective in accounting for the uncertainty in y in a regression-based analysis.

6.6.9 Identification of Non-Random Patterns

The three tests described in the preceding section attempt to identify departures from monotonic trends. An even less restrictive approach to identifying influential variables is to determine if the scatterplot for the points $(x_k, y_k), k = 1, 2, \ldots, m$, appears to be random conditional on the marginal distributions for x and y. Specifically, the χ^2 test can be used to indicate if the pattern appearing in a scatterplot appears to be nonrandom (Wagner, 1995; Kleijnen and Helton, 1999a, Section 7; Shortencarier and Helton, 1999). For convenience, the χ^2 test for nonrandom patterns will be denoted as a test for statistical independence (SI).

With the χ^2 test, the values for the sampled variable (i.e., the x-values on the abscissa) are divided into classes (Figure 6.20). As in Section 6.7, let $q, q = 1, 2, \ldots, n_X$, denote the individual classes into which the values of x have been divided; let $\mathscr{X}_q$ designate the set such that $k \in \mathscr{X}_q$ only if x_k belongs to class q; and let n_{X_q} equal the number of elements contained in $\mathscr{X}_q$ (i.e., the number of x_k associated with class q). Similarly, the values for the dependent variable (i.e., the y values on the ordinate) are also divided into classes (Figure 6.20). For notational convenience, let $p, p = 1, 2, \ldots, n_Y$, designate the individual classes into which the values of y are divided; let $\mathscr{Y}_p$ denote the set such that $k \in \mathscr{Y}_p$ only if y_k belongs to class p; and let n_{Y_p} equal the number of elements contained in $\mathscr{Y}_p$ (i.e., the number of y_k associated with class p). Typically, the classes $\mathscr{X}_q$ and $\mathscr{Y}_p$ are defined by ordering the x_k and y_k, respectively, and then requiring the individual classes to have similar numbers of elements (i.e., the n_{X_q} are approximately equal for $q = 1, 2, \ldots, n_X$, and the n_{Y_p} are approximately equal for $p = 1, 2, \ldots, n_Y$).

The partitioning of x and y into n_X and n_Y classes in turn partitions (x, y) into $n_X\, n_Y$ classes (Figure 6.20), where (x_k, y_k) belongs to class (p, q) only if x_k belongs to class q of the x values (i.e., $k \in \mathscr{X}_q$) and y_k belongs to class p of the y values (i.e., $k \in \mathscr{Y}_p$)). For notational convenience, let O_{pq} denote the set such that $k \in O_{pq}$ only if $k \in \mathscr{X}_q$ (i.e., x_k is in class q of x-values)

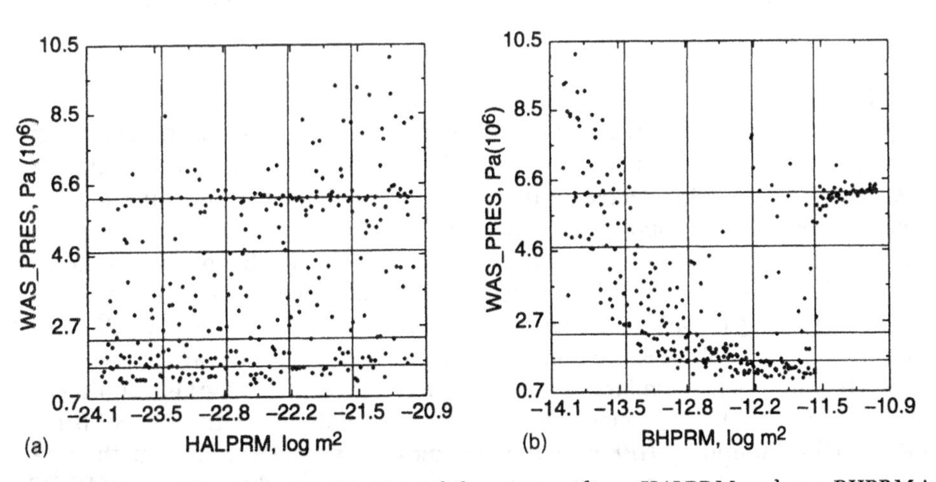

Figure 6.20 Examples of the partitioning of the ranges of $x = HALPRM$ and $x = BHPRM$ into $n_X = 5$ classes and the range of $y = WAS\_PRES$ at 10 000 years in Figure 6.18(a) into $n_Y = 5$ classes.

and also $k \in \mathcal{Y}_p$ (i.e., y_k is in class p of y-values), and let $n_{O_{pq}}$ equal the number of elements contained in O_{pq}. Further, if x and y are independent, then

$$n_{E_{pq}} = \frac{n_{Y_p}}{m}\frac{n_{X_q}}{m}m = \frac{n_{Y_p}n_{X_q}}{m} \tag{6.6.53}$$

is an estimate of the expected number of observations (x_k, y_k) that should fall in class (q, p).

The following statistic can be defined:

$$T = \sum_{q=1}^{n_X}\sum_{p=1}^{n_Y}\frac{\left(n_{O_{pq}} - n_{E_{pq}}\right)^2}{n_{E_{pq}}}. \tag{6.6.54}$$

Asymptotically, T follows a χ^2 distribution with $(n_X - 1)(n_Y - 1)$ degrees of freedom when x and y are independent (Conover, 1980, p. 156). Thus, the probability $\text{prob}_{\chi^2}[\tilde{T} > T|(n_X - 1)(n_Y - 1)]$ of obtaining a value of $\tilde{T}$ that exceeds T when x and y are independent is given by (6.6.51).

The preceding probability provides a way to identify scatterplots that appear to display a significant relationship (i.e., pattern) involving the x and y variables on the abscissa and ordinate. In particular, $\text{prob}_{\chi^2}[\tilde{T} > T|(n_X - 1)(n_Y - 1)]$ is the probability that a larger value of the statistic would occur owing to chance variation (i.e., a p-value). A small p-value indicates that, under the assumptions of the test, an outcome equal to or greater than the observed value of the statistic is unlikely to occur due to chance. Thus, the implication is that the pattern in the scatterplot arose from some underlying relationship involving x and y rather than from chance alone.

As an example, a ranking of variable importance based on p-values for $y = WAS\_PRES$ and a 5×5 grid (Figure 6.20) is given in Table 6.13 under the heading SI: 5×5. The most important variable is *BHPRM*, which is consistent with the well-defined pattern in the corresponding scatterplot in Figure 6.20(b). In contrast, this pattern is completely missed by the regression analyses in Table 6.11. The next most important variable is *HALPRM*, with an effect that is discernible but rather weak in the corresponding scatterplot (Figure 6.20a). In particular, the dependent variable tends to increase as *HALPRM* increases, but with much noise around this trend. After *BHPRM* and *HALPRM*, small possible effects are indicted for *WGRCOR* and *ANHPRM*. After *BHPRM*, *HALPRM*, *WGRCOR*, and *ANHPRM*, the p-values increase rapidly (Table 6.13), and there is little reason to believe that the ordering of the remaining variables on the basis of their p-values is due to anything other than chance. Similar variable rankings were also obtained in the analyses with CMNs, CMDs, and CLs (Table 6.12).

The χ^2 statistic for identifying nonrandom patterns is based on superimposing grids on the scatterplots under consideration (Figure 6.20). As a result, the outcome of such an analysis can depend on the grid selected for use. In particular, different grids can lead to different orderings of variable importance, although the identification of strong patterns is relatively insensitive to reasonable grid selections (i.e., grids that do not have an excessive number of cells relative to the number of points in the scatterplots under consideration). As an example, a ranking of variable importance based on p-values for $y = WAS\_PRES$ and a 10×10 grid is given in Table 6.13 under the heading SI: 10×10. The rankings with 5×5 and 10×10 grids produce similar but not identical results, with both grids resulting in the identification of *BHPRM* as the most important variable and the identification of *BHPRM*, *HALPRM*, *WGRCOR*, and *ANHPRM* as the four most important variables. Both analyses suggest that none of the remaining variables have a discernible effect on *WAS_PRES*. Similar robustness is also present in the analyses of y with CMNs, CMDs, and CLs (Kleijnen

Table 6.13 Comparison of variable rankings with the χ^2 statistic for repository pressure at 10 000 years after a drilling intrusion at 1000 years (i.e., $y = WAS\_PRES$ at 10 000 years in Figure 6.18a) obtained with a maximum of five classes of x-values (i.e., $n_X = 5$) and analytic determination of p-values with variable rankings obtained with (i) a maximum of ten classes of x-values (i.e., $n_X = 10$) and analytic determination of p-values and (ii) a maximum of five classes of x-values (i.e., $n_X = 5$) and Monte Carlo determination of p-values (Kleijnen and Helton, 1999a, Table 23; for omitted results, see Kleijnen and Helton, 1999c, Table 10.23).

Variable	**SI: 5 × 5** [b]		**SI: 10 × 10** [c]		**SIMC: 5 × 5** [d]	
name [a]	**Rank**	***p*-value**	**Rank**	***p*-value**	**Rank**	***p*-value**
BHPRM	1.0	0.0000	1.0	0.0000	1.5	0.0000
HALPRM	2.0	0.0002	4.0	0.0082	1.5	0.0000
WGRCOR	3.0	0.0002	2.0	0.0028	3.0	0.0002
ANHPRM	4.0	0.0049	3.0	0.0032	4.0	0.0033
SHRGSSAT	5.0	0.0698	22.0	0.8482	5.0	0.0699
SHBCEXP	6.0	0.1010	15.0	0.3495	6.0	0.0989
WGRMICI	7.0	0.1985	11.0	0.1646	7.0	0.2013
ANHBCVGP	8.0	0.2427	14.0	0.3398	8.0	0.2380
—	—	—	—	—	—	—
SHPRMHAL	24.0	0.9064	24.0	0.8863	24.0	0.9102
SHPRMCON	25.0	0.9898	20.0	0.5316	25.0	0.9933

[a] Twenty-five variables from Table 6.1 included in analysis, with (i) *ANHCOMP* and *HALCOMP* not included because of the −0.99 rank correlations within the pairs (*ANHPRM*, *ANHCOMP*) and (*HALPRM*, *HALCOMP*) and (ii) *BPCOMP*, *BPINTPRS*, *BPPRM*, and *BPVOL* not included because brine pocket properties are not relevant to the E2 intrusion under consideration.
[b] Variable rankings and p-values obtained with a maximum of five classes of x values (i.e., $n_X = 5$), five classes of y values (i.e., $n_Y = 5$), and analytic determination of p-values (see (6.6.54) and (6.6.51)). Discrete variables (e.g., *WMICDFLG*, which has only three distinct values) are divided into less than n_X classes when they have less than n_X distinct values.
[c] Variable rankings and p-values obtained with a maximum of ten classes of x-values (i.e., $n_X = 10$), ten classes of y-values (i.e., $n_Y = 10$), and analytic determination of p-values.
[d] Variable rankings and p-values obtained with a maximum of five classes of x-values (i.e., $n_X = 5$), five classes of y-values (i.e., $n_Y = 5$), and Monte Carlo determination of p-values.

and Helton, 1999a, Table 24; for comparisons with additional variables, see Kleijnen and Helton, 1999a, Tables 8, 14, and 19).

The p-values used to identify important variables in Table 6.13 are calculated with statistical assumptions that are not fully satisfied. In particular, the sample from the xs consists of three pooled LHSs rather than a random sample (see (6.3.9)). A Monte Carlo simulation can be used to assess if the use of formal statistical procedures to determine p-values is producing misleading results. Specifically, a large number of samples (10 000 in this example) of the form

$$(x_k, y_k), \qquad k = 1, 2, \ldots, 300, \tag{6.6.55}$$

can be generated by pairing the 300 values for x (i.e., the 300 values for the particular x under consideration contained in the samples indicated in (6.3.9)) with the 300 predicted values for y (i.e., the 300 values for y that resulted from the use of the sample elements indicated in (6.3.9)). The specific pairing algorithm used was to randomly and without replacement assign an x value to each y value, which is similar to bootstrapping (Efron and Tibshirani, 1993) except that the sampling is being performed without replacement. This random assignment was repeated 10 000 times to produce 10 000 samples of the form in (6.6.55). Each of the 10 000 samples can be used to calculate the value of the χ^2 statistic.

The resulting empirical distribution of the χ^2 statistic can then be used to estimate the *p*-value for the χ^2 statistic actually observed in the analysis. Comparison of the *p*-value obtained from (6.6.51) with the *p*-value obtained from the empirical distribution provides an indication of the robustness of the variable rankings with respect to possible deviations from the assumptions underlying the formal statistical procedure in (6.6.51). As indicated by comparing the results in columns SI: 5×5 and SIMC: 5×5 in Table 6.13, the analytical determination of *p*-values in (6.6.51) and the just-described Monte Carlo determination of *p*-values are producing similar results. Thus, at least in this example, the variable rankings are not being adversely impacted by the use of (6.6.51). Similar comparisons were also obtained in the analyses for y with CMNs, CMDs and CLs (Kleijnen and Helton, 1999a, Table 24; for comparisons with additional variables, see Kleijnen and Helton, 1999a, Tables 8, 14, and 19). Monte Carlo procedures of the type described in this paragraph provide a useful alternative to the use of formal statistical procedures to calculate *p*-values.

6.7 SUMMARY

Sampling-based methods for uncertainty and sensitivity analysis have a number of desirable properties, including:

(i) conceptual simplicity;
(ii) ease and flexibility in adaptation to specific analysis situations;
(iii) stratification over the range of each uncertain variable;
(iv) direct estimation of distribution functions to characterize the uncertainty in model predictions;
(v) availability of a variety of sensitivity analysis techniques.

The results of sampling-based uncertainty and sensitivity analyses are conditional on the distributions assigned to the uncertain (i.e., sampled) variables. Thus, care must be used in assigning these distributions. A possibility is to carry out multiple iterations of an analysis. The first iteration could be performed with rather crude distribution assumptions to determine the most important variables. Then, additional resources could be focused on characterizing the uncertainty in these variables before a second iteration of the analysis is carried out. Under certain conditions, the effects of the changed distributions can be determined without rerunning the model under consideration (Iman and Conover, 1980; Beckman and McKay, 1987).

A number of techniques for sensitivity analysis have been described. However, many additional techniques for analyzing multivariate data exist that could be productively applied in a sampling-based sensitivity analysis. In particular, there are undoubtedly many pattern recognition techniques that could be successfully adapted for use in sensitivity analysis.

Sampling-based uncertainty and sensitivity analyses are usually performed for two reasons: first, to determine the uncertainty in model predictions (e.g., to ascertain if model predictions fall within some region of concern), and second, to determine the dominant variables in giving rise to the uncertainty in model predictions (e.g., to identify the variables on which limited research funds should be concentrated). However, there is also a third reason to perform a sampling-based uncertainty and sensitivity analysis, namely, to verify that the model under consideration is operating correctly. Owing to the concurrent variation of many model inputs and the efficacy of sensitivity analysis procedures in identifying the effects of model inputs on individual model predictions, sampling-based uncertainty and sensitivity analysis procedures provide a powerful tool for model and analysis quality assurance.

Sampling-based sensitivity analysis procedures are based on identifying patterns in a mapping between model inputs and predictions. Different procedures are predicated on the identification of different types of patterns. Thus, a procedure will not perform well if the mapping under consideration does not contain the type of pattern that that particular procedure seeks to identify. As a result, a good sensitivity analysis strategy is to use several different procedures that seek to identify different types of patterns. With this approach, there is a reasonable chance that each important model input will be identified by at least one of the procedures.

Sensitivity analysis provides a way to identify the model inputs that most affect the uncertainty in model predictions. However, sensitivity analysis does not provide an explanation for such effects. This explanation must come from the analysts involved and, of course, be based on the mathematical properties of the model under consideration. An inability to develop a suitable explanation for the effects of a particular model input is often indicative of an error in the development of the model or the implementation of the analysis.

This chapter has emphasized the propagation and analysis of the effects of subjective uncertainty as characterized by the distributions in (6.1.3). However, many large analyses involve two distinct sources of uncertainty: stochastic or aleatory uncertainty, which arises because the system under study can behave in many different ways, and subjective or epistemic uncertainty, which arises from an inability to specify an exact value for a quantity that is assumed to have a constant value within a particular analysis (Helton, 1994, 1997; Helton and Burmaster, 1996; Cullen and Frey, 1998). An example of such an analysis is the US Nuclear Regulatory Commission's reassessment of the risk from commercial nuclear reactors in the USA (commonly referred to as the NUREG-1150 analysis after its report number), where stochastic uncertainty arose from the many possible accidents that could occur at the power plants under study and subjective uncertainty arose from the many uncertain quantities required in the estimation of the probabilities and consequences of these accidents (US NRC, 1990–91; Breeding *et al.*, 1992b; Helton and Breeding, 1993). Numerous other examples also exist (e.g., PLG, 1982, 1983; Øvreberg *et al.*, 1992; Payne *et al.*, 1992; McKone, 1994; Allen *et al.*, 1996; Price *et al.*, 1996).

In such analyses, importance sampling (Section 6.3.2) is often used to propagate the effects of stochastic uncertainty through the use of event trees, and random or Latin hypercube sampling (Sections 6.3.1 and 6.3.3) is used to propagate the effects of subjective uncertainty. In the NUREG-1150 analysis, stochastic uncertainty was propagated through the use of event trees, and subjective uncertainty was propagated through the use of Latin hypercube sampling. The 1996 WIPP PA, from which the example involving two-phase fluid flow is taken (Section 6.1), used random sampling to propagate the effects of stochastic uncertainty and Latin hypercube sampling to propagate the effects of subjective uncertainty (Helton *et al.*, 1998, 1999; Helton, 1999). The NUREG-1150 analysis and the 1996 WIPP PA provide examples of large and complex analyses that used sampling-based methods for uncertainty and sensitivity analysis.

7

Reliability Algorithms: FORM and SORM Methods

Jeff D. Cawlfield

University of Missouri at Rolla, USA

7.1 INTRODUCTION

7.1.1 Why Use a Reliability Approach?

As noted in the introduction to this book, there are times when the entire suite of probabilistic outcomes are not necessarily of interest; rather, there may be one particular mode of failure (or reliability) that is of concern. Examples from engineering include failure of a beam or girder under loading, 'failure' of a groundwater contaminant remediation scheme, and the reliability of a foundation under earthquake loading. In each of these cases, there will be input parameters describing the loads and the material characteristics, and each parameter may be treated as an uncertain variable. There will be a mathematical model that includes those input parameters and that can be solved analytically or numerically to provide a result—perhaps in the form of a factor of safety, perhaps in the form of a stress or strain, or a result framed in some other fashion. That result can be compared to a threshold level such that if the stress exceeds the threshold there is failure, or if the contaminant level exceeds the threshold there is failure, or if the foundation displaces an amount greater than the threshold there is failure.

Reliability algorithms are often very efficient in providing an approximate evaluation of the probability of failure in these cases. And, of great interest here, these algorithms also directly provide sensitivity measures that incorporate the statistical characteristics of the input parameters as well as the particular notion of failure that is specific to any given case. Little, if any, additional computational effort is required to obtain the sensitivity measures; they are evaluated in the optimization scheme within the reliability algorithm. The

Sensitivity Analysis. Edited by A. Saltelli *et al.*

algorithms that are most commonly used are known as FORM and SORM (first-order and second-order reliability methods).

7.1.2 Conceptual Basis (What Are FORM and SORM?)

The first-order (or second-order) reliability methods, referred to as FORM and SORM, may also be called limit-state methods, and there are probably other names that will arise in the literature. The historical highlights and necessary theory will be reviewed shortly, but at this point it is instructive to consider conceptually what FORM and SORM can and cannot do. For convenience, let us use the acronym FORM generically to refer to all reliability algorithms unless the second-order method is of particular interest.

FORM seeks to find a point in the space of all possible realizations of the uncertain variables that is, in some sense, most likely to lead to failure. Let each uncertain variable be X_i, where there are n total uncertain variables in a model, and the vector $\mathbf{X}$ is a vector of all n variables. Failure is defined by the analyst using a function of the $\mathbf{X}$ variables that is called a performance function $g(\mathbf{X})$; mathematically, failure is the exceeding of some threshold output value when a model is run. FORM utilizes an optimization scheme to find the 'most likely failure point', using the input parameters to the model and the performance function that defines failure. Once this most likely failure point (often called the design point) has been determined, a first-order surface is fit at that point in order to conveniently evaluate an approximate probability of failure.

FORM does not generate a large number of random realizations in order to fully evaluate the statistical distribution of all possible outcomes, but, rather, it tries to find the most likely point leading to a particular outcome using optimization (which typically requires only a few iterations, rather than thousands of realizations). FORM does not provide a complete statistical portrait of all possible model outputs, but only gives an approximate probability of failure for a very specific performance function. However, that may be all the analyst needs in many applications. And FORM directly provides sensitivity measures related to each uncertain variable, whereas a full Monte Carlo analysis might require significant additional calculations before yielding sensitivity measures.

7.1.3 Sensitivity Measures Obtained from FORM and SORM

Optimization algorithms typically utilize some form of derivative evaluations in order to move efficiently toward the most likely failure point as required for FORM. Therefore, as a natural part of the FORM algorithm, an optimization scheme will be invoked that will require derivatives of the performance function with respect to each uncertain input variable. These derivatives are by themselves sensitivity measures, albeit in a crude sense, but, as will be shown, they can be utilized with other FORM calculations and characteristics to provide very useful and sophisticated sensitivity measures that incorporate the statistical nature of the uncertain variables, correlation, and the specific form of the performance function. Much recent work, to be discussed later, focuses on the nature and robustness of the various sensitivity measures and how FORM sensitivity measures compare to MC and other approaches. Local sensitivity measures, global sensitivity measures, and other issues of importance are the focus of recent and ongoing research related to FORM techniques. The reader is referred to Chapter 5, which discusses the evaluation of derivatives by a variety of methods.

7.2 BRIEF REVIEW OF RELIABILITY ALGORITHMS

The reader can find a wealth of journal articles and textbook chapters that provide detailed accounts of the reliability approach (for a variety of discussions and applications, see, e.g., Madsen *et al.*, 1986; Der Kiureghian and Liu, 1986; Cawlfield and Wu, 1993). Historical highlights and necessary theory are provided here—but only in brief. The interested reader is encouraged to delve into the references for detailed theoretical presentations.

7.2.1 Early Theoretical Development

Mean-value approaches

Early reliability approaches were based on using the mean of all uncertain variables in a model as the design point. Cornell (1972) gave perhaps the earliest application of what has come to be called the mean-value first-order second-moment method (MVFOSM). In this method, the model (solved either analytically or numerically) is evaluated using first-order approximations for the mean and variance of the performance function. Mathematically, this treatment is straightforward, except perhaps for the need to evaluate the derivative of the performance function with respect to each uncertain variable. This requirement is ubiquitous for all reliability algorithms. The derivatives may be obtained analytically for some simple mathematical models, but often they must be evaluated using finite-difference techniques, or more sophisticated adjoint sensitivity equations.

The MVFOSM method provides a reliability index, which is defined as the ratio of the mean of the performance function to the variance of the performance function (evaluated at the mean point). In the literature, this index is often symbolized as

$$\beta_{MVFOSM} = \frac{g(\mathbf{M})}{\sum_{i=1}^{n}\sum_{j=1}^{n}\frac{\partial g}{\partial X_i}\frac{\partial g}{\partial X_i}\rho_{ij}\sigma_i\sigma_j}$$

In this formulation $\mathbf{M}$ is the vector of mean values for the $\mathbf{X}$ variables, and the correlation coefficients and standard deviations are given by ρ_{ij} and σ_i, respectively. The larger the reliability index, the less likely is the probability of failure. If the analyst wishes to assume a normal distribution, the index can be used to estimate the probability of failure by using the approximation $p_f \approx \Phi(-\beta)$, where $\Phi(\cdot)$ is the standard normal probability function.

Hasofer–Lind limit-state surface technique

Hasofer and Lind (1974) recognized that the MVFOSM technique did not provide unique values of β, even if equivalent forms of the performance function are specified. They proposed a design point other than the mean as a more consistent measure of reliability. This is the basis of the FORM technique—that an optimization point (the design point) can be found that is, in some sense, the most likely failure point. This point is unique, irrespective of equivalent formulations of the performance function, and a surface fit through this point can be used to divide the space of all possible outcomes into two regions: a region of failure and a region of safety. Again, the analyst might be willing to assume a normal probability distribution over the space of all possible outcomes, such that the probability of failure might be evaluated if the surface separating failure from safety can be defined or approximated (later, we discuss the nature of these surfaces in the FORM and SORM techniques).

The original uncertain variables $\mathbf{X}$ are transformed into standard variates $\mathbf{Y}$, and the optimization scheme can be most efficiently applied in the standard variate space. Extensions of the Hasofer and Lind method have been appearing over the last two decades or so. Specific optimization algorithms have been proposed, as have techniques for incorporating correlation and probability distribution function information. But fundamentally the technique is unchanged—find the design point that most effectively separates the space of all possible outcomes into two regions: a region of failure and a region of safety. In order to find that point, optimization algorithms are invoked and derivatives are required, which can be utilized for sensitivity measures. Also, if the analyst can assume something about the nature of the joint probability density function, a probability of failure can be evaluated.

7.2.2 Example Application

Rather than encumber the discussion with equations and additional theory at this point, let the reader consider an example, which will, it is hoped, shed light on the nature of the performance function, uncertain variables in a specific model, derivative calculations, and the reliability index. The following is based on an example used by Der Kiureghian (1985) to illustrate the nature of the FORM techniques.

For simplicity, and so the problem can be illustrated in two dimensions, consider a nonlinear performance function with two random variables only:

$$g(\mathbf{X}) = \tfrac{1}{2}(X_2)^2 - X_1.$$

The random variable X_1 has a mean of 100.0 and a variance of 400.0. The random variable X_2 has a mean of 20.0 and a variance of 25.0. The correlation between the two variables is 0.50. By convention, the performance function is constructed such that failure is defined as $g(\mathbf{X}) < 0$. Since only the sign of $g(\mathbf{X})$ is important, an infinite number of equivalent $g(\mathbf{X})$ functions may be formulated. Recall that the MVFOSM technique will give different results for equivalent $g(\mathbf{X})$ formulations, while the Hasofer–Lind technique (and FORM) will give a consistent reliability index no matter what equivalent $g(\mathbf{X})$ form is used.

The MVFOSM method utilizes first-order equations for the mean and variance of $g(\mathbf{X})$, evaluated at the mean points. The MVFOSM equations result in a reliability index of 1.09 for this example.

The Hasofer–Lind technique (a basic portion of the FORM technique) seeks a design point that is the most likely point in the $\mathbf{X}$ space that will lead to $g(\mathbf{X}) = 0$. Actually, the problem is transformed into standard space, and the design point is that point most likely leading to $G(\mathbf{Y}) = 0$. An optimization algorithm is used to find the design point, which is that point on the limit-state surface in the standard space that is closest to the origin (the limit-state surface is defined by $G(\mathbf{Y}) = 0$). The techniques used in this example for transformation of correlated variables into uncorrelated standard space are those given by Der Kiuerghian and Liu (1986). Essentially, a rotation matrix $\mathbf{\Gamma}$ is constructed from the covariance matrix utilizing eigenvalue techniques or Cholesky decomposition (a number of $\mathbf{\Gamma}$ matrices are possible). Then the transformation from $\mathbf{X}$ to $\mathbf{Y}$ space is obtained in the usual fashion (but multiplied by the rotation matrix) as follows:

$$\mathbf{Y} = \mathbf{\Gamma}\mathbf{D}^{-1}(\mathbf{X} - \mathbf{M}),$$

where **M** is the vector of means, and **D** is the matrix of standard deviations. Results for the example under consideration give the design point and β as follows:

$$X_1 = 93.7, \qquad X_2 = 13.6, \qquad \beta = 1.314.$$

It is interesting to note that, because of the correlation, the most likely value leading to failure is below the mean for both variables.

What about the sensitivity measures? In the **X** space, we might use the traditional derivatives of the performance function with respect to each X_i. But in the **Y** space we have more useful sensitivity measures that can be utilized with some of the statistical characteristics of the uncertain variables to give us measures of the sensitivity of the performance function with respect to equally likely changes in the standard variables **Y**. In the optimization scheme, the direction cosines α_i obtained at the design point in the **Y** space are often used as sensitivity factors; for the example under consideration, the direction cosines are

$$\alpha_1 = -0.69, \qquad \alpha_2 = 0.72.$$

Here the analyst would conclude there was little difference between the **X** variables based on the alpha sensitivity factors (because the magnitude of the α-values are very similar). However, the direction cosines at the design point will differ, depending on whether the rotation matrix in the transformation is obtained using eigenvalue techniques or Cholesky decomposition. For example, if a Cholesky decomposition approach is used to develop the rotation matrix for the example here, the reliability index is unchanged, but the direction cosines become

$$\alpha_1 = -0.23, \qquad \alpha_2 = -0.97.$$

Now the analyst might conclude that uncertain variable number 2 is much more important, because it has much higher magnitude alpha sensitivity than uncertain variable number 1. Thus some inconsistency arises in the procedure. A better sensitivity measure is provided by Der Kiureghian and Liu (1986), which they call the gamma sensitivity vector. The gamma sensitivities are obtained in standard space, and are measures of the sensitivity of the performance function (and thus the reliability index) to *equally likely* changes in the uncertain variables **X**.

The rotation matrix is utilized in the calculation of each individual gamma sensitivity as follows:

$$\gamma = \frac{\mathbf{\Gamma}^{\mathrm{T}}\boldsymbol{\alpha}}{|\mathbf{\Gamma}^{\mathrm{T}}\boldsymbol{\alpha}|}.$$

$\boldsymbol{\alpha}$ is the vector of direction cosines at the design point, which are the individual derivatives of the reliability index with respect to each standard variate y_i. Of course, derivatives of the reliability index are formulated as a function of derivatives of the performance function (see Wu and Cawlfield, 1992).

For the example under consideration, the gamma sensitivity values are as follows:

$$\gamma_1 = 0.28, \qquad \gamma_2 = 0.96.$$

These gamma sensitivity measures do not depend on the specific technique used to create the rotation matrix, and, therefore, the inconsistency in the alpha sensitivities mentioned earlier is not a problem. In this case, the gamma sensitivity measures indicate that, for equally likely changes in the two uncertain variables, changes in uncertain variable 2 are much more important to the reliability index and the probability of failure than are changes in uncertain variable number 1.

The entire FORM process may be listed in step-by-step fashion as follows:

1. Formulate the problem in terms of a performance function $g(\mathbf{X})$.
2. Transform the problem and the original variables $\mathbf{X}$ into standard space and in terms of standard variates $\mathbf{Y}$ so that the performance function is now in standard space, $G(\mathbf{Y})$. The transformation is carried out with the equation given above, utilizing the rotation matrix $\mathbf{\Gamma}$, which is a function of the covariances modified to account for the marginal distribution information (if available). Refer to Der Kiureghian and Liu (1986) for specific details of modifying and constructing this rotation matrix.
3. An optimization algorithm is then applied to find the point y^* that gives $G(\mathbf{Y}) = 0$ and is closest to the origin in standard space. The distance between y^* and the origin is called the reliability index, and it is this index that can be used to estimate the probability that $G(\mathbf{Y}) < 0$. The optimization algorithm can be one of many; in general, the optimization algorithm will require that partial derivatives of the function $G(\mathbf{Y})$ be calculated with respect to each $\mathbf{Y}$ variable, and these partial derivatives are used to evaluate the direction cosines α, which can then be used to evaluate the gamma sensitivities γ.

7.2.3 Application of FORM and SORM to a Nonlinear Test Case: Chemical Kinetics

In Section 2.9 in Chapter 2, a test case involving chemical kinetics is presented that may be analyzed using FORM and SORM in order to illustrate the interpretation of the sensitivity measures over a suite of trial values.

The time-dependent analytical solution to the chemical kinetics case is given as

$$Y(t) = \left(2x_1 \exp\left(-\frac{x_2}{T}t\right) + \frac{1}{x_3}\right)^{-1},$$

and the uncertain variables are all uniformly distributed as

$$X_1 \sim U(8.97 \times 10^6, 3.59 \times 10^7), \qquad X_2 \sim U(0, 1000), \qquad X_3 \sim U(1.0, 1.2).$$

Assume that $T = 300$ K is a constant. Of interest here are the gamma sensitivity values, how they relate to the probability of exceeding some target value of $\mathbf{Y}$, and how the FORM and SORM procedure compares with a Monte Carlo analysis. The performance function is set up as $g(\mathbf{X}) = \text{target} - Y(t)$; therefore, we shall evaluate the probability that $Y(t) > \text{target}$. To begin, let's look at the case where $t = 1.0 \times 10^{-8}$. The mean value $E[Y(t)]$ is given as 0.98.

We use the CALREL program (see the Software Appendix) and look first at a suite of target values and investigate the FORM, SORM, and MC estimates for probability of exceeding the target, and the gamma sensitivity information. Table 7.1 shows results. The results show that the gamma sensitivity values change in relative importance depending upon whether we are interested in a low- or high-probability event. For example, there is a high probability of $Y(t)$ exceeding a value of 0.80, and that result is most sensitive to changes in uncertain variable number 2 (which has gamma sensitivity of 0.89). However, as the target of interest is changed, the relative importance of uncertain variable 2 as opposed to variable 3 changes. For a very low-probability event, for example the probability that $Y(t)$ exceeds 1.18, the result is most sensitive to changes in uncertain variable number 3 (with a gamma sensitivity of

Table 7.1 Probability of exceeding target and gamma sensitivities from FORM and SORM.

Target	FORM	SORM	MC	γ_1	γ_2	γ_3
0.80	0.8731	0.9063	0.9120	0.40	0.89	0.23
0.95	0.6403	0.6477	0.6470	0.31	0.88	0.34
1.10	0.1918	0.1258	0.1290	0.29	0.63	0.75
1.15	0.0485	0.0240	0.0245	0.34	0.55	0.75
1.18	0.0030	0.0009	0.0007	0.44	0.51	0.73

Table 7.2 Probability of exceeding target and gamma sensitivities from FORM and SORM.

Target	FORM	SORM	MC	γ_1	γ_2	γ_3
0.003	0.8733	0.9045	0.9050	0.41	0.91	0.00
0.005	0.7392	0.7740	0.7760	0.35	0.93	0.00
0.040	0.1646	0.1461	0.1377	0.51	0.86	0.00
0.080	0.0358	0.0221	0.0216	0.64	0.76	0.01

0.73). It is interesting to note that the gamma sensitivity for uncertain variable number 1 is relatively constant throughout the suite of target values.

Also of interest is the comparison between FORM, SORM, and MC estimates of probability. As expected for this nonlinear performance function, the FORM results for the probability estimate are not as accurate as SORM or MC (which are very close to each other); this is especially true for the extreme values when either very high- or very low-probability events are being investigated. The gamma sensitivity measures, of course, are not affected by whether FORM or SORM is used. However, the analyst is cautioned that the FORM probability estimates may be very approximate for some cases.

Now let us consider the same chemical kinetics case, but at a much different time $t = 1 \times 10^{-5}$, when the expected value for $Y(t)$ is $E[Y(t)] = 0.02$. All other assumptions are as in the previous discussion. Results for this case are given in Table 7.2. These results show that the probability estimate is most sensitive to changes in uncertain variable number 2 throughout the range of the target values investigated. However, the relative difference between gamma values between uncertain variables numbers 2 and 1 becomes smaller as more extreme values of probability are investigated. Perhaps most interesting is the result that uncertain variable number 3 is not important in this case for $t = 1 \times 10^{-5}$, whereas it was an important uncertain variable for the case for $t = 1 \times 10^{-8}$. This does not mean that the parameter is unimportant—it does mean that the uncertainty of this parameter can be neglected and that it can be treated as a deterministic variable for this particular value of t.

7.3 A REVIEW OF APPLICATIONS

7.3.1 Structural Reliability

Madsen *et al.* (1986) present a number of applications of the reliability approach, in its various forms, to structural safety problems. Der Kiureghian and Taylor (1983) provide an early application incorporating numerical modelling with FORM, and they suggest a number of techniques to improve the efficiency and computer storage requirements when large structural systems are encountered. A number of computer codes and general application shells are available for performing FORM and SORM, and most have been formulated with structural engineering and reliability in mind: two that are commonly used are the CALREL code (Liu *et al.*, 1989) and PROBAN (Madsen, 1988). The CALREL code is used by Cawlfield and Wu (1993), Hamed *et al.* (1995), and Piggott and Cawlfield (1996), as described in other sections of this chapter.

An innovative application of the PROBAN code to various offshore oil field design and operation activities is presented by Bysveen *et al.* (1990). They give a brief overview of applications of FORM and SORM to such diverse problems as project economics, tension leg oil platform design, and geotechnical concerns such as the bearing capacity of offshore rig foundations. Of particular interest in this study is the use of the sensitivity factors when analyzing tolerances on mispositioning of the anchor tethers for an offshore rig. The engineers are required to specify tolerances on mispositioning (that is, anchoring the rig at slightly inaccurate locations—thereby changing the loads in the tethers). The paper describes an approach where a FORM analysis provides sensitivity information concerning the probability of failure as influenced by mispositioning tolerances; then the consequences of relaxing the tolerances (which influences construction cost) are analyzed in order to optimize the project economics.

7.3.2 Subsurface Hydrology

Early applications of the reliability approach to subsurface hydrology include Sitar *et al.* (1987), Cawlfield and Sitar (1988), Schanz and Salhotra (1992), Wu and Cawlfield (1992), and Jang *et al.* (1994). In these papers, the FORM and SORM techniques were applied to various hypothetical cases of subsurface flow and transport that invoked analytical and numerical solution techniques. In these works, FORM and SORM were shown to be efficient and reasonably accurate techniques for estimating probabilities associated with groundwater flow and transport. Comparisons with Monte Carlo simulation were favorable with respect to accuracy of the probability calculations.

Hamed *et al.* (1995) present the FORM technique as especially useful as a screening tool in their work for investigating the effects of parameter uncertainty on the response of a groundwater transport model with reactive and non-reactive solutes. Hamed *et al.* (1995) also show good agreement between Monte Carlo and reliability approaches, and they note that for very low-probability events the reliability approach is much more computationally efficient. Hamed *et al.* (1996) extend the approach using numerical modelling and a system probability estimate that allows the analysis of joint probability of failure at several wells in an aquifer (failure is defined as exceeding a threshold contaminant concentration). Sensitivity information was used to determine which grid block in the numerical model had the greatest influence on failure probability. Thus, the numerical modeller can use sensitivity information to determine where additional information (perhaps from field work or laboratory studies) would be most beneficial in terms of reducing the uncertainty in the probabilistic model result.

The FORM and SORM techniques have recently been used to explore unsaturated flow and transport, which involves highly nonlinear solutions. Piggott and Cawlfield (1996) explored the sensitivity results for one-dimensional transport in the vadose zone, and concluded that, out of the many uncertain parameters that influence the problem, the saturated water content was usually the most important. The reliability algorithm occasionally has difficulty converging on a design point for the highly nonlinear problems studied by Piggott and Cawlfield, and others have noticed the same difficulties under certain situations. Typically the convergence problems arise because of local minimums encountered in the algorithm; this problem can usually be overcome with selection of an appropriate initial point other than the mean point.

Xiang and Mishra (1997) studied multiphase flow and transport, and state that the determination of the design point was computationally intensive for these types of highly nonlinear problems. They recommend that the analyst reduce the number of random parameters, perhaps by preliminary modelling to determine the most important parameters, in general, over a suite of problem scenarios. They also suggest that adjoint techniques for determining partial derivatives will be more efficient when the number of performance functions (for a systems reliability problem) are significantly smaller than the number of random parameters (they reference the work by RamaRao and Mishra, 1996). Xiang and Mishra (1997) also recommend that the performance function and the random parameters must be properly scaled in order to avoid numerical problems during the application of the reliability optimization algorithm.

7.3.3 A Geotechnical Example Focused on Bayesian Updating

Luckman *et al.* (1987) utilize a FORM approach to study failure of soil slopes. Their presentation focuses particularly on utilizing the FORM technique for back-analysis and determination of parameter values. Back-analysis, or inverse analysis, takes into account that the slope has failed, and then attempts to predict the values of the parameters that likely exist and led to such failure. Luckman *et al.* (1987) note that the FORM design point is, in some sense, the most likely failure point. They present a very logical and straightforward technique for incorporating knowledge of failure, sensitivity information, and the design point into a Bayesian approach for updating prior parameter statistics.

Luckman *et al.* (1987) also note that, although a soil slide tends to occur (at least in the idealized setting) along a well-defined failure point that involves only some of the soil, if spatial correlation exists (and it usually does) then the design point will provide information about the most likely soil parameters and load values even in areas some distance from the sliding plane. Examples in this work illustrate how engineering judgement, test data, knowledge of failure, and model uncertainty might all be incorporated in the analysis.

7.4 SUMMARY OF RECENT THEORETICAL ADVANCES

7.4.1 Comparison of First- and Second-Order Approaches (FORM and SORM) and Monte Carlo simulation

When approached with a problem involving some sort of failure or reliability analysis, the scientist is faced with an array of possible choices. Typically, Monte Carlo simulation (MC) is the first line of attack. Why not always use MC methods? Twenty years ago, the motivation was to find alternatives to MC that were more computationally efficient in order to save on computer time and man-hours. Today computers are incredibly fast, and even running thousands of realizations using MC techniques is usually not particularly burdensome. However, there are still cases that favor an approach other than MC; for example, when extremely low-probability events are of interest (because an enormous number of realization may be necessary for statistical validity in an MC simulation).

Another case where FORM may offer advantages over MC is the focus of this book—sensitivity analysis. Other chapters describe techniques for obtaining sensitivity measures during a MC simulation, but these techniques may often be computationally burdensome and are perhaps ill-suited for some problems where extremely low-probability events are of interest. FORM can provide an efficient technique for obtaining probabilistic sensitivity measures over a wide range of problem types and statistical characteristics.

A number of authors have compared the efficiencies of MC versus FORM and SORM techniques over the past few years. Jang *et al.* (1994) applied the methods to one- and two dimensional models of contaminant transport in porous media under saturated conditions, and found that FORM overestimates probabilities when the hydraulic conductivity and dispersivity of the porous media are heterogeneous. Their SORM and MC results were essentially identical. Their conclusion was that FORM or SORM was the method of choice when a single probability of failure is desired, and the number of uncertain variables was not extremely large. For cases where the number of uncertain variables became quite large, MC simulation was perhaps more efficient for calculating a failure probability. However, note that this efficiency question as posed in Jang *et al.* (1994) did not address the calculation of sensitivity measures. Cawlfield *et al.* (1997) note that, although FORM may overestimate the probability of failure (owing to the use of a first-order approximation of the failure surface), the sensitivity results are not affected because they are only a function of the design point and the partial derivatives, not the type of failure surface approximation.

7.4.2 Convergence Problems Associated with FORM and SORM

Xiang and Mishra (1997) have studied the use of FORM, SORM, and MC for evaluating the probability associated with multiphase flow in porous media. This application is concerned with flow at the USA's proposed high level nuclear waste disposal facility at Yucca Mountain, Nevada. They also concluded that FORM and SORM give similar results to MC and are often more computationally efficient. Their mathematical models are extremely nonlinear, and, therefore, provide a rather sophisticated test of the optimization algorithms utilized for FORM and SORM. They do caution that as the probability of failure approaches the extreme values of 0.0 or 1.0, the optimization algorithm may have difficulty converging on a design point. That is certainly a limitation that must be considered when applying FORM techniques with extremely nonlinear solutions—the optimization algorithm must actually find the global design point. The optimization technique, convergence tolerance, and other numerical solution characteristics must be carefully constructed in order to avoid instabilities and other difficulties.

Contaminant transport in porous media provides an illuminating example of one potential pitfall in the FORM approach related to the interaction of the performance function and the optimization algorithm. If the performance function is formulated such that failure is defined as the exceeding of some threshold concentration, and if the contaminant source is a single pulse or spill, there may be two times at which the actual $g = 0$ (the limit state) is achieved. Early in the transport process, the concentration will increase until the threshold is reached, and later, after the peak concentration has passed, the concentration will decrease and the threshold will be achieved again. If the performance function is not carefully constructed to take into account the time of occurrence of the concentration threshold, or perhaps formulated in terms of the total mass cumulatively passing a point (or some other monotonic function), the optimization scheme might converge on the wrong design point.

Piggot and Cawlfield (1996) and Cawlfield *et al.* (1997) also utilized FORM to study nonlinear solutions; in this case, unsaturated flow and transport. Their results indicate that FORM provides very useful sensitivity information, but, as with the Xiang and Mishra (1997) conclusions, the optimization algorithm may have difficulty converging on a design point. Alternative optimization techniques may be used, even if derivatives are not integral to the technique. For example, perhaps a genetic algorithm might be used to find the design point. Once a design point has been obtained, a single evaluation of derivatives at that design point could be used to evaluate the typical sensitivity measures revealed by FORM.

7.4.3 Local versus Global Sensitivity and Probabilistic Dissonance

Cooke and van Noortwijk (1999) present a comparison of MC and FORM results for evaluating the probability of failure for overtopping of dikes. They are particularly interested in the concept of global versus local sensitivity measures and how they might compare between the MC and FORM methods. Cooke and Van Noortwijk present an important new concept, which they call 'local probabilistic dissonance' in their paper. This behavior is explained as follows: the performance function $Z = g(\mathbf{X})$ might be strictly increasing as a function of one variable X_i, but for some specific value or neighborhood of Z, the conditional distribution of the variable X_i is decreasing in Z. Cooke and van Noortwijk point out that local sensitivity measures (such as a partial derivative) and, indeed, global sensitivity measures may both be a little misleading in such cases. Their example results show that local probabilistic sensitivity measures and FORM results may be significantly different! They introduce a graphical procedure for inspecting the joint distribution and the role of individual variables using what they call conditional percentile cobweb plots.

7.5 CONCLUDING REMARKS

FORM and SORM techniques provide an efficient method for estimating the probability associated with engineering component failure or reliability, and have been used in a number of applications throughout the realm of engineering and science. They are approximate methods, however, and provide a single estimate of probability for a particular level of failure, rather than a complete suite of simulation statistics as provided by Monte Carlo simulation. The FORM and SORM sensitivity information is extremely valuable for judging the importance of each uncertain variable and any deterministic parameters as related to the probabilistic outcome. Of particular usefulness are the gamma sensitivity measures, which evaluate the sensitivity of the probability estimate to equally likely changes in each uncertain variable. The gamma sensitivity measures are able to incorporate the effects of correlation and marginal distributions, which provides a more complete evaluation of sensitivity as compared with more traditional deterministic sensitivity measures.

8

Variance-Based Methods

Karen Chan, Stefano Tarantola and Andrea Saltelli

European Commission, Joint Research Centre, Ispra, Italy

Ilya M. Sobol'

Russian Academy of Sciences, Moscow, Russia

8.1 INTRODUCTION

This chapter is devoted to a class of global sensitivity analysis (SA) techniques that are known as variance-based methods. In particular, we focus on three such methods: correlation ratio, Sobol', and Fourier amplitude sensitivity test (FAST).

The idea of using variance as an indicator of importance for input factors is not new, and, in fact, it can be shown to underlie the regression-based methods (see Chapter 6). Other variance-based techniques such as the standardized regression coefficients (SRC), correlation coefficients (Pearson), and partial correlation coefficients (PCC) rely on the assumption that the output and input factors are near-linearly related, and their rank equivalents such as the standardized rank regression coefficients (SRRC), Spearman correlation, and partial rank correlation coefficients (PRCC) rely on the assumption that output and input are near-monotonically related. These techniques are described in Chapter 6.

Correlation ratios (McKay, 1995) and 'importance measures' (Hora and Iman, 1986) are derived from a simple description of uncertainty using probability distributions, and are based on the conditional variance of the model output. Although starting from quite a different setting, as will be shown in Section 8.2, both methods arrive at the same quantity, $\mathrm{Var}_X[E(Y|x)]$, as an indicator of importance for an input factor x. This quantity can be shown to be equivalent to the Sobol' and FAST first-order sensitivity indices. McKay (1997) shows that the regression-based methods are special cases of this class of variance-based methods.

The Fourier amplitude sensitivity test (FAST), created in the 1970s by Cukier, Schaibly and others, and further developed by Koda and McRae, offers a SA method that is independent of any assumptions about the model structure, and works for monotonic and non-monotonic

Sensitivity Analysis. Edited by A. Saltelli *et al.*

models. The core feature of FAST is that it explores the multidimensional space of the input factors by a search curve that scans the entire input space. Some variations of the basic scheme of FAST are also known; an example is given by the Walsh amplitude sensitivity procedure (WASP) (Pierce and Cukier, 1981), a method for discrete models where the factor variation is intrinsically two-valued. Saltelli *et al.* (1999b) propose a new FAST technique, which uses a new Fourier transformation function, and a resampling plan. Also introduced in that article is the use of the new FAST technique (the extended FAST) to compute the so-called 'total sensitivity indices (TSI)'.

The Sobol' sensitivity indices, an original extension of DOE to the world of numerical experiments first published in 1990, are similar to FAST, in the sense that the total variance of the model output is assumed to be made up of terms of increasing dimensionality. Sobol' indices are superior to the original FAST in that the computation of the higher interaction terms is very natural and is similar to the computation of the main effects. Each effect (main or otherwise) is computed by evaluating a multidimensional integral via a Monte Carlo (MC) method.

In this chapter, some sampling schemes usually associated with the methods are also described (see Section 8.5). Several analytical test cases are used to illustrate the methods, and to compare the techniques. Saltelli *et al.* (2000) offers a review of applications of these methods.

8.2 CORRELATION RATIOS/IMPORTANCE MEASURES

In this section, a simple measure of importance of an input factor x, or a set of inputs S_x based on the so-called *variance of the conditional expectation* (VCE) of prediction, is described. This variance-based technique stems from a simple expression of uncertainty using probability distributions.

Let us assume that we are interested in describing the importance of a single input x. The importance of x with regard to (prediction) uncertainty can be assessed by considering the conditional probability distributions of output y conditioned on x. This is because the marginal distribution of Y can be written in terms of the conditional distribution of Y given x, namely

$$p_Y(y) = \int p_{Y|x}(y|x)p_X(x)\,dx.$$

The notion that the importance of an input is related to how well it controls the model prediction is reasonable. Intuitively, x is important if fixing its value substantially reduces the (conditional) prediction variance relative to the marginal prediction variance.

The prediction variance is written in terms of conditional variance, without any assumptions about the functional relation between y and x. Hence, it is reasonable to use various (conditional) prediction variance ratios as appropriate measures of importance. We shall describe one such ratio in this section, but first we define the general setting within this class of (conditional) variance-based measures.

8.2.1 Prediction Variance

Given a set of k input factors $\mathbf{x}$, we define the general analysis model

$$y = E(Y|\mathbf{x}) + \varepsilon,$$

where $\boldsymbol{\varepsilon}$ denotes a vector of error terms, with the assumption that $E(\boldsymbol{\varepsilon}) = 0$ and $\mathrm{Cov}[E(Y|\mathbf{x}), \varepsilon] = \Sigma$. In numerical experiments, ε will be omitted. Note that no specific assumptions are made about the mathematical form of the conditional expectation $E(Y|\mathbf{x})$ or the covariance matrix Σ. In the standard linear regression analysis model, $E(Y|\mathbf{x})$ is assumed to be of the form $\mathbf{x}^{\mathrm{T}}\boldsymbol{\beta} = \sum_{i=1}^{k} x_i \beta_i$, where the β_i are the regression coefficients, which can be estimated by the standard least squares method.

The prediction variance of Y can be written as

$$\mathrm{Var}(Y) = \mathrm{Var}_{\mathbf{X}}[E(Y|\mathbf{x})] + E_{\mathbf{X}}(\mathrm{Var}[Y|\mathbf{x}]), \tag{8.1}$$

where

$$\mathrm{Var}_{\mathbf{X}}[E(Y|\mathbf{x})] = \int [E(Y|\mathbf{x}) - E(Y)]^2 p_{\mathbf{x}}(\mathbf{x})\, d\mathbf{x},$$

$$E_{\mathbf{X}}(\mathrm{Var}[Y|\mathbf{x}]) = \int\int [y - E(Y|\mathbf{x})]^2 p_{Y|\mathbf{x}}(y)\, dy\, p_{\mathbf{x}}(\mathbf{x})\, d\mathbf{x},$$

$$E(Y|\mathbf{x}) = \int y p_{Y|\mathbf{x}}(y)\, dy.$$

The two components of the variance decomposition of (8.1) are called the variance of the conditional expectation (VCE) and the residual part, respectively. The first term is the variance of the conditional expectation of Y, conditioned on $\mathbf{x}$, which is a suitable measure of the importance of $\mathbf{x}$ since it looks at the constituent parts in (8.1) to reveal the ways in which they relate to $\mathbf{x}$. The importance of $\mathbf{x}$ relates to how well $\mathbf{x}$ drives or controls y, that is, how well $E[Y|\mathbf{X} = \mathbf{x}]$ mimics y. In particular, if the total variation in y is matched by the variability in $E[Y|\mathbf{X} = \mathbf{x}]$ as $\mathbf{x}$ varies, this imples that $\mathbf{x}$ could be a very important set of inputs. That variation is measured by $\mathrm{Var}_{\mathbf{X}}[E(Y|\mathbf{x})]$. The second term is described as an error or residual term, measuring the remaining variability in y that is due to other unobserved inputs or other unknown sources of variation when $\mathbf{x}$ is fixed. In an example given later for a two-input-factor case, the residual term is equivalent to a measure of the effects of the complement of x and the interaction, if any, between X and the other factors.

8.2.2 Definition

The magnitude of VCE relative to prediction variance is measured by

$$\eta^2 = \frac{\mathrm{Var}_{\mathbf{X}}[E(Y|\mathbf{x})]}{\mathrm{Var}[Y]}, \tag{8.2}$$

and is termed the *correlation ratio* (McKay, 1995).

Prior to McKay (1995), the correlation ratio measure had been studied and applied by many investigators, including Kendall and Stuart (1979), Krzykacz (1990), and Hora and Iman (1986). In particular, derived from a slightly different setting (see the discussion of the resampling-based method below), Hora and Iman (1986) used $\sqrt{\mathrm{Var}\,[E(Y|X_i)]}$ in the analysis of fault trees, assuming a linear polynomial approximation for the conditional expectation of Y, and named it an *importance measure.* For numerical robustness reasons, Iman and Hora (1990) proposed the following to measure the importance of x_i:

$$\frac{\mathrm{Var}_{X_i}[E(\log Y|X_i)]}{\mathrm{Var}\,[\log Y]},$$

where Var_{X_i} denotes variance over all possible values of X_i and $E[\log Y|X_i]$ is estimated using linear regression. Saltelli *et al.* (1993) discussed a modified version of Hora and Iman (1986), which was later found to coincide with the method of Sobol' (1990b, 1993).

The idea of the correlation ratio can be extended to the partial correlation ratio, paralleling the partial correlation coefficient in linear models (for further details, see McKay, 1995). Rank-transformed versions of the correlation ratio or importance measures are discussed in McKay and Beckman (1994), Saltelli and Sobol' (1995), and Homma and Saltelli (1996). A brief description of the estimation of η^2 is given in the following section.

8.2.3 Computational Issues

By writing $\mathbf{x}$ as X_i and from (8.2), the correlation ratio for an input factor x_i is estimated by a ratio of two estimators, namely

$$\hat{\eta}^2 = \frac{\widehat{\mathrm{VCE}}(X_i)}{\widehat{\mathrm{Var}}[Y]}.$$

Hence, to estimate the correlation ratio for each input factor, we need to estimate $\mathrm{VCE}(X_i)$ separately for each input and $\mathrm{Var}[Y]$.

The following two subsections describe methods for estimating the correlation ratios or VCE for individual input factors. First is that of McKay (1995), who uses the sums of squares derived from analysis of variance to estimate η^2. The sampling plan is based on Latin hypercube sampling (LHS) (McKay *et al.*, 1979); see Chapter 6 for a description of the LHS methods. The second method is that of Saltelli *et al.* (1993), which provides a more effective computation of the VCE.

Variance-components-based method

Estimation of variance components for all of the individual inputs can be achieved by using a single r-LHS of size m with r replicates. An LHS of size m for k inputs is denoted by the matrix

$$\mathbf{D}_o = [\mathbf{X}_1 \quad \mathbf{X}_2 \quad \cdots \quad \mathbf{X}_k]. \tag{8.3}$$

Each column $\mathbf{X}_i$ is a m-dimensional vector containing values x_{ji}, for $j = 1, \ldots, m$, sampled from equiprobable intervals and randomized as to the position in the vector. A design matrix $\mathbf{D}$ for r replicates in an r-LHS_m is given by

$$\mathbf{D} = \begin{bmatrix} \mathbf{D}_1 \\ \mathbf{D}_2 \\ \vdots \\ \mathbf{D}_r \end{bmatrix},$$

where $\mathbf{D}_l = [\tilde{\mathbf{X}}_1^{(l)} \quad \tilde{\mathbf{X}}_2^{(l)} \quad \cdots \quad \tilde{\mathbf{X}}_k^{(l)}]$, for $l = 1, 2, \ldots, r$, and each $\tilde{\mathbf{X}}_i^{(l)}$ is an independent permutation of the rows of the vector $\mathbf{X}_i$ given in Equation (8.3), for $i = 1, \ldots, k$. The model outputs are denoted by $\{y_{jl}\}$, for $j = 1, \ldots, m$ and $l = 1, 2, \ldots, r$. In other words, each $\mathbf{D}_l$ yields a vector of m output values $\mathbf{y}_l$. The estimator of the output variance is given by

$$\widehat{\mathrm{Var}}[Y] = \frac{1}{mr} \sum_{j=1}^{m} \sum_{l=1}^{r} (y_{jl} - \bar{y})^2,$$

where $\bar{y} = (1/mr)\sum_{j=1}^{m}\sum_{l=1}^{r} y_{jl}$ is the grand mean. This estimator is termed the total mean square, namely total sum of squares (TSS) divided by the total sample size, mr.

McKay (1995) shows that an estimate of $\mathrm{VCE}(X_i)$ is given by

$$\widehat{\mathrm{VCE}}(X_i) = \frac{1}{m}\sum_{j=1}^{m}(\bar{y}_{j.} - \bar{y})^2 - \frac{1}{mr^2}\sum_{j=1}^{m}\sum_{l=1}^{r}(y_{jl}^{(i)} - \bar{y}_{j.})^2,$$

where $\{y_{jl}^{(i)}\}$ is obtained by fixing the m entries of the ith column, corresponding to the factor X_i, in all r replicates to the values taken from the ith column of the matrix $\mathbf{D}_o$.

Hence, the estimate of η^2 for X_i is

$$\hat{\eta}^2(X_i) = \frac{\frac{1}{m}\sum_{j=1}^{m}(\bar{y}_{j.} - \bar{y})^2 - \frac{1}{mr^2}\sum_{j=1}^{m}\sum_{l=1}^{r}(y_{jl}^{(i)} - \bar{y}_{j.})^2}{\frac{1}{mr}\sum_{j=1}^{m}\sum_{l=1}^{r}(y_{jl} - \bar{y})^2}$$

$$= \frac{r\sum_{j=1}^{m}(\bar{y}_{j.} - \bar{y})^2 - \frac{1}{r}\sum_{j=1}^{m}\sum_{l=1}^{r}(y_{jl}^{(i)} - \bar{y}_{j.})^2}{\sum_{j=1}^{m}\sum_{l=1}^{r}(y_{jl} - \bar{y})^2}. \tag{8.4}$$

The estimator in (8.4) has been adjusted for bias induced by the sample design, but has the undesirable property that it can yield negatives values. An example is given in Figure 8.1, in which the model is the Legendre polynomial (see Section 2.9 in Chapter 2) with two factors X_1 and X_2. The solid lines are analytical values of the first-order index of the two factors. The replicates are chosen to be $r = 4, 8, 16$, and 32; and the sample sizes are $m = 8, 16, 32$, and 64.

McKay (1997) gives a different formula for estimating η^2, namely

$$\hat{\eta}^2_{\text{bias}}(X_i) = \frac{r\sum_{j=1}^{m}(\bar{y}_{j.} - \bar{y})^2}{\sum_{j=1}^{m}\sum_{l=1}^{r}(y_{jl} - \bar{y})^2},$$

and shows that this estimator produced bias estimates, but by increasing the number of replicates the estimates slowly converge to the analytical values. This suggests that an accurate estimate of the expectation of Y, conditioned on x_i, is needed; which is intuitively appropriate since, as explained earlier, the importance of X_i relates to how well $E(Y|X_i)$ mimics y.

This method for estimating η^2 for all the factors is computationally expensive, since it requires $rm \times (k+1)$ model evaluations, including the mr model evaluations to estimate $E(Y)$. Another method that can be used to estimate VCE is that of Saltelli *et al.* (1993), who elaborate a computation scheme suggested by Ishigami and Homma (1990) to estimate the importance measure of factor X_i, denoted by I_i. In order to describe the estimation procedure, it is necessary to reproduce the derivation of I_i.

Resampling-based method

Let $Y = f(\mathbf{x})$ be the function of a set of input factors $\mathbf{X}$ and assume the joint probability distribution of $\mathbf{X}$, $p(\mathbf{x})$, to be the product of the marginal distributions of the X_i, i.e. the X_i

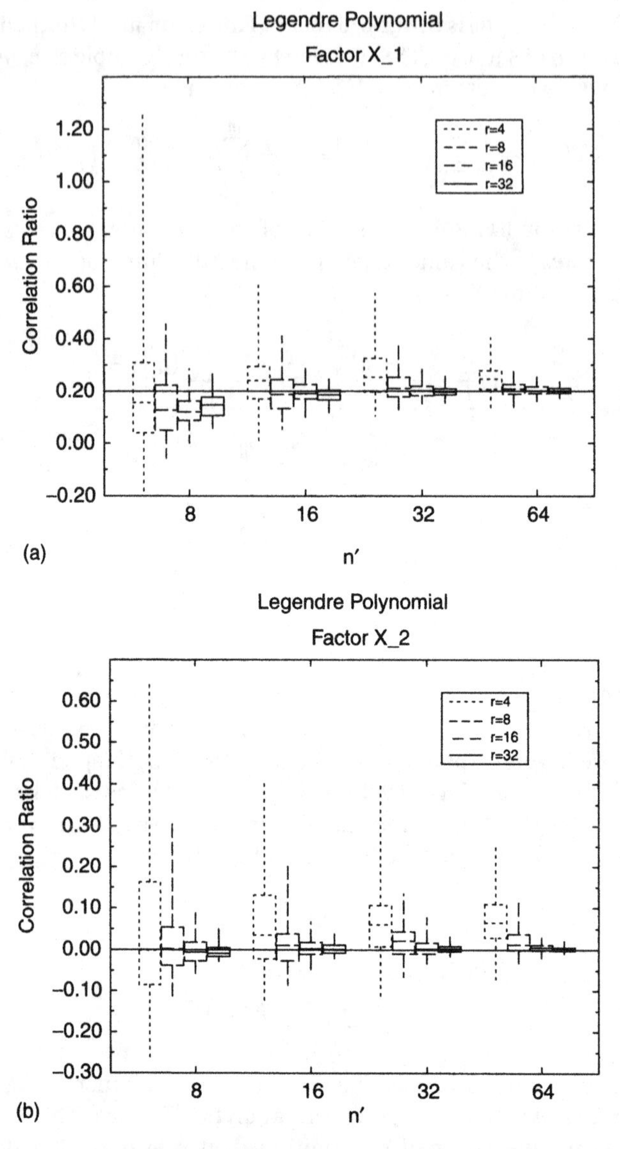

Figure 8.1 Boxplots of 100 estimates of the correlation ratios for (a) factor X_1 and (b) factor X_2 against the sample sizes m, with four replicates, $r = 4, 8, 16$, and 32. Horizontal lines are the respective analytical values, i.e. $S_1 = 0.2$ and $S_2 = 0.0$.

are independent; then the unconditional and conditional (on X_i) variances of Y are given by

$$\text{Var}(Y) = \int_{\Omega^k} f(\mathbf{x})^2 \prod_{j=1}^{k} p_j(x_j)\, dx_j - [E(Y)]^2 \tag{8.5}$$

and

$$\begin{aligned}\text{Var}(Y|X_i) &= \int \dots \int [y(x_i) - E(Y|X_i)]^2 \prod_{j \neq i} p_j(x_j)\, dx_j \\ &= \int \dots \int y(x_i)^2 \prod_{j \neq i} p_j(x_j)\, dx_j - [E(Y|X_i)]^2,\end{aligned} \tag{8.6}$$

respectively, where

$$E(Y|X_i) = \int \dots \int y(x_i) \prod_{j \neq i} p_j(x_j)\, dx_j$$

and

$$y(x_i) = f(X_1, \dots, X_{i-1}, x_i, X_{i+1}, \dots, X_k).$$

The dependence of the variance $\mathrm{Var}(Y|X_i)$ upon a specific value x_i can be eliminated by averaging $\mathrm{Var}(Y|X_i)$ according to $p(x_i)$, to yield

$$E[\mathrm{Var}(Y|X_i)] = \int \mathrm{Var}(Y|X_i) p_i(x_i)\, dx_i \tag{8.7}$$

Substituting (8.6) in (8.7) gives

$$E[\mathrm{Var}(Y|X_i)] = \int f(\mathbf{x})^2 \prod_{j=1}^{k} p_j(x_j)\, dx_j - \int [E(Y|X_i)]^2 p_i(x_i)\, dx_i. \tag{8.8}$$

By comparing Equations (8.5) and (8.8), we obtain the following relation:

$$\mathrm{Var}(Y) - E[\mathrm{Var}(Y|X_i)] = U_i - [E(Y)]^2, \tag{8.9}$$

where

$$U_i = \int [E(Y|X_i)]^2 p_i(x_i)\, dx_i. \tag{8.10}$$

Hora and Iman (1986) use

$$I_i = \sqrt{\mathrm{Var}(Y) - E[\mathrm{Var}(Y|X_i)]} = \sqrt{U_i - [E(Y)]^2}$$

as the importance measure. The right-hand side of Equation (8.9) is simply VCE, as defined earlier. Since $E(Y)$ is constant, i.e. it does not depend on X_i, the factor ranking is in fact based on the values of U_i. That is, factor l is more important than factor l' if $U_l > U_{l'}$. Hence, the importance measure is related to the computation of the integral in (8.10).

If there is no analytical solution to the integral in (8.10), a MC method can be applied. The number of simulations required for the computation of the average $E(Y|X_i)$ and the evaluation of the integral over all possible values of x_i is hence Mn, where M is the mesh size used to compute the integral in (8.10) and n is the sample size in evaluating $E(Y|X_i)$ for a specified value of x_i. This method is too expensive and impractical. Saltelli *et al.* (1993) adapt a computation scheme suggested by Ishigami and Homma (1989, 1990), and propose to evaluate U_i by

$$U_i = \frac{1}{n} \sum_{l=1}^{n} y_l y'_l, \tag{8.11}$$

where $y_l = f_l(\mathbf{x})$ is the lth sample output obtained from the base sample matrix $\{x_{ij}; i = 1, \dots, k, j = 1, \dots, n\}$ and $y'_l = f_l(x'_1, x'_2, \dots, x'_{i-1}, x_i, x'_{i+1}, x'_k)$ is the lth sample output evaluated from the resampling matrix for factor X_i; that is the y'_l are computed by resampling all the factors but the ith one. This scheme reduces the total number of model evaluations to $n(k+1)$, and is the same as that used in estimating the Sobol' first-order indices (see Section 8.3). As will be shown in an example, the measure of importance of individual input factors based on the correlation ratios or measures of importance are equivalent to the Sobol' and FAST first-order sensitivity indices.

8.3 METHOD OF SOBOL'

8.3.1 Sensitivity Indices

This variance-based MC method is due to Sobol' (1990b) (in Russian—Sobol' (1993) in English). In order to describe his approach, let us define the input factor space Ω^k as a k-dimensional unit cube, i.e. the region

$$\Omega^k = (\mathbf{x} | 0 \leqslant x_i \leqslant 1; \quad i = 1, \ldots, k).$$

The main idea behind the Sobol' approach is to decompose the function $f(\mathbf{x})$ into summands of increasing dimensionality, namely

$$f(x_1, \ldots, x_k) = f_0 + \sum_{i=1}^{k} f_i(x_i) + \sum_{1 \leqslant i < j \leqslant k} f_{ij}(x_i, x_j) + \cdots + f_{1,2,\ldots,k}(x_1, \ldots, x_k). \qquad (8.12)$$

In his earlier work, published in 1967, Sobol' based the decomposition (8.12) on multiple Fourier Haar series; in the 1990s, he offered a more general representation of the same decomposition using multiple integrals, and we summarize it as follows.

For (8.12) to hold, f_0 must be a constant, and the integrals of every summand over any of its own variables must be zero, i.e.

$$\int_0^1 f_{i_1,\ldots,i_s}(x_{i_1}, \ldots, x_{i_s})\, dx_{i_k} = 0 \quad \text{if } 1 \leqslant k \leqslant s. \qquad (8.13)$$

A consequence of (8.12) and (8.13) is that all the summands in (8.12) are orthogonal, i.e. if $(i_1, \ldots, i_s) \neq (j_1, \ldots, j_l)$, then

$$\int_{\Omega^k} f_{i_1,\ldots,i_s} f_{j_1,\ldots,j_l}\, d\mathbf{x} = 0. \qquad (8.14)$$

Since at least one of the indices will not be repeated, the corresponding integral will vanish due to (8.13). Another consequence is that

$$f_0 = \int_{\Omega^k} f(\mathbf{x})\, d\mathbf{x}.$$

Sobol' (1993) showed that the decomposition (8.12) is unique and that all the terms in (8.12) can be evaluated via multidimensional integrals, namely

$$f_i(x_i) = -f_0 + \int_0^1 \ldots \int_0^1 f(\mathbf{x})\, d\mathbf{x}_{\sim i},$$

$$f_{ij}(x_i, x_j) = -f_0 - f_i(x_i) - f_j(x_j) + \int_0^1 \ldots \int_0^1 f(\mathbf{x})\, d\mathbf{x}_{\sim(ij)},$$

with the convention that $d\mathbf{x}_{\sim i}$, $d\mathbf{x}_{\sim(ij)}$ denote integration over all variables except x_i, and x_i and x_j, respectively. Analogous formulae can be obtained for the higher-order terms. The variance-based sensitivity indices develop very naturally from this scheme; the total variance D of $f(\mathbf{x})$ is defined to be

$$D = \int_{\Omega^k} f^2(\mathbf{x})\, d\mathbf{x} - f_0^2, \tag{8.15}$$

while partial variances are computed from each of the terms in (8.12) as

$$D_{i_1,\ldots,i_s} = \int_0^1 \cdots \int_0^1 f^2_{i_1,\ldots,i_s}(x_{i_1},\ldots,x_{i_s})\, dx_{i_1} \ldots dx_{i_s} \tag{8.16}$$

where $1 \leqslant i_1 < \cdots < i_s \leqslant k$ and $s = 1,\ldots,k$. By squaring and integrating (8.12) over Ω^k, and by (8.14), we obtain

$$D = \sum_{i=1}^{k} D_i + \sum_{1 \leqslant i < j \leqslant k} D_{ij} + \cdots + D_{1,2,\ldots,k}. \tag{8.17}$$

Hence, the sensitivity measures $S_{i_1,\ldots,i_s}$ are given by

$$S_{i_1,\ldots,i_s} = \frac{D_{i_1,\ldots,i_s}}{D} \quad \text{for } 1 \leqslant i_1 < \cdots < i_s \leqslant k, \tag{8.18}$$

where S_i is called the *first-order sensitivity index* for factor x_i, which measures the main effect of x_i on the output (the fractional contribution of x_i to the variance of $f(\mathbf{x})$), S_{ij}, for $i \neq j$, is called the second-order sensitivity index which measures the interaction effect (the part of the variation in $f(\mathbf{x})$ due to x_i and x_j that cannot be explained by the sum of the individual effects of x_i and x_j), and so on. The decomposition in (8.17) has the useful property that all the terms in (8.18) sum to 1; that is,

$$\sum_{i=1}^{k} S_i + \sum_{1 \leqslant i < j \leqslant k} S_{ij} + \cdots + S_{1,2,\ldots,k} = 1.$$

It can be noted that Equation (8.17) is similar to the ANOVA decomposition scheme (Archer *et al.*, 1997), apart from the missing residual error term. Furthermore, the terms defined in (8.18) are very similar to the 'effects' that are computed in experimental design theory (see Chapter 3).

Apart from the FAST method described in Section 8.4, decompositions similar to (8.12) and (8.17) are discussed in Cotter (1979) and Sacks *et al.* (1989a). Sacks *et al.* (1989a), in particular, plotted the individual $f_{i_1,\ldots,i_s}$ terms in (8.12) and used the plots to investigate the influence of the various factors on the output. This kind of analysis can give a valuable insight into the problem, but it becomes impractical for systems with many factors, especially when the interaction terms are to be computed. A different high-dimensional model representation (HDMR) is discussed in Chapter 9.

8.3.2 Total Effect Indices

The total sensitivity index (TSI) is defined as the sum of all the sensitivity indices involving the factor in question. For example, suppose that we have three factors in our model, the total effect of factor 1 on the output variance, denoted by TS(1), is given by $\mathrm{TS}(1) = S_1 + S_{12} + S_{13} + S_{123}$, where S_1 is the first-order sensitivity index for factor 1, S_{1j} is the second-order sensitivity index for the two of factors 1 and $j\,(\neq 1)$, i.e. the interaction between factors 1 and $j\,(\neq 1)$, and so on.

This measure is derived from a notion of Sobol', whose problem was that of 'freezing' the unimportant factors to their midpoint (Sobol', 1993). The k factors are divided into two subsets, and we treat each subset as a new factor. For example, $\mathbf{x}$ can be partitioned into $\mathbf{v}$ and $\mathbf{w}$, where $\mathbf{v}$ contains factors from x_1 to x_t, and $\mathbf{w}$ the remaining $k-t$ factors. Let consider a generalized ANOVA decomposition of $f(\mathbf{x})$,

$$f(\mathbf{x}) = f_0 + f_1(\mathbf{v}) + f_2(\mathbf{w}) + f_{12}(\mathbf{v}, \mathbf{w}),$$

with

$$\int f_1\, d\mathbf{v} = \int f_2\, d\mathbf{w} = \int f_{12}\, d\mathbf{v} = \int f_{12}\, d\mathbf{w} = 0,$$

and

$$D_{\mathbf{v}} = \int f_1^2\, d\mathbf{v}, \qquad D_{\mathbf{w}} = \int f_2^2\, d\mathbf{w}, \qquad D_{\mathbf{vw}} = \int f_{12}^2\, d\mathbf{v}\, d\mathbf{w}.$$

Then, the total variance of the output is given by

$$D = D_{\mathbf{v}} + D_{\mathbf{w}} + D_{\mathbf{vw}}.$$

If $D_{\mathbf{v}}/D$ is high (say 0.8 or higher) then the factors in $\mathbf{w}$ can be fixed; otherwise we need to compute the total effect of $\mathbf{v}$ on the output, which can be measured by

$$D_{\mathbf{v}}^{\mathrm{tot}} = D_{\mathbf{v}} + D_{\mathbf{vw}} = D - D_{\mathbf{w}}.$$

Hence, the total sensitivity index for $\mathbf{v}$ is defined as

$$TS(\mathbf{v}) = D_{\mathbf{v}}^{\mathrm{tot}}/D.$$

The above approach can be extended to measure the individual influence of all factors, including the interaction effects between the factors (Homma and Saltelli, 1996). It is achieved by partitioning $\mathbf{x}$ into $\mathbf{x}_{\sim i}$ and x_i; then one can compute with just one MC integral the total effect term $\mathrm{TS}(i)$:

$$\mathrm{TS}(i) = S_i + S_{i(\sim i)} = 1 - S_{\sim i},$$

where $S_{\sim i}$ is the sum of all the $S_{i_1,\ldots,i_s}$ terms that do not include the index i, i.e. the total fractional variance complement to factor x_i, $D_{\sim i}$. Thus, the total contribution of factor x_i to the total output variation is given by

$$\mathrm{TS}(i) = 1 - \frac{D_{\sim i}}{D}. \tag{8.19}$$

Note that the computation of the $\mathrm{TS}(i)$ does not provide a complete characterisation of the system (since this could only be achieved by computing all the $2^k - 1$ sensitivity indices), but it is much more reliable than the first-order indices, in order to investigate the overall effect of each single factor on the output variable.

The rank transformation described in Chapter 6 can also be used when sensitivity indices are employed. In this case, Saltelli and Sobol' (1995) show that the main result of the rank transformation is to increase the relative weight of the first-order terms, representing the linear effects, at the expense of the higher-order ones. As a consequence, the influence of those factors whose 'total effect' mostly arises from their interactions with other factors, may be overlooked. Although rank transformation is a very useful tool in uncertainty and sensitivity analysis, being able to cope with nonlinear and non-monotonic models, the limitations linked to this simplification of the original model have to be kept in mind.

8.3.3 Alternative Sensitivity Estimates

Jansen *et al.* (1994) proposed a new measure to compute the main effects, namely

$$D_i^J = D - \tfrac{1}{2}E[f(X_i, \mathbf{X}_{\sim i}) - f(X_i, \mathbf{X}'_{\sim i})]^2, \tag{8.20}$$

and the following to compute the total effect:

$$D_i^{J\text{tot}} = \tfrac{1}{2}E[f(X_i, \mathbf{X}_{\sim i}) - f(X'_i, \mathbf{X}_{\sim i})]^2. \tag{8.21}$$

In the other words, Jansen *et al.* (1994) used the mean-square difference to compute the indices whereas the Sobol' indices use the product. This mean-square difference has also discussed in Šaltenis and Dzemyda (1982). Jansen (1996a,b) showed that

$$\tfrac{1}{2}E[f(X_i, \mathbf{X}_{\sim i}) - f(X_i, \mathbf{X}'_{\sim i})]^2 = D - \mathrm{Cov}[f(X_i, \mathbf{X}_{\sim i}), f(X_i, \mathbf{X}'_{\sim i})]. \tag{8.22}$$

Also, it can be shown that

$$\tfrac{1}{2}E[f(X_i, \mathbf{X}_{\sim i}) - f(X'_i, \mathbf{X}_{\sim i})]^2 = D - \mathrm{Cov}[f(X_i, \mathbf{X}_{\sim i}), f(X'_i, \mathbf{X}_{\sim i})]. \tag{8.23}$$

The two covariances in Equations (8.22) and (8.23) are equivalent to those of the Sobol' first-order and total partial variances, since they can be computed by a MC method, namely

$$\int f(x_i, \mathbf{x}_{\sim i}) f(x_i, \mathbf{x}'_{\sim i})\, dx_i\, d\mathbf{x}_{\sim i}\, d\mathbf{x}'_{\sim i} - f_0^2 = D_i$$

and

$$\int f(x_i, \mathbf{x}_{\sim i}) f(x'_i, \mathbf{x}_{\sim i})\, dx_i\, dx'_i\, d\mathbf{x}_{\sim i} - f_0^2 = D_{\sim i},$$

respectively. Furthermore, it can be shown that

$$\mathrm{Var}[D_i^{J\text{tot}}] \leqslant \mathrm{Var}[D_i^{\text{tot}}], \tag{8.24}$$

while

$$\mathrm{Var}[D_i^J] \geqslant \mathrm{Var}[D_i]. \tag{8.25}$$

8.3.4 Computational Issues

One attractive feature of the Sobol' indices is that the integrals in (8.15) and (8.16) can be computed with the same kind of MC integral. Hence, the MC estimates of f_0, D, and D_i are given by the following formulae:

$$\hat{f}_0 = \frac{1}{n}\sum_{m=1}^{n} f(\mathbf{x}_m), \tag{8.26}$$

$$\hat{D} = \frac{1}{n}\sum_{m=1}^{n} f^2(\mathbf{x}_m) - \hat{f}_0^2, \tag{8.27}$$

$$\hat{D}_i = \frac{1}{n}\sum_{m=1}^{n} f(\mathbf{x}_{(\sim i)m}^{(1)}, x_{im}^{(1)}) f(\mathbf{x}_{(\sim i)m}^{(2)}, x_{im}^{(1)}) - \hat{f}_0^2. \tag{8.28}$$

In Equations (8.26)–(8.28), n is the number of samples generated to obtain the MC estimates, $\mathbf{x}_m$ is a sampled point in Ω^k, and

$$\mathbf{x}_{(\sim i)m} = (x_{1m}, x_{2m}, \ldots, x_{(i-1)m}, x_{(i+1)m}, \ldots, x_{km}).$$

The superscripts (1) and (2) in (8.28) indicate that we are using two sampling data matrices for $\mathbf{x}$. Both matrices have dimension $n \times k$. Hence, (8.28) says that in computing $\hat{D}_i$, we multiply values of f corresponding to $\mathbf{x}$ from matrix (1) by values of f computed using a different matrix (2), but for the ith column, which is kept constant. This may offer an intuitive justification of why S_i should be high if x_i is an important factor: if this is the case, in fact, high f values are multiplied by high f values. When x_i is unimportant, low and high f values tend to be randomly multiplied by each other, and the resulting S_i is lower. Note that the formulation of the MC estimate of D_i given in (8.28) and of U_i given in (8.11) are the same.

Formulae similar to (8.28) can be derived for the partial variances of higher order (see Homma and Saltelli, 1996). A drawback of the method is that a separate MC integral is needed to compute any effect, be it of first or higher order. Counting also the set of model evaluations needed to obtain $\hat{f}_0$, a total of 2^k MC integrals are needed for a full characterization of the system, which is far too many unless k is low.

The integral that is needed for the computation of $S_{\sim i}$ is estimated by the MC integral

$$\hat{D}_{\sim i} + \hat{f}_0^2 = \frac{1}{n}\sum_{m=1}^{n} f(\mathbf{x}_{(\sim i)m}^{(1)}, x_{im}^{(1)}) f(\mathbf{x}_{(\sim i)m}^{(1)}, x_{im}^{(2)}), \tag{8.29}$$

where the superscripts (1) and (2) are defined as above. Hence, the estimated TSI is given by

$$\widehat{\mathrm{TS}}(i) = 1 - \frac{\hat{D}_{\sim i}}{\hat{D}}, \tag{8.30}$$

where $\hat{D}_{\sim i}$ is given in (8.29) and $\hat{D}$ is given in (8.27). The number of MC integrals to be performed is now only equal to $k+1$, that is, the number of factors plus one (for $\hat{f}_0$).

Homma and Saltelli (1996) suggested the following correction term to improve the Sobol' sensitivity indices estimate:

$$\frac{1}{n}\sum_{m=1}^{n} f(\mathbf{x}_m^{(1)}) f(\mathbf{x}_m^{(2)}) - \hat{f}_0^2,$$

which is applied to all the terms $D_{i_1 \ldots i_s}$ in Equation (8.16). With this correction term, an extra n model evaluations are needed. In the example below, we have used this correction term for computing the first order indices.

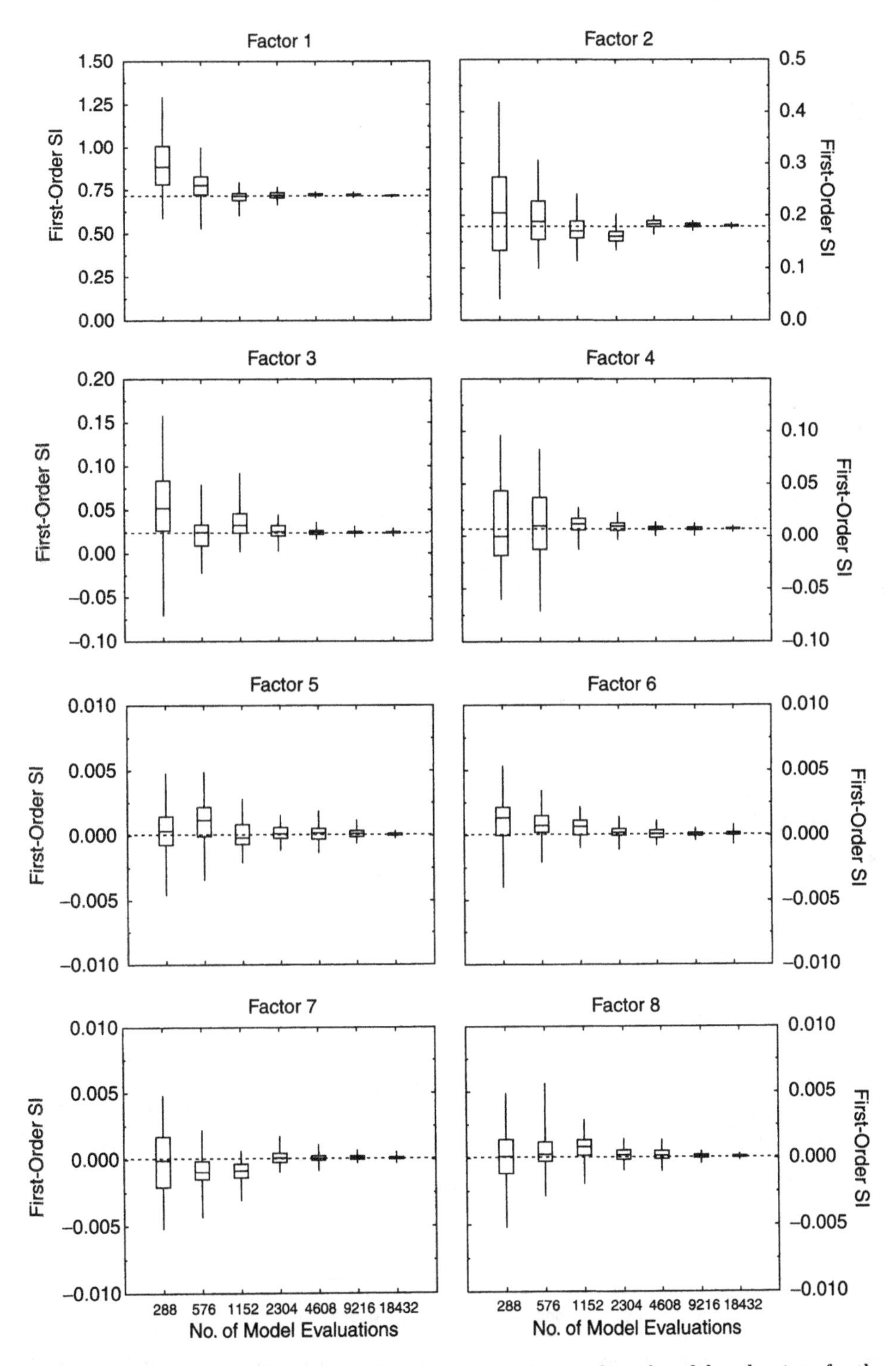

Figure 8.2 Boxplots of 100 first-order estimates against the number of model evaluations for the g-function test case with eight factors; analytical values of the sensitivity indices are shown by dotted lines.

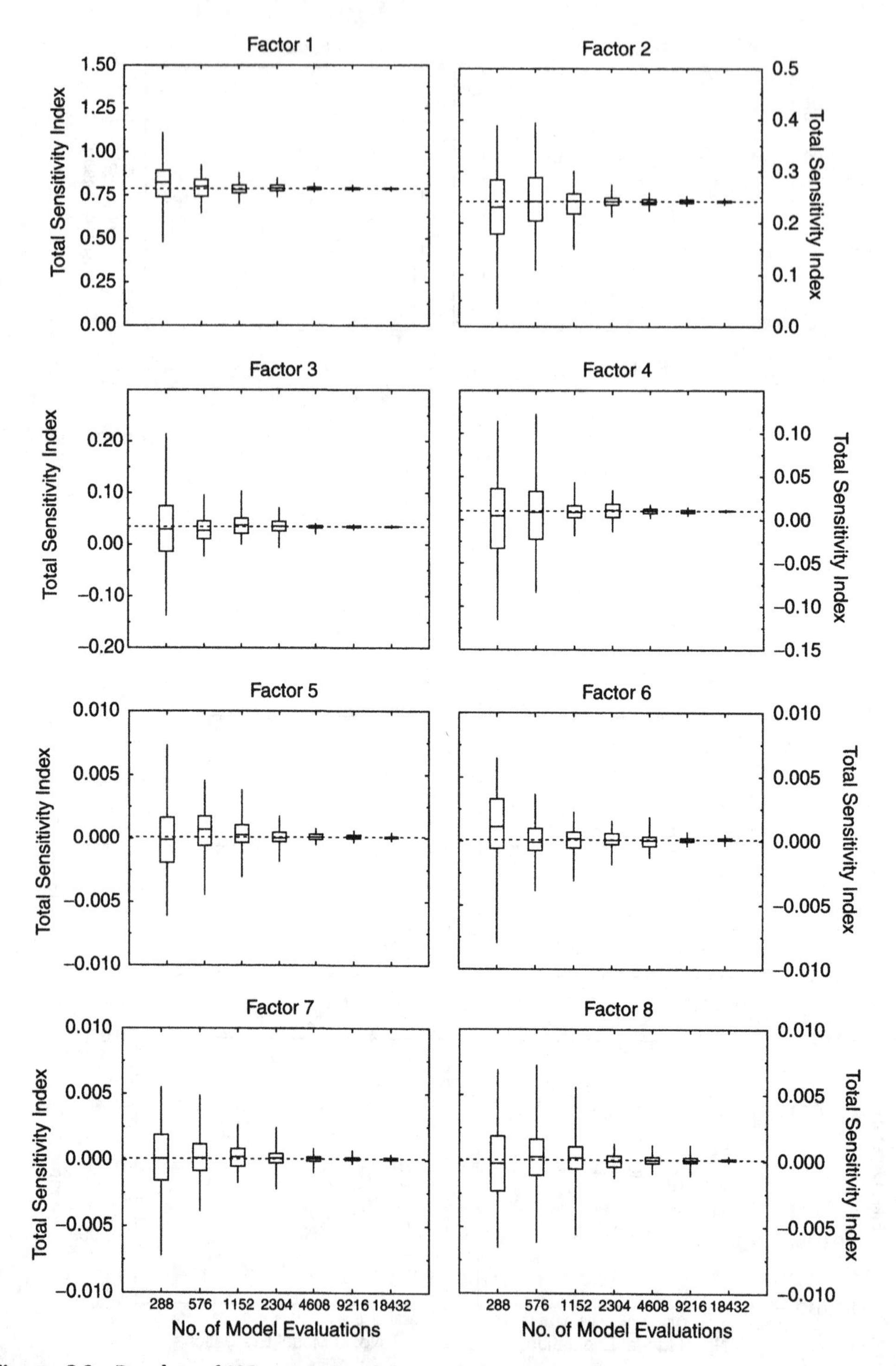

Figure 8.3 Boxplots of 100 estimates of the total sensitivity index against the number of model evaluations for the *g*-function test case with eight factors; analytical values of the sensitivity indices are shown by dotted lines.

Figures 8.2 and 8.3 show an example of the Sobol' method applied to an analytical test case, the g-function (Section 2.9 in Chapter 2). Here we have used eight factors, with the parameter a_i associated with each factor being $\{0, 1.0, 4.5, 9, 99, 99, 99, 99\}$. This choice implies that the first factor is the most important, the next factor is the second most important, ..., and the last four factors are equally unimportant. One hundred estimates of the first-order and total indices are computed with different sample sizes, i.e. $\{32, 64, 128, \ldots, 2048\}$; the replicates are obtained by the scrambled Sobol' LP_τ method (see Section 8.5). The boxplots of their summary statistics, for each factor, are plotted against the number of model evaluations, and are shown in Figures 8.2 and 8.3, respectively. The analytical values of the indices for each factor are plotted as dotted lines. First of all, for both set of indices, the estimates converge to the analytical values and the precision of the estimates increases as the number of model evaluations increases.

At small sample size, the estimates of the SIs and the TSIs vary greatly around the analytical values, and at times can take on negative values. This negative-value phenomenon is due to numerical computation, and if this happens then we usually set the index to zero. Also, at very low sample size, the Sobol' first-order indices can produce values greater than 1 (see factor 1 in Figure 8.2, for example).

The random data matrix mentioned previously is generated using quasi-random numbers for the computation of MC integrals. Quasi-random numbers are characterized by an enhanced convergence (Sobol', 1990a) under certain limitations (Davis and Rabinowitz, 1984; Bratley and Fox, 1988; Sobol' *et al.*, 1992). Note that other sampling strategies, such as Latin hypercube sampling (LHS) (see Chapter 6), can be used to compute sensitivity indices, but Homma and Saltelli (1995) found that LP_τ sequences performed better than those they have studied. The Sobol' LP_τ and other related sampling schemes are described in Section 8.5.

8.4 THE FAST METHOD

The Fourier amplitude sensitivity test (FAST) is a procedure that has been developed for uncertainty and sensitivity analysis (Cukier *et al.*, 1973, 1975, 1978; Schaibly and Shuler, 1973). This procedure provides a way to estimate the expected value and variance of the output variable and the contribution of individual input factors to this variance. An advantage of FAST is that the evaluation of sensitivity estimates can be carried out independently for each factor using just a single set of runs because all the terms in a Fourier expansion are mutually orthogonal.

The main idea behind the FAST method is to convert the k-dimensional integral in $\mathbf{x}$ into a one-dimensional integral in s by using the transformation functions G_i for $i = 1, \ldots, k$, namely

$$x_i = G_i(\sin \omega_i s), \tag{8.31}$$

where $s \in (-\pi, \pi)$ is a scalar variable and $\{\omega_i\}$ is a set of integer angular frequencies.

For properly chosen ω_i and G_i, the expectation of Y can be approximated by

$$E(Y) \doteq \frac{1}{2\pi} \int_{-\pi}^{\pi} f(s)\, ds, \tag{8.32}$$

where $f(s) = f(G_1(\sin \omega_1 s), \ldots, G_k(\sin \omega_k s))$.

By using the properties of Fourier series (Cukier *et al.*, 1973), an approximation of the variance of Y is given by

$$\begin{aligned}\mathrm{Var}(Y) &\doteq \frac{1}{2\pi}\int_{-\pi}^{\pi} f^2(s)\,ds - [E(Y)]^2 \\ &\approx \sum_{j=-\infty}^{\infty}(A_j^2 + B_j^2) - (A_0^2 + B_0^2) \\ &\approx 2\sum_{j=1}^{\infty}(A_j^2 + B_j^2),\end{aligned} \tag{8.33}$$

where A_j and B_j are the Fourier coefficients and are defined as follows:

$$A_j = \frac{1}{2\pi}\int_{-\pi}^{\pi} f(s)\cos js\,ds \tag{8.34}$$

and

$$B_j = \frac{1}{2\pi}\int_{-\pi}^{\pi} f(s)\sin js\,ds. \tag{8.35}$$

The expressions in (8.32) and (8.33) provide a means to estimate the expected value and variance associated with Y.

Application of the FAST method involves defining the ω_i and G_i, and evaluating the original model at a sufficient number of points to allow numerical evaluation of the integrals in (8.34) and (8.35). In Section 8.4.1, we shall describe some choices for G_i.

8.4.1 FAST Sampling

A suitable transformation G_i given in (8.31) should provide a uniformly distributed sample for each factor x_i, $\forall i = 1, 2, \ldots, k$ in the unit cube Ω^k. A number of transformations have been proposed; see Table 8.1 for examples.

Transformation (A) was proposed by Cukier *et al.* (1973), and is plotted in Figure 8.4(a), with $\bar{x}_i = e^{-5}, \bar{\nu}_i = 5$, and $\omega = 11$. Transformation (B), suggested by Koda *et al.* (1979), is plotted in Figure 8.4(c), with $\bar{x}_i = \frac{1}{2}, \nu_i = 1$, and $\omega = 11$. Saltelli *et al.* (1999b) proposed transformation (C), which is a set of straight lines oscillating between 0 and 1 (Figure 8.4(e)).

For a given transformation, the search curve oscillates over the range of s. As s varies, all the factors change simultaneously and their range of uncertainty is systematically

Table 8.1 Various transformations used in Equation (8.31).

Transformation	G_i	Reference
(A)	$x_i = \bar{x}_i e^{\bar{\nu}_i \sin \omega_i s}$	Cukier *et al.* (1973)
(B)	$x_i = \bar{x}_i(1 + \bar{\nu}_i \sin \omega_i s)$	Koda *et al.* (1979)
(C)	$x_i = \frac{1}{2} + \dfrac{1}{\pi}\arcsin(\sin \omega_i s)$	Saltelli *et al.* (1999b)
(D)	$x_i = \frac{1}{2} + \dfrac{1}{\pi}\arcsin[\sin(\omega_i s + \varphi_i)]$	Saltelli *et al.* (1999b)

Note: $\bar{x}_i$ denotes the nominal value of the factor x_i; $\bar{\nu}$ denotes the endpoints that define the estimated range of uncertainty of x_i; s takes values between $-\frac{1}{2}\pi$ and $\frac{1}{2}\pi$; and φ_i is a random phase-shift parameter taking values in $[0, 2\pi)$.

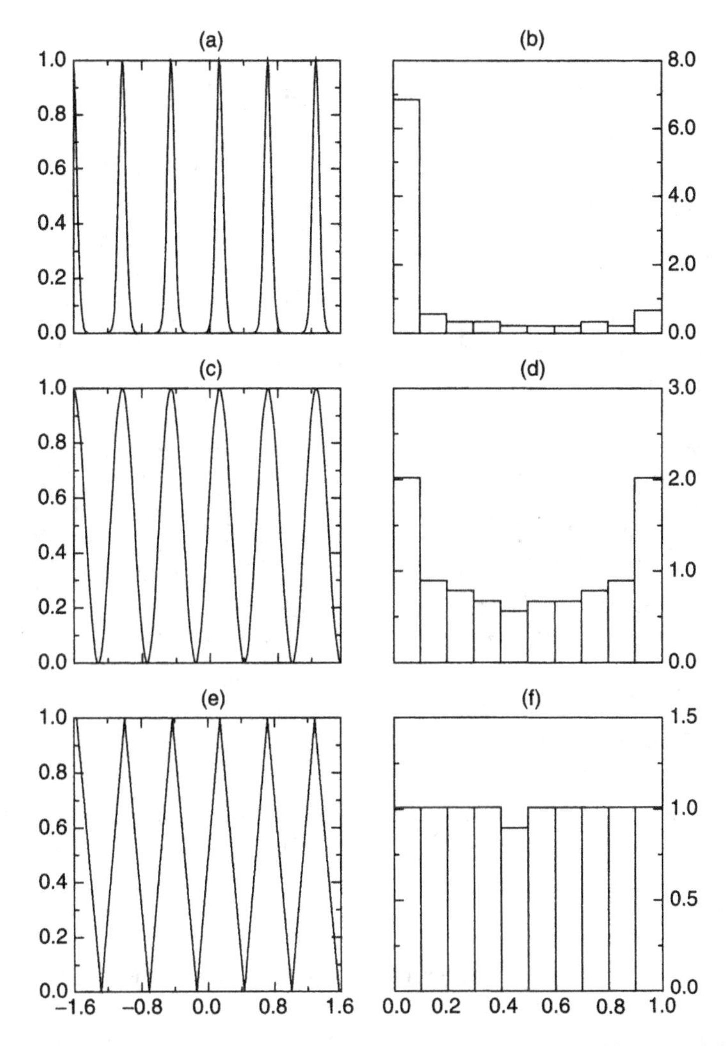

Figure 8.4 (a), (c), (e) Plots of three transformation functions given in Table 8.1 and (b), (d), (f), their respective empirical distributions.

explored. The curve drives arbitrarily close to any point **x** of the input domain *if and only if* a set of incommensurate ω_i frequencies is used. A set of frequencies is said to be *incommensurate* if none of them may be obtained as a linear combination of the other frequencies with integer coefficients. If this is the case then we say that the curve is space-filling.

The respective histograms of 89 sample points based on these three transformations are plotted in Figures 8.4(b), 8.4(d), and 8.4(f). Clearly, transformations (A) and (B) fail to provide uniformly distributed samples, in comparison with transformation (C). Note that transformation (C) is a special case of a more general differential form, proposed by Cukier *et al.* (1978), namely

$$\pi(1 - x_i^2)^{1/2} p_i(G_i) \frac{dG_i(x_i)}{dx_i} = 1,$$

where $p_i(\cdot)$ is the pdf of X_i.

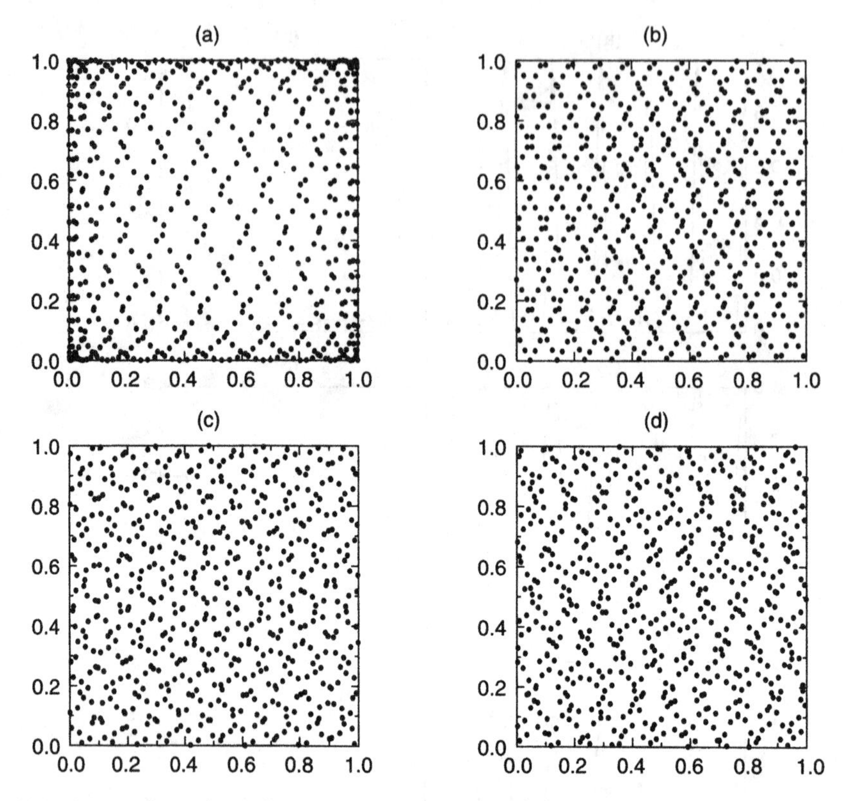

Figure 8.5 Scatterplots of sampling points in a two-factor case based on (a) transformation (B), (b) transformation (C), and (c,d) transformation (D) with one and two resampling curves (shown in (c) and (d), respectively).

Figures 8.5(a) and 8.5(b) illustrate differences between transformation (B) and transformation (C) in a two-factor case with $\{\omega_1, \omega_2\} = \{11, 21\}$. As can be seen, the sample points obtained by the latter transformation are more uniformly spread in the unit square than those generated by transformation (B).

Saltelli *et al.* (1999b) further proposed a random-phase-shift version of transformation (C), which is a curve defined by a set of parametric equations (transformation (D) in Table 8.1). The advantage of transformation (D) is that the starting point of the search curve can be anywhere within Ω^k, as shown in Figure 8.5(c), hence providing replicate samples of the same transformation. Figure 8.5(d) shows an example obtained by using two sampling curves with the same number of sample points.

8.4.2 First-Order Indices (Classical FAST)

The computation of the first-order indices is well documented. They are computed by evaluating the A_j and B_j for the fundamental frequency ω_i, for $i = 1, 2, \ldots, k$ and its higher harmonics—a periodic function with period $2\pi/\omega_0$ has non-zero spectral components at the fundamental frequency ω_0 and at all its higher harmonics $2\omega_0, 3\omega_0, \ldots$—denoted by $p\omega_i$ for $p = 1, 2, \ldots$. Provided that the ω_i are integers, the contribution to total variance of Y by

X_i can be approximated by

$$D_{\omega_i} \approx 2\sum_{p=1}^{\infty}(A_{p\omega_i}^2 + B_{p\omega_i}^2).$$

Furthermore, the Fourier amplitudes decrease as p increases; hence, we may further approximate D_{ω_i} by

$$\hat{D}_{\omega_i} = 2\sum_{p=1}^{M}(A_{p\omega_i}^2 + B_{p\omega_i}^2), \tag{8.36}$$

where M is the maximum harmonic we consider, and is usually taken to be 4 or 6 (Cukier *et al.*, 1975).

The ratios $\hat{D}_{\omega i}/\hat{D}^{\text{FAST}}$, denoted by S_i^{FAST}, provide a way to rank individual factors on the basis of their contribution to the variance of Y, D^{FAST}, given in Equation (8.33). Saltelli and Bolado (1998) showed that S_i^{FAST} is equivalent to the Sobol' sensitivity indices of the first order, S_i.

The computation of the partial variances $D_{\omega i}$ also requires the choice of the frequencies ω_i and the number of sample points. It can be shown that the minimum sample size (Saltelli *et al.*, 1999b) required to compute $D_{\omega i}$ is

$$N_s = 2M\omega_{\max} + 1,$$

where $\omega_{\max}$ is the maximum frequency amongst the set of ω_i. Cukier *et al.* (1975) derived an empirical algorithm to choose a set of ω_i for a given number of factors; but the algorithm is somewhat restricted as the number of factors increases. In Section 8.4.4, we shall describe an automated algorithm to select the ω_i.

8.4.3 Total Indices (Extended FAST)

The computation of the total indices using FAST was proposed by Saltelli *et al.* (1999b). The basic idea behind the computation of the total indices by the FAST method is to consider the frequencies that do not belong to the set $\{p_1\omega_1, p_2\omega_2, \ldots, p_k\omega_k\}$, for $p_i = 1, 2, \ldots, \infty$ and $\forall i = 1, 2, \ldots, k$. These frequencies contain information about the residual variance

$$D - \sum_{i}^{k} D_i$$

that is not accounted for by the first-order indices, that is, including the interactions between the factors at any order.

We assign a frequency ω_i for the factor X_i and a set of almost identical frequencies, but different from ω_i, to all the remaining factors, denoted by $\omega_{\sim i}$. As in Section 8.3.2, we use $\sim i$ to represent 'all but i'. Then, by evaluating the spectrum at the frequencies $\omega_{\sim i}$ and their higher harmonics $p\omega_{\sim i}$, we can compute the partial variance $D_{\sim i}$:

$$\hat{D}_{\sim i}^{\text{FAST}} = 2\sum_{p=1}^{M}(A_{p\omega_{\sim i}}^2 + B_{p\omega_{\sim i}}^2). \tag{8.37}$$

$\hat{D}_{\sim i}$ is a measure including all the effects of any orders that do not involve the factor X_i. As in the Sobol' indices, the total indices TS(i) are computed by using the formula given in Equation (8.19), namely $\text{TS}(i) = 1 - \hat{D}^{\text{FAST}}_{\sim i}/\hat{D}^{\text{FAST}}$.

8.4.4 Choice of Frequencies ω_i

Usually, a high value is assigned to ω_i and a low one to all the $\omega_{\sim i}$. For example, Figure 8.6 shows a plot of an artificial spectrum with $\omega_i = 20$ and $\omega_{\sim i} = 1$. The components at $\{p\omega_i\} = \{20, 40, 60, \ldots\}$ contribute to the computation of D_i, and decrease rapidly in amplitude as p increases. The partial variance of $X_{\sim i}$, $D_{\sim i}$, can be estimated from the first few spectral components, since the higher harmonics of $\omega_{\sim i}$ usually converge to zero after a few terms. The information about the term $D_{i(\sim i)}$, which measures the interaction between X_i and $X_{\sim i}$, can be observed at all the other frequencies within $[1, M\omega_{\max}]$, in particular around each harmonic $p\omega_i$ for $p = 1, 2, \ldots, M$, where M is the number of terms in the partial variances summation (see Equations (8.36) and (8.37)), and is usually set to 4, and $\omega_{\max}$ is the maximum frequency amongst the ω_i.

This approach has two advantages: (i) for each factor X_i, only two frequencies are needed; and (ii) the problem of interference is avoided. Interference is a problem when information provided by the frequencies corresponding to D_i and $D_{\sim i}$ is mixed; hence, the problem could lead to overestimating all the D_i. Saltelli *et al.* (1999b) proposed an automated algorithm to select the frequencies for ω_i and $\omega_{\sim i}$. The algorithm is as follows:

First, we set the maximum allowable frequency for the complementary factors $X_{\sim i}$, i.e.

$$\max\{\omega_{\sim i}\} = \frac{\omega_i}{2M},$$

where M was defined previously. Then, the frequencies for the complementary set are chosen so that the entire range $[1, \max\{\omega_{\sim i}\}]$ is covered, according to the following:

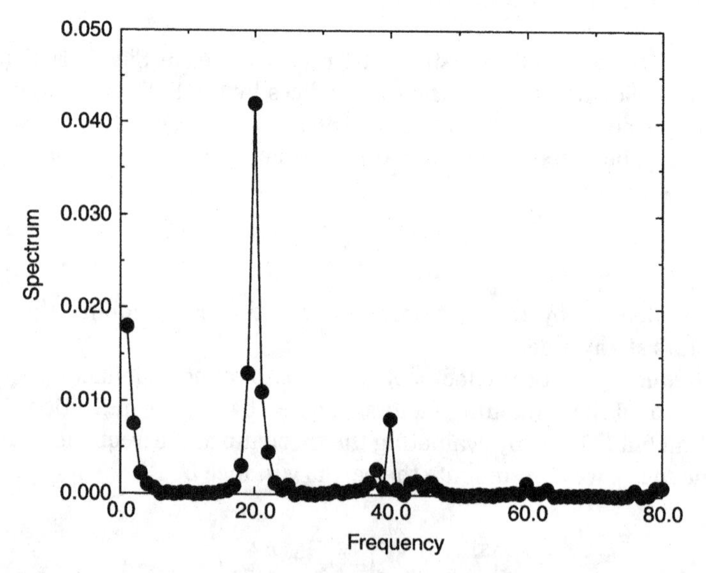

Figure 8.6 A plot of the spectrum against frequency, with $\omega_i = 20$ and $\omega_{\sim i} = 1$.

Table 8.2 Examples of complementary frequencies obtained by the automated algorithm at different sample sizes. The assumed number of factors is 8 and the factor of interest is X_4.

Sample size N_s	Frequency ω_4 for factor X_4	Complementary frequencies		Step size
		$\max\{\omega_{\sim 4}\}$	$\{\omega_1, \omega_2, \omega_3, \omega_5, \omega_6, \omega_7, \omega_8\}$	
65	8	1	$\{1, 1, 1, 1, 1, 1, 1\}$	0
129	16	2	$\{1, 2, 1, 1, 2, 1, 2\}$	1
257	32	4	$\{1, 2, 3, 1, 2, 3, 4\}$	1
513	64	8	$\{1, 2, 3, 5, 6, 7, 8\}$	1
1025	128	16	$\{1, 3, 5, 9, 11, 13, 15\}$	2
2049	256	32	$\{1, 5, 9, 17, 21, 25, 29\}$	4
4097	512	64	$\{1, 9, 17, 33, 41, 49, 57\}$	8
8193	1024	128	$\{1, 17, 33, 65, 81, 97, 113\}$	16

- the step between two consecutive frequencies must be as large as possible, and
- the number of factors to which the same frequency is assigned must be as low as possible.

This means that $D_{\sim i}$ can easily be estimated from the first few spectral components. For example, Table 8.2 shows the choice of frequencies for a eight-factor case for different sample sizes N_s. Here, we use N_s to denote the sample size used in FAST, since it is slightly different from that used in Sobol'.

Figures 8.7 and 8.8 show an example of the extended FAST method applied to the g-function test case with eight factors. The parameters a_i associated with the input factors are chosen to be $\{0, 1.0, 4.5, 9, 99, 99, 99, 99\}$. One hundred estimates, obtained using different starting points, of the first-order and total indices are computed. The boxplots of their summary statistics, for each factor, are plotted against the number of model evaluations, and are shown in Figures 8.7 and 8.8, respectively. The analytical values of the indices for each factor are plotted with dotted lines.

For both sets of indices, the estimates converge to the analytical values and the precision of the estimates increases, as the number of model evaluations increases. The FAST indices do not have the problems of the indices being negative or being greater than one, as were found in the Sobol' indices. A different set of model evaluations is needed for the estimation of each $\mathrm{TS}(i), \forall i = 1, \ldots, n$. However, with the same set of model evaluations (i.e. at no extra computational cost) the extended FAST can compute both $\mathrm{TS}(i)$ and S_i; unlike the Sobol', where a separate set of model evaluation is needed to compute the first-order and the total indices. Note that the minimum value of ω_i, for $i = 1, 2, \ldots, k$, is 8! Hence, the lowest sample size that FAST can use is 64.

8.4.5 Choice of resampling size N_r

Recall that the search curve introduced by Saltelli *et al.* (1999b) has the capability of producing replicates by specifying a new value for φ_i (see Table 8.1). For a given sample size N_s, we can choose ω_i and N_r such that the ratio ω_i/N_r lies between 16 and 64 (see Figure 8.9). This is because, if ω_i is low and N_r is large, the sampling over each curve is too sparse; on the other hand, a high ω_i and small N_r implies dense sampling over a small number of closed paths.

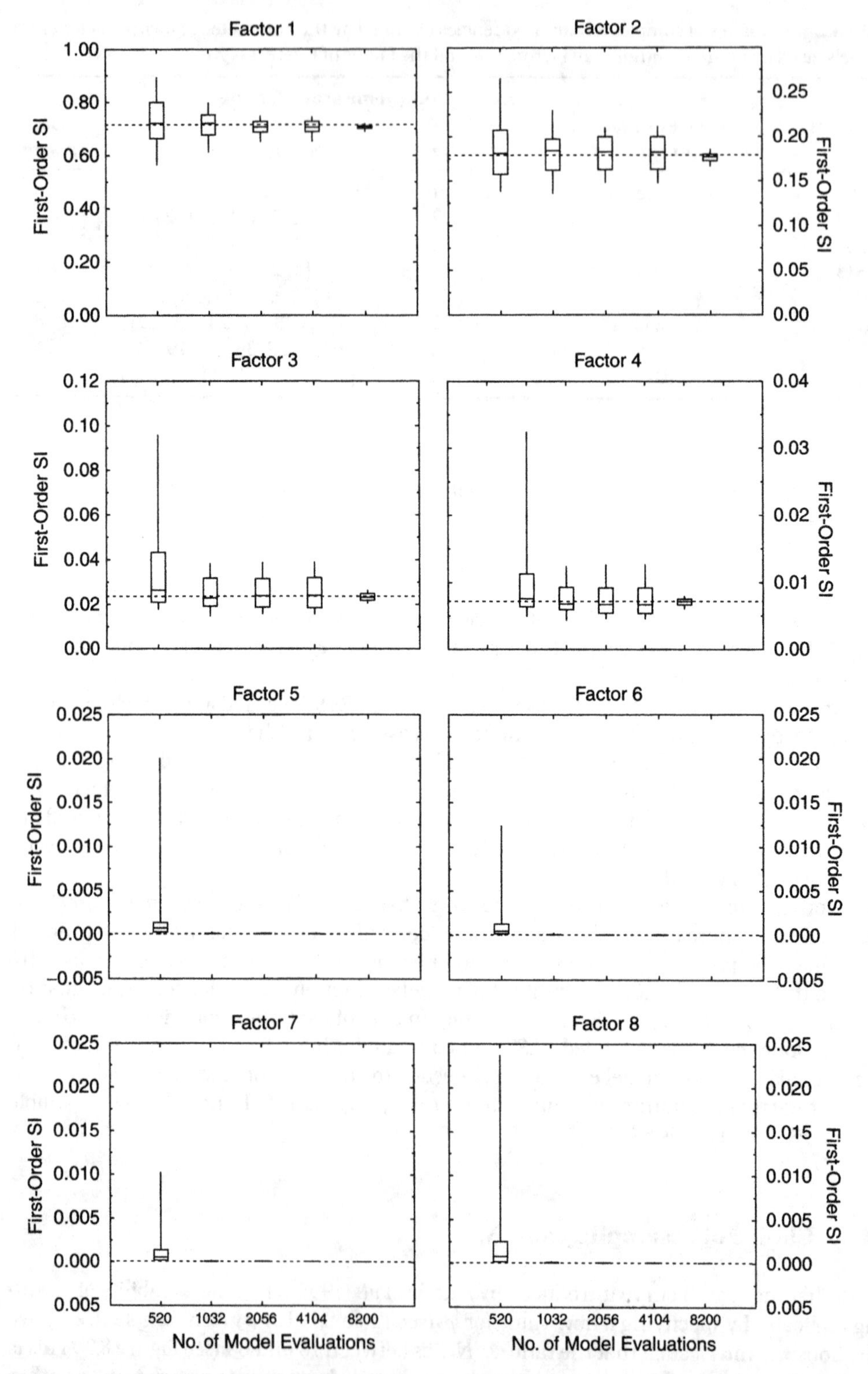

Figure 8.7 Boxplots of 100 first-order estimates, obtained by the extended FAST, against the number of model evaluation for the g-function test case with eight factors; analytical values of the sensitivity indices are shown by dotted lines.

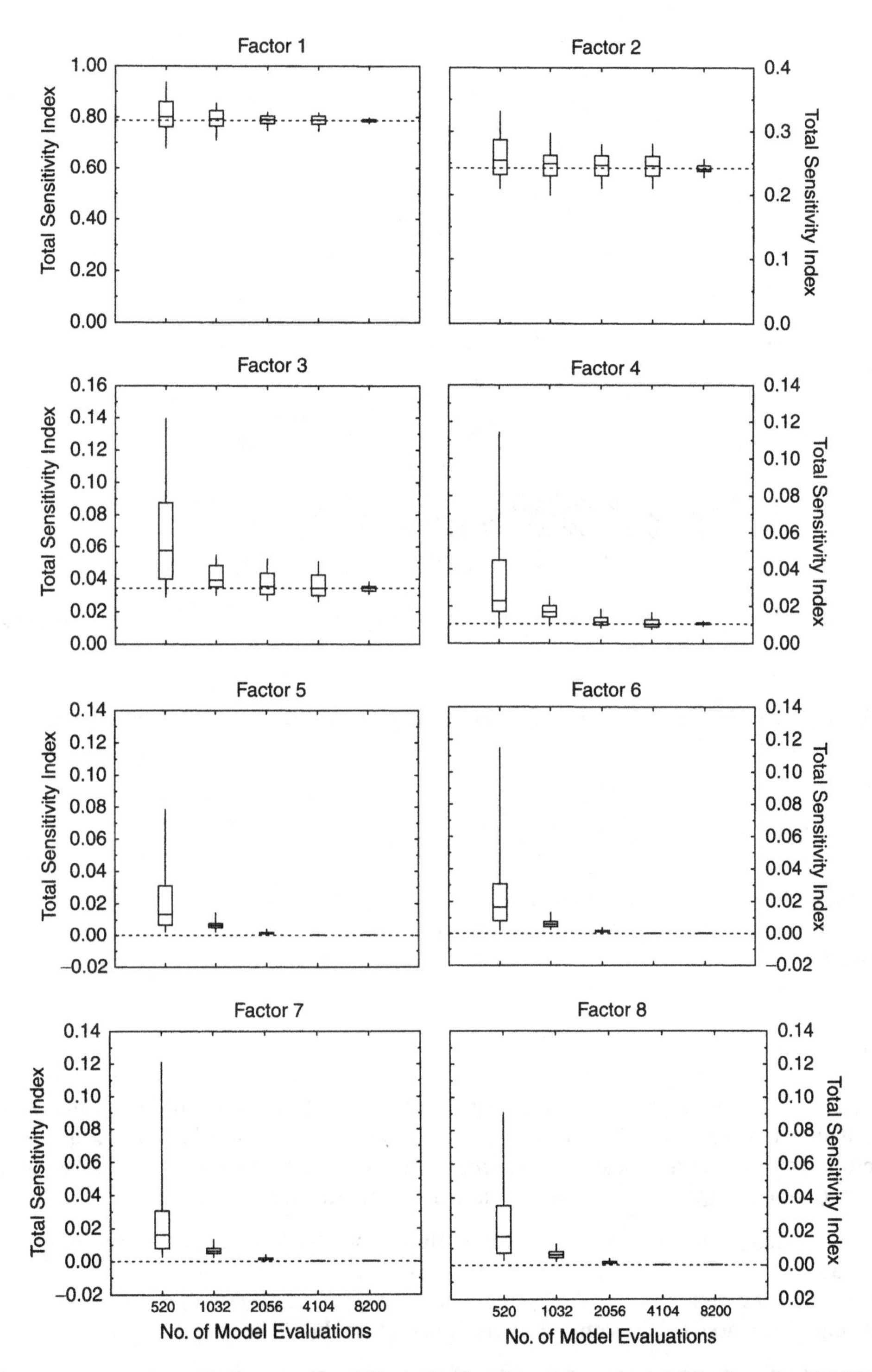

Figure 8.8 Boxplots of 100 estimates of the total sensitivity index, obtained by the extended FAST, against the number of model evaluation for the *g*-function test case with eight factors; analytical values of the sensitivity indices are shown by dotted lines.

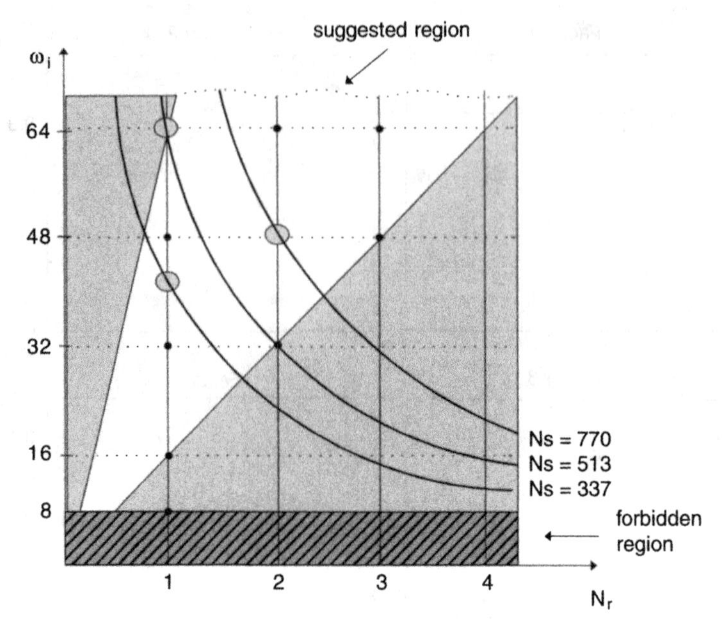

Figure 8.9 Plot of the recommended region for choosing the frequency for factor X_i and resampling size N_r. Note that $\omega_i \geqslant 8$ and $N_s \in \mathbb{Z}$.

However, at low sample size, we could be forced to select values that lie outside the recommended region.

8.5 SAMPLING STRATEGIES

In this section, we describe three sampling schemes, namely Sobol' LP_τ, winding stairs (WS), and combined scrambled Sobol' and winding stairs (SS–WS) to compute the Sobol' sensitivity indices.

8.5.1 LP_τ sampling

Sequences of LP_τ vectors represent a strategy to produce sample points uniformly distributed in a unit cube, and are essentially quasi-random sequences that are defined as sequences of points that have no intrinsic random properties. Generally, sequences of quasi-random vectors, $Q_1, Q_2, \ldots, Q_n$, should fulfill the following requirements (Sobol', 1994):

- The uniformity of the distribution is to be optimal when the length of the sequence tends to infinity.
- Uniformity of vectors $Q_1, Q_2, \ldots, Q_n$ should be observed for fairly small n.
- The algorithm used for the computation of the vectors should be simple.

Quasi-random sequences are used in place of random points to guarantee convergence of estimates in the classical sense. The use of points of LP_τ sequences usually results in better convergence when employed in numerical integration, instead of random points in the MC algorithm with finite constructive dimension. In a MC algorithm, the constructive dimension is the number of random numbers to be generated for each trial. The LP_τ sequences were introduced by Sobol' in 1966 and obey the requirements listed above.

The theory underlying the Sobol' LP_τ sequences is given in Sobol' (1967, 1976) and the algorithm to generate the sequences has been coded in FORTRAN77 and C (Bratley and Fox, 1988; Sobol' *et al.*, 1992). These programs are widely available, and hence we shall not describe the algorithm here. However, note that there is a maximum number of columns allowed; 52 is the highest constructive dimension for which LP_τ is better than a random number generator. For readers who wish to know more about the theoretical development of the LP_τ sequences, a good summary description can be found in Bratley and Fox (1988).

With as few as 128 points, the uniformly regular filling-in feature of the LP_τ sequence can clearly be seen (Figure 8.10a), compared with the purely random placement of pseudo-random points (Figure 8.10b). The pattern of quasi-random points changes as the number of points increases, but the property of regularity remains (Figure 8.11). However,

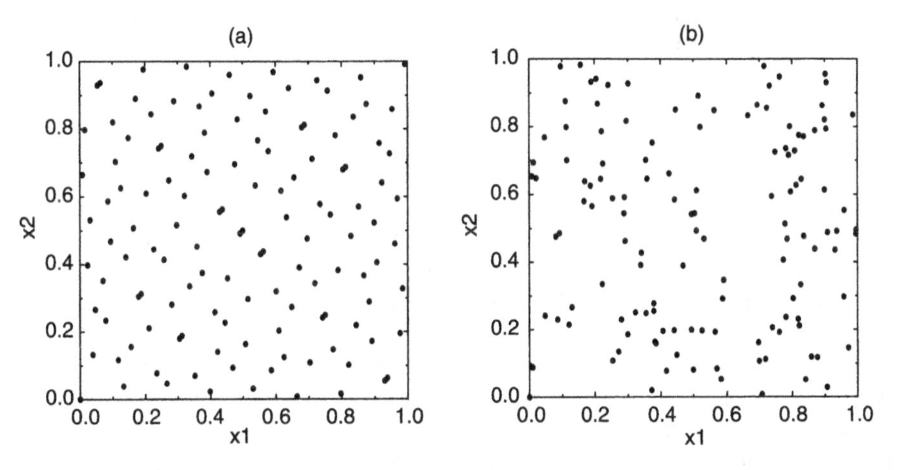

Figure 8.10 (a) Filling-in points generated by LP_τ sequences, and (b) by pseudo-random algorithm. The number of points is 128.

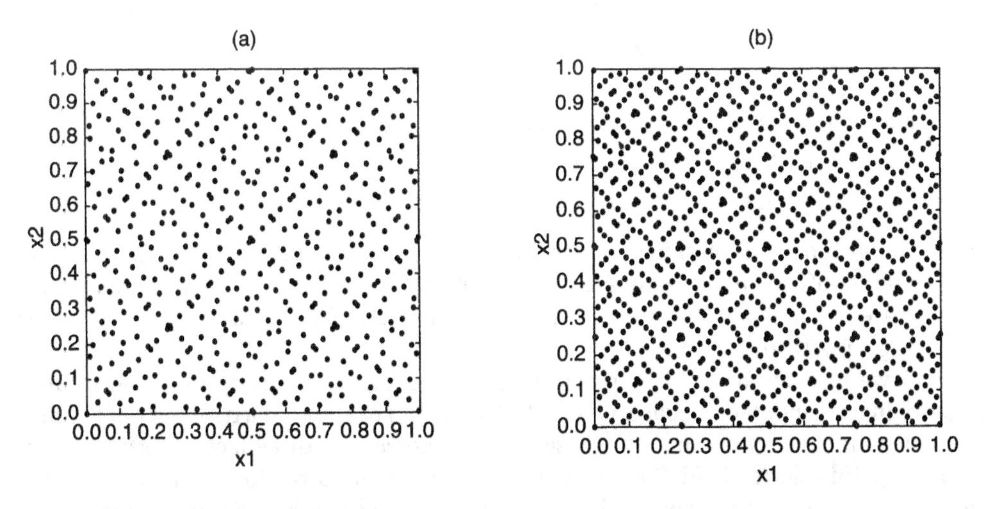

Figure 8.11 Patterns of LP_τ sequences with (a) 512 points and (b) 1024 points.

sequences that are intended for numerical computation must satisfy the following additional property

$$\int_0^1 \cdots \int_0^1 g(x_1, x_2, \ldots, x_k)\, dx_1 \ldots dx_k = \lim_{n\to\infty} \frac{1}{n} \sum_{j=1}^{n} g(Q_j),$$

where $g(x_1, x_2, \ldots, x_k)$ is an arbitrary integrable function.

The main reason why investigators favor the use of quasi-random points is the enhanced rate of convergence. The approximate error, for large n, can approach $1/n$, compared with the error of standard MC methods, which is of the order of $1/\sqrt{n}$. Thus, without changing the computation algorithm but merely replacing the random numbers with coordinates of quasi-random points, we can improve our results considerably.

Sequences of LP_τ points have been used successfully as nodes for multidimensional integration (Sobol', 1990a; Sobol' and Shukhman, 1995; Homma and Saltelli, 1995), search points in global optimization, trial points in multicriteria decision making (Sobol', 1992) and quasi-random points for quasi-MC algorithms (Sobol', 1990a).

8.5.2 Winding Stairs

The winding stairs method (Jansen *et al.*, 1994) consists of calculating y after each drawing of a new value for an individual factor X_i, for $i = 1, 2, \ldots, k$. Each new value is drawn at random from the marginal distribution of X_i, p_i. In other words, a sequence of sample points is generated by changing one input factor value at a time. Hence, the new input points are sampled in a fixed *cyclic* order.

The output is evaluated after each sample input point is generated, yielding a sequence of outputs y_l, for $l = 1, 2, \ldots, N$, where N is the total number of model evaluations. We arrange the sequence of N output values into k columns and $r + 1$ rows, where $r(= N/k)$ is the number of turns to repeat the cyclic order. For example, the following output matrix shows a set of output $y_1, y_2, \ldots, y_{15}$ for $k = 3$ and $r = 4$, and their corresponding input points (the input factor values that are sampled after each sample point is obtained, are highlighted in **bold**).

$$\begin{bmatrix} y_1 & y_2 & y_3 \\ y_4 & y_5 & y_6 \\ y_7 & y_8 & y_9 \\ y_{10} & y_{11} & y_{12} \\ y_{13} & y_{14} & y_{15} \end{bmatrix} = \begin{bmatrix} f(x_{11}, x_{21}, x_{31}) & f(x_{11}, \mathbf{x_{22}}, x_{31}) & f(x_{11}, x_{22}, \mathbf{x_{32}}) \\ f(\mathbf{x_{12}}, x_{22}, x_{32}) & f(x_{12}, \mathbf{x_{23}}, x_{32}) & f(x_{12}, x_{23}, \mathbf{x_{33}}) \\ f(\mathbf{x_{13}}, x_{23}, x_{33}) & f(x_{13}, \mathbf{x_{24}}, x_{33}) & f(x_{13}, x_{24}, \mathbf{x_{34}}) \\ f(\mathbf{x_{14}}, x_{24}, x_{34}) & f(x_{14}, \mathbf{x_{25}}, x_{34}) & f(x_{14}, x_{25}, \mathbf{x_{35}}) \\ f(\mathbf{x_{15}}, x_{25}, x_{35}) & f(x_{15}, \mathbf{x_{26}}, x_{35}) & f(x_{15}, x_{26}, \mathbf{x_{36}}) \end{bmatrix}.$$

The entries within each column are independent of each other (e.g. y_1, y_4, y_7, y_{10}, and y_{13}), since the values of each input factor are all different. However, the consecutive points within each row are not independent, in the sense that the points differ in a single input factor value (e.g. y_1, y_2, y_3, and so on).

In total, we generate $k(r + 1)$ input points, and the sequence of these input points forms a trajectory in the input factor space. Figure 8.12 shows an example of a trajectory of the first 32 sample points generated by the WS sampling scheme in a three-factor case; the total number of points plotted is 63 (i.e. $r = 20$). Note that a pseudo-random number generator is used to obtain each input factor value. The main advantage of the WS sampling scheme is the multiple use of the model evaluations (for further details, see Chan *et al.*, 2000).

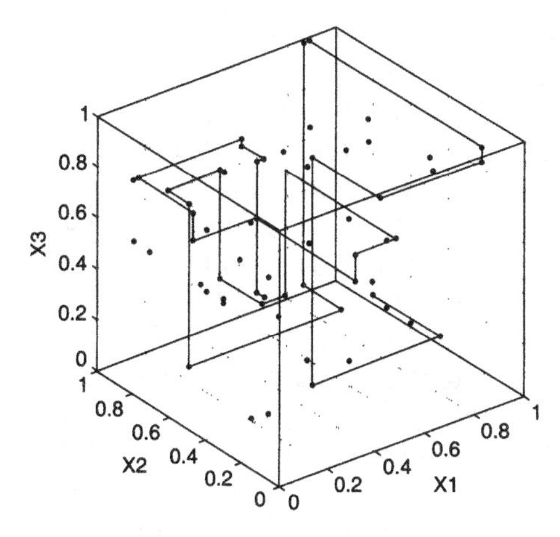

Figure 8.12 Plots of 63 winding stairs sample points and the path of the first 26 points for three factors.

8.5.3 Combined Scrambled Sobol' and Winding Stairs

In the WS sampling scheme, new sample points are obtained by drawing a value at random for each input factor. The combined scrambled Sobol' and winding stairs (SS–WS) makes use of the regularity feature of the Sobol' LP_τ sequences to obtain a set of WS samples. The procedure involves the construction of the so-called scrambled Sobol' sequences (Owen, 1997). Owen (1997) showed that, by scrambling the values of each coordinate of the Sobol' sequences independently and in a restricted way, the resulting sequences retain the equidistributed properties of the original samples. Recall that the individual coordinates of the Sobol' sample lie inside the interval [0,1]. The values of each individual coordinate are scrambled as follows:

Step 1: Divide the interval [0,1] into two subintervals, each containing the same number of values.
Step 2: Interchange the values in these two halves with probability 0.5.
Step 3: Divide each of these subintervals into two halves, each containing the same number of values.
Step 4: Repeat Step 2 independently for each subinterval.
Step 5: Continue until there are only two elements in each subinterval.

Hence, the scrambled Sobol' LP_τ data matrix contains coordinates of the original LP_τ sequences in a restricted order. The combined SS–WS sample points are generated by replacing input factor values, one at a time, with the coordinates from the corresponding scrambled matrix. An example of three factors with 16 sample points is given in Figure 8.13, which shows the transitions of the data matrix obtained from the original Sobol' to the combined SS–WS input matrix. Jansen (1999) suggested that the rows of the scrambled data matrix should be randomized before performing WS, but our experience suggests that this is not necessary.

Figures 8.14(a) and 8.14(c) show plots of 256 Sobol' LP_τ sample points and the random scrambled Sobol' points, respectively. The combined SS–WS points constructed from the scrambled Sobol' points by changing one input factor value at a time are plotted in Figure 8.14(d). Note that the resulting number of points generated using the combined SS–WS method is proportional to the number of factors, i.e. kn; hence, we have only plotted the first

$$\begin{bmatrix} 0.5000 & 0.5000 & 0.5000 \\ 0.2500 & 0.7500 & 0.2500 \\ 0.7500 & 0.2500 & 0.7500 \\ 0.1250 & 0.6250 & 0.8750 \\ 0.6250 & 0.1250 & 0.3750 \\ 0.3750 & 0.3750 & 0.6250 \\ 0.8750 & 0.8750 & 0.1250 \\ 0.0625 & 0.9375 & 0.6875 \\ 0.5625 & 0.4375 & 0.1875 \\ 0.3125 & 0.1875 & 0.9375 \\ 0.8125 & 0.6875 & 0.4375 \\ 0.1875 & 0.3125 & 0.3125 \\ 0.6875 & 0.8125 & 0.8125 \\ 0.4375 & 0.5625 & 0.0625 \\ 0.9375 & 0.0625 & 0.5625 \\ 0.0313 & 0.5313 & 0.4063 \end{bmatrix} \rightarrow \begin{bmatrix} 0.5000 & 0.6875 & 0.5625 \\ 0.1250 & 0.5000 & 0.2500 \\ 0.7500 & 0.8125 & 0.8750 \\ 0.4375 & 0.1875 & 0.8125 \\ 0.6250 & 0.8750 & 0.3750 \\ 0.0313 & 0.5625 & 0.6875 \\ 0.9375 & 0.3125 & 0.1250 \\ 0.3125 & 0.3750 & 0.6250 \\ 0.5625 & 0.6250 & 0.1875 \\ 0.1875 & 0.7500 & 0.7500 \\ 0.8125 & 0.2500 & 0.4063 \\ 0.3750 & 0.5313 & 0.3125 \\ 0.6875 & 0.4375 & 0.9375 \\ 0.0625 & 0.1250 & 0.0625 \\ 0.8750 & 0.9375 & 0.5000 \\ 0.2500 & 0.0625 & 0.4375 \end{bmatrix} \rightarrow \begin{bmatrix} 0.5000 & 0.6875 & 0.5625 \\ 0.5000 & 0.5000 & 0.5625 \\ 0.5000 & 0.5000 & 0.2500 \\ 0.1250 & 0.5000 & 0.2500 \\ 0.1250 & 0.8125 & 0.2500 \\ 0.1250 & 0.8125 & 0.8750 \\ 0.7500 & 0.8125 & 0.8750 \\ 0.7500 & 0.1875 & 0.8750 \\ 0.7500 & 0.1875 & 0.8125 \\ 0.4375 & 0.1875 & 0.8125 \\ 0.4375 & 0.8750 & 0.8125 \\ 0.4375 & 0.8750 & 0.3750 \\ 0.6250 & 0.8750 & 0.3750 \\ 0.6250 & 0.5625 & 0.3750 \\ 0.6250 & 0.5625 & 0.6875 \\ \vdots & \vdots & \vdots \end{bmatrix}$$

Original Sobol′ LP_τ Scrambled Sobol′   Combined SS–WS

Figure 8.13 A diagram of the transitions from the original Sobol′ data matrix to the scrambled Sobol′ and then to the combined SS–WS. Three factors and 16 points.

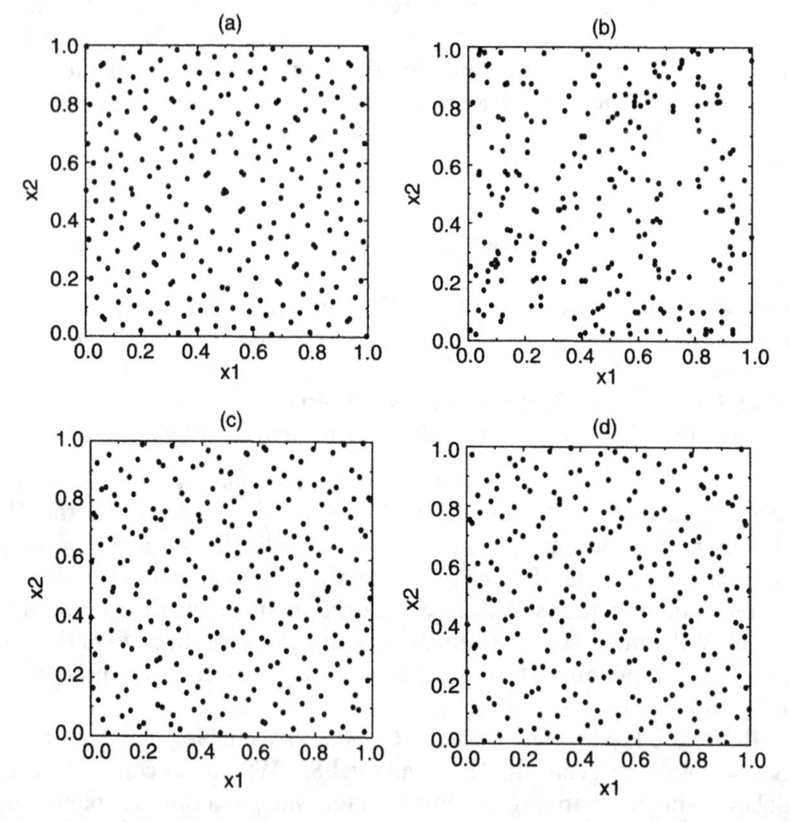

Figure 8.14 Plots of 256 sample points obtained by (a) Sobol′ LP_τ sequences, (b) winding stairs, (c) scrambled Sobol′, and (d) combined scrambled Sobol′ and winding stairs.

256 points here. For comparison, we also include the sample points generated by the original WS (Figure 8.14b). As can be seen, both scrambled and combined SS–WS sample points retain some regularity feature of the Sobol' LP_τ.

All of these sampling schemes can be used to compute both the Sobol' and Jansen global sensitivity measures. The Sobol' indices have always been associated with the LP_τ sequences, while the WS sampling scheme with the Jansen sensitivity measure has only been investigated recently. A brief observation of their relative merits can be found in Chan *et al.* (2000) and in Section 8.6. The usefulness of the combined SS–WS is still to be established. In Section 8.6, we shall draw our conclusions and provoke some thoughts on the methods that we have presented here.

8.6 DISCUSSION

A quantitative SA method should offer sensitivity measures for the various input factors that sum to one. Some techniques capture the additive (first-order) effects only, and can thus be considered quantitative as long as the model is additive. On the other hand, even computing all effects (that is, all the terms in the development in Equation (8.12)) represents a problem when a model contains many input factors. If a model has k factors, the total number of terms (including the first-order ones) is as high as $2^k - 1$. This difficulty affects in theory both experimentalists preparing an experimental design and modellers, although for the latter the problem is more acute. Models are mainly used to perform impossible or difficult experiments. With models, one wants to explore more factors than can be controlled in an experiment carried out under laboratory conditions, or in the field. Similarly, factors are varied on a wider scale in a numerical experiment than they are in a physical one. Unfortunately, increasing the number of factors and their ranges of variation will result in more and larger interactions. This suggests that not only should a generally useful SA method account for interactions, but also, in order to be of practical use, it should avoid this 'dimensionality curse'. In other words, the following property is desirable:

Efficiency property: The sensitivity measures should account for all effects in a computationally affordable way.

A further step can be made to complete the process of making the analysis of model sensitivity more agile and easier to interpret. This is based on the observation that in complex models with several submodels, or modules, or simply with sets of variables pertaining to different logical levels, one might desire to decompose the uncertainty according to subgroups.

Simplicity property: Sets or subgroups of inputs can be treated as single entities (factors).

By grouping the input factors, the computational cost is reduced, and a more compact display of the results is obtained, at the cost of losing the importance measure of individual factors within the groups.

Finally, the last but not least important property, especially when dealing with nonlinear models, is the following:

Model independence property: The level of additivity or linearity of the model does not influence the accuracy of the method.

The variance-based methods described in this chapter can be considered as a quantitative method for global SA. They can all compute the 'main effect' contribution of each input factor to the output variance. It is a matter of preference whether one uses the term 'importance measure' or 'correlation ratio', or others. However, it has been shown that the Sobol' and FAST techniques are more efficient than others. Furthermore, both Sobol' and FAST can compute the 'total sensitivity index', which is a more accurate measure of the effect of a factor on the model output, since it takes into account all interaction effects involving that factor.

In terms of efficiency, FAST performs better, since it can compute both first-order and total indices with a single set of model evaluations, The extended FAST method uses kN_s model evaluations, where $N_s = (8\omega_i + 1)N_r$ is the sample size; whereas the Sobol' method requires $n(2k + 1)$ model evaluations. For example, Table 8.3 shows the total number of model evaluations require to compute both the first-order and total indices for the Sobol' and the extended FAST methods, with $k = 8$. Also, shown in Table 8.3 is the total number of model evaluations needed if the winding stairs sampling scheme is used to compute the Sobol' indices. The WS sampling scheme was designed to make multiple use of model evaluations. Hence, with a single set of model evaluations, it can also compute both the first-order and total indices. As can be seen from Table 8.3, WS reduces N by more than half. But there is concern over the accuracy of the estimates yielded by the WS sampling scheme because of this reduction.

Chan *et al.* (2000) has found that Jansen's total effect measure (Equation (8.20)) together with the WS sampling scheme is superior to the Sobol' (with LP_τ). However, there is a reason to suspect that this superiority is due to the use of the square difference, as shown in Equation (8.24), rather than the WS sampling scheme. This issue will be the subject of further investigation. As to the scrambled SS–WS, the effectiveness and accuracy are being investigated. Rabitz (see Chapter 9) suggests an alternative efficient method to compute sensitivity indices provided that the development in Equation (8.12) stops at low-dimensionality terms.

Finally, the Sobol' and FAST methods are model-independent, unlike the SR(R)Cs (Chapter 6), for instance, which are known to be inadequate for non-monotonic models. The total sensitivity indices are computationally more expensive than the correlation/regression measures and the screening tests (see the Morris method and the iterated fractional factorial design (IFFD) in Chapter 4). The Morris method, for instance, is especially effective in its capacity to discriminate, at low sample size, among non-influential, linearly influential, and influential by way of either nonlinear or interaction effects. The total sensitivity indices

Table 8.3 Number of model evaluations, N, required to compute both sets of first-order and total sensitivity indices with $k = 8$ for the Sobol', the extended FAST, and the WS sampling methods.

	Number of model evaluations *N*		
Sample size *n*	**Sobol'**	**FAST ($N_s = n + 1$)**	**WS ($r = n$)**
32	544	–	256
64	1088	520	512
128	2176	1032	1024
256	4352	2056	2048
512	8704	4104	4096
1024	17408	8200	8192
2048	34816	16392	16384

offer something more at a higher price, namely, their capacity to rank quantitatively the factors, based on all effects, be they additive or not.

8.7 LAST REMARKS ON ANOVA DECOMPOSITION

The correlation ratios are familiar to statisticians, and are well-known in 'design of experiments' (DOE) textbooks (Box *et al.*, 1978b). The DOE techniques also decompose the response into terms of increasing dimensionality (main effects, two-way interactions, three-way interactions, etc.). The same decompositions are seen in ANOVA studies, the variance being decomposed into partial variances of increasing dimensionality; these type of decompositions are called ANOVA-like decompositions (see Archer *et al.*, 1997). A variance decomposition used by Cox (1982) consists of a sum of terms depending on subsets of inputs of size 1, 2, and so forth. That is,

$$\mathrm{Var}[Y] = \sum_{i}^{k} V_i + \sum_{i<j} V_{ij} + \sum_{i<j<l} V_{ijl} + \cdots + V_{12\ldots k},$$

where

$$\begin{aligned}
V_{ijl\ldots} &= \mathrm{Var}[Z_{ijl\ldots}], \qquad 1 \leqslant i < j < l < \ldots \leqslant k, \\
Z_i &= E[Y|X_i], \qquad i = 1, \ldots, k, \\
Z_{lj} &= E\left[Y - \sum_{i=1}^{k} Z_i | X_l, X_j\right], \qquad 1 \leqslant l \leqslant j \leqslant k, \\
Z_{ljm} &= E\left[Y - \sum_{i=1}^{k} Z_i - \sum_{p<q} Z_{pq} | X_l, X_j, X_m\right], \qquad 1 \leqslant l < j < m \leqslant k,
\end{aligned}$$

and so forth.

As noted by McKay (1995), the first summation in the decomposition can be seen as the VCEs, and the subsequent terms involve variances of prediction residuals. The expansion can certainly be used as an importance indicator; however, it requires the inputs be statistically independent. Furthermore, as noted by Archer *et al.* (1997), decomposition similar to the above and the decompositions in (8.12) and (8.17) are given in Efron and Stein (1981).

9

Managing the Tyranny of Parameters in Mathematical Modelling of Physical Systems

Herschel Rabitz and Ömer F. Alış

Princeton University, USA

9.1 INTRODUCTION

The physical models of various phenomena and their resultant mathematical structure often contain many input variables. Such models act to represent real world or laboratory processes, and this paper will generally use the world 'model' to also encompass working directly with the laboratory variables and observations. In some cases, the system variables serve as controls to be set in the laboratory, while in other circumstances, they are internal model variables whose values may be uncertain, calling for a multivariate statistical analysis. In either case, a general desire is to deduce the detailed structure of the n-dimensional variable space in order to identify regions of special impact on the model or observed output.

Sensitivity analysis in general terms aims to reveal the relationship between the model inputs and the outputs. There are two different goals in this respect:

1. Deducing the contribution of the input variable uncertainties upon the output uncertainties: Monte Carlo-based regression–correlation measures (Helton, 1993), the Fourier amplitude sensitivity test (FAST) (Helton, 1993), the Sobol' sensitivity estimates (Sobol' 1993; Saltelli and Sobol', 1995), among other methods, serve this purpose. The remainder of this book is largely focused on this topic.
2. Deducing the detailed mapping of the input variable space upon the output: previously this latter goal was typically explored by a perturbation analysis of the model output around a nominal value employing gradient based sensitivity analysis.

Although one may perform the uncertainty analysis in (1) without directly examining the full-space mapping in (2), the latter map is fundamental to overall model analysis, and it can

Sensitivity Analysis. Edited by A. Saltelli *et al.*

be used for many additional purposes. For this reason, this paper will focus on goal (2), including a demonstration that the statistical aim in (1) can be very efficiently attained by a special form of analysis in (2).

Many current applications require analyses with either goal (1) or (2) above calling for operations beyond the capabilities of local gradient methods (Rabitz, 1989). Without any *a priori* physical assumptions on the nature of the model input–output relationship, construction of a full-space analysis would be NP-complete (i.e., the computational complexity of the problem is an exponential function of the dimension of the model, which makes the construction of the mapping intractable) with computationl complexity scaling exponentially as $\sim s^n$, where n is the number of variables and s is the number of sample values for each variable. This problem is sometimes referred to as the *curse of dimensionality*, i.e. without any regularization or other form of simplification, learning the input–output function mapping from its sample values is computationally NP-complete in the dimension of its input variables. Stone (1982) showed that, using a local polynomial regression, one can achieve a rate of convergence $\epsilon_N = N^{-p/(2p+n)}$, with N being the number of sample points and p the degree of smoothness of the function (e.g. $p = 2$ implies that the function is twice-differentiable). It is easy to see how the curse of dimensionality appears. Without a high degree of smoothness ($p \ll n$), $\epsilon_N \approx N^{-p/n}$, and one needs $\sim s^{n/p}$ sample points to approximate the function to a resolution of $1/s$. The high-dimensional model representations (HDMR) introduced in Rabitz *et al.* (1999) aim to show that a dramatic reduction in this scaling is often expected to arise due to the presence of only low-order correlations amongst the input variables having a significant impact upon the output.

The model output is assumed to be a *function* taking values over an n-dimensional Euclidean space $\mathbb{R}^n$. The HDMR expansions introduced here are especially useful for the purpose of representing the outputs of a physical system when the number of input variables is large. The resulting mathematical representation might be further used for sensitivity analysis or uncertainty analysis of the model output. The hierarchical form of the HDMR expansions is suited for this purpose: the HDMR expansions are based on exploiting the correlated effects of the input variables, which are naturally created by the input-output mapping. The term 'correlation' employed here is generally distinct from that employed in statistics, since the input variables often will not be random. We assume that the model output(s) is rationally behaved in terms of the input variables (i.e., the output is a well-defined function, but rapid or even discontinuous behavior would still be permitted). The high dimensionality of the input space and the expense of performing model calculations or experiments often prevent a full sampling of the input space (i.e., s^n computer simulations or experiments). The notion of 'high' dimensionality is system-dependent, with some situations being considered high for $n \approx 3$–5 while others will only reach that level for $n \gg 10$ or more.

With the comments above as background, we seek a fast algorithm that can circumvent the apparent exponential difficulty of the high-dimensional mapping problem. The HDMR expansions (Rabitz *et al.*, 1999) can be written in the following form, for $f(x) \equiv f(x_1, x_2, \ldots, x_n)$ representing the mapping between the input variables $x_1, x_2, \ldots, x_n$ and output defined on the domain $\Omega \subset \mathbb{R}^n$:

$$f(x) \equiv f_0 + \sum_i f_i(x_i) + \sum_{i<j} f_{ij}(x_i, x_j) + \cdots + f_{12\cdots n}(x_1, x_2, \ldots, x_n). \tag{9.1}$$

Here f_0 denotes the zeroth-order effect, which is a constant everywhere in the domain Ω. The function $f_i(x_i)$ gives the effect associated with the variable x_i acting independently, although generally nonlinearly, upon the output f. The function $f_{ij}(x_i, x_j)$ describes the cooperative effects of the variables x_i and x_j, and higher-order terms reflect the cooperative effects of increasing numbers of variables acting together to impact upon f. The last term $f_{12...n}(x_1, \ldots, x_n)$ gives any residual dependence of all the variables locked together in a cooperative way to influence the output f. If there is no cooperation between the input variables, then only zeroth-order and first-order terms will appear in the expansion. However, even to first order, the expansion is not a linear superposition, since $f_i(x_i)$ could have an arbitrary dependence on x_i. The notion of 0th-, 1st-, 2nd-order, etc. in the HDMR expansion should not be confused with the terminology of a Taylor series; the HDMR expansion is exact and always of finite order. The HDMR expansion is a very efficient formulation of the physical output if higher-order variable correlations are weak, permitting the physical model to be captured by the first few lower-order terms. The resultant computational effort to determine the expansion functions will scale polynomically with n rather than the traditional view of it being exponential. The choice of input variables can be important, but evidence suggests that often no special effort is required to find rapid convergence of the HDMR expansion *for well-defined physical systems*. Typically, physical input variables are chosen to have a distinct role that aids in the convergence of the HDMR. In some cases, a suitable transformation of the variables may be employed to simplify the analysis.

The HDMR has a structure analogous to the many-body expansions used in molecular physics (Schatz, 1989) to represent potential surfaces created by a system of atoms. Generally two-body terms dominate and rarely are terms beyond third-order significant. The many-body expansions can be viewed as a special case of a HDMR that rapidly converges for particular physical reasons. Similar rapidly convergent cluster expansions are utilized in statistical mechanics (Hill, 1987). The HDMR concept rests on suggesting that a similar rapid loss of correlation is expected to arise under more general physical conditions. Perhaps the best evidence for this conjecture lies in statistics, where rarely do more than input covariances play a significant role. The latter behavior can depend on the dynamic range of the input variables, but the observed lack of higher-order correlations appears to be rather generic. Conversely, an HDMR may not be of practical utility (i.e., high-order terms play a role) for arbitrary mathematical functions, although Equation (9.1) is always an exact representation.

In order to understand the effects of each of the terms in the HDMR expansion, projection operators $F_{i_1 i_2 \ldots i_l}$ may be introduced so that $P_{i_1 i_2 \ldots i_l} f = f_{i_1 i_2 \ldots i_l}$ determines a particular term in the HDMR expansion. The orthogonality of the projection operators ensures that the terms within and between each order give unique correlated information about the variables contributing to the output function $f(x)$. The sum of the full set of projection operators $\{P_{i_1 i_2 \ldots i_l}\}$ provides a resolution of the identity operator and different sets of projection operators give rise to distinct HDMR expansions. The choice of a particular HDMR expansion depends on the application and the nature of any constraints for sampling the input variables $x_1, \ldots, x_n$. The input space is assumed to be a normed vector space furnished with an inner product expressed in terms of a suitable measure.

Attempts at approximating multivariate functions by linear or nonlinear superpositions of functions has a long history, and some relevant cases are mentioned here. *Projection pursuit* algorithms (Friedman and Stuetzle, 1981b; Huber, 1981; Diaconis and Shahshahani, 1984; Stone, 1985) approximate the multivariate function $f(x)$ in the form

$$f(x) \equiv f(x_1, x_2, \ldots x_n) = \mu + \sum_{i=1}^{K} f_i\left(\sum_{k=1}^{n} \beta_{ik} x_k\right), \tag{9.2}$$

where $\boldsymbol{\beta}_i \equiv [\beta_{i1} \ldots \beta_{in}]$ represent the projection directions and μ is taken to be the average of the function. The parameter vectors $\boldsymbol{\beta}_i$ and functions f_i are estimated from the data. One may view (9.2) as a special case of (9.1) using linear combinations of the original variables and truncating the expansion to the first order. *Multilayer perceptrons* (MLPs) (Parker, 1985) used in artificial neural networks approximate the multivariate function $f(x)$ in the form

$$f(x) = h\left(\sum_{i=1}^{K} \alpha_i g\left(\sum_{k=1}^{n} \beta_{ik} x_k\right)\right), \tag{9.3}$$

where h and g are arbitrary nonlinear functions (i.e., this is a MLP with a single *hidden layer*). Such a *learning network* is trained with a given set of inputs and output values, and a MLP approximation seeks to find the scalars α_k and the vectors $\boldsymbol{\beta}_i$ by least-squares minimization. *Radial basis functions* (Poggio and Girosi, 1990) have been used to approximate $f(x)$ as a nonlinear function of its input variables under a regularization criteria. This approach under certain conditions leads to the expansion

$$f(x) = \sum_{i=1}^{K} \beta_i f_i(\|x - u_i\|) + \mathscr{P}(x) \tag{9.4}$$

where the u_i are centers similar to the knots of splines, the β_i are constants, the f_k are a chosen set of *radial basis functions* (e.g., a Gaussian), and $\mathscr{P}$ is a polynomial. The coefficients are then fitted to data using least-squares minimization. Although the representations above are useful for particular applications, there is no general rule of thumb to choose one over another. There is no widely accepted procedure for determining the number of hidden layers, number of functions, etc. The parameters of the networks are obtained by least-squares minimization, and if the objective functional is not globally convex and has many local minima, this may result in a nonunique representation.

All of the above representations are inspired by a theorem of Kolmogorov (Lorentz *et al.*, 1996) which states that a multivariate function defined on the unit cube $K^n = [0, 1]^n$ can be represented in the following way:

$$f(x_1, x_2, \ldots, x_n) = \sum_{q=1}^{2n+1} g(\lambda_1 \phi_q(x_1) + \cdots + \lambda_n \phi_q(x_n)), \tag{9.5}$$

i.e., any multivariate function can be written as a linear superposition of univariate functions. Although the functions ϕ_q are continuous, they are highly nonsmooth and their practical utility for approximation/interpolation is very limited (Girosi and Poggio, 1989). The HDMR technique aims to represent multivariate functions arising in physical contexts rather than for arbitrary function interpolation. As argued earlier, in most physical systems it is natural to expect only very low-order correlations to exist amongst the input variables for their action upon the output function $f(x)$. There is a predisposition towards this behavior in describing physical systems, as one naturally chooses the variables to act as independently as possible. In light of the theorem of Kolmogorov, it appears that although the natural physical variables in typical physical systems are not perfect in the sense of Equation (9.5), the additional low-order correlations are easily managed.

There is no unique decomposition of the model output $f(x_1, x_2 \ldots, x_n)$ in the form of Equation (9.1). This richness of the HDMR expansions may be exploited for a specific representation objective. For example, in the case of uncertainty analysis of the model output (e.g., an analysis of the variance of the output), one should choose the component functions in the HMDR so that they represent the independent contributions of input variables to the overall uncertainty of the output. Such an ANOVA–HDMR used in statistics can measure the importance of the input variable variance upon the variance of the output. In this case, each component function is a random quantity uniquely contributing to the overall variance of the output. The hierarchical formulation of ANOVA–HDMR allows for the identification of how each input variable or group of input variables determine the variance of the output. In this way, ANOVA–HDMR also supplies a nonlinear sensitivity analysis of the model output. A drawback of traditional ANOVA–HDMR is the need for Monte Carlo simulations to compute the component functions (Sobol', 1993; Homma and Saltelli, 1996). Cut-HDMR is a different HDMR expansion using a specific sampling of the model output to generally provide a computationally more efficient representation than ANOVA–HDMR. If an uncertainty analysis of the model output is of interest then a cut-HDMR can be easily converted into ANOVA–HDMR of the output. This approach is generally computationally more efficient than the direct route of computing the ANOVA–HDMR of the model output, and this point is discussed in this chapter.

The chapter is organized as follows. Section 9.2 gives the general formulation of the HDMR expansions in terms of projection operators where the input variables reside in $\mathbb{R}^n$. Sections 9.2.1 and 9.2.2 presents ANOVA–HDMR and cut-HDMR respectively. Section 9.2.3 compares the computational costs of computing these two HDMRs. In Section 9.3, the illustration of cut-HDMR is presented with its application to several test examples. Section 9.4 presents a general discussion about further potential applications of the HDMR tools. Concluding remarks and future perspectives are given in Section 9.5.

9.2 HIGH-DIMENSIONAL MODEL REPRESENTATIONS

We assume that the input–output relationship of a physical model is represented by a real, scalar function $f(x) \equiv f(x_1, x_2, \ldots, x_n)$ defined on the unit cube $K^n = \{(x_1, x_2, \ldots, x_n) : 0 \leqslant x_i \leqslant 1,\ i = 1, 2, \ldots, n\}$. The extension to an output vector is readily apparent as each vector (function) component can be treated separately. $f(x)$ belongs to a linear vector space of functions denoted by X. A measure μ on Borel subsets of K^n is defined so that $\{K^n, \mathscr{B}(K^n), \mu)\}$ becomes a measure space (with $\mathscr{B}(K^n)$ denoting the Borel σ-algebra on K^n). We consider the subspace X to consist of all integrable functions with respect to μ. We further stipulate that μ is a product measure with unit mass, and has a density, i.e.,

$$\begin{aligned} d\mu(x) &\equiv d\mu(x_1, \ldots, x_n) = \prod_{i=1}^{n} d\mu_i(x_i), \\ &\int_K^1 d\mu_i(x_i) = 1, \\ d\mu(x) &= g(x)\, dx = \prod_{i=1}^{n} g_i(x_i)\, dx_i, \end{aligned} \tag{9.6}$$

where $g_i(x_i)$ is the marginal density of the input x_i.

The inner product $\langle .,. \rangle$ on X induced by the measure μ is defined as follows

$$\langle f, h \rangle \equiv \int_{K^n} f(x)h(x)\, d\mu(x) \qquad f(x), h(x) \in X. \tag{9.7}$$

Two functions $f(x)$ and $h(x)$ will be called *orthogonal* if $\langle f, h \rangle = 0$. Note that functions $f(x)$ and $h(x)$ may depend on different sets of components of the input vector x. The norm $\| \cdot \|_X$

on X induced by the above inner product is defined as follows:

$$\|f(x)\|_X \equiv (\langle f, f\rangle)^{1/2} \equiv \left[\int_{K^n} f^2(x)\, d\mu(x)\right]^{1/2} \tag{9.8}$$

We now define the following decomposition of X into subspaces whose mutual intersections are empty:

Definition $\mathscr{V}_0, \{\mathscr{V}_i\}, \{\mathscr{V}_{ij}\}_{i<j} \cdots, \mathscr{V}_{12\ldots,n} \subset X$ are defined as follows:

$$\begin{aligned}
\mathscr{V}_0 &\equiv \{f \in X : f = C, \text{ where } C \in \mathbb{R} \text{ is a constant}\},\\
\mathscr{V}_i &\equiv \{f \in X : f = f_i(x_i) \text{ is a univariate function of the input } x_i\\
&\qquad \text{with } \int_{K^1} f_i(x_i)\, d\mu_i(x_i) = 0\},\\
\mathscr{V}_{ij} &\equiv \{f \in X : f = f_{ij}(x_i, x_j) \text{ is a bivariate function of the inputs } x_i,\ x_j\\
&\qquad \text{with } \int_{K^1} f_{ij}(x_i, x_j)\, d\mu_k(x_k) = 0, k = i, j\},\\
\mathscr{V}_{i_1\ldots i_l} &\equiv \{f \in X : f = f_{i_1\ldots i_l}(x_{i_1}, x_{i_2} \ldots, x_{i_l}) \text{ is an } l-\text{ variate function of the inputs } x_{i_1,\ldots,x_{i_l}}\\
&\qquad \text{with } \int_{K^1} f_{i_1\ldots i_l}(x_{i_1,\ldots,x_{i_l}}) d\mu_k(x_k) = 0, k = i_1, \ldots, i_l\},\\
&\ \vdots\\
\mathscr{V}_{12\ldots n} &\equiv \{f \in X : f = f_{12\ldots n}(x_1, x_2, \ldots, x_n) \text{ is an } n\text{-variate function of all inputs}\\
\text{with } &\qquad \int_{K^1} f_{12\ldots n}(x_1, x_2, \ldots, x_n)\, d\mu_k(x_k) = 0, k = 1, \ldots, n\}.
\end{aligned} \tag{9.9}$$

The integral null property introduced above in any subspace $\mathscr{V}_{i_1 i_2\ldots i_l}$ serves to assure that the functions are orthogonal:

$$\langle f_{i_1\ldots i_s}, f_{j_1\ldots j_p}\rangle = 0 \tag{9.10}$$

for at least one index differing in $\{i_1, \ldots, i_s\}$ and $\{j_1, \ldots, j_p\}$, and s may be the same as p.
The following proposition can be deduced immediately from the above definition:

Lemma 9.1. X is the direct sum of the subspaces defined above, i.e.,

$$X = \mathscr{V}_0 \oplus \sum_i \mathscr{V}_i \oplus \sum_{i<j} \mathscr{V}_{ij} \oplus \ldots \oplus \sum_{i_1<i_2\ldots<i_l} \mathscr{V}_{i_1 i_2\ldots i_l} \cdots \oplus \mathscr{V}_{12\ldots n}, \tag{9.11}$$

where $\oplus$ denotes the direct sum operator. The corollary to this proposition is that $f(x) \in X$ can be written as

$$f(x) = f_0 + \sum_i f_i(x_i) + \sum_{i<j} f_{ij}(x_i, x_j) + \cdots + f_{12\ldots n}(x_1, x_2, \ldots, x_n). \tag{9.12}$$

The decomposition (9.11) of X is unique subject to the choice of measure in (9.6), which in turn implies that the expansion (9.12) is unique.

Lemma 9.2. Equation (9.11) suggests the following family of projection operators defined from X into one of the subspaces above:

$$\begin{aligned} f_0 &\equiv P_0 f(x) = \mathbf{M} f(x), \\ f_i(x_i) &\equiv P_i f(x) = \mathbf{M}^i f(x) - P_0 f(x), \\ f_{ij}(x_i, x_j) &\equiv P_{ij} f(x) = \mathbf{M}^{ij} f(x) - P_i f(x) - P_j f(x) - P_0 f(x), \\ &\vdots \\ f_{i_1 \ldots i_l}(x_{i_1}, \ldots, x_{i_l}) &\equiv P_{i_1 \ldots i_l} f(x) = \mathbf{M}^{i_1 \ldots i_l} f(x) - \sum_{j_1 < \ldots < j_{l-1} \subset \{i_1, \ldots, i_l\}} P_{j_1 \ldots j_{l-1}} f(x) \\ &\quad - \sum_{j_1 < \ldots < j_{l-2} \subset i_1, \ldots, i_l} P_{j_1 \ldots j_{l-2}} f(x) - \cdots - \sum_j P_j f(x) - P_0 f(x). \end{aligned} \tag{9.13}$$

These operators are chosen so that the variational problem

$$\min_u \|f(x) - u\|_X, \qquad u \in \mathscr{V}_0 \oplus \sum_i \mathscr{V}_i \oplus \sum_{i<j} \mathscr{V}_{ij} \oplus \cdots \oplus \sum_{i_1 < i_2 \ldots < i_l} \mathscr{V}_{i_1 \ldots i_l}, \tag{9.14}$$

is minimized with

$$u = \left(P_0 + \sum_i P_i + \sum_{i<j} P_{ij} + \cdots + \sum_{i_1 < i_2 \ldots < i_l} P_{i_1 i_2 \ldots i_l} \right) f(x). \tag{9.15}$$

Importantly, each of these functions $f_{i_1, \ldots, i_l}$ is determined separately in a sequence starting from the lower members and moving up.

Returning again to the collection of projection operators, we observe that they have the following properties:

1. Idempotency:

$$P^2_{i_1 2 \ldots i_l} \equiv P_{i_1 i_2 \ldots i_l} \neq 0, \qquad i_1 < i_2 < \cdots < i_l \subset \{1, 2, \ldots, n\}. \tag{9.16}$$

2. Orthogonality:

$$P_{i_1 i_2 \ldots i_l} P_{j_1 j_2 \ldots j_k} \equiv \begin{cases} 0 & (l \neq k), \\ P_{i_1 i_2 \ldots i_l} \delta_{i_1 j_1} \delta_{i_2 j_2} \ldots \delta_{i_l j_l} & (l = k), \end{cases} \tag{9.17}$$

with δ_{ij} representing the Kronecker delta function. In (9.16) and (9.17), composition of projection operators is understood to imply an integration over the input variable space.
3. Resolution of the identity:

$$P_0 + \sum_i P_i + \sum_{i<j} P_{ij} + \cdots + P_{12 \ldots n} = \mathbf{1}, \tag{9.18}$$

where $\mathbf{1}$ denotes the identity operator.

Corollary 9.1. The operators defined by

$$Q_l = P_0 + \sum_i P_i + \sum_{i<j} P_{ij} + \cdots + \sum_{i_1 < i_2 < \ldots < i_l \subset \{1, 2, \ldots, n\}} P_{i_1 i_2 \ldots i_l} \tag{9.19}$$

are also projections.

Corollary 9.2. An important property of the HDMR is that if a set of output-model functions obey a set of linear-superposition conversation laws (e.g., conservation of mass), then their HDMR expansion to any order also obey these conservation law(s) order-by-order. Given the output functions $\{f^1(x), f^2(x), \ldots, f^N(x)\}$, suppose that they obey the following m conservation laws:

$$\sum_{j=1}^{N} w_{kj} f^j(x) = c_k, \qquad k = 1, 2, \ldots m, \tag{9.20}$$

where w_{kj} are constants. Then these conservation laws are obeyed by the $Q_l f^j(x)$, i.e,

$$\sum_{j=1}^{N} w_{kj} Q_l f^j(x) = c_k, \qquad k = 1, 2, \ldots m, \quad l = 0, 1, 2, \ldots N. \tag{9.21}$$

This result is obtained by using the following identities:

$$\sum_{j=1}^{N} w_{kj} P_{i_1 \ldots i_l} f^j(x) = \begin{cases} 0 & (P_{i_1 \ldots i_l} \neq P_0), \\ c_k & (P_{i_1 \ldots i_l} = P_0). \end{cases} \tag{9.22}$$

Corollary 9.3. The properties of the projection operators assure the following:

1. The projection of f to form $f_{i_1 i_2 \ldots i_l}$ is unique given the operator $P_{i_1 i_2 \ldots i_l}$.
2. The functions $f_{i_1 i_2 \ldots i_l}$ and $f_{j_1 j_2 \ldots j_k}$ are independent and orthogonal provided that at least one member of each of $\{i_1, i_2, \ldots, i_l\}$ and $\{j_1, j_2, \ldots, j_k\}$ differ from one another.
3. The HDMR expansion has a finite number of terms that exactly represent $f(x)$.

The demand that $P_{i_1 i_2 \ldots i_l} \neq 0$ assures that each order of correlation in $f(x)$ is allowed to be naturally identified. The HDMR expansion functions $f_0, f_i(x_i), f_{ij}(x_{ij})$, etc. are understood to be particular types of correlation functions in relation to the properties of the projection operators as explained later.

The HDMR expansion may be understood in terms of $\mathbf{1} \cdot f(x) = f(x)$, where $\mathbf{1}$ is the unit operator. We may represent $\mathbf{1}$ many ways, such as in terms of tensor products of orthogonal functions in each of the variables x_i. The latter representations of $f(x)$ are complete, but they contain an infinite number of terms. The key to the HDMR is the choice of $\mathbf{1}$ as a hierarchy of projections into subspaces of increasing dimensions, playing on the natural expectation of rapidly diminishing contributions from the higher spaces corresponding to high-order correlations amongst the input variables. An important property of HDMR expansions is that they will all converge at the same order. If a choice of the measure μ for a function f gives a HDMR that converges at order L with an error $\mathcal{O}(\epsilon)$, then another HDMR defined by the measure μ' will also converge at order L with an error $\mathcal{O}(\epsilon)$. However, the expansion functions $f_{i_1 \ldots i_l}(\cdot), l \leqslant L$, in each case will be different. In this sense, the correlation interpretation of the function $f_{i_1 \ldots i_l}(\cdot)$ is associated with the measure used to define their corresponding HDMR. The equivalence of one converged HDMR with respect to any other may be exploited to calculate a convenient one and convert it to another more suitable HDMR for a particular application (see Figure 9.1 below).

The HDMR expansion (9.1) is used in statistics as the ANOVA decomposition (Scheffe, 1959; Efron and Stein, 1981) of a multivariate statistical function $f(x_1, \ldots, x_n)$ that depends on independently distributed random variables $x_1, \ldots, x_n$. This particular form of HDMR is treated in Section 9.2.1. Owing to the orthogonality of the individual component functions,

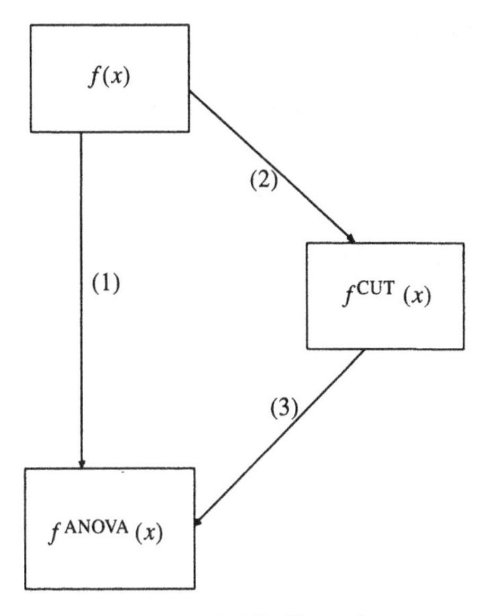

Figure 9.1. Two equivalent routes to ANOVA–HDMR. The indirect route through cut-HDMR is more efficient.

the variance of f will be equal to the sum of variances of each of the individual random variables on the right-hand side of Equation (9.1). Usually the few lowest-order terms have significant contributions to the overall variance of f. HDMR expansions can be used in a broader sense, since the input variables need not be random and one may be interested in more than the variance or some finite moment of the output. They may be used for representing the input–output mapping over the operating region of the input variables. As such, they are multivariate approximation/interpolation schemes as well as a means to analyze the relevant statistics of a random output. The measure μ defined over the input variable space does not necessarily assume randomness of the input variables. It is rather a 'weight' chosen to give different HDMR expansions distinct characteristics that may be useful for a certain representation problem. For example, the ANOVA–HDMR in Section 9.2.1 is useful for measuring the contribution of the variance of individual component functions to the overall variance of the output. On the other hand, a cut-HDMR expansion (Section 9.2.2) is an exact representation of the output $f(x)$ along the hyperplanes passing through a reference point. Thus, the choice of the particular HDMR is suggested by what is desired to be known about the output and is also dictated by the amount of available data.

One may show that a compact formulation of the projection operators is given by

$$
\begin{aligned}
P_{i_1\ldots i_l} f(x) &= \int_{K^n} K_{i_1\ldots i_l}(x; x') f(x')\, dx', \\
K_0(x; x') &= \prod_{j=1}^{n} g_j(x'_j), \\
K_{i_1\ldots i_l}(x; x') &= \prod_{j\in\{i_1,\ldots,i_l\}} [\delta(x_j - x'_j) - g_j(x'_j)] \prod_{j\in\{i_1,\ldots,i_l\}} g_j(x'_j).
\end{aligned}
\tag{9.23}
$$

From the analysis above, it is evident that the overall form of the HDMR expansion is uniquely defined once the projection operators are specified. For illustration, we present two classes of HDMRs with the projection operators specified below.

9.2.1 ANOVA–HDMR Expansion

In this case, X is defined as the space of square-integrable functions on K^n. The measure μ is taken as the ordinary Lebesgue measure

$$d\mu(x) = dx = dx_1 dx_2 \ldots dx_n \tag{9.24}$$

With this choice, the action of the projection operators in the ANOVA–HDMR is given by

$$\begin{aligned}
f_0(x) &\equiv P_0 f(x) = \int_{K^n} f(x)\,dx,\\
f_i(x_i) &\equiv P_i f(x) = \int_{K^{n-1}} f(x) \prod_{j\neq i} dx_j - P_0 f(x),\\
f_{ij}(x_i, x_j) &\equiv P_{ij} f(x) = \int_{K^{n-2}} f(x) \prod_{k\notin\{i,j\}} dx_k - P_i f(x) - P_j f(x) - P_0 f(x),\\
&\vdots\\
f_{i_1\ldots i_l}(x_{i_1}, \ldots, x_{i_l}) &\equiv P_{i_1\ldots i_l} f(x) = \textstyle\int_{K^{n-l}} f(x) \displaystyle\prod_{k\in\{i_1,\ldots,i_l\}} dx_k - \sum_{j_1<\cdots<j_{l-1}\subset\{i_1,i_2,\ldots i_l\}} P_{j_1\ldots j_{l-1}}\\
&\quad - \sum_{j_1<\cdots<j_{l-1}\subset\{i_1,i_2,\ldots,i_l\}} P_{j_1\ldots j_{l-2}} f(x) - \ldots\\
&\quad - \sum_j P_j f(x) - P_0 f(x),
\end{aligned} \tag{9.25}$$

where the kernel functions are given by the expressions

$$\begin{aligned}
K_0(x; x') &= 1,\\
K_{i_1\ldots i_l}(x; x') &= \prod_{j\in\{i_1,\ldots,i_l\}} [\delta(x_j - x'_j) - 1].
\end{aligned} \tag{9.26}$$

The expansion based on these integrals is a multivariate representation of the model output. It is mainly used (Scheffe, 1959; Efron and Stein, 1981; Sobol', 1993; Saltelli and Sobol', 1995; Homma and Saltelli, 1996) for statistical analysis. If the input consists of independently distributed uniform random variables (corresponding to the ordinary Lebesgue measure above), then the component functions will be uncorrelated and the overall variance of the ANOVA–HDMR can be written as follows:

$$D \equiv \mathbf{E}(f - f_0)^2 = \sum_i D_i + \sum_{i<j} D_{ij} + \cdots + \sum D_{12\ldots n}, \tag{9.27}$$

where the individual variances $D_{i_1,\ldots,i_l}$ are given by

$$D_{i_1\ldots i_l} = \int_{K^l} (f_{i_1\ldots i_l})^2 dx_{i_1} \ldots dx_{i_l} \tag{9.28}$$

Global sensitivity indices based on these variances are defined as (Sobol', 1993; Saltelli and Sobol', 1995; Homma and Saltelli, 1996)

$$S_{i_1\ldots i_l} = \frac{D_{i_1\ldots i_l}}{D}, \tag{9.29}$$

where $S_{i_1 \dots i_l}$ is the fractional contribution of the input set $\{x_{i_1,\dots,}x_{i_l}\}$ to the variance of the output. Although an ANOVA–HDMR might be used as a multivariate representation of the output, it is most useful as a sensitivity or uncertainty analysis of the model output. A thorough discussion of ANOVA–HDMR is contained in Chapter 8.

A significant drawback of employing ANOVA–HDMR is the need to compute the above integrals to extract each component function for systems with high dimensions $n \gg 10$. These high-dimensional integrals would likely need to be carried out by Monte Carlo integration, and a large number of sample points will be required to attain good accuracy. The computation of the sensitivities $S_{i_1 \dots i_l}$ requires the generation of $N \times 2n$ uniformly distributed random numbers and the number of model evaluations needed for Lth-order ANOVA–HDMR is given by

$$N \times \sum_{i=0}^{L} \frac{n!}{(n-i)!i!}, \tag{9.30}$$

where N is the sample size for the computation of each integral in the Lth-order ANOVA–HDMR expansion. Even if $L \ll n$, the cost of such an analysis can be very high, since reliable results will often call for $N \gg 10^3$. To circumvent this difficulty, a computationally more efficient cut-HDMR expansion will be introduced in Section 9.2.2 for more general model representation and reexpression into the ANOVA–HDMR form. Finally, although ANOVA–HDMR is often costly, if only the total contribution of each variable x_i to the variance is sought, then a simplification is possible using Monte Carlo sampling methods (Homma and Saltelli, 1996).

9.2.2 Cut-HDMR Expansion

In this case, X is defined as the space of functions taking finite value at the point $y = (y_1, y_2, \dots, y_n)$. The measure μ is taken as the Dirac measure located at the point $y = (y_1, y_2, \dots, y_n)$, i.e.

$$d\mu(x) = \prod_{i=1}^{n} \delta(x_i - y_i)\, dx_i \tag{9.31}$$

The point y will be called the 'cut' center. Cut-HDMR is an expression of the function $f(x)$ as a superposition of its values on lines, planes, and hyperplanes of higher orders passing through the cut center y. The exploration of the output surface $f(x)$ may be global, and the value of y is irrelevant if the expansion is taken out to convergence.

The component functions of $f(x)$ obtained within cut-HDMR are given as follows. The notation $f^{i_1 \dots i_l}(x_{i_1}, \dots, x_{i_l})$ stands for the function $f(x)$ with all the remaining variables, besides $x_{i_1}, \dots, x_{i_l}$ of the input vector, set to y, e.g. $f^i(x_i)$ stands for $f(y_1, \dots, y_{i-1}, x_i, y_{i+1}, \dots, y_n)$.

$$\begin{aligned}
f_0 &= P_0 f(x) = f(y), \\
f_i(x_i) &= P_i f(x) = f^i(x_i) - P_0 f(x), \\
f_{ij}(x_i, x_j) &= P_{ij} f(x) = f^{ij}(x_i, x_j) - P_i f(x) - P_j f(x) - P_0 f(x), \\
&\vdots \\
f_{i_1 \dots i_l}(x_{i_1}, \dots, x_{i_l}) &= P_{i_1 \dots i_l} f(x) = f^{i_1 \dots i_l}(x_{i_1}, \dots, x_{i_l}) - \sum_{\{j_1,\dots,j_{l-1}\} \subset \{i_1,\dots,i_l\}} P_{j_1 \dots j_{l-1}} f(x) \\
&\quad - \sum_{\{j_1,\dots,j_{l-2}\} \subset \{i_1,\dots,i_l\}} P_{j_1 \dots j_{l-2}} f(x) - \dots - \sum_j P_j f(x) - P_0 f(x),
\end{aligned} \tag{9.32}$$

with the kernel functions given by the expressions

$$
\begin{aligned}
K_0(x; x') &= \prod_{j=1}^{n} \delta(x'_j - y_j), \\
K_{i_1 \ldots i_l}(x; x') &= \prod_{j \in \{i_1, \ldots, i_l\}} [\delta(x_j - x'_j) - \delta(x'_j - y_j)] \prod_{j \notin \{i_1, \ldots, i_l\}} \delta(x'_j - y_j)
\end{aligned}
\tag{9.33}
$$

The cut-HDMR at Lth order with the cut center y is exact along the hypersurfaces of dimension L that pass through the point y. Although the dependence of cut-HDMR on the cut center y makes it appear like a local approximation, it is globally accurate if the approximation is taken out to a level where the high-order interactions are negligible. The computational or experimental cost of generating the cut-HDMR up to the Lth order, when it is used for interpolation purposes, is given by

$$
\sum_{i=0}^{L} \frac{n!}{(n-i)!\,i!}(s-1)^i, \tag{9.34}
$$

where s is the number of sample points taken along each axis. If convergence of the cut-HDMR expansion occurs at $L \ll n$, then the sum above is dominated by the Lth term, and considering $s \gg 1$ we get full space resolution at the computational labor of $(ns)^L/L!$. This result is in stark contrast with the conventional view of exponential scaling $\sim s^n$. Applications indicate that one may expect $L \approx 1$–3 for good quality results.

9.2.3 Comparison of ANOVA–HDMR and cut-HDMR

Several relationships exist between ANOVA–HDMR and cut-HDMR. The first of these concerns the relation between the component functions of ANOVA–HDMR and those of cut-HDMR given by

$$
f^{\text{ANOVA}}_{i_1,\ldots,i_l}(x_{i_1}, \ldots, x_{i_l}) = \int_{K^n} f^{\text{cut}}_{i_1,\ldots,i_l}(x_{i_1}, \ldots, x_{i_l})dy, \tag{9.35}
$$

where the integrand is a cut-HDMR expansion function implicitly understood to depend on y. Recall that $f^{\text{cut}}_{i_1 \ldots i_l}$ is exact along the hyperplanes through the cut center y. Thus, $f^{\text{ANOVA}}_{i_1 \ldots i_l}$ is good on average throughout the subvolumes K^l, $l \leqslant n$.

Computation of the ANOVA–HDMR expansion of f involves multidimensional Monte Carlo integrations of f, which may be costly. If the objective is to compute the ANOVA–HDMR component functions, then an efficient two-step approach can be envisioned as follows (note that computation of cut-HDMR functions requires that the function $f(x)$ be sampled at special points, i.e., it can be computed if one has control over the input variables):

1. Compute the cut-HDMR component functions and approximate the function $f(x)$ by

$$
f(x) \approx f^{\text{cut}}(X) = f^{\text{cut}}_0 + \sum_i f^{\text{cut}}_i(x_i) + \cdots. \tag{9.36}
$$

2. Compute the ANOVA–HDMR expansion of $f^{\text{cut}}(x)$ Assuming that Lth-order cut-HDMR approximates $f(x)$ to good accuracy, then $f^{\text{cut}}(x)$ will consist of functions of at most L variables. This situation is an illustration of the more general comparative convergence behavior of different HDMRs mentioned earlier. When $L \ll n$, quadrature techniques can

be used for the integrations of the function $f(x)$ to compute the ANOVA–HDMR functions in a computationally accurate and efficienty way.

A graphical representation of the above scheme is shown in Figure 9.1.

If the cut-HDMR approximates $f(x)$ to good accuracy at the Lth order, then the computation route (2) + (3) will give a very close approximation to the true $f^{\text{ANOVA}}(x)$ and the computational effort required to construct $f^{\text{ANOVA}}(x)$ will be much less. The computational cost of generating the Lth-order cut-HDMR is given by Equation (9.34). In contrast, direct computation of Lth-order ANOVA–HDMR calls for the generation of $N \times 2n$ random numbers, and the number of model evaluations for the Lth-order ANOVA–HDMR is given by Equation (9.30). Thus, the ratio of the computational cost of cut-HDMR to that of ANOVA–HDMR is given by

$$\rho \equiv \frac{\sum_{i=0}^{L} \frac{n!}{(n-i)!\,i!} \frac{(s-1)^i}{N}}{\sum_{i=0}^{L} \frac{n!}{(n-i)!\,i!}}. \tag{9.37}$$

The condition $(s-1)^L < N$ ensures that the ratio ρ is less than 1. Assuming that cut-HDMR converges at low order ($L \approx 1$–3), the ratio will satisfy $\rho \ll 1$, observing that typically $s \approx 10$ and $N \gg 10^3$. Computation of ANOVA–HDMR using the route (3) after obtaining $f^{\text{cut}}(x)$ is straightforward, involving integrals of dimensions no larger than L. Since L is assumed to be small on physical grounds, the integrations can be carried out very accurately via quadrature methods with only a few points, while using Monte Carlo integration directly in ANOVA–HDMR to the same accuracy will call for a very large number of points N. This argument is illustrated in the following example.

The model is an analytical expression of three variables (Homma and Saltelli, 1996)

$$f(x_1, x_2, x_3) = \sin x_1 + 7(\sin x_2)^2 + 0.1(x_3)^4 \sin x_1, \tag{9.38}$$

where the joint probability density of the inputs are given by

$$\begin{aligned} p(x_1, x_2, x_3) &= \prod_{i=1}^{3} p_i(x_i), \\ p_i(x_i) &= \begin{cases} \dfrac{1}{2\pi} & (-\pi \leqslant x_i \leqslant \pi), \\ 0 & (x_i < \pi, x_i > \pi). \end{cases} \end{aligned} \tag{9.39}$$

Suppose that the goal is to compute the global sensitivity indices $S_{i_1 i_2 \ldots i_l}, l \leqslant 3$, defined in Equation (9.29). A sample size of 1024 is used for the evaluation of each of the integrals appearing in the ANOVA–HDMR expansion of $f(x_1, x_2, x_3)$, so that one needs 1024×7 evaluations. Note that the term $f_{123}^{\text{ANOVA}}(x_1, x_2, x_3)$ is determined by subtracting all the other component functions from the output function. The exact sensitivities and their approximation by Monte Carlo simulations are given respectively, in the first and second columns of Table 9.1. Alternatively, we tested the route (2) + (3) to compute the sensitivities $S_{i_1 i_2 \ldots i_l}, l \leqslant 3$. The cut-HDMR expansion of the output to second order was computed, and then the ANOVA–HDMR expansion of the output was obtained using this expansion. The necessary integrations were computed by Gaussian-quadrature integrations, and cut-HDMR was evaluated at these points. The results shown in Table 9.1 are for $s = 11$ quadrature points, which give values good to at least four decimal places.

Table 9.1 Comparison of routes in Figure 9.1 to ANOVA.

Variables	Exact	Route (1)	Route (2) + (3) (11 points)
x_1	0.3138	0.3230	0.3138
x_2	0.4424	0.4390	0.4424
x_3	0.0	0.0078	0.0
x_1, x_2	0.0	0.0	0.0
x_1, x_3	0.2436	0.2354	0.2436
x_2, x_3	0.0	0.0	0.0
x_1, x_2, x_3	0.0	0.0	0.0

The ratio ρ, which represents the computational savings gained by the computation route (2) + (3), is given by

$$\rho = \frac{\sum_{i=0}^{2} \frac{3!}{(3-i)!\,i!} 10^i}{1024 \times 7} = 0.046. \tag{9.40}$$

At $s = 8$, the accuracy of route (2) + (3) is the same as for route (1) in the table, but in this case the savings by route (2) + (3) are even better at $\rho = 0.023$. In general, the savings above arise because step (1) calls for integrations over f of dimensions $n, n-1, \ldots, n-L$. In contrast, step (2) calls for no integrations, and step (3) only calls for integrations of dimensions 1, 2, , . . . , L and we expect that $L \ll n$. It is evident that the dimension of the integrals to compute via steps (2) and (3) is the complement of those via step (1).

9.3 EXAMPLES

For illustration of the capabilities of cut-HDMR, we give the following examples. These were taken from a set of test cases given in Chapter 2. In each case, the 'model' is actually an analytic function, but for HDMR analysis purposes, we treat it as a source of input–output information to be processed. In this regard, an important point is that only Example 9.4 is a true physical system, while the others are arbitrary mathematical constructions. This comment is important, since we expect HDMR to be broadly applicable to physically derived systems. Examples 9.1 and 9.2 are exactly representable by cut-HDMR at first order, and they are included only to confirm that the procedure properly identifies this behavior.

For the outputs in the examples below, the cut-HDMR was carried out at most to second order. The cut centers used for the construction of the expansions were chosen arbitrarily. Cubic spline interpolation was used for constructing first-order cut-HDMR functions and bicubic splines were used for interpolation of second-order functions. Two criteria of quality were considered for the cut-HDMR approximation $f^{\text{cut}}(x)$ in representing the the output function $f(x)$. They are defined as follows:

1. $\|f(x) - f^{\text{cut}}(x)\|_{L^\infty} = \sup_{x \in \Omega} \|f(x) - f^{\text{cut}}(x)\|$,
2. $\|f(x) - f^{\text{cut}}(x)\|_{L^2} = \int_\Omega [f(x) - f^{\text{cut}}(x)]^2 dx$,

where Ω is the operating region of the input variables. Note that when the domain Ω is bounded in $\mathbb{R}^n$, convergence in the norm $\|\cdot\|_{L^\infty}$ implies convergence in the norm $\|\cdot\|_{L^2}$.

Example 9.1. The model (Model 1 in Section 2.9.1 in Chapter 2) is linear:

$$f(x) = \sum_{i=1}^{k} x_i, \qquad k = 3, \tag{9.41}$$

and x_i are uniformly distributed in the range $(\bar{x}_i - \sigma_i, \bar{x}_i + \sigma_i)$, with $\bar{x}_i = 3^{i-1}$ and $\sigma_i = 0.5\bar{x}_i$.

We have chosen the cut center $y = (1, 3, 9)$ and have taken 10 samples in each input variable's direction. As expected, the second-order cut-HDMR functions have been found to be close to zero (not exactly zero, owing to interpolation errors) and may be discarded. The number of samples necessary to construct first-order cut-HDMR in this case is 28. The errors in both norms are zero, neglecting the interpolation errors.

Example 9.2. The model (Model 4 in Section 2.9.2 in Chapter 2) is an additive nonlinear function

$$f(x) = x_1 + x_2^4, \tag{9.42}$$

where x_1 and x_2 are uniformly distributed over the two possible ranges

(a) $(0 \leqslant x_1 \leqslant 1)$ and $(0 \leqslant x_2 \leqslant 1)$;
(b) $(0 \leqslant x_1 \leqslant 5)$ and $(0 \leqslant x_2 \leqslant 5)$.

This is again a case where cut-HDMR is exact at first order. However, in order to simulate the situation where we do not know the form of the function in advance, we carried out a cut-HDMR to second order. In both ranges, we have chosen the cut center as $y = (0.5, 0.5)$.

(a) The second-order functions were found to be close to zero, and were discarded. Ten sample values in each input variable were used to construct the first-order cut-HDMR, requiring a total of 19 samples of the function. In this small range, the component function $f_1(x_1)$ was more pronounced, i.e., it has a bigger contribution to the output.
(b) The independent effect $f_2(x_2)$ of the input x_2 is more pronounced in this domain.

We again used 19 samples of the output to construct the first-order cut-HDMR.

Example 9.3. The model (Model b in Section 2.9.3 in Chapter 2) is a nonlinear analytical case:

$$f(x) = \frac{x_2^4}{x_1^2}, \tag{9.43}$$

where x_i are uniformly distributed in the range $(1 - \epsilon, 1 + \epsilon)$ with $\epsilon = 10^{-1}$. The cut center was chosen as $y = (1, 1)$. Five samples of the output were taken in each input variable, with the overall number of samples being 9. A first order cut-HDMR was constructed and the following error norms were computed; the respective norms of the function $f(x)$ are also given for comparison:

$$\|f(x) - f^{\text{cut}}(x)\|_{L^\infty} = 0.1, \quad \|f(x)\|_{L^\infty} = 1.8,$$
$$\|f(x) - f^{\text{cut}}(x)\|_{L^2} = 3 \times 10^{-5}, \quad \|f(x)\|_{L^2} = 0.045,$$

The L^2 norm indicates that the average error over the domain is 0.07%, while the L^∞ norm shows that the worst point on the domain is in error by 6%.

Example 9.4. The model (Model 11 in Section 9.9.3 in Chapter 2) corresponds to a nonlinear chemical reaction

$$A + A \xrightarrow{k} \text{products}, \tag{9.44}$$

with rate constant $k = x_1 e^{-x_2/T}$. The equation governing the concentration f of species A is

$$\frac{df}{dt} = -2kf^2, \tag{9.45}$$

with some initial condition x_3. The following ranges for the inputs x_1, x_2, and x_3 were chosen:

$$x_1, x_2 \in [2.5, 7.5], \qquad x_3 \in [1.0, 1.2] \tag{9.46}$$

The rate constant parameters are taken with broad uncertainty. We have also assigned $T = 5$. The cut-HDMR representation was carried out for the mapping $f(t) = f(x_1, x_2, x_3)$, with $t = 0.5$. The functional form of the output is given by

$$f(x_1, x_2, x_3) = \frac{1}{(0.5x_1 e^{-x_2/5} + 1/x_3)}. \tag{9.47}$$

The cut center is $y = (5, 5, 1)$. A first-order cut-HDMR provided satisfactory accuracy with the error norm

$$\|f(x) - f^{\text{cut}}(x)\|_{L^2} = 0.026, \qquad \|f(x)\|_{L^2} = 1.5,$$

indicating an overall cut-HDMR prediction error of 2%. In a second experiment, we chose a broader range

$$x_1, x_2 \in [2.5, 12.5], \qquad x_3 \in [1.0, 1.2] \tag{9.48}$$

of rate constant uncertainty, and computed the following error for the first-order cut-HDMR

$$\|f(x) - f^{\text{cut}}(x)\|_{L^2} = 0.818, \qquad \|f(x)\|_{L^2} = 26.$$

This corresponds to an overall error of just 3%.

Example 9.5. This case is a nonlinear computational mode with six input variables (similar to Model 10 in Section 2.9.3 in Chapter 2):

$$f(x) = \beta_0 + \sum_{i}^{6} \beta_i w_i + \sum_{i<j}^{6} \beta_{ij} w_i w_j + \sum_{i<j<l}^{6} \beta_{ijl} w_i w_j w_l + \sum_{i<j<l<s}^{6} \beta_{ijls} w_i w_j w_l w_s, \tag{9.49}$$

where $w_i = 2(x_i - 0.5)$, except for $i = 3, 5$, where $w_i = 2[1.1x_i/(x_i + 0.1) - 0.5]$. The coefficients have the values

$$\begin{aligned} \beta_0 &= 10, \\ \beta_i &= +20 \quad \text{for} \quad i = 1, \ldots, 4, \\ \beta_{ij} &= -15 \quad \text{for} \quad i < j \leqslant 4, \\ \beta_{ijl} &= -10 \quad \text{for} \quad i < j < l \leqslant 4, \\ \beta_{ijls} &= +5 \quad\ \ \text{for} \quad i < j < l < s \leqslant 4. \end{aligned} \tag{9.50}$$

The remaining first- and second-order coefficients are generated independently from a normal distribution with zero mean and unit standard deviation; the remaining third- and fourth-order coefficients are set to zero. The domain Ω of the input variables is $K^6 = [0,1]^6$.

Cut-HDMR was applied to the above model at second order. We carried out different experiments with different cut points and different coefficients of third- and fourth-order polynomials. As the error criterion, we have computed the following quantity for an ensemble of 1000 uniformly distributed points in K^6:

$$\boldsymbol{\varepsilon} \equiv \left\{ i = 1,\ldots,1000 : \left| \frac{f^i(x) - f^i_{\text{cut}}(x)}{f^i(x)} \right| \right\} \tag{9.51}$$

(i.e., $i = 1,\ldots,1000$ here). However, this criterion can be misleading at the points where the model output is close to zero. In fact, we observed large outliers in these 1000-sample Monte Carlo runs. We redefined the error vector $\boldsymbol{\varepsilon}$ again such that large values are not permitted. Since the frequency of these large values is also important, they were recorded too, i.e. we consider the following redefined error vector $\hat{\boldsymbol{\varepsilon}}$:

$$\hat{\boldsymbol{\varepsilon}} \equiv \{\epsilon_i : \epsilon_i < 1\} \tag{9.52}$$

and its length Length $[\hat{\boldsymbol{\varepsilon}}]$. In Tables 9.2–9.5, we first indicate the cut center for the experiment being used. The mean, median, and the standard deviation (SD) about the mean of the above vector $\hat{\boldsymbol{\varepsilon}}$ corresponding to different values of third- and fourth-order coefficients are also shown. The length of the vector $\hat{\boldsymbol{\varepsilon}}$ is also given, since it is a measure of the domain where the relative error is smaller than one. Note that for all of the experiments,

Table 9.2 HDMR accuracy. Cut center $y = (0.5, 0.5, 0.5/6, 0.5, 0.5/6, 0.5)$.

	$\beta_{ijl} = -10, \beta_{ijls} = +5$	$\beta_{ijl} = -1, \beta_{ijls} = +1$
Mean $[\hat{\boldsymbol{\varepsilon}}]$	0.1310	0.0406
Median $[\hat{\boldsymbol{\varepsilon}}]$	0.0785	0.0117
SD $[\hat{\boldsymbol{\varepsilon}}]$	0.1652	0.0957
Length $[\hat{\boldsymbol{\varepsilon}}]$	950	994

Table 9.3 HDMR accuracy. Cut center $y = (1,1,1,0,0,0)$.

	$\beta_{ijl} = -10, \beta_{ijls} = +5$	$\beta_{ijl} = -1, \beta_{ijls} = +1$
Mean $[\hat{\boldsymbol{\varepsilon}}]$	0.2199	0.0349
Median $[\hat{\boldsymbol{\varepsilon}}]$	0.1389	0.049
SD $[\hat{\boldsymbol{\varepsilon}}]$	0.2233	0.0805
Length $[\hat{\boldsymbol{\varepsilon}}]$	908	982

Table 9.4 HDMR accuracy. Cut center $y = (0.278, 0.068, 0.711, 0.184, 0.915, 0.175)$.

	$\beta_{ijl} = -10, \beta_{ijls} = +5$	$\beta_{ijl} = -1, \beta_{ijls} = +1$
Mean $[\hat{\boldsymbol{\varepsilon}}]$	0.1204	0.0313
Median $[\hat{\boldsymbol{\varepsilon}}]$	0.0555	0.068
SD $[\hat{\boldsymbol{\varepsilon}}]$	0.1692	0.0737
Length $[\hat{\boldsymbol{\varepsilon}}]$	936	993

Table 9.5 HDMR accuracy. Cut center $y = (0.684, 0.394, 0.946, 0.433, 0.185, 0.982)$.

	$\beta_{ijl} = -10, \beta_{ijls} = +5$	$\beta_{ijl} = -1, \beta_{ijls} = +1$
Mean $[\hat{\varepsilon}]$	0.0827	0.0195
Median $[\hat{\varepsilon}]$	0.0286	0.026
SD $[\hat{\varepsilon}]$	0.1308	0.0540
Length $[\hat{\varepsilon}]$	968	990

the coefficients of the monomials have the same fixed value. The inputs w_i are all zero at the first cut center given below.

In the following experiments, we have chosen cut centers as random in K^6.

The mean and the standard deviation of the vector $\hat{\epsilon}$ are the most meaningful quantities for measuring the quality of the second-order cut-HDMR approximation. The accuracy of the approximation always improves when the coefficient of the third-order and fourth-order terms are decreased, which is understandable since a second-order cut-HDMR consists of only independent actions of the inputs and their pairwise cooperative effects. When the coefficients are held constant, convergence to the true model output occurs at the same order for the different cut centers. It was observed that all the independent action terms, i.e. $f_i(x_i)$, were significant and pair-interaction terms of the other inputs with the inputs x_5 and x_6 were found to be insignificant. Ten sample points were taken for each input variable, totalling 1270 evaluations of the model output to get a second-order cut-HDMR. The overall error of the second-order cut-HDMR from the tables is globally about 10–20%, which is quite good considering that an equivalent sampling of the six-dimensional space to the same resolution would require 10^6 points. This corresponds to an order of 10^3 savings of effort (clearly, the function is simple in this case, but other more complex physical systems have also indicated similar savings).

9.4 APPLICATIONS OF HDMR

The HDMR technique is a tool to enhance modelling where the interest centers on the input–output relationships. There is a broad family of applications to exploit HDMR capabilities. Presently, applications have been made to chemical kinetics (Shorter *et al.*, 1999), radiative transport (Shorter *et al.*, 2000) materials discovery (Shim and Rabitz, 1998; Rabitz and Shim, 1999), and statistical analysis (Saltelli and Hjorth, 1995). The particular applications of HDMR discussed below are not exhaustive, and they should be regarded as representative.

9.4.1 Fully Equivalent Operational Models

All the HDMR applications in Sections 9.4.2–9.4.7 are associated with exploring the role and relationships amongst the model variables. An application HDMR, to some degree underlying the ability to execute the other applications in Sections 9.4.2–9.4.7, is model component replacement by highly efficient equivalent forms. This operation takes advantage of the fact that complex models are typically broken into various components (e.g., involving chemistry, mechanical coupling, mass transport, etc.). These components of an overall model are often treated by numerical splitting techniques, thus isolating them for efficient replacement with equivalent HDMRs. Some components (e.g., radiation transport in weather modelling) can be exceedingly costly contributions to the computational effort, since they correspond to high-overhead operations that are repeated many times in the course of model execution. In some applications, the model 'component' may be the entire model. The input-output space can be of high dimension n, but typically it is expected to have systematic structure. Thus, one could envision 'learning' the model input–output behavior through the observation of model runs for subsequent encapsulation of the information into an HDMR expansion. Once the expansion functions f_0, f_i, f_{ij}, etc. are learned, they may be reused as a basis to predict output behavior at any other point x in the space called upon in additional execution of the model. The HDMR expansion obtained in this way corresponds to a fully equivalent operational model (FEOM), which could replace the original one (i.e., the model component(s)). The logic will be most appropriate for model components that involve very large numbers of computational operations that are repeated many times in executing the overall model. The computational savings using an FEOM can be dramatic (Rabitz *et al.*, 1999). A study on atmospheric radiation transport treating water vapor and temperature as input column functions led to a computational saving by a factor of the order of 10^3 for the atmospheric heating rate, with errors no larger than a few percent at all altitudes (Shorter *et al.*, 2000). In an additional study (Shorter *et al.*, 1999), an FEOM acted as an integrator of a set of coupled ordinary differential equations describing atmospheric chemical kinetics. In this case, the FEOM variables $(x_1, x_2, \ldots, x_N)$ consisted of the initial conditions for the system of equations and the outputs were the species at 24 hours later. High-quality results were produced over an interval of many years by repeated use of the FEOM from one day to the next at a computational savings of the order of 10^3 over the original code while maintaining very good accuracy. In general, the method may be applied to an autonomous set of differential equations of the form

$$\begin{aligned} \dot{x} &= h(x), \\ x(t_0) &= x_0. \end{aligned} \tag{9.53}$$

After an appropriate discretization over the vector of input variables x_0, the dynamics of the above system can be approximated by a discrete map of the form

$$x_{m+1} = H(x_m), \quad m = 0, 1, 2, \ldots, \tag{9.54}$$

or equivalently

$$x_m = H^m(x_0), \quad m = 1, 2, \ldots, \tag{9.55}$$

where m indicates the time discretization. Importantly, the map can be discretized on steps much larger than traditional integrator time-steps (cf. the 24-hour HDMR steps mentioned above, while the GEAR integration took about 2000 steps over one day). The map H can be approximated by a cut-HDMR representation and it can be repeatedly used as an integrator of the dynamical system. The behavior of the HDMR approximation to the discrete map when the system exhibits chaotic or possibly other types of unstable behavior is an open question.

FEOMs have potentially broad applicability in many areas, and importantly the FEOM replacement of a model component (e.g., radiation transport) can be handled as a simple swap for an existing routine without alteration of the remainder of the code. The original model component would likely still be retained, but only called upon if a new required input point x fell outside of the regime explored in generating the FEOM or if the FEOM prediction is estimated to have significant errors. In this case, the output from the new model component run would also serve to enhance the FEOM for its further use. Following this logic, the creation of a FEOM for model component replacement could be performed as a background operation while the parent code is being exercised in a normal fashion. The latter approach would correspond to generating a FEOM from a space of variables x chosen with a probability distribution dictated by the physical model.

9.4.2 Identification of Key Variables and their Interrelationships

The terms in the HDMR expansion can be regarded as the generalized sensitivities of the output function $f(x_1, x_2, \ldots, x_n)$ with respect to groups of variables. Each of the functions $f_i(x_i), f_{ij}(x_i, x_j), \ldots$ reveals a unique contribution of the variables separately or cooperatively on the output f. These comments apply to any of the HDMR expansions, and the choice of the expansion can give different physical interpretation and quantitative measures of the variable cooperativity. The choice of HDMR expansion to employ will often be guided by physical considerations. For example, if a clear reference state y exists, then the cut-HDMR is natural for variable analysis in relation to y. If statistical variance analysis is needed, then ANOVA–HDMR is appropriate (Sobol', 1993; Saltelli and Sobol', 1995), as discussed in Section 9.4.3. An example in a study of atmospheric radiation transport (Shorter *et al.*, 2000) involved a cut-HDMR expansion of the output atmospheric heating rate with respect to the input column densities of chemical species. Discretization of the functions led to 62 input variables, but the HDMR expansions revealed that many variables were important acting independently and only a modest number of pair-correlated contributions existed. Techniques from traditional gradient-based sensitivity analysis can reveal similar relationships (Rabitz, 1989), but only over small operating uncertaintly or scenario variations around a nominal set of conditions. In contrast, the HDMR expansion puts no restrictions on the magnitude, shape, or variable range of the individual expansion functions, to reveal the true underlying variable relationships.

9.4.3 Global Uncertainty Assessments

Expressing the output uncertainty in terms of model input uncertainty has been a topic of prime interest in virtually all areas of modeling. Monte Carlo sampling (Shreider, 1967) and perturbative sensitivity analysis (Rabitz, 1989) are the traditional approaches to these problems. In both cases, a probability measure is defined on $\mathbb{R}^n$, and the goal is to calculate the mean $\bar{f} = \mathbf{E}f$ and standard deviation $\sigma_f = \mathbf{E}(f - \bar{f}^2)$ of the output f, as well as to reveal the contributions from the various input variables and their interrelationships. One may show that the ANOVA–HDMR expansion has a direct statistical correlation interpretation (Sobol', 1993; Saltelli and Sobol', 1995). Each term of the expansion is associated with a particular contribution to the variance of the output. Orthogonality of the individual terms ensures this behavior. The overall variance σ_f can be written as follows:

$$\begin{aligned}
\sigma_f &= \sum_i \sigma_i + \sum_{i<j} \sigma_{ij} + \cdots, \\
\sigma_i &= \int f_i^2(x_i)\, d\mu(x_i), \\
\sigma_{ij} &= \int f_{ij}^2 d\mu_i(x_i)\, d\mu_j(x_j), \\
&\ \vdots
\end{aligned} \tag{9.56}$$

Monte Carlo integration (Shreider, 1967) would be a means for evaluating the above integrals. Illustrations of these uncertainty analyses have been made for applications to environmental modelling (Saltelli and Sobol', 1995). The information gained from the decomposition of the variance σ_f into its subcomponents σ_i, σ_{ij}, etc. can be most valuable for attaining a physical understanding of the origins of uncertainty.

9.4.4 Quantitative Risk Assessment

Environmental, industrial, and economic modelling are often performed for the ultimate purpose of providing an assessment of the risk associated with some action that is subject to remediation. There is a serious need for more quantitative and efficient means for performing these assessments. For a quantitative risk assessment, it is generally necessary to split the original set of input variables into two components $(x_i, \ldots, x_s; y_1, \ldots, y_r; r + s = N)$, where the set $\{x_i\}$ will be referred to as scenario variables under human control (e.g., industrial emissions, etc.) and the set $\{y_j\}$ correspond to all other model variables (e.g., chemical rates, transport coefficients, mechanical properties, etc.) that are present and subject to some degree of uncertainty. Typically risk is associated with identifying whether the output exceeds (or goes below) a critical value $f > f_c$. The risk R is defined as the probability $P(f > f_c)$ for this event to occur while simultaneously taking into account the uncertainty amongst the model variables $\{y_i\}$. Thus, the risk is defined as

$$R = \int d\mu(x)\, d\upsilon(y) H(f(x; y) - f_c), \tag{9.57}$$

and the variance of the risk is

$$\sigma_R = -R^2 + \int d\upsilon(y) d\mu(x)\, d\mu(x') H(f(x; y) - f_c) H(f(x'; y) - f_c), \tag{9.58}$$

where the distribution of the variable group y is given by υ and $H(z)$ is the Heaviside function. We may take special advantage of the HDMR expansion in evaluating the risk and the variance around the risk in a quantitative fashion. These tasks are facilitated by the ability to rapidly evaluate $f(x; y)$. In addition, it will be possible to determine the portion of the scenario variables $(x_1, \ldots, x_r)$ that contribute independently or in a correlated fashion to the risk (cf. Section 9.4.2). The analysis will not only provide the risk R, but also a quality assurance on the risk through its variance σ_R due to the model variables $\{y_j\}$ and their uncertainty.

9.4.5 Inverse Problems

Information on the input variables of an input–output relationship is often incomplete. If the objective is to gain knowledge about these unknown input variables from the available output data, then the task becomes an inverse problem. This problem is typically ill-posed (Tikhonov and Arsenin, 1977) in the sense that there are usually more unknowns than available measurements. In addition, measurements (i.e., function(al)s of the outputs) will inevitably contain errors, and this further contaminates the quality of the information to be extracted. These issues are typically dealt with through regularization based on a local linear sensistivity analysis mapping between the data-model deviations and the sought-after input variables (Tikhonov and Arsenin, 1977; Ho and Rabitz, 1993). The linearization is an approximation to the true model, and can lead to algorithmic instabilities and even a false solution. For inversion applications, the HDMR expansion needs to be extended to consider the case of multiple outputs $f^l(x_1, x_2, \ldots, x_n)$ labeled by the index $l = 1, 2, \ldots$. This index corresponds to distinct observations (e.g., chemical species as well as their spatial locations and/or temporal behavior). The forward problem consists of expressing $f^1, f^2, \ldots$ in terms of the input $x_1, x_2, \ldots, x_n$. The inverse problem is to map the data back to identify the input variables. Reversing the roles of the fs and xs, one can construct an inverse-HDMR expansion for xs in terms of fs. A regularization functional also can be introduced if there is *a priori* knowledge of the input variables to be inverted. An open question is how to express the HDMR expansion of the inverse map in terms of the HDMR expansion of the forward map and analyze the relation between the convergence behaviors of the two expansions.

9.4.6 Financial and Econometric Applications

For many financial applications, one wishes to relate (regress) a response y_t to several variables $x_{1t}, x_{2t}, \ldots, x_{lt}$ (t signifies the time component of the financial or econometric data). Parametric regression models have the structure $y_t = f(\Theta; x_{1t} \ldots x_{lt})$, where the form of the function f is known and the parameter set Θ is to be estimated by satisfying some optimization criteria. Nonparametric regression of y_t takes the form $y_t = f(x_{1t} \ldots x_{lt})$, with an unknown function f. Modelling with nonlinear nonparametric regression is computationally very intensive owing to the curse of dimensionality mentioned before. Sampling the input space is of exponential cost in the number of input variables. Various learning algorithms have been devised to model the output to good accuracy while being computationally cheaper than a full sampling of the space. Nonparametric regressive schemes include *average derivative estimators* (Stoker, 1986), *artificial neural networks* (White, 1992), *radial basis function interpolators* (Poggio and Girosi, 1990), and *projection pursuit regression* (Friedman and Stuetzle, 1981). The main disadvantage of these nonlinear models is that there is no widely accepted procedure for choosing one specific model over another. *A priori* knowledge of the system to be modelled is crucial to the success of the model. HDMR is a nonparametric nonregressive model, and as such it has advantages for direct applications to problems in finance and time series econometrics. One possible application is the modelling of the price of financial securities that are dependent on several factors, i.e., parameters of the stochastic differential equation governing the evolution of the price, interest rate, time to maturity of the security, and other factors specific to the pay-off structure of the particular security.

9.4.7 Laboratory Applications of HDMR

In many cases, laboratory or field experiments are performed for the explicit purpose of exploring broad regions of the input variable space. Applications of this type abound in many areas, including industrial processes, environmental studies, materials discovery, engineering control, etc. The most challenging and interesting of these problems arises when the dimension n of the space of inputs is large. This task may be recognized as the learning of the output mapping $f: \Omega \rightarrow \mathbb{R}$, where Ω is the space of input values. Traditional statistical sampling techniques, including factorial design, are inadequate, since they provide no guidance based on the system behavior. In this case, the notion of the 'model' in the HDMR sense is understood to mean the realization of laboratory or field experiments and representing them in a logical fashion to provide an efficient and thorough sampling for analysis of the results. Given the above objective, the HDMR expansion provides an ideal approach. When the HDMR expansion converges at low orders, the original NP-complete problem is reduced to one of polynomial scaling in the dimension of the input space. As experiments can be exceedingly expensive to perform, this saving can be critical. One particular application of the cut-HDMR has been made for semiconductor electronic band gap as the observable material property, with the input variables being the material compositions. It was shown that the band gap of the material $Ga_\alpha In_{1-\alpha} P_\beta As_{1-\beta}$ could be reliably described as a function of α and β by a first order cut-HDMR centered at $\alpha = 0, \beta = 0$, corresponding to laboratory band gap input variables for the materials $Ga_\alpha In_{1-\alpha}As$ and $InP_\beta As_{1-\beta}$. In this case, $n = 2$, and this is a modestly 'high'-dimensional semiconductor system. Although each application will have its own features, the domain of combinatorial synthesis poses an intriguing opportunity, since in most realistic cases, one has $n \gg 1$ and there will very likely be correlated behavior amongst the input materials or chemical variables. One could envision active feedback between the ongoing experiments and the development of an HDMR to guide the subsequent experiments to rapidly converge on the desired library of materials or compounds for application purposes (Rabitz and Shim, 1999).

9.5 CONCLUSIONS

This chapter has introduced a family of nonparametric multivariate approximation/interpolation schemes for physically based functions with a large number of input variables. Among the various applications of the multivariate interpolation theory, interest in this book centers mostly on the sensitivity and uncertainty analysis aspects. A specific form of the high-dimensional model representations (HDMR) reveals the correlations among the input variables as reflected upon the model output and the nature of the metric in the variable space. Each component function in an HDMR gives the specific contribution of an input or a set of inputs to the output. The HDMRs are useful if they can represent the output to good accuracy at sufficiently low orders. As argued in this chapter for most well-defined physical systems, high-order correlated behavior of the input variables is expected to be weak and HDMRs are designed to capture this effect. It is interesting that the greatest body of evidence supporting this statement comes from the multivariate statistical analysis of many systems where rarely more than covariances are needed to capture the physical behavior.

General HDMR expansions are similar in form to the ANOVA expansion used in statistics to analyze the variance of a physical quantity. HDMR is a hierarchical description of the multivariate function $f(x)$ by a sum of component functions of fewer variables. HDMR does not assume an *a priori* parametric form, and as such it is not a fitting algorithm like other regressive multivariate approximation schemes. As shown in Section 9.2, the HDMR functions are optimal with respect to a suitably defined norm, with the optimality criterion being a quadratic cost functional. However, the HDMR avoids an explicit optimization (fitting) procedure to represent the output function $f(x)$. For example, the cut-HDMR approach only calls for evaluation of the output on 'specified' lines, planes, and hyperplanes of higher dimensions to construct the expansion (9.1).

The high computational complexity of representing multivariate functions was one of the motivations of the present work. Even if the ultimate aim is the sensitivity or uncertainty analysis of the model output, this complexity of the representation problem is usually the main concern. Without *a priori* information on the nature of the output, multivariate function approximations suffer from the *curse of dimensionality*. Sampling in the n-dimensional space scales exponentially with the dimension n. It is well known that the number of samples needed to achieve a given degree of accuracy depends on the dimension of the input vector and the smoothness of the function to be approximated. Without any regularization of the output function, one needs s^n sample points to approximate/interpolate the function to a resolution of $1/s$. Therefore, for physical systems with many input variables, a blind reconstruction of the output function $f(x)$ from its sample values is virtually impossible. In this respect, one should consider the storage and computational limitations in constructing $f(x)$. If the time required to obtain a single value of the output is sufficiently short, then representation of the multivariate output may not be difficult. However, when this time is of the order of minutes or longer for a computational model, then representation of the output in high dimensions becomes impossible unless some regularization on the output is imposed. The regularization implicitly imposed by the HDMR approximation is that the high-order correlated effects of the inputs upon the output are negligible. Similar comments apply for representing the input–output behavior of experimental systems, where the curse of dimensionality can be even more critical.

A number of multivariate approximation schemes exist to circumvent this curse of dimensionality. They are inspired by a theorem of Kolmogorov, which states that any multivariate function can be written as a superposition of functions of a suitable set of variables. In the introduction, we mentioned three of these approximation schemes, collectively known as *learning networks*. Each of these schemes can be preferred over another for a specific problem; however, there are no general rules that determine which one is suitable for a specific problem. This comment comes as no surprise, since there is no universal approximation scheme that will succeed in representing every function. The HDMR expansions are based on a general ansatz that for physical systems the cooperative effect of a group of input variables upon the output decreases as the number of inputs in this group increases. With this ansatz, the first few lowest-order cooperative terms are often enough to approximate the output to good accuracy. The computational advantage gained by HDMR (when it converges at low order) is significant. Assuming that a second-order HDMR is an accurate representation of the output, the computational complexity scales quadratically with the number of input variables. Although a thorough analysis of the convergence behavior of the HDMR expansion is not given in this chapter, numerical tests tried on a number of systems suggest that low-order terms will represent the output to good accuracy. The systems include chemical kinetics modelling (Shorter *et al.*, 1999), radiative transport modelling (Shorter *et al.*, 2000), solid state material modelling and experiments (Rabitz and Shim, 1998; Shim and Rabitz, 1998), as well as testing on various mathematically defined tasks that we have addressed in this chapter. We emphasize that HDMR is designed to deal with well-defined physical systems where the input variables are rationally chosen to have a nominally specific action or a role. The latter circumstance is natural in setting up physical systems, but no such guidance exists for arbitrary mathematical functions, notwithstanding the theorem of Kolmogorov. These comments are important to keep in mind when exploiting HDMRs.

In Section 9.2, we introduced the general formulation of HDMR and presented two distinct HDMR expansions. The ANOVA–HDMR useful in statistics called for the computation of multidimensional integrals, which is quite prohibitive in high dimensions at good accuracy. Although Monte Carlo integration is viable in high dimensions, one still needs a large number of samples values to carry out the necessary integrations. The cut-HDMR expansion does not necessitate the computation of any multidimensional integrals. It uses the sample values of the output on lines, planes, and hyperplanes of higher dimensions passing through a reference point, and constructs the function from these values according to the formulae (9.32). A very important result connecting ANOVA–HDMR and cut-HDMR is that the ANOVA–HDMR approximation $f^{\text{ANOVA}}(x)$ of the function $f(x)$ can be obtained in a much more efficient manner from $f^{\text{cut}}(x)$ taken to convergence. In this case, the ANOVA–HDMR of $f^{\text{cut}}(x)$ is a good approximation to $f^{\text{ANOVA}}(x)$, and computing the former expansion is often expected to be much less intensive compared with that of the latter, since the dimension of the integrals to be computed is smaller. Given the low dimension of the integrals, accurate quadrature techniques could be very efficient. In this case, the sampling of the cut-HDMR subspaces could be performed at the particular quadrature points. A graphical illustration of this method is given in Figure (9.1).

New HDMR expansions with distinct character can be generated by changing the measure μ in the formulae (9.13). The measure μ in this respect acts as a 'weight' to give more importance to certain regions of the input space. Coordinate transformations of the original input variables also generate new HDMR expansions. Kolmogorov's theorem implies that there is a coordinate system in which the HDMR expansion is exact at first-order. However, an *a priori* analysis of the output data must be generated first to learn which coordinate system is the best one. This task is as difficult as learning the output function, and as a first-order approximation one can search for optimal linear coordinate transformations, which is the basis for the projection pursuit regression technique.

As a multivariate approximation/interpolation scheme, HDMR has a broad variety of applications. It can also be used in assessing the importance of the individual and cooperative effects of input variables. Selected applications are mentioned in Section 9.4. This list is not exhaustive, and should be regarded as representative. The relative ease of employing HDMRs should aid in their various future applications.

ACKNOWLEDGMENTS

The authors acknowledge support from NASA and NSF. One of the authors would like to acknowledge numerous conversations and developments on HDMR with Drs J. Shorter and K. Shim.

10

Bayesian Sensitivity Analysis

David Ríos Insua

Universidad Rey Juan Carlos, Madrid, Spain

Fabrizio Ruggeri

CNR–IAMI, Milano, Italy

Jacinto Martín

Universidad Extremadura, Cáceres, Spain

10.1 SENSITIVITY ANALYSIS IN BAYESIAN ANALYSIS: AN INTRODUCTION

> Subjectivists should feel obligated to recognize that any opinion (especially the initial one) is only vaguely acceptable. So it is important to know not only the exact answer for an exactly specified initial problem, but also what happens on changing, within a reasonable neighborhood, the assumed initial opinion. (De Finetti, as quoted by Dempster (1975).)

By contrast with other chapters in this book, which focus more on inference and/or prediction, we shall emphasise decision-making aspects. We are interested in choosing an action from a feasible set. We shall assume that such a decision-making problem is under uncertainty, and shall aim at solving it within the Bayesian framework; see French and Ríos Insua (2000). This essentially implies:

(a) modelling beliefs over states with a probability distribution, which in the presence of additional information, is updated via Bayes' formula;
(b) modelling preferences over consequences by a utility function;
(c) determining the alternative of maximum expected utility.

Typical reasons for undertaking sensitivity analysis in a quantitative model, as have been discussed in other chapters, apply in our context. In addition to them, since inputs to a decision analysis encode the decision maker's (DM's) judgements, the DM would ask

Sensitivity Analysis. Edited by A. Saltelli *et al.*

for means to determine the implications, and possible inconsistencies, of his/her judgements. This is an especially important point in our setting since the DM's judgements will evolve through analysis. The assessment of beliefs and preferences is a difficult task—even more so in the case of several decision makers and/or experts. Some would even argue that priors and utilities can never be quantified exactly, i.e., without error, especially in a finite amount of time. In fact, within the Bayesian framework, sensitivity analysis has been the subject of recent research efforts; see Ríos Insua and Ruggeri (2000) for a thorough survey. In this chapter, we shall describe some of these, with emphasis on recent developments in sensitivity analysis (SA) methods and concepts specific to Bayesian analysis.

More formally, in a decision-making context, we assume that we have to choose among a set $\mathcal{A}$ of alternatives a. For that, we assess the prior beliefs on a state variable $\theta \in \Theta$ in a prior distribution with density π_0. They are updated to the posterior distribution, with density $\pi_0(\cdot|x)$, where x is the result of an experiment with likelihood $l_x(\theta)$ over a sample space x. We associate a consequence $c \in \mathscr{C}$ to each pair (a, Θ). Preferences over consequences are modelled with a utility function u_0, and we associate with each alternative a its posterior expected utility:

$$T(u_0, \pi_0, l_x, a) = \frac{\int_\Theta u_0(a, \theta) l_x(\theta)\, d\pi_0(\theta)}{\int_\Theta l_x(\theta)\, d\pi_0(\theta)}.$$

We maximize, in a, $T(u_0, \pi_0, l_x, a)$, as a way of obtaining the optimal alternative $a^*(u_0, \pi_0, l_x)$.

However, the assessment of u_0 and π_0 and the choice of model l_x may be far from simple, and we shall need tools to check their impact on the optimal alternative and its expected utility, the model output in our case. This has been widely acknowledged in the Bayesian arena, leading to a vast literature on Bayesian robustness and sensitivity analysis; see reviews by Berger (1990, 1994b), Wasserman (1992), and Ruggeri (1994). Note that these studies have concentrated mainly on inference issues, and therefore on sensitivity to the prior.

We illustrate the sensitivity approach with a case study taken from Rios Insua *et al.* (1999) concerning forecasting labour accidents.

Case study (robustness to prior)

Given the accident history D_k of a company (number of workers, number of accidents, ...) prior to period k of length t_k, as expressed in Table 10.1, we want to forecast the number X_k of accidents of the company in period k. We consider the Poisson model

$$X_k|\theta, n_k \sim \text{Po}(n_k\theta),$$

where n_k is the (known) number of workers in period k and θ is the accident rate.

In a Bayesian approach, the parameter θ will be treated as a random variable whose distribution is assessed according to prior beliefs. Data, modelled through the likelihood function, will update that prior distribution, via Bayes' theorem, and inferences and decisions are based upon the updated (posterior) distribution of θ. See Berger (1985) for a thorough presentation of the Bayesian approach.

In this case, we elicited from experts the following quartiles for θ : 0.38, 0.58, and 0.98. For mathematical convenience and flexibility, we chose a gamma prior,

$$\pi(\theta|D_1) \propto \exp(-a\theta)\theta^{p-1}I[0, \infty)(\theta),$$

Table 10.1 Accident data.

Month	No. of accidents	No. of workers	Month	No. of accidents	No. of workers	Month	No. of accidents	No. of workers
Jan. 88	8	350	Jan. 89	8	391	Jan. 90	1	300
Feb. 88	5	352	Feb. 89	6	392	Feb. 90	3	286
Mar. 88	3	342	Mar. 89	3	376	Mar. 90	9	303
Apr. 88	1	334	Apr. 89	9	367	Apr. 90	1	309
May 88	4	344	May 89	5	356	May 90	5	340
Jun. 88	2	343	Jun. 89	8	340	Jun. 90	2	358
Jul. 88	3	357	Jul. 89	7	344	Jul. 90	8	359
Aug. 88	4	364	Aug. 89	4	342	Aug. 90	7	365
Sep. 88	8	358	Sep. 89	5	330	Sep. 90	10	376
Oct. 88	7	374	Oct. 89	5	316	Oct. 90	9	378
Nov. 88	6	382	Nov. 89	5	310	Nov. 90	5	401
Dec. 88	3	381	Dec. 89	3	293	Dec. 90	–	–

with parameters adapted to fit such quantiles, namely $\hat{p} = 1.59$ and $\hat{a} = 2.22$. With such a choice, the posterior distribution for θ before period k is given by

$$\pi(\theta|D_k) \propto \exp\left[-\left(a + \sum_{i=1}^{k-1} n_i\right)\theta\right] \theta^{\sum_{i=1}^{k-1} r_i + p - 1} I_{(0,\infty)}(\theta),$$

i.e. a gamma distribution with parameters $a + \sum_{i=1}^{k-1} n_i$ and $\sum_{i=1}^{k-1} r_i + p$. Simple computations lead to the predictive distribution and the predictive expected value E^p and variance V^p of the number of accidents. Suppose we focus our interest on the predictive expected value

$$E^p = n_k \frac{\sum_{i=1}^{k-1} r_i + p}{\sum_{i=1}^{k-1} n_i + a}$$

and compute it for the data in Table 10.1.

Since, apart from fitting a few quantiles, the gamma choice may be arbitrary, we may undertake an alternative robust approach based on the available information (the quartiles), define the class of all priors compatible with it (i.e., all distributions with such quantiles), and investigate, as shown in the following sections, the effect of such imprecision on inferences and decisions.

The rest of the chapter describes such a robust Bayesian approach. We start by reviewing sensitivity to the prior in Section 10.2. In Section 10.3, we discuss additional difficulties brought in by decision-making aspects to sensitivity analysis; by means of examples, we show potential problems in undertaking simplified approaches, which lead us to review foundational issues in Section 10.4. These provide a framework for sensitivity analysis. Stability theory, described in Section 10.5, provides a qualitative theory for sensitivity analysis; however, the required conditions are difficult to check in practice. This leads us to introduce and look for nondominated alternatives, as the basic solution concept when there is imprecision in the inputs to a Bayesian analysis. In cases in which there are several nondominated alternatives, and they are different in terms of expected utility, we may need to extract additional information from decision makers to reduce imprecision, as we describe in Section 10.7. Section 10.8 is devoted to maximin concepts, viewed from a robust Bayesian perspective, as a way of choosing a robust solution. We end up with a brief discussion on topics for further research. An appendix includes most technical details. Appropriate topological and measure theoretical conditions will be assumed throughout.

The rest of this introduction provides other notation to be used. All through the paper, we shall assume that the model is fixed (but see the discussion); hence dependence on l_x will be dropped in the posterior expected utility $T(u_0, \pi_0, a)$. In Section 10.2, we shall fix u_0 and a, designating $g(\theta) = u_0(a, \theta)$ and

$$T_g(\pi) = \frac{\int_\Theta g(\theta) l_x(\theta)\, d\pi(\theta)}{\int_\Theta l_x(\theta)\, d\pi(\theta)}, \tag{10.1}$$

assuming that T_g is well defined. We shall also use the notation $T_g(\pi) = N_\pi / D_\pi$, with

$$N_\pi = \int_\Theta g(\theta)\, l_x(\theta)\, d\pi(\theta), \quad D_\pi = \int_\Theta l_x(\theta)\, d\pi(\theta).$$

When describing differences in posterior expected utility between two alternatives a and b, we shall use

$$T^{ab}(u, \pi) = \frac{\int_\Theta [u(a, \theta) - u(b, \theta)] l_x(\theta)\, d\pi(\theta)}{\int_\Theta l_x(\theta)\, d\pi(\theta)}.$$

We shall sometimes use the operator

$$N^{ab}(u, \pi) = \int_\Theta [u(a, \theta) - u(b, \theta)] l(x|\theta)\, d\pi(\theta).$$

$\mathscr{P}$ will designate the class of all probability measures over Θ, endowed with an appropriate σ-algebra.

10.2 SENSITIVITY TO THE PRIOR

Early papers on Bayesian robustness are discussed in Berger (1984), who justifies the need for a robust Bayesian viewpoint because of a belief in the assumption that 'prior distributions can never be quantified or elicited exactly (i.e., without error), especially in a finite amount of time'. Nowadays, there is a different perception among Bayesians about robustness and sensitivity. Most Bayesians accept the robust approach because of the wealth of papers published in the early 1990s, devoted to laying foundations for Bayesian robustness, developing computational techniques and sophisticated mathematical tools, and applying robust ideas in inference and decision making. Specialized conferences have been held, and papers presented therein are collected in volumes such as those edited by Berger (1994a) and Berger *et al.* (1996). As a result of such flourishing in recent years, we may say that the field has reached maturity and practical applications are appearing; see e.g. Greenhouse and Wasserman (1996), Sargent and Carlin (1996), and Ríos Insua *et al.* (1999). In this section, we give a brief review of sensitivity to the prior analyses. This will typically be a task to perform when there is little imprecision in the utility function (say a standard loss is adopted) but there is imprecision in the prior. There are two reasons to start with this case: it is the simplest and most thoroughly studied one, and it will provide many insights for the general problem. We distinguish between global and local sensitivity analysis. Note, however, that these concepts differ slightly from those usually considered in the SAMO literature.

10.2.1 Global Sensitivity Analysis

Most robust Bayesian research deals with *global sensitivity analysis*: the elicitation process ends with specifying some features (e.g., quantiles, symmetry, unimodality, . . .) of the prior measure, and all measures compatible with such information are considered. Given a class Γ of prior measures, global sensitivity analysis is typically interested in the range of a posterior functional of interest $T_g(\pi)$ as π ranges in the class Γ, i.e. $\sup_{\pi\in\Gamma} T_g(\pi) - \inf_{\pi\in\Gamma} T_g(\pi)$. The search for suprema and infima of functionals is therefore relevant in the global sensitivity approach. This global sensitivity measure, *range*, is generally regarded as being relatively easy to interpret, and as having reasonable asymptotic properties, as compared, for instance, to some local sensitivity measures. For related details, see Berger (1994b) and Gustafson and Wasserman (1995). The usual interpretation of the range is that the posterior quantity is robust (to deviations from the prior) when the range is small, and is not robust when the range is large.

Berger (1990) gives a broad review of the most important prior classes used in global sensitivity analysis: conjugate, neighborhoods (mainly ε-contaminations), approximately specified moments, sub-sigma field (mainly quantile), and density ratio classes. We shall illustrate the basic ideas in global sensitivity analysis with the generalized moments class (Betrò *et al.*, 1994), which covers the realistic case in which the elicitation process enables us to specify constraints on the generalised moments of the prior measure: we consider compatible with our prior knowledge all priors in the class

$$\Gamma_M = \left\{ \pi \in \mathcal{P} : \int_\Theta H_i(\theta)\, d\pi(\theta) \leqslant \alpha_i, i = 1, \ldots, n \right\},$$

where H_i are π-integrable functions and $\alpha_i, i = 1, \ldots, n$, are fixed real numbers. Among others, this class contains as a particular case the class of quantiles, which we use later. It often turns out that extrema of posterior functionals are attained at a smaller class, typically, the extreme points of the original class Γ. Specifically, Betrò *et al.* (1994) proved that, under generalized moment conditions, $\sup \pi \in \Gamma_M\, T_g(\pi)$ is achieved by a discrete distribution in Γ_M concentrated in, at most, $n + 1$ points. Betrò and Guglielmi (1996) provided a software code for such an approach, based on interval analysis and semi-infinite linear programming, allowing for fast and reliable algorithms.

While the interpretation of the range as a sensitivity measure is intuitively reasonable and appealing, it does have shortcomings. For one, it is very much problem-dependent. For instance, it is not clear what is meant by 'large' and 'small', since the range will be scale-dependent. More importantly, the range, as a measure of sensitivity to prior misspecification of the prior, does not take account of the uncertainty within a single posterior. It is therefore of interest to take into account the posterior standard deviation (with respect to a single prior), the 'within posterior uncertainty', when assessing the size of a *range*, which is a measure of the 'between posterior uncertainty'.

Ruggeri and Sivaganesan (2000) defined the global sensitivity measure

$$R = \frac{[T_g(\pi) - T_g(\pi_0)]^2}{V^\pi},$$

where V^π is the posterior variance of $g(\theta)$ with respect to the prior π. This measure, called *relative sensitivity*, is, in fact, a scaled version of the *range*, and Ruggeri and Sivaganesan argued that it overcomes the above-mentioned concerns, besides having other properties.

10.2.2 Local Sensitivity Analysis

Suppose a unique prior π is elicited, but we allow for small changes in the concentration of the probability, due to errors in the elicitation process. Hence, we consider measures that are 'functionally close' to π, and study the behavior of prior or posterior functionals under infinitesimal departures from π. Recent work on local sensitivity considers either the prior probability measure embedded in an ε-contaminated class of priors, computing Fréchet derivatives of posterior functionals (Diaconis and Freedman, 1986; Ruggeri and Wasserman, 1993; Srinivasan and Truszczynska, 1990), or density-based classes of priors, computing the derivative of the supremum (and the infimum) of functionals over classes of priors shrinking towards a given prior probability measure (Ruggeri and Wasserman, 1995). We illustrate some of the ideas with the example of the case study from Section 10.1.

Case study (continued)

To check whether the hypothesis about a gamma distribution is critical, we undertake a global robustness analysis with respect to the class

$$\Gamma_{\mathbf{p}} = \{\pi : \pi(0, 0.38] = 0.25, \pi(0.38, 0.58] = 0.25, \pi(0.58, 0.98] \\ = 0.25, \pi(0.98, \infty) = 0.25\}.$$

Interest will focus on the predictive mean $E(X_k|D_k)$. Applying a result in Ruggeri (1990), we find that

$$\frac{\overline{E}(X_k|D_k)}{n_k} = \frac{\sup_{\pi\in\Gamma_{\mathbf{p}}} E(X_k|D_k)}{n_k} = \sup_{\theta_1\in A_1,\theta_2\in A_2,\theta_3\in A_3,\theta_4\in A_4} \frac{\sum_{j=1}^{4} \theta_j^{s+1}\exp(-\theta_j r)}{\sum_{j=1}^{4} \theta_j^{s}\exp(-\theta_j r)},$$

with s the sum of accidents and r the sum of workers, before period k, and $A_1 = (0, 0.38]$, $A_2 = (0.38, 0.58], A_3 = (0.58, 0.98], A_4 = (0.98, \infty)$. A similar expression may be found for $\underline{E}(X_k|D_k) = \inf_{\pi\in\Gamma_{\mathbf{p}}} E(X_k|D_k)$. The corresponding optimization problems may be solved with the aid of a MATHEMATICA package, with the results shown in Table 10.2.

Note that additional data produced little variation in the accident range, because of the small annual variation of s/r. These results suggest that the class $\Gamma_{\mathbf{p}}$ is not robust in this example: expected forecasts may vary widely. When robustness is lacking, one should be interested in reducing the imprecision, through elicitation of additional information from the expert. We use the sensitivity measure (10.A.1) from the appendix to this chapter which is decomposed in terms relating to the four intervals A_i defining the class $\Gamma_{\mathbf{p}}$, with the results shown in Table 10.3. For example, if $\varepsilon_0 = 0.2$, the imprecision in the forecast for 1989 is 5.49 accidents, which might be considered too big. In all years, the supremum is due to the first quartile (A_1). Hence, it is worth eliciting additional information about this quartile from the expert.

Table 10.2 Bounds on scaled expected accidents.

Year	$\underline{E}[X_k\|D_k]/n_k$	$\bar{E}[X_k\|D_k]/n_k$
1988	0.05	0.58
1989	0.05	0.58
1990	0.05	0.58

Table 10.3 Local sensitivity.

Year	$\sup_{\delta\in\Delta_{\varepsilon_0}} \|\dot{T}^{\pi}_{g}\|$	$\hbar_1, \hbar_2, \hbar_3, \hbar_4$
1988	$25.1561\varepsilon_0$	$25.1561, -2.18\times 10^{23}, -3.24\times 10^{-69}, 0$
1989	$27.4553\varepsilon_0$	$27.4553, 8.59\times 10^{-19}, 2.07\times 10^{-57}, 6.24\times 10^{-153}$
1990	$23.2267\varepsilon_0$	$23.2267, -5.18\times 10^{-83}, -3.23\times 10^{-223}, 0$

Table 10.4 Bounds on revised scaled expected accidents.

Year	$\underline{E}[X_k\|D_k]/n_k$	$\bar{E}[X_k\|D_k]/n_k$
1988	0.15	0.24
1989	0.15	0.23
1990	0.15	0.23

Suppose the interval is split into three subintervals, $A_{11} = [0, 0.15), A_{12} = [0.15, 0.2)$, and $A_{13} = [0.2, 0.38)$, with probabilities $0.1, 0.05$, and 0.1, respectively, so that the same gamma prior is kept. Then, for example, for 1989, the new value of $\sup_{\sigma\in\Delta_{\varepsilon_0}} |\dot{T}^{\pi}_{g}|$ is $7.7623\varepsilon_0$, which is a considerable reduction. Upper and lower bounds of the predictive expectation are computed with respect to this new class, as shown in Table 10.4.

Observe the reduction in ranges from Table 10.2 to Table 10.4, suggesting much more robustness in the new class.

This example shows how local sensitivity, even if arising from a different framework, can be actually used in a global robustness analysis.

10.3 ISSUES IN GENERAL SENSITIVITY ANALYSIS

The previous section concentrated on sensitivity to the prior issues. Since we are more interested in decision-making problems, we move on to more general sensitivity analyses based on both the prior and the utility. This problem is not easy: simple examples show that new questions arise when undertaking general sensitivity analysis.

(a) *It is not enough to study changes in output by trying some other pairs of utilities and probabilities.*

Example 10.1 Suppose we have a decision problem summarized by the following decision table:

	θ_1	θ_2
a	c_1	c_2
b	c_2	c_3

Let $\pi(\theta_1) = \pi_1$. Suppose we have no data and we have assessed that $0.4 \leqslant \pi_1 \leqslant 0.6$, $0 \leqslant u(c_1) \leqslant 0.5, 0.25 \leqslant u(c_2) \leqslant 0.75$, and $0.5 \leqslant u(c_3) \leqslant 1$. We consider the following four utility–probability pairs associated with the bounds on utilities and probabilities:

π_1	utility (u)	$T(u, \pi_1, a)$	$T(u, \pi_1, b)$
0.4	$u(c_1) = 0, u(c_2) = 0.25, u(c_3) = 0.5$	0.15	0.4
0.4	$u(c_1) = 0.5, u(c_2) = 0.75, u(c_3) = 1$	0.65	0.9
0.6	$u(c_1) = 0, u(c_2) = 0.25, u(c_3) = 0.5$	0.1	0.35
0.6	$u(c_1) = 0.5, u(c_2) = 0.75, u(c_3) = 1$	0.6	0.85

Since, in all four cases, $T(u, \pi_1, a) < T(u, \pi_1, b)$, we could conclude that alternative b is preferable to alternative a. Note, though, that this is not true: if, e.g., we take $\pi = 0.6, u(c_1) = 0.5, u(c_2) = 0.25, u(c_3) = 0.5$, then $T(u, \pi_1, a) = 0.4$ and $T(u, \pi_1, b) = 0.35$.

(b) *Partial sensitivity studies are not sufficient: a problem may be insensitive to changes in utility and changes in probability, but sensitive to simultaneous changes in utility and probability.*

Example 10.2 Consider Example 10.1 again. If we fix $\pi_1 = 0.5$ and study sensitivity with respect to the utility function, we have

$$\begin{aligned} T(u, \pi_1, a) - T(u, \pi_1, b) &= 0.5u(c_1) + 0.5u(c_2) - 0.5u(c_2) - 0.5u(c_3) \\ &= 0.5[u(c_1) - u(c_3)]. \end{aligned}$$

Since $u(c_1) \leqslant u(c_3)$ for all values of u, we can conclude that the problem is robust against changes in u. Analogously, if we consider $u(c_1) = 0.25, u(c_2) = 0.5$, and $u(c_3) = 0.75$, we have

$$T(u, \pi_1, a) - T(u, \pi_1, b) = \pi_1 0.25 + (1 - \pi_1)0.5 - \pi_1 0.5 - (1 - \pi_1)0.75 = -0.25.$$

Then, we have robustness against changes in probabilities. However, we cannot conclude that the problem is robust; see Example 10.1.

(c) *When performing sensitivity analysis, there are cases in which expected utility may change a lot, with virtually no change in the optimal action, even if the utility is fixed.*

Example 10.3 (based on Kadane and Srinivasan, 1994) Suppose there are no data and the class of priors is

$$\Gamma = \{\pi = (1 - \varepsilon)\pi_0 + \varepsilon\pi', \quad \pi_0 \sim \mathcal{N}(0, 1) \quad \textit{and} \quad \pi' \in \Gamma_0\},$$

with

$$\Gamma_0 = \{\text{all symmetric distributions with mean} \quad 0\},$$

whereas $\mathscr{A} = \mathbb{R}$ and the consequences are $(a, \theta) = a - \theta$. Assume the utility function is $u(c) = -c^2$; for estimation problems under square loss, the optimal action is the posterior mean. In this case, $\forall \pi \in \Gamma$, the optimal action is 0. However, if we consider the distributions

$$p_n(x) = \begin{cases} \frac{1}{2} & \text{if } x = -n, \\ \frac{1}{2} & \text{if } x = n, \end{cases}$$

then $T(u, p_n, 0) = (\varepsilon - 1) - \varepsilon n^2$, and the expected utility of 0 would tend to $-\infty$, without changes in the optimal alternative.

(d) *There are cases in which the optimal alternative varies widely, but the maximum expected utility does not practically change.*

Example 10.4 (based on Kadane and Srinivasan, 1994) Suppose there are no data and the class of priors is

$$\Gamma = \{\pi_\lambda \sim \mathcal{U}\{[0, 1-\lambda] \cup [2-\lambda, 3]\}, \lambda \in [-0.01, 0.01]\}$$

$\mathcal{A} = \mathbb{R}$ and the consequences are $(a, \theta) = a - \theta$. Suppose further that the utility function is $u(c) = |c|$; thus, for an estimation problem under absolute error loss, the optimal action is the posterior median. Hence, the optimal action is 1 when $\lambda \in [-0.01, 0)$, 2 when $\lambda \in (0, 0.01]$, and the interval [1,2] when $\lambda = 0$. On the other hand, we have $T(u, \pi_\lambda, 1) = -1 - \frac{1}{2}\lambda$ when $\lambda \in [-0.01, 0)$, $T(u, \pi_\lambda, 2) = -1 + \frac{1}{2}\lambda$ when $\lambda \in (0, 0.01]$, and $T(u, \pi_\lambda, a) = -1$ when $\lambda = 0$ and $a \in [1, 2]$. Therefore, the optimal action changes greatly, but the expected utility varies in the range $[-1, -0.995]$.

(e) *Big changes in expected utility do not necessarily correspond to big changes in consequences of interest.*

See Example 10.5.

(f) *Standard global Bayesian robustness studies, as described in Section 10.2, based, e.g., on ranges of expected utilities of actions, may not be sufficient within a decision-theoretic perspective.*

Example 10.5 Suppose we have the classes of priors and utilities

$$\Gamma = \{\pi \sim \mathcal{E}xp(\theta), 1 < \theta < 2\},$$
$$\mathcal{U} = \{u, \text{increasing in } (0, \infty)\}$$

Actions are a and b, with $(a, \theta) = \frac{1}{2}\theta$ and $(b, \theta) = \theta$. In this case, $u(a, \theta) < u(b, \theta), \forall_u, \theta$, so that $T(u, \pi, a) < T(u, \pi, b), \forall_u \in \mathcal{U}, \pi \in \Gamma$, and the decision is robust. If we consider the range of the posterior expected utility of b, as in Example 10.3, we can prove that $\sup T(u, \pi, b) = \infty$. We have that

$$\mathcal{U}_0 = \{u_n(c) = c^n, n = 1, 2, \ldots\} \subset \mathcal{U}.$$

Taking $\pi_0 \sim \mathcal{E}xp(1), T(u_n, \pi_0, b) = n$, and the problem does not seem robust. Obviously, changes in utility do not correspond to changes in consequences.

10.4 FOUNDATIONAL ISSUES

The above issues suggest that we should be careful when addressing sensitivity analysis questions within a decision-theoretic framework. Standard methods may lead to errors. To set up an appropriate framework for general sensitivity analysis, we reconsider the foundations of the robust Bayesian approach. Although this question has not been settled completely, a number of results (see Girón and Ríos, 1980; Ríos Insua, 1990; Walley, 1991; Seidenfeld *et al.*, 1995; Nau *et al.*, 1998) lead to essentially the same conclusion: imprecise beliefs and preferences may be modelled by a class of priors and a class of utility functions, so that preferences among alternatives may be represented by inequalities for the corresponding posterior expected utilities. More technically, under certain conditions, we are able to model beliefs by a class Γ of priors π and preferences by a class $\mathcal{U}$ of utility functions u, so that we find alternative a at most as preferred as $b(a \preceq b)$ if and only if $T(u, \pi, a) \leqslant T(u, \pi, b)$, $\forall_u \in \mathcal{U}, \pi \in \Gamma$.

These results have two basic implications. On the one hand, they provide a qualitative framework for sensitivity analysis in decision making: they describe under which conditions (mainly imprecision in preferences and/or beliefs described by quasi-orders, plus other conditions close to the standard Bayesian axioms; see French and Rios Insua, 2000) we may undertake the standard sensitivity analysis approach of perturbing the initial probability–utility assessments, within some reasonable constraints. On the other hand, they point to the basic solution concept associated with the robust approach, indicating, therefore, the basic computational objective in sensitivity analysis, as long as we are interested in decision-analytic issues: that of nondominated solution. To wit, we say that an alternative a dominates another alternative b, denoted $b \prec a$, if $b \preceq a$ and $\neg(a \preceq b)$, or in terms of posterior expected utilities, $T(u, \pi, b) \leqslant T(u, \pi, a), \forall u \in \mathcal{U}, \pi \in \Gamma$, with strict inequality for a pair (u, π). We then say that an alternative a is nondominated if there is no other alternative in $\mathcal{A}$ dominating it.

The foundations of the robust Bayesian approach therefore set the stage for sensitivity analysis within a decision-making perspective. We first deal with qualitative aspects describing a theory of stability, and then provide a strategy to compute nondominated alternatives.

10.5 STABILITY THEORY

According to the subjective view of Savage (1954), inputs must be subjectively elicited, but being aware that prior opinions and desires cannot be exactly represented by priors, likelihoods and utilities. The effects of imprecise elicitations have been considered from different approaches, often interlaced, such as Bayesian sensitivity and stability theory.

Stability theory (Kadane and Chuang, 1978; Chuang, 1984; Salinetti, 1994; Kadane and Srinivasan, 1994) studies the convergence of decisions a_n, 'nearly' optimal for the pair (π_n, u_n), to a_0 when π_n and u_n (the approximately specified opinions and desires) converge, respectively, to π_0 and u_0 (the 'true' ones). Stability is strongly related to the continuity of the posterior expected utility functional.

Different notions of stability have been introduced, and are reviewed in, e.g., Kadane and Srinivasan (1994), who considered a bounded continuous likelihood function l_x, a sequence of utility functions u_n converging (in some topology) to u_0, and a sequence of priors π_n converging weakly to π_0 (denoted by $\pi_n \Longrightarrow \pi_0$)

Definition 10.1 The decision problem (u_0, l_x, π_0) is *strongly stable I (SSI)* if for any sequence $\pi_n \Longrightarrow \pi_0$ and $u_n \to u_0$

$$\lim_{\varepsilon \downarrow 0} \limsup_{n \to \infty} \left[\int u_n(\theta, a_0(\varepsilon)) l_x(\theta)\, d\pi_n(\theta) - \sup_{a \in \mathcal{A}} \int u_n(\theta, a) l_x(\theta)\, d\pi_n(\theta) \right] = 0 \tag{10.2}$$

for every $a_0(\varepsilon)$ such that

$$\int u_0(\theta, a_0(\varepsilon)) l_x(\theta)\, d\pi_0(\theta) \geqslant \sup_{a \in \mathcal{A}} \int u_0(\theta, a) l_x(\theta)\, d\pi_0(\theta) - \varepsilon.$$

Definition 10.2 The decision problem (u_0, l_x, π_0) is *weakly stable I (WSI)* if (10.2) holds for a particular choice $a_0(\varepsilon)$.

Because of the practical difficulty in checking the conditions of the previous definitions, some effort has been devoted in seeking sufficient conditions ensuring strong stability. Kadane and Srinivasan (1994) present some of them, based upon the works by Salinetti (1994). Those results may be seen in Chuang (1984), Kadane and Srinivasan (1994), and Kadane *et al.* (1997). We present the following

Theorem 10.1 Let (u_0, l_x, π_0) be a finite decision problem such that $u_0(\theta, a)l_x(\theta)$ is bounded for each a and π_0 ({discontinuity points of $u(\cdot, a)$}) $= 0$. Under uniform convergence of utilities, (u_0, l_x, π_0) is SSI.

Further definitions and results are presented in Section 10.A.2 in the appendix to this chapter.

10.6 COMPUTATION OF NONDOMINATED ALTERNATIVES

We now consider the fundamental problem of approximating the nondominated set. Except under very tight structural conditions, we may only hope to provide some type of approximation scheme. The one we suggest follows the intuitive idea of discretizing the set of alternatives and, by means of a pairwise comparison method, determining the set of nondominated alternatives within the discrete approximation.

1. Sample $A_n = \{a_1, \ldots, a_1\} \subset \mathcal{A}$.
2. Choose $(u, \pi) \in \mathcal{U} \times \Gamma$ and relabel the as as $(e_1, \ldots, e_n)$ so that $T(u, \pi, e_{i+1}) \geqslant T(u, \pi, e_i), \forall i$.
3. Let $d(i) = 0, \forall i$.
4. For $i = 1$ to $n - 1$
    If $d(i) = 0$
        For $j = i + 1$ to n
            If $d(j) = 0$
                If $e_j \preceq e_i$ then $d(j) = 1$
                    Else, if $(T(u, \pi, e_i) = T(u, \pi, e_j)$ and $e_i \preceq e_j)$
                        then $d(i) = 1$ and next i.
5. Let $N_n = \{e_i : d(i) = 0\}$.

N_n is the approximation to the nondominated set. Under mild conditions, we may prove that N_n converges to the nondominated set (Bielza *et al.*, 1998) as n goes to infinity.

Note that in step 4, we need a procedure to check whether an alternative a dominates an alternative b. This leads us to study the operator T^{ab} introduced before, since

$$\text{if} \quad \inf_{(u,\pi) \in \mathcal{U} \times \Gamma} T^{ab}(u, \pi) \geqslant 0 \quad \text{then } b \preceq a. \tag{10.3}$$

We are actually interested in knowing whether the infimum (10.3) is positive. This will be useful, since sometimes we are only able to bound that supremum. Hence, ignoring the common denominator, we could use, instead of T^{ab}, the operator N^{ab}.

There is little more we can say in general about such problems, there remains only the issue of studying it for important cases, i.e. for specially important classes of utilities and priors. We illustrate the ideas with an example.

Table 10.5 Estimation of the infimum for $a = 1$ and $b = 0$.

n	I_n	$\widehat{EE}_n$	$I_n + 2\widehat{EE}_n$
500	−0.000 80	0.001	—
1000	−0.0010	0.000 8	—
5000	0.000 34	0.000 41	−0.000 48
10 000	0.000 18	0.000 25	−0.000 32
20 000	0.000 4	0.000 15	0.0001

Example 10.6 Suppose we have a problem with $\mathscr{A} = \{a \in [0, 1]\}$ and $c(a, \theta) = 1.1 + (\theta - 0.1)a$. Moreover, we have parametric classes for preferences and beliefs:

$$\mathscr{U}_\omega = \{u_\omega(c) = 1 - e^{-\omega c}, \quad \text{with} \quad \omega \in [1, 2.5]\},$$
$$\Gamma = \{\pi(\theta|\mu) \sim \mathscr{N}(\mu, 0.1), \quad \text{with} \quad \mu \in [0.12, 0.15]\}.$$

For two alternatives $a, b \in \mathscr{A}$, we have,

$$b \preceq a \Longleftrightarrow \sup_{\omega,\mu} \int [1 - e^{-\omega(1.1+(\theta-0.1)b)} - (1 - e^{-\omega(1.1+(\theta-0.1)a)})]\pi(\theta|\mu)\, d\theta \leqslant 0. \tag{10.4}$$

To illustrate the method to compute such supremum (see Section 10.A.3 in the appendix to this chapter), we consider a normal distribution with mean 1 and standard deviation 0.1 as the importance sampling distribution. Then

$$\pi(\theta|\mu) = (\sqrt{2\pi 0.1})^{-1} \exp\left[-\frac{1}{2}\left(\frac{\theta - \mu}{0.1}\right)^2\right],$$
$$h(\theta) = (\sqrt{2\pi 0.1})^{-1} \exp\left[-\frac{1}{2}\left(\frac{\theta}{0.1}\right)^2\right],$$
$$\frac{\pi(\theta|\mu)}{h(\theta)} = \exp[-50(\mu^2 - 2\mu\theta)].$$

Fix a and b. We have to compute, as an approximation to the infimum,

$$\inf_{\omega,\mu}\left[\frac{1}{n}\sum_{i=1}^{n}(e^{-\omega(\theta_i - 0.1)^b} - e^{-\omega(\theta_i - 0.1)^a} e^{-50(\mu^2 - 2\mu\theta_i)})\right],$$

where $\theta_1, \ldots, \theta_n$ is iid from $\mathscr{N}(0, 0.1)$. Table 10.5 shows the results for $b = 0$ and $a = 1$.

Then b dominates a. We apply the discretization scheme, considering only a sample size of 500. The values of d are represented in Figure 10.1. We can conclude that the nondominated set is $[0.8, 1]$. It is easy to prove analytically that this is the actual non-dominated set.

10.7 EXTRACTING ADDITIONAL INFORMATION

It may happen that there are several nondominated alternatives and differences in expected utilities are non-negligible. If such is the case, we should look for additional information

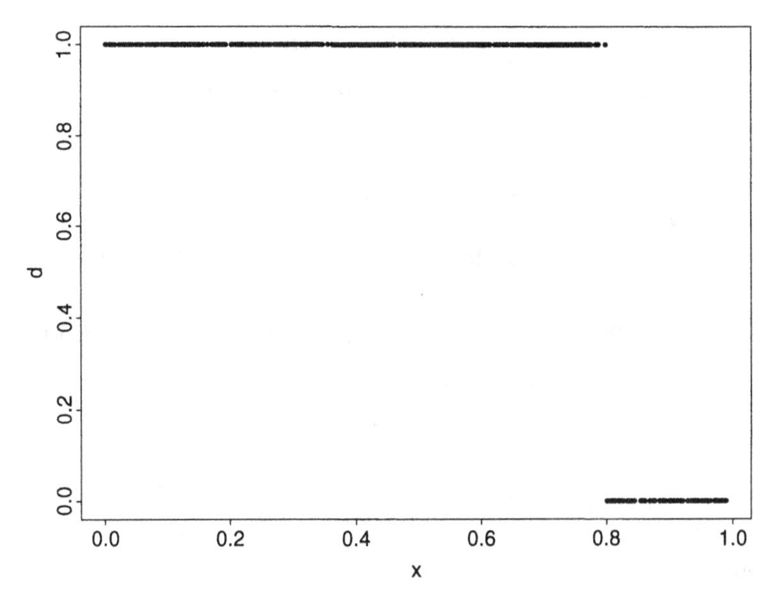

Figure 10.1 Dominated ($d = 1$) and nondominated ($d = 0$) points.

that would help us to reduce the classes, and, perhaps, reduce the nondominated set. We shall describe tools based on functional derivatives to elicit additional information. See Ruggeri and Wasserman (1993) and Martin and Rios Insua (1996) for additional details. First, we need the following definition, which describes perturbations of a current assessment preserving preferences.

Definition 10.3 (u, π) is ε-robust for $b \preceq a$ within $\mathcal{U} \times \Gamma$ if $b \preceq a$ whenever we use $(tv + (1-t)u, tq + (1-t)\pi) \in \mathcal{U} \times \Gamma$ in place of (u, π), where $(v, q) \in \mathcal{U} \times \Gamma$ and $t < \varepsilon$.

When $\varepsilon = 1$, the information $b \preceq a$ is robust. We again use $N^{ab}(u, \pi)$ rather than $T^{ab}(u, \pi)$. We first compute the Fréchet derivative of $N^{ab}(u, \pi)$.

Theorem 10.2 Let $\dot{N}^{ab}_{u\pi}(\cdot, \cdot)$ be the Fréchet derivative of N^{ab} with respect to (u, π). Then

$$\dot{N}^{ab}_{u\pi}(m, \delta) = N^{ab}(u, \delta) + N^{ab}(m, \pi).$$

We shall use such Fréchet derivatives to bound the ε- robustness of a preference over a pair of alternatives, as shown in Section 10.A.4 of the appendix to this chapter. A simple example illustrates the potential of the tools there described in guiding modelling efforts.

Example 10.7 Let $\Theta = \{\theta_1, \theta_2, \theta_3\}$. We consider two options a and b, meaning entering or not in a gamble. Consequences for these alternatives are $(a, \theta_1) = -1000$, $(a, \theta_2) = 0$, and $(a, \theta_3) = 3000$, and $(b, \theta_i) = 0$ for $i = 1, 2, 3$. Suppose we have assessed

$$\begin{aligned}
\mathcal{U}_K &= \{u : 0 \leqslant u(-1000) \leqslant 0.1 \leqslant u(0) \leqslant 0.9 \leqslant u(3000) \leqslant 1\}, \\
\Gamma_Q &= \{\pi : 0.3 \leqslant p_1 \leqslant 0.4, \leqslant p_2 \leqslant 0.6, 0.1 \leqslant p_3 \leqslant 0.2, \\
&\qquad \text{with} \quad p_i = \pi(\theta_i), \quad i = 1, 2, 3\}.
\end{aligned}$$

Initially, we adopt a utility function u_0 and a prior π_0 such that

$$u_0(-1000) = 0, \qquad u_0(0) = 0.25, \qquad u_0(3000) = 1,$$
$$\pi_0(\theta_1) = \tfrac{1}{3}, \qquad \pi_0(\theta_2) = \tfrac{1}{2}, \qquad \pi_0(\theta_3) = \tfrac{1}{6}.$$

Suppose, also, that $l_x(s_i) = \frac{1}{3}, i = 1, 2, 3$, for the observed x, Then $T(u_0, \pi_0, a) = 0.296$ and $T(u_0, \pi_0, b) = 0.25$, so that $b \prec a$.

Let us compute the ε-robustness for this information within $\mathcal{U}_K$ and Γ_Q. First, we compute ε_{u_0}:

$$I_1 = 0\pi_0(S_a(-1000)) + 0.9[\pi_0(S_a(0)) - \pi_0(S_b(0))] + 0.9\pi_0(S_a(3000))$$
$$= 0 - 0.45 + 0.15 = -0.3$$

and

$$\varepsilon_{u_0} = \frac{T^{ab}(u_0, \pi_0)}{T^{ab}(u_0, \pi_0) - I_1} = \frac{0.046}{0.046 + 0.3} = 0.133$$

Now, we compute ε_{π_0}. We have,

$$H_1 = \tfrac{1}{3}\min\{p_1(0 - 0.25) + p_2(0.25 - 0.25) + p_3(1 - 0.25)\}$$

such that

$$0.4 \leqslant p_1 \leqslant 0.6,$$
$$0.3 \leqslant p_2 \leqslant 0.4,$$
$$0.1 \leqslant p_3 \leqslant 0.2,$$

so that $H_1 = -0.025$. Therefore,

$$\varepsilon_{\pi_0} = \frac{N^{ab}(u_0, \pi_0)}{N^{ab}(u_0, \pi_0) - H_1} = \frac{0.015}{0.015 + 0.025} = 0.375.$$

In this case, we verify that for $\delta = \varepsilon(q - \pi_0), \| \delta \| \leqslant 0.2\varepsilon$, and for $m = \varepsilon(v - u)$, $\| m \| \leqslant 0.8\varepsilon$. Therefore, $K = 4 \times 0.2 \times 0.8 \times \frac{1}{3} = 0.213$. We then have

$$J_1 = \tfrac{1}{3}I_1 + H_1 - 2N^{ab}(u_0, \pi_0) = -0.1 - 0.025 - 0.030 = -0.155$$

and $2\sqrt{KN^{ab}(u_0, \pi_0)} = 0.115$: we are in case (a) of Corollary 10.1. So, (u_0, π_0) is ε-robust for $b \preceq a$ within $\mathcal{U}_K \times \Gamma_Q$ with $\varepsilon \in [\varepsilon_1, \varepsilon_2]$ and

$$\varepsilon_1 = \frac{J_1 + \sqrt{J_1^2 + 4KN^{ab}(u_0, \pi_0)}}{2K} = 0.08,$$

$$\varepsilon_2 = \min\left\{\frac{\sqrt{N^{ab}(u_0, \pi_0)}}{K}, \varepsilon_u, \varepsilon_\pi\right\} = \min\{0.26, 0.133, 0.375\} = 0.133.$$

ε_1 is small, suggesting sensitivity. Let us see how can we increase it, preserving the same utility function u_0 and the same prior π_0. ε_1 is increasing in J_1. Therefore, we must look for the most influential value in J_1. Note that the term relating to u_0 is the smallest (-0.1), so we need to center our efforts on it. Also, the smallest term in I_1 is that referred to the utility of $0(-0.45)$. Therefore, we should investigate a possible reduction in the upper bound of $u(0)$.

10.8 MAXIMIN SOLUTIONS

The approach proposed in previous sections may be summarized as follows: at a given stage of analysis, we elicit information on the decision maker's beliefs and preferences, and consider the class of all priors and utilities consistent with such information; we approximate the set of nondominated solutions; if these alternatives do not differ too much in expected utility, we may stop the analysis; otherwise, we need to gather additional information, possibly with tools described in Section 10.7.

It is conceivable that, in such a context, at some stage we might not be able to gather additional information, yet there are several nondominated alternatives with very different expected utilities. In these situations, maximin solutions may be useful as a way of selecting a single robust solution. We associate with each alternative its smallest expected utility; we then suggest the alternative with maximum smallest expected utility. Specifically, an alternative $a_M \in \mathcal{A}$ is $\mathcal{U} \times \Gamma$-maximin if $\min_{u\in\mathcal{U},\pi\in\Gamma} T(u, \pi, a_M) = \max_{a\in\mathcal{A}}, \min_{u\in\mathcal{U},\pi,\in\Gamma} T(u, \pi, a)$. Maximin (and related concepts such as minimax or minimax regret-solutions) are usually introduced as *ad hoc* concepts. We view them as heuristics that allow us to choose a nondominated alternative, in cases where we are not able to gather additional information, as the following results show

Proposition 10.1 If the set of $\mathcal{U} \times \Gamma$-maximin alternatives is finite, one of them is nondominated.

As a corollary, if there is a unique $\mathcal{U} \times \Gamma$-maximin alternative, it is nondominated. Interestingly enough, some of these minimax concepts have axiomatic foundations compatible with those described for robust Bayesian analysis in Section 10.6.

Computation of maximin alternatives is not simple in general cases, but, again, we may appeal to discretization schemes. Once we have a discrete sample from the set of alternatives, we compute the corresponding nondominated set, and then the minimum posterior expected utility of each alternative (this is a conventional global sensitivity analysis problem), picking that with the maximum minimum. Again, under mild conditions, we may prove convergence of the approximate maximin solution to the actual maximin solution.

10.9 DISCUSSION

We have given a brief outline of approaches to sensitivity analysis when undertaking a Bayesian analysis. As we have seen, this is a rich field with many recent developments. Relevant applications may be seen in Berger *et al.* (1996) and Rios Insua and Ruggeri (2000). As we have described, we believe that sensitivity analysis is extremely important in Bayesian analysis, to the point that we support a Sensitivity Analysis-based approach to such analyses, as outlined in Section 10.8.

There are still many open issues. We have mentioned some fundamental results, yet most of them deal with the so-called simple case; extensions to the general continuous case would be desirable. We have described some ideas on calibration of sensitivity measures in Ruggeri and Sivaganesan (1997); other alternatives may be seen in McCulloch (1989) and Weiss (1996), yet there is still room for further calibration approaches, especially as far as decision making is concerned. We have introduced procedures to check dominance with several important classes, yet many others should be considered. Other sensitivity analysis approaches as illustrated in this book are also relevant in our context; some ideas may be seen in various articles in the special issue on sensitivity analysis of the *Journal of Multi-criteria Decision Analysis*. We have not paid attention to model robustness, but this is also an important problem when undertaking general sensitivity analyses. Some ideas are described in Berger (1994b). Last but not least, examples and case studies following our approach are still sparse.

ACKNOWLEDGMENTS

This work was partially supported by grants from CICYT, NATO, CNR and CAM.

10.A APPENDIX

10.A.1 Sensitivity to the Prior

We consider local sensitivity. Once we have fixed $\pi \in \mathscr{P}$, suppose we compute the posterior expectation for measures in $\Delta_{\mathscr{M}}(\pi) = \{\pi + \delta : \delta \in \mathscr{M}\}$, where $\mathscr{M}$ is a subset of Δ containing all signed measures δ with $\delta(\Theta) = 0$. Δ is a normed, linear space with norm given by $\| \delta \| = d(\delta, 0)$, where $d(P, Q) = \sup_{A \in B(\Theta)} |P(A) - Q(A)|$ is the total variation metric, and $\mathscr{B}(\Theta)$ the Borel σ-field. We view T_g as a nonlinear operator taking Δ into $\mathbb{R}$.

$\dot{T}_G^{\pi}$ will be the Fréchet derivative of the operator T_g, that is, a linear map on Δ satisfying $\dot{T}_g^{\pi}(\delta) = T_g(\pi + \delta) - T_g(\pi) + o(\| \delta \|)$. $\dot{T}_g^{\pi}$ measures how a small change in π affects the posterior expectation. The norm of $\dot{T}_g^{\pi}$ over $\mathscr{M}$ is defined by $\| \dot{T}_g^{\pi} \|_{\mathscr{M}} = \sup_{\delta \in \mathscr{M}} |\dot{T}_g^{\pi}| / \| \delta \|$. We consider it as a sensitivity measure of the posterior expectation as the prior ranges in $\mathscr{M}$. Let $\Delta_{\not{p}}$ be the set of all signed measures $\delta = \varepsilon(q - \pi)$, where $q \in \mathscr{P}$ and $\varepsilon \in (0, 1]$. Under such a choice of $\mathscr{M}$, it follows that $\Delta_{\mathscr{M}}(\pi) = \{(1 - \varepsilon)\pi + \varepsilon q\}$ is an ε-contamination class of priors.

Both Fréchet derivatives and their norms may be computed as described in the following results by Ruggeri and Wasserman (1993).

Theorem 10.A.1 $\dot{T}_g^{\pi}(\delta) = \{D_\pi\}^{-1}(N_\delta -_\rho D_\delta)$, where $\rho = T_g(\pi)$.

Define $h(\theta) = l_x(\theta)(g(\theta) - \rho), \bar{h} = \sup_{\theta \in \Theta} h(\theta), \underline{h} = \inf_{\theta \in \Theta} h(\theta)$. Then we have the following theorem.

Theorem 10.A.2

$$\| \dot{T}_g^{\pi} \|_{\Delta_{\not{p}}} = \{D_\pi\}^{-1}\{\bar{h} - \underline{h}\}.$$

Ruggeri and Wasserman (1993) provide results for specific classes including the quantile class $\Gamma_{\mathbf{p}}$,

$$\Gamma_{\mathbf{p}} = \{\pi : \pi(A_i) = p_i, i = 1, \ldots, n\},$$

where $(A_1, \ldots, A_n)$ is a partition of Θ, and $\sum_{i=1}^{n} p_i = 1$.

A related result, which we use in Example 10.1, due to Martîn and Ríos Insua (1996), shows that

$$\sup_{\delta \in \Delta_{\varepsilon_0}} |\dot{T}_g^{\pi}| = \frac{\varepsilon_0}{D_\pi} \max\left\{ \sum_{i=1}^{k} p_i \bar{h}_i, - \sum_{i=1}^{k} p_i \underline{h}_i \right\}, \qquad (10.A.1)$$

with

$$\Delta_{\varepsilon_0} = \{\delta \in \Delta : \delta = \varepsilon(q - \pi), \text{with} \quad q \in \Gamma_{\mathbf{p}}, \varepsilon \leqslant \varepsilon_0\}.$$

and

$$\underline{h}_i = \inf_{\theta \in A_i} h(\theta),$$
$$\bar{h}_i = \sup_{\theta \in A_i} h(\theta).$$

Note that, except for an infinitesimum, $\sup_{\delta \in \Delta_{\varepsilon_0}} |\dot{T}_g^\pi|$ gives the maximum difference between the expected value obtained with π and that obtained with $(1 - \varepsilon)\pi + \varepsilon q, \forall \varepsilon \leqslant \varepsilon_0, q \in \Gamma_Q$.

10.A.2 Stability Theory

There are several other definitions of stable problems

Definition 10.A.1 The decision problem (u_0, l_x, π_0) is *strongly stable II (SSII)* if for any sequence $\pi_n \Longrightarrow \pi_0, q_n \Longrightarrow \pi_0, u_n \to u_0$ and $w_n \to u_0$.

$$\lim_{\varepsilon \downarrow 0} \limsup_{n \to \infty} \left[\int u_n(\theta, a_{q_n}(\varepsilon)) l_x(\theta)\, d\pi_n(\theta) - \sup_{a \in A} \int u_n(\theta, a) l_x(\theta)\, d\pi_n(\theta) \right] = 0 \qquad (10.A.2)$$

for every $a_{q_n}(\varepsilon)$ such that

$$\int w_n(\theta, a_{q_n}(\varepsilon)) l_x(\theta)\, dq_n(\theta) \geqslant \sup_{a \in A} \int w_n(\theta, a) l_x(\theta)\, dq_n(\theta) - \varepsilon.$$

Definition 10.A.2 The decision problem (u_0, l_x, π_0) is *weakly stable II (WSII)* if (10.A.2) holds for a particular choice $a_{q_n}(\varepsilon)$.

It can easily be seen that SSII implies SSI, while the converse is not true, in general, as shown by examples in Chuang (1984) and Kadane and Srinivasan (1994). Sufficient conditions ensuring the equivalence of SSI and SSII are presented in Kadane and Srinivasan (1994), including the one showing that equivalence holds for concave utility functions.

The following is another important result presented in Kadane and Srinivasan (1994).

Theorem 10.A.3 Let (u_0, l_x, π_0) be a finite decision problem such that π_0 ({discontinuity points of $u(\cdot, a)$}) $= 0$. Then (u_0, l_x, π_0) is SSII (and SSI, of course).

Further convergence results may be seen in Kadane *et al.* (1997).

10.A.3 Nondominated Alternatives

We describe the problem of checking dominance using parametric classes. If $\pi_\lambda(\theta)$ denotes the density function, we call

$$\Gamma_\lambda = \{P_\lambda : P_\lambda \text{ has df } \pi_\lambda(\cdot), \lambda \in \Lambda\}$$

the *parametric class of priors*. Typical examples are the class of normal distributions with constraints on mean and standard deviation or a family of exponentials with constraints on the parameter. This class is fairly popular in robust Bayesian analysis; see e.g. Berger (1994a). As far as the utility function is concerned, we shall consider the class

$$\mathcal{U}_\Omega = \{u_\omega : \text{utility function depending on } \omega \in \Omega\}.$$

For example, it could be a class of exponential utility functions with risk aversion coefficient in a certain range. Other important classes can be derived from results in Bell (1995).

In this case, to check dominance, we have to compute

$$\inf_{\Lambda,\Omega} \int [u_\omega(a,\theta) - u_\omega(b,\theta)] l_x(\theta)\pi_\lambda(\theta) d\theta = I. \tag{10.A.3}$$

If we have a primitive for those integrals for each u_ω, π_λ, we shall have a nonlinear programming problem. If we lack a primitive, we may use a Monte Carlo importance sampling strategy to solve (10.A.3) approximately.

Suppose that Λ and Ω are compact and, for each $\theta, u_\omega(a,\theta), u_\omega(b,\theta)$ and $\pi_\lambda(\theta)$ and continuous functions of ω and λ respectively, and measurable functions of θ for (λ, ω). Assume also that, for a certain density $h(\theta)$,

$$\left| [u_\omega(b,\theta) - u_\omega(a,\theta)] \frac{l_x(\theta)\pi_\lambda(\theta)}{h(\theta} \right| \leqslant g(\theta),$$

and g is integrable with respect to the probability measure associated with h, which will play the role of an importance sampling function. Then, we are under the conditions of a uniform version of the strong law of large numbers (Jennrich, 1969), and, for almost every sequence $(\theta_1, \theta_2, \ldots) \sim h$,

$$\frac{1}{n}\sum_{i=1}^{n} [u_\omega(b,\theta_i) - u_\omega(a,\theta_i)] \frac{l_x(\theta_i)\pi_\lambda(\theta_i)}{h(\theta_i)} \to_n \int [u_\omega(b,\theta) - u_\omega(a,\theta)] l_x(\theta)\pi_\lambda(\theta)\, d\theta,$$

uniformly $\forall(\lambda,\omega) \in \Lambda \times \Omega$ Hence, for any ε, there is n such that, for almost all $(\theta_1, \theta_2, \ldots) \sim h$,

$$\sup\left\{\left| \int [u_\omega(b,\theta) - u_\omega(a,\theta)] l_x(\theta)\pi_\lambda(\theta) d\theta - \frac{1}{n}\sum_{i=1}^{n} [u_\omega(b,\theta_i) - u_\omega(a,\theta_i)] \frac{l_x(\theta_i)\pi_\lambda(\theta_i)}{h(\theta_i)} \right|\right\} \leqslant \varepsilon.$$

This suggests the following strategy:

1. Generate $\theta_1, \ldots, \theta_n \sim h$.
2. Solve

$$\inf_{\lambda,\omega} \frac{1}{n}\sum_{i=1}^{n} [u_\omega(b,\theta_1) - u_\omega(a,\theta_i)] \pi_\lambda(\theta_i) \frac{l_x(\theta_i)}{h(\theta_i)} = I_n.$$

We have that $I_n \to I$ almost surely. Note, though, that in this and other cases, we are interested in the sign of the infimum, rather than the infimum, so it may be cheaper to proceed as follows:

1. Compute $I_n = \Psi_n(\lambda_n, \omega_n)$.
2. Estimate $\sqrt{\text{Var}(\Psi(\lambda_n, \omega_n))} = \widehat{\text{EE}}_n$.

3. If $I_n - 2\widehat{\mathrm{EE}}_n > 0$, then $b \prec a$. If $I_n + 2\widehat{\mathrm{EE}}_n < 0$, then stop. Otherwise, if $\widehat{\mathrm{EE}}_n$ is small enough, then stop. Otherwise, resample (unless a certain maximum sample size is attained).

10.A.4 Additional Information

We use the following results for Example 10.7.

Theorem 10.A.4 (a) If

$$-\inf_{\mathscr{U}\times\Gamma} \hat{N}(v,\pi') > 2\sqrt{KN^{ab}(u,\pi)} \tag{10.A.4}$$

then (u,π) is ε-robust for $b \preceq a$ within $\mathscr{U} \times \Gamma$ with $\varepsilon \in [\varepsilon_1, \varepsilon_2]$ and

$$\varepsilon_1 = \min\left\{1, \frac{\inf_{\mathscr{U}\times\Gamma} \hat{N}(v,\pi') + \sqrt{(\inf_{\mathscr{U}\times\Gamma} \hat{N}(v,\pi'))^2 + 4KN^{ab}(u,\pi)}}{2K}\right\},$$

$$\varepsilon_2 = \min\left\{1, \sqrt{\frac{N^{ab}(u,\pi)}{k}}, \varepsilon_u, \varepsilon_\pi\right\},$$

with

$$K = 4\sup|l_x(\theta)|,$$

$$\hat{N}(v,\pi') = N^{ab}(v,\pi) + N^{ab}(u,\pi') - 2N^{ab}(u,\pi),$$

$$\varepsilon_u = \min\left\{1, \frac{T^{ab}(u)}{T^{ab}(u) - \inf_{\omega\in\mathscr{U}} T^{ab}(\omega)}\right\},$$

$$\varepsilon_\pi = \min\left\{1, \frac{N^{ab}(\pi)}{N^{ab}(\pi) - \inf_{\pi'\in\Gamma} N^{ab}(\pi')}\right\}.$$

(b) If (10.A.4) does not hold, then (u,π) is ε-robust for $b \preceq a$ within $\mathscr{U} \times \Gamma$, for $\varepsilon \in [\varepsilon_1, \varepsilon_3]$ and

$$\varepsilon_3 = \min\{1, \varepsilon_u, \varepsilon_\pi\}.$$

We may further specify the above result for classes $\mathscr{U}_k$ and Γ_Q and extension of $\Gamma_{\mathbf{p}}$ defined by

$$\Gamma_Q = \{\pi \in \mathscr{P} : \underline{p}_i \leqslant \pi(A_i) \leqslant \bar{p}_i, i = 1,\ldots,n\}.$$

Corollary 10.A.1 Suppose that Θ or C is discrete and that

$$J_1 = I_n D(\pi) + H_1 - 2N^{ab}(u,\pi) < -2\sqrt{KN^{ab}(u,\pi)} \tag{10.A.5}$$

holds, with

$$I_1 = T^{ab}(\omega,\pi),$$

$$H_1 = \min \sum_{i=1}^{n} p_i \inf_{\theta\in A_i} \{[u(a,\theta) - u(b,\theta)] l_x(\theta)\}$$

subject to

$$\underline{p}_i \leqslant p_i \leqslant \bar{p}_i, \quad i = 1, \ldots, n,$$

$$\sum_{i=1}^{n} p_i = 1,$$

$$K = 4 \max_{j=1,\ldots,k} \{c_j - c_{j-1}\} \sup |l_x(\theta)|,$$

and ω defined as follows: let $S_a(c) = \{\theta \in \Theta : (a, \theta) = c\}$ and $S_b(c) = \{\theta \in \Theta : (b, \theta) = c\}$; for each $c \in C_i, i = 1, \ldots, k$,

$$\omega(c) = \begin{cases} v_i & \text{if } \pi(S_a(c)|x) \leqslant \pi(S_b(c)|x), \\ v_{i-1} & \text{if } \pi(S_a(c)|x) > \pi(S_b(c)|x). \end{cases}$$

Then, (u, π) is ε-robust for $b \preceq a$ within $\mathscr{U}_K \times \Gamma_Q$, with $\varepsilon \in [\varepsilon_1, \varepsilon_2]$ and

$$\varepsilon_1 = \min\left\{1, \frac{J_1 + \sqrt{(J_1)^2 + 4KN^{ab}(u, \pi)}}{2K}\right\},$$

$$\varepsilon_2 = \min\left\{1, \sqrt{\frac{N^{ab}(u, \pi)}{K}}, \varepsilon_u, \varepsilon_\pi\right\},$$

with

$$\varepsilon_\pi = \min\left\{\frac{N^{ab}(\pi)}{N^{ab}(\pi) - H_1}\right\},$$

$$\varepsilon_u = \min\left\{1, \frac{T^{ab}(u, \pi)}{T^{ab}(u, \pi) - I_1}\right\}.$$

11
Graphical Methods

Roger M. Cooke

Delft University of Technology, The Netherlands

Jan M. van Noortwijk

HKV Consultants, Lelystad, The Netherlands

11.1 INTRODUCTION

'A picture is worth a thousand words' is true in many areas of statistical analysis and in modelling. Graphs contribute to the formulation and construction of conceptual models and facilitate the examination of underlying assumptions. Data visualization is an area of considerable scientific challenges, particularly when one is faced with high-dimensional problems characteristic of sensitivity analysis.

A literature search reveals very little in the way of theoretical development for graphical methods in sensitivity and uncertainty analysis, apart from certain reference books (e.g. Cleveland, 1993). Perhaps it is the nature of these methods that one simply 'sees' what is going on. Cleveland studies the visualization of univariate, bivariate, and general multivariate data. Our focus, however, is not on visualizing data as such, but rather visualization to support uncertainty and sensitivity analysis. The main sources for graphical methods are software packages. Standard graphical tools such as scatterplots and histograms are available in almost all packages, but the more challenging multidimensional visualization tools are less widely available.

In this chapter, we first choose a simple problem for illustrating generic graphical techniques. The virtue of a simple problem is that we can easily understand what is going on, and therefore we can appreciate what the various techniques are and are not revealing. Then, in subsequent sections, we shall discuss more generic techniques that can be used for more complex problems. The generic techniques discussed here are tornado graphs, radar plots, generalized reachable sets, matrix and overlay scatterplots, and cobweb plots. Simple scatterplots have already been presented in Chapter 2 and Chapter 6 and will not be discussed separately here. Where appropriate, the software producing the plots and/or analysis will be indicated.

Sensitivity Analysis. Edited by A. Saltelli *et al.*

In the last sections of this chapter, having grasped what graphical techniques can do, it is instructive to apply them to three real problems where we do not immediately 'see' what is going on. First, a problem concerning dike ring reliability is used to illustrate the use of cobweb plots in identifying local probabilistically important parameters. The second problem, from internal dosimetry, illustrates the use of radar plots to scan a very large set of parameters for important contributors to the overall uncertainty. Finally, a study of uplifting and piping failure mechanisms is used to illustrate the use of scatterplots and coplots in steering a Monte Carlo sampling routine. The detailed discussion of these problems can be found in the original reports; in this chapter, we concentrate rather on the usefulness of the different graphical techniques in providing insights into the model behavior.

11.2 A SIMPLE PROBLEM

The following problem[1] serves to illustrate the generic techniques. Suppose we are interested in how long a car will start *after* the headlights have stopped working. We build a simple reliability model of the car consisting of three components:

- the battery (bat),
- the headlight lampbulb (bulb),
- the starter motor (strtr).

The headlight fails when either the battery or the bulb fail. The car's ignition fails when either the battery or the starter motor fail. Thus, considering bat, bulb, and strtr as life variables:

$$\text{headlite} = \min(\text{bat}, \text{bulb}),$$
$$\text{ignitn} = \min(\text{bat}, \text{strtr}).$$

The variable of interest is then

$$\text{ign-head} = \text{ignitn} - \text{headlite}.$$

Note that this quantity may be either positive or negative, and that it equals zero whenever the battery fails before the bulb and before the starter motor.

We shall assume that bat, bulb, and strtr are independent exponential variables with unit expected lifetime. The question is, which variable is most important to the quantity of interest, ign-head?

We shall now present some more specialized graphical tools, using this simple example to illustrate their construction and interpretation.

11.3 TORNADO GRAPHS

Tornado graphs are simply bar graphs where (in this case) the rank correlation between the factors and the model response is arranged vertically in order of descending absolute value. The spreadsheet add-on Crystal Ball (1996) (see the software appendix) performs uncertainty analysis and gives graphic output for sensitivity analysis in the form of tornado graphs (without using this designation). After selecting a 'target forecast variable', in this

[1]This has been developed for the Cambridge Course for Industry, Dependence Modelling and Risk Management (Cambridge Course for Industry, 1998).

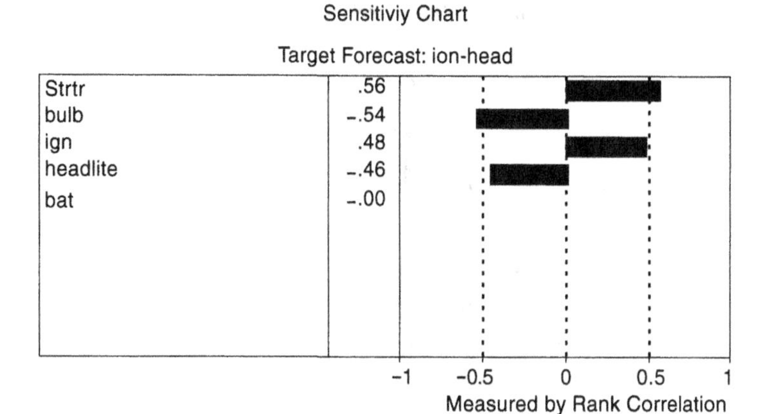

Figure 11.1 Tornado graph.

case ign-head, Crystal Ball shows the rank correlations of other input variables and other forecast variables as in Figure 11.1.

The values, in this case, the rank correlation coefficients, are arranged in decreasing order of absolute value. Hence, the variable strtr with rank correlation 0.56 is first, and bulb with rank correlation -0.54 is second, and so on. When influence on the target variable, ign-head, is interpreted as rank correlation, it is easy to pick out the most important variables from such graphs. Note that bat is shown as having rank correlation 0 with the target variable ign-head. This would suggest that bat was completely unimportant for ign-head. Obviously, any of the other global sensitivity measures mentioned in Chapter 2 could be used instead of rank correlation (see also Kleijnen and Helton, 1999a,b).

11.4 RADAR GRAPHS

Continuing with this simple example, another graphical tool, radar graphs, is introduced. A radar graph provides another way of showing the information in Figure 11.1. Figure 11.2 shows a radar graph made in EXCEL (1995) by entering the rank correlations from Figure 11.1. An identical picture could be produced in other general software.

Each variable corresponds to a ray in the graph. The axis on each ray spans the absolute value of the correlation values $(-0.6, 0.6)$, and the value of the rank correlation for each factor is then plotted on the corresponding ray and connected. The variable with the highest rank correlation is plotted furthest from the midpoint, and the variable with the lowest rank correlation is plotted closest to the midpoint. Thus bulb and headlite are plotted closest to the midpoint, and srtr and ignitn plotted furthest. The discussion of the internal dosimetry problem in Section 11.9 shows that the real value of radar plots lies in their ability to handle a large number of variables.

11.5 GENERALIZED REACHABLE SETS

The method of generalized reachable sets (Bushenkov *et al.*, 1995) involves approximating and visualizing multidimensional sets given implicitly by a mapping. Although most often

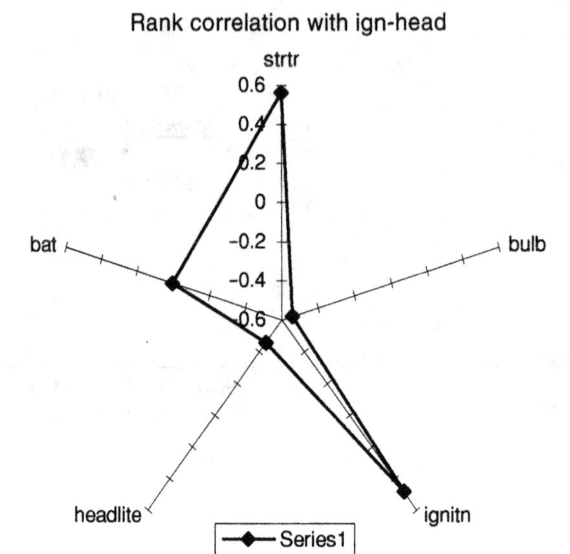

Figure 11.2 Radar graph.

employed in multicriteria analysis, its visualization facility as implemented in the MS WINDOWS software package VISAN can be readily applied to a problem such as the one involving ign-head. The method enables the visualization of the influence of two variables on a target variable (ign-head) by shaded contours (when implemented, the shades are colored but in this case we use a simple gray scale). In the present case, the functional relations are extracted by sampling (bat, bulb, strtr), computing ign-head, and clustering the results. The clustering introduces numerical inaccuracies.

Figure 11.3 shows shading of the value of bulb, with i-h (ign-head) on the vertical axis, and bat on the horizontal axis. The shading at the point $|i - h| = 1.015$, bat $= 1.973$) represents

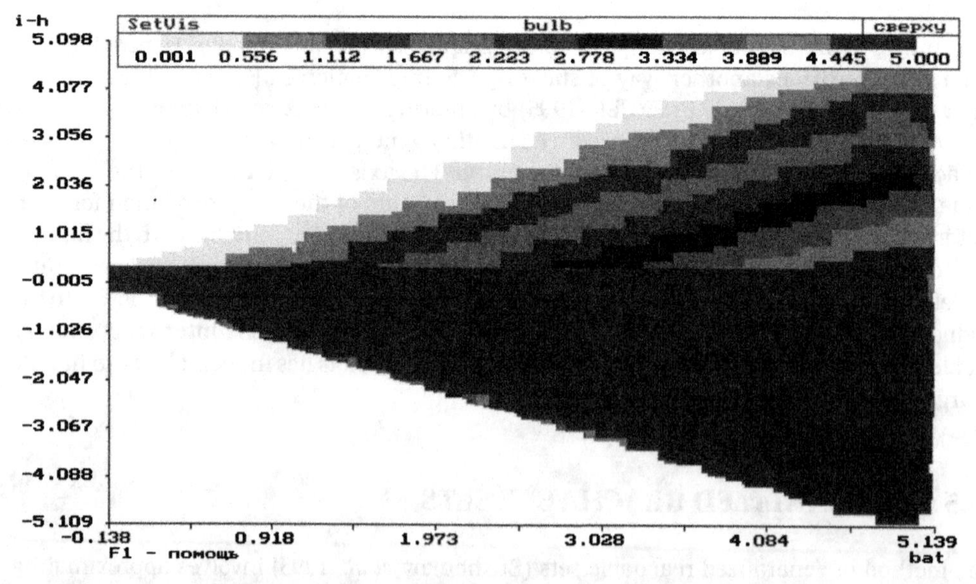

Figure 11.3 Relationships between bulb, bat, and ign-head.

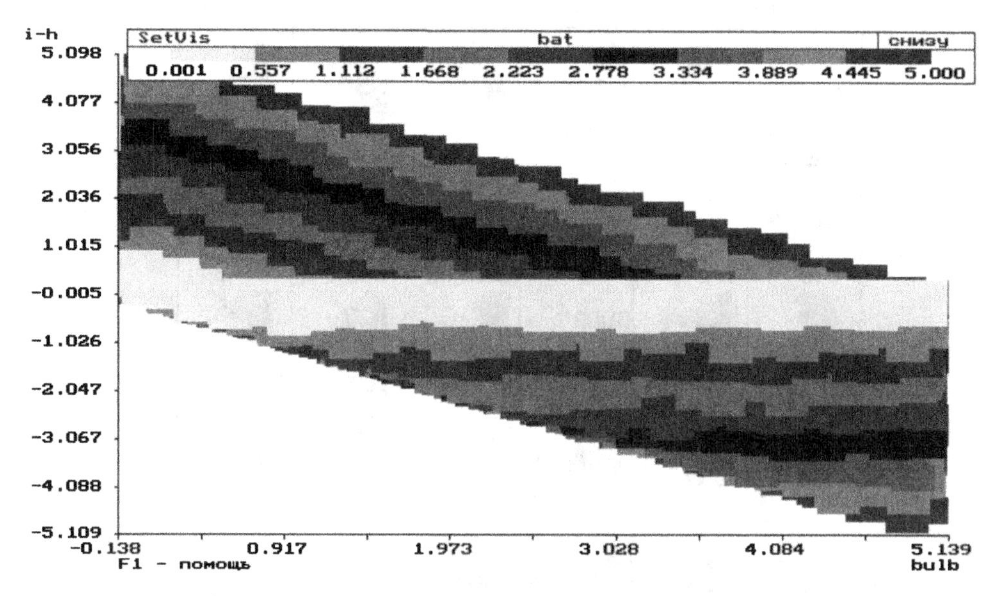

Figure 11.4 Alternative visualization of relationships between ign-head and bat and bulb.

the range of values of bulb found in the sample. We see that |i-h| < bat, which indeed is obvious from the definitions. Further, i-h is positive only if bulb < bat (the values of bat for i-h < 0 and 4.445 < bulb < 5 have overwritten these values for bulb < 4.445). From the definitions, we see if bat = 1.973 and i-h = 1.015, then bulb cannot be larger than 1.973 − 1.015 = 0.958. Figure 11.3 shows bat between 1.112 and 1.667. Figure 11.4 shows the same information, except that the roles of bulb and bat have been reversed. If we now take bulb = 1.973, then we see i-h increases in bat if i-h > 0 and decreases in bat if i-h < 0 (this may be difficult to see in black and white reproduction). In spite of numerical inaccuracies, these graphs provide gradient-type information; we can see what combinations of bat and bulb lead to increasing or decreasing values of ign-head. The technique can deal with up to five variables by constructing two-dimensional arrays of figures, but the interpretation of these figures is not easy. This illustrates some of the difficulties in visualising multidimensional data. Other graphical presentations that would prove useful here would include contour plots; however, they do not easily generalise to more than three variables.

11.6 MATRIX AND OVERLAY SCATTERPLOTS

The simple scatterplot has also been the subject of some development for use in multivariate cases. Such extensions include the matrix scatterplot and the idea of overlaying multiple scatterplots on the same scale. Many statistical packages support these (and other) variations of scatterplots. For example, SPSS (1997) provides a matrix scatterplot facility. Simulation data produced by UNICORN (Cooke, 1995) for 1000 samples has been read into SPSS to produce the matrix scatterplot shown in Figure 11.5. We see pairwise scatterplots of the variables in our problem. The first row, for example, shows the scatterplots of ign-head and, respectively, bat, bulb, strtr, ignitn and headlite. The matrix scatterplot is symmetrical, and we need only focus on the top right plots, since they are replicated in the lower left.

Figure 11.5 summarizes the relationships between each pair of variables, in the problem (15 such individual plots). Let (a,b) denote the scatterplot in a row a and column b; thus, (1,2)

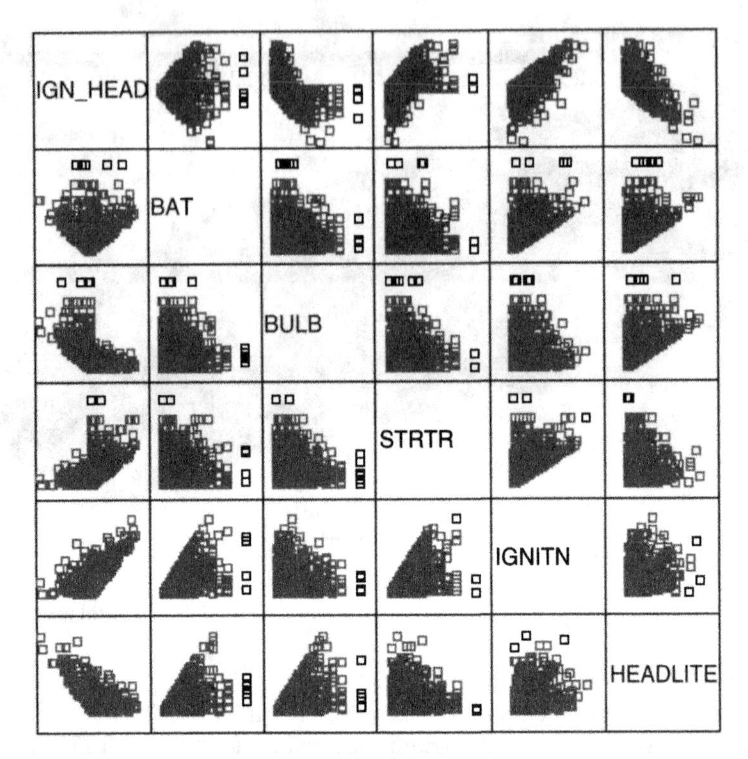

Figure 11.5 Matrix scatterplot.

denotes the second plot in the first row with ign-head on the vertical axis and bat on the horizontal axis. Note that (2,1) shows the same scatterplot, but with bat on the vertical and ing-head on the horizontal axes.

Figure 11.5 shows that the value of bat can say a great deal about ign-head. Thus, if bat assumes its lowest possible value, then the values of ign-head are severely constrained. This reflects the fact that if bat is smaller than bulb and strtr, then ignitn = headlite and ign-head = 0. From (1,3), we see that large values of bulb tend to associate with small values of ign-head; if bulb is large, then the headlight may live longer than the ignition, making ign-head negative. Similarly, (1,4) shows that large values of strtr are associated with large values of ign-head. These latter facts are reflected in the rank correlations of Figure 11.1.

In spite of the above remarks, the relation between *rank* correlations depicted in Figure 11.1 and the scatterplots of Figure 11.5 is not direct. Thus, bat and bulb are statistically independent, but if we look at (2,3), we might infer that high values of bat tend to associate with low values of bulb. This, however, is an artifact of the simulation. There are very few very high values of bat, and, as these are independent of bulb, the corresponding values of bulb are not extreme. If we had a scatter plot of the *rank* of bat with the *rank* of bulb, then the points would be uniformly distributed on the unit square.

Continuing with the developments for simple scatterplots, we consider the idea of overlaying separate plots (on the same scale) and using different symbols to convey additional information; see Figure 11.6 for an example. Ign-head is depicted on the vertical axis, and the values for bat, bulb, and strtr are shown as squares, triangles, and diamonds, respectively. Figure 11.6 is a superposition of plots (1,2), (1,3), and (1,4) of Figure 11.5. However, Figure 11.6 is more than just a superposition. Inspecting Figure 11.6 closely, we see that there are always a square, triangle, and diamond corresponding to each realized value on

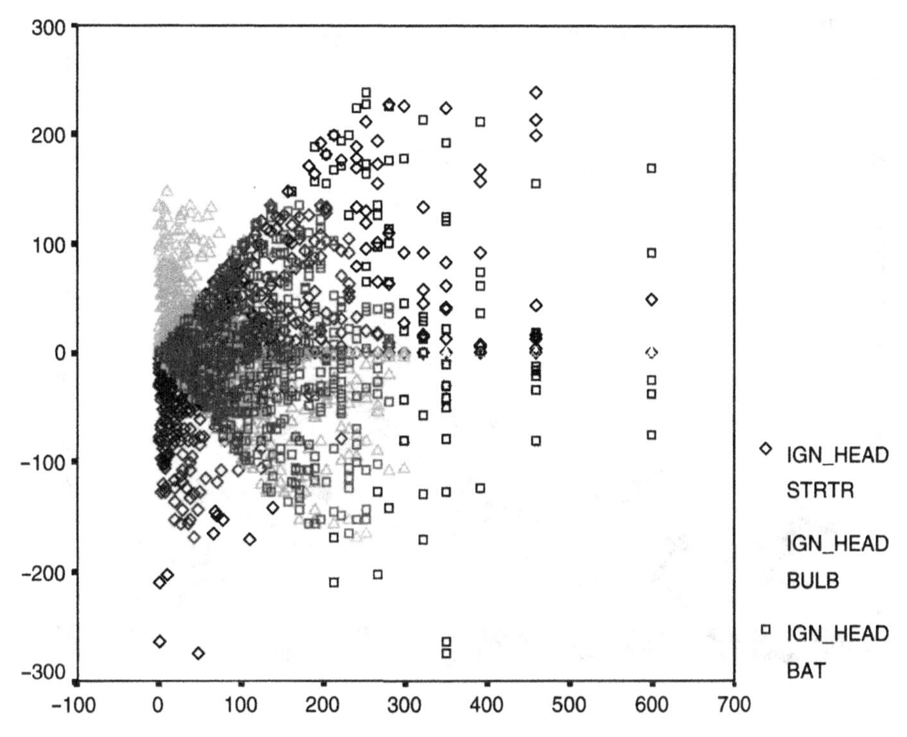

Figure 11.6 Overlay of three scatterplots.

the vertical axis. Thus, at the very top there is a triangle at ign-head = 238 and bulb slightly greater than zero. There are a square and diamond also corresponding to ign-head = 238. These three points correspond to the same sample. Indeed, ign-head attains its maximum value when strtr is very large and bulb is very small. If a value of ign-head is realized twice, then there will be two triples of squares–triangle–diamonds on a horizontal line corresponding to this value, and it is impossible to resolve the two separate data points. For ign-head = 0, there are about 300 realizations.

The joint distribution underlying Figure 11.5 is six-dimensional. Figure 11.5 does not show this distribution, but rather shows 30 two-dimensional (marginal) projections from this distribution. Figure 11.6 shows more than a collection of two-dimensional projections, since we can sometimes resolve the individual data points for bat bulb, strtr, and ign-head, but it does not enable us to resolve all data points. The full distribution can, however, be shown in cobweb plots.

11.7 COBWEB PLOTS

The uncertainty analysis program UNICORN (see the software appendix) contains a graphical feature that enables interactive visualization of a moderately high dimensional distribution. Our sample problem contains six random variables. Suppose we represent the possible values of these variables as parallel vertical lines (Wegman, 1990). One sample from this distribution is a six-dimensional vector. We mark the six values on the six vertical lines and connect the marks by a jagged line. If we repeat this 200 times, we get Figure 11.7. The

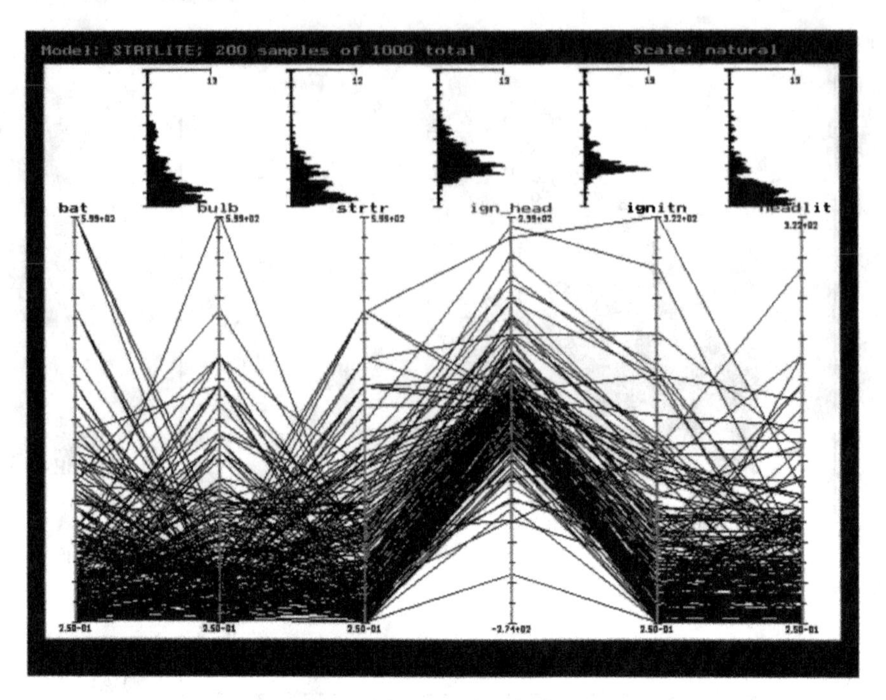

Figure 11.7 Cobweb plot.

number of samples (200) is chosen for black and white reproduction. Onscreen, the lines may be color-coded according to the leftmost variable: e.g. the bottom 25% of the axis is yellow, the next 25% is green, then blue, then red. This allows the eye to resolve a greater number of samples, and greatly aids visual inspection. From the cobweb plot, we can recognize the exponential distributions of bat, bulb, and strtr. Ignitn and headlite, being the minima of independent exponentials, are also exponential. Ign-head has a more complicated distribution. The graphs at the top are the 'cross-densities'; they show the density of line-crossings midway between the vertical axes. The role of these in depicting dependence becomes clear when we transform the six variables to ranks or percentiles, as in Figure 11.8.

A number of striking features emerge when we transform to the percentile cobweb plot. First of all, there is a curious hole in the distribution of ign-head. This is explained as follows. On one-third of the samples, bat is the minimum of $\{\text{bat, bulb, strtr}\}$. On these samples, ignitn = headlite and ign-head = 0. Hence, the distribution of ign-head has an atom at zero with weight 0.33. On one-third of the samples, strtr is the minimum, and on these samples, ign-headlite is negative, while on the remaining third, bulb is the minimum and ign-head is positive. Hence, the atom at zero means that the percentiles 0.33 up to 0.66 are all equal to zero. The first positive number is the 67th percentile.

Note the cross-densities in Figure 11.8. One can show the following for two adjacent continuously distributed variables X and Y in a (unconditional) percentile cobweb plot:[2]

- If the rank correlation between X and $Y = 1$, then the cross-density is uniform.
- If X and Y are independent (rank correlation 0), then the cross-density is triangular.

[2] These statements are easily proved with a remark by Tim Bedford of the Department of Mathematics of Delft University of Technology. Notice that the cross-density is the density of $X + Y$, where X and Y are uniformly distributed on the unit square. If X and Y have rank correlation 1, then $X + Y$, is uniform [0,2]; if they have rank correlation -1 then $X + Y = 1$; if X and Y are independent, then the mass for $X + Y = a$ is proportional to the length of the segment $x + y = a$.

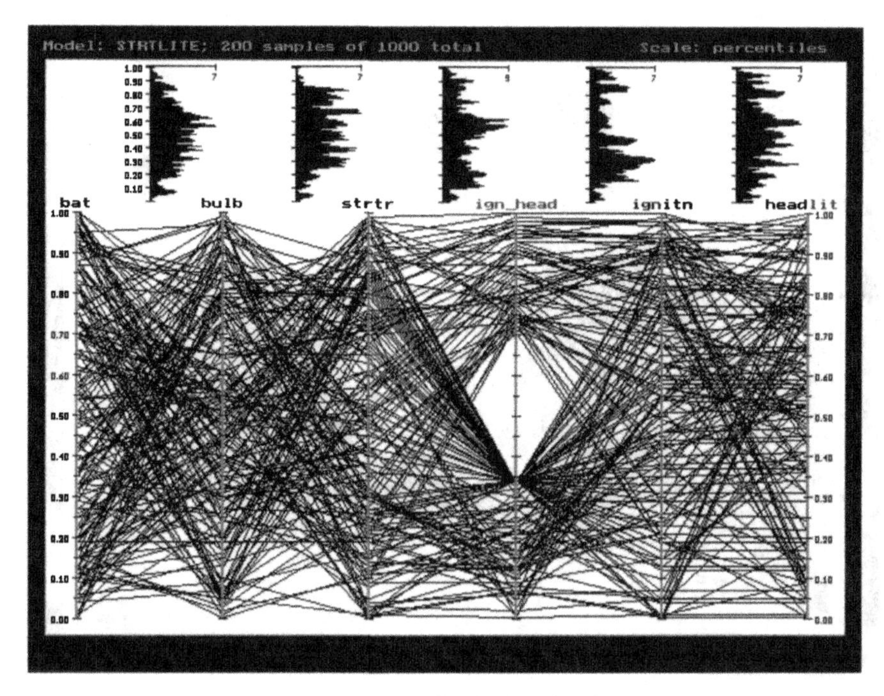

Figure 11.8 Cobweb plot of ranks.

- If the rank correlation between X and $Y = -1$, then the cross-density is a spike in the middle.

Intermediate values of the rank correlation yield intermediate pictures. The cross-density of ignitn and headlite is intermediate between uniform and triangular; and the rank correlation between these variables is 0.42.

Cobweb plots support interactive conditionalization; that is, the user can define regions on the various axes and select only those samples that intersect the chosen region. Figure 11.9 shows the result of contitionalizing on ign-head $= 0$. Notice that if ign-head $= 0$, then bat is almost always the minimum of $\{$bat,bulb,strtr$\}$, and ignitn is almost always equal to headlite. This is reflected in the conditional rank correlation between ignitn and headlite being almost equal to 1. We see that the conditional correlation as in Figure 11.9 can be very different from the unconditional correlation of Figure 11.8. From Figure 11.9, we also see that bat is almost always less than bulb and strtr.

Cobweb plots allows us to examine local sensitivity. Thus, supposing that ign-head is very large, we can ask what values the other variables should take. The answer is reached simply by conditionalizing on high values of ign-head. Figure 11.10 shows conditionalization on high values of ign-head, while Figure 11.11 conditionalizes on low values (the number of unconditional samples has been increased to 1000).

If ign-head is high, then bat, strtr, and ignitn must be high. If ign-head is low, then bat, bulb, and headlite must be high. Note that bat is high in both cases. Hence, we should conclude that bat is very important both for high values and for low values of ign-head. This conclusion differs from what we would have drawn if we considered only the rank correlations of Figures 11.1 and 11.2.

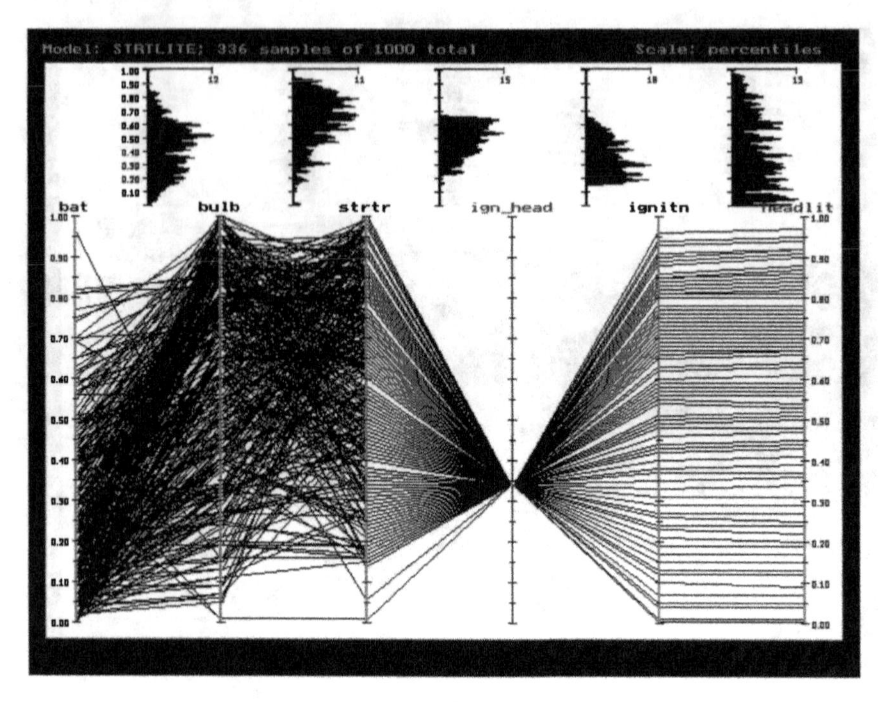

Figure 11.9 Conditioned cobweb plot.

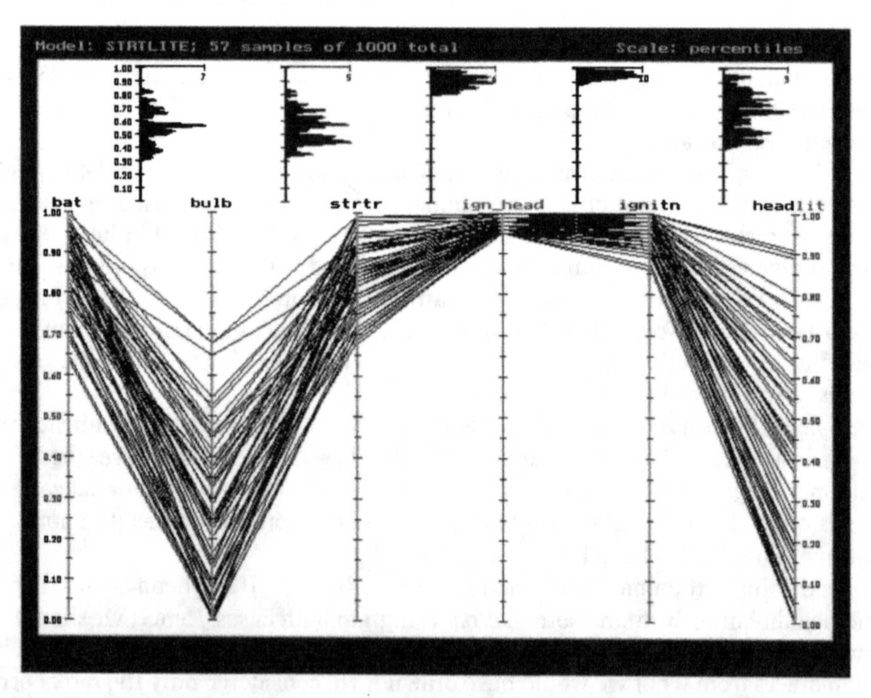

Figure 11.10 Conditional cobweb plot (high values for ign-head).

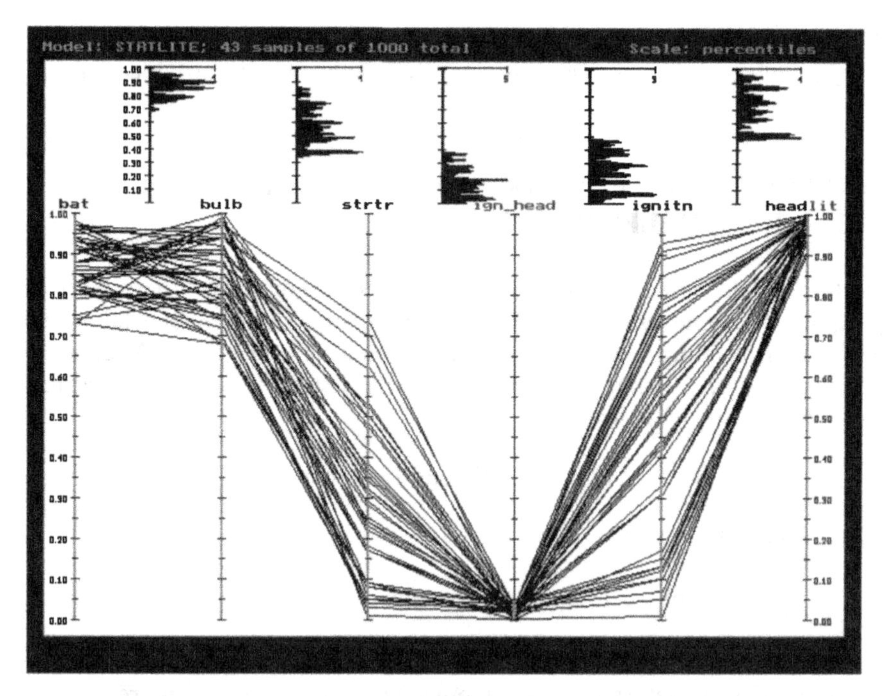

Figure 11.11 Conditional cobweb plot (low values of ign-head).

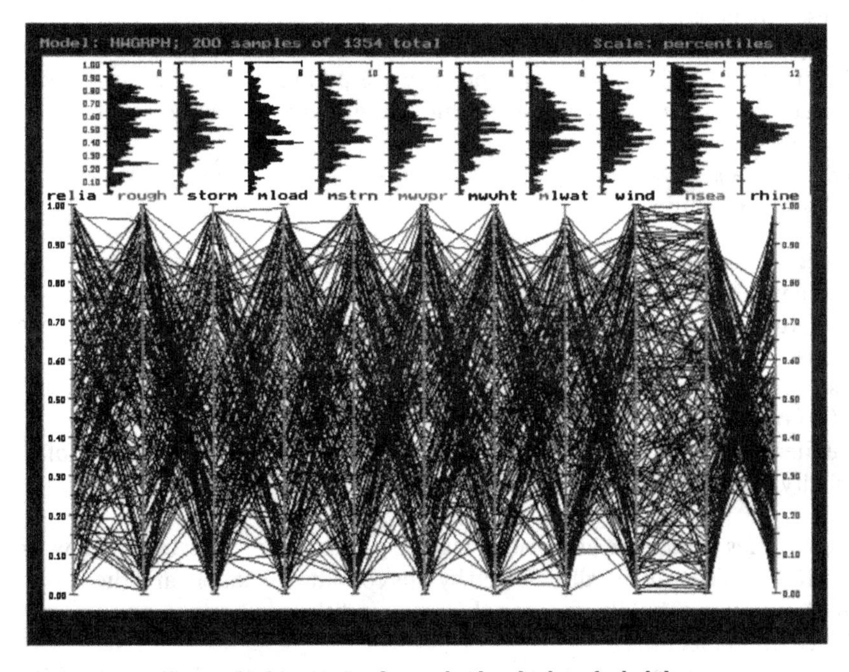

Figure 11.12 Unconditional cobweb plot of reliability.

These facts can also be readily understood from the formulae themselves. Of course, the methods come into their own in complex problems where we cannot see these relationships immediately from the formula. The graphical methods then draw our attention to patterns, which we must then seek to understand. The following three sections illustrate graphical methods used in anger—that is, for real, problems where our intuitive understanding of the many complex relationships is assisted by the graphical tools introduced earlier.

11.8 COBWEB PLOTS FOR LOCAL SENSITIVITY: DIKE-RING RELIABILITY

In this section, we discuss a recent application in which graphical methods were used to identify important parameters in complex uncertainty analyses. This application concerns the uncertainty in dike ring reliability, and was discussed in Cooke and van Noortwijk (1997, 1999). The dike ring in question is built up of more than 30 dike sections. The reliability of each dike section *i* is modelled as

$$\text{Reliability}_i = \text{Strength}_i - \text{Load}_i.$$

The reliability of the dike ring is

$$\text{relia} = \text{Reliability}_{\text{ring}} = \min\{\text{Reliability}_i\}.$$

The dike ring fails when relia <0. Figure 11.12 shows the unconditional percentile cobweb plot for relia and 10 explanatory variables. From left to right, the variables are:

- roughness ('rough');
- storm length ('storm');
- model factors for load, strength, significant wave period, significant wave height, and local water level ('mload', 'mstrn', 'mwvpr', 'mwvht', and 'mlwat', respectively);
- wind ('wind');
- North Sea ('nsea');
- Rhine discharge ('rhine').

For a further discussion of these variables and their role in determining reliability, we refer to Cooke and van Noortwijk (1999).

Note from the cross-densities that nsea and rhine are negatively correlated, whereas nsea and wind are positively correlated.

Figures 11.13, 11.14, and 11.15 show the results of conditionalizing, respectively, on the upper 5% of relia, the 25- to 30-percentiles, and the bottom 5%. Failure occurs at the 2-percentile; hence, Figure 11.15 shows conditionalization to 'dangerous' values of the dike ring reliability. We make the following observations:

- Very large values of relia are associated with very high values of mstrn and low values of storm; other variables are hardly affected by this conditionalization, and their conditional distributions are practically equal to their unconditional distributions (i.e. uniformly distributed over percentiles).
- For values of relia around the 30-percentile, mstrn must be rather low; other variables are hardly affected by the conditionalization.
- For dangerous values of relia, nsea and wind must be high; other variables are unaffected by the conditionalization.
- For dangerous values of relia, the correlations wind–nsea and nsea–rhine are *weaker* than in the unconditional sample.

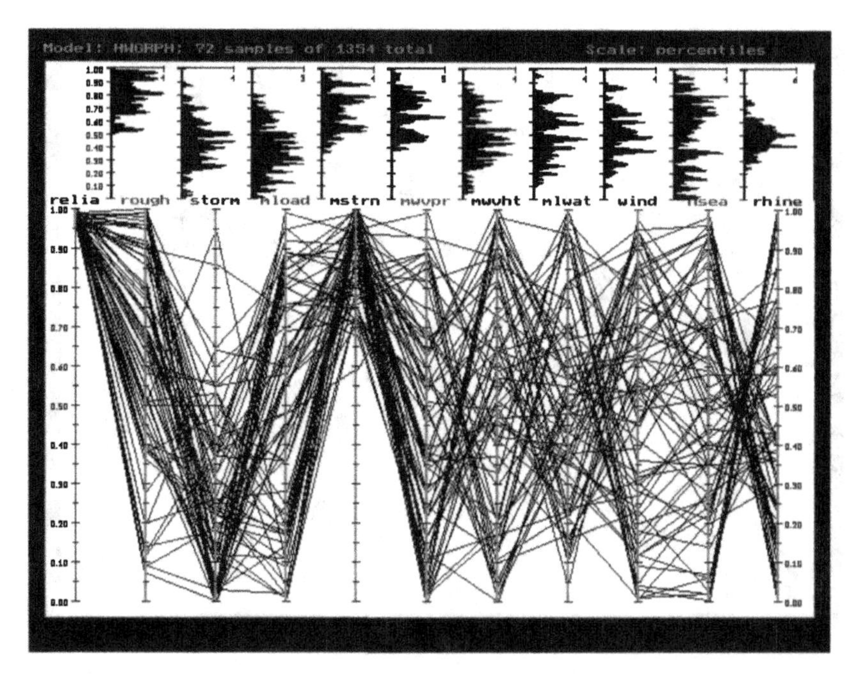

Figure 11.13 Conditional cobweb plot (upper 5% of relia).

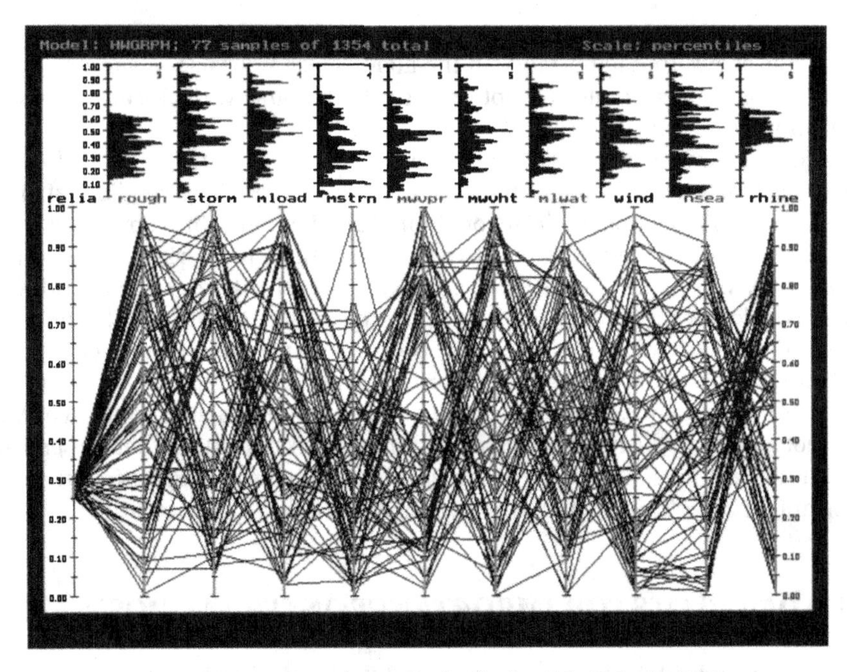

Figure 11.14 Conditional cobweb plot (25–30% of relia).

We interpret 'unaffected by the conditionalization' in this context as 'unimportant for the values of "relia" on which we conditionalize'. For example, knowing that reliability is very high, we should expect that storm length is low and the model factor for strength is very high. With regard to other variables, knowing that relia is very high does not affect our knowledge about what values these variables might take.

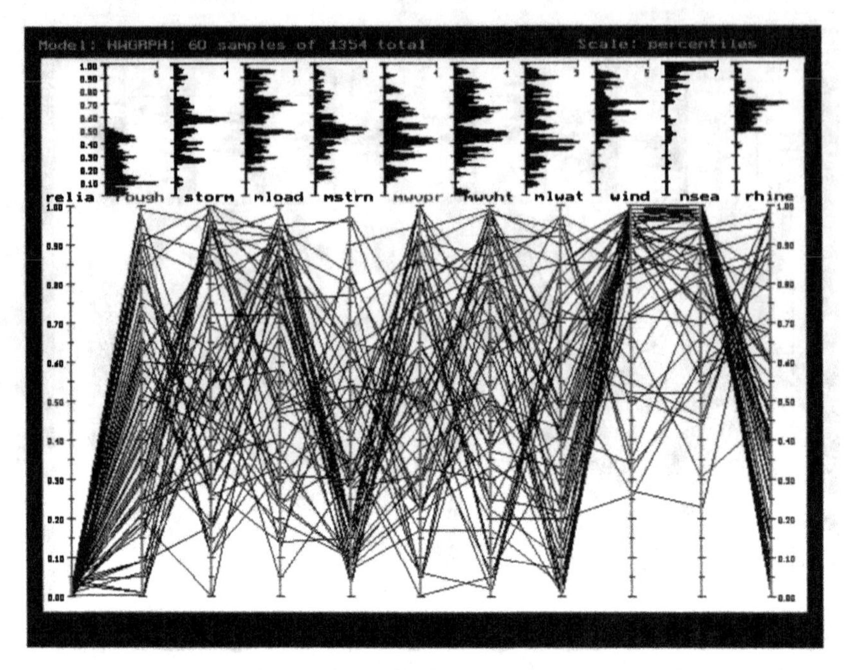

Figure 11.15 Conditional cobweb plot (lower 5% of relia).

This example shows that importance in the above sense is local. The variables that are important for high values of relia are not necessarily the same variables as those that are important for very low values of relia.

Intuitively, when we conditionalize an output variable on a given range of values, those input variables that are important *for the given range* are those whose conditional distributions differ most markedly from their unconditional distributions. One convenient measure of this is the derivative of the conditional expectation of the input variable (Cooke and van Noortwijk, 1999). More precisely, the local probabilistic sensitivity measure (LPSM) of variable X for variable Z when Z takes the value z is proportional to the rate of change of the expectation of X conditional on $Z = z$. In the special case that Z is a linear combination of independent normals, LPSM(X, z) is just the product moment correlation between X and Z (i.e., it does not depend on z). If LPSM(X, z) = 0, then the conditional expectation of X given $Z = z$ does not change in the neighborhood of z, which is taken to mean that X is probabilistically not important for $Z = z$. Conversely, large absolute values of LPSM(X, z) suggest that X is important for $Z = z$. These notions are applied in the following example.

11.9 RADAR PLOTS FOR IMPORTANCE: INTERNAL DOSIMETRY

An ongoing joint study of the European Union and the US Nuclear Regulatory Commission aims to quantify the uncertainty for accident consequence models for nuclear power plants, based on structured expert judgement (Goossens *et al.*, 1997). The number of uncertain variables is on the order of 500. Not all of these can be considered in the Monte Carlo uncertainty analysis. Roughly 150 of the 'most important' variables must be selected for the Monte Carlo analysis. In this study, there are a large number of output variables. In the example discussed here, there are six output variables, corresponding to collective dose to six important

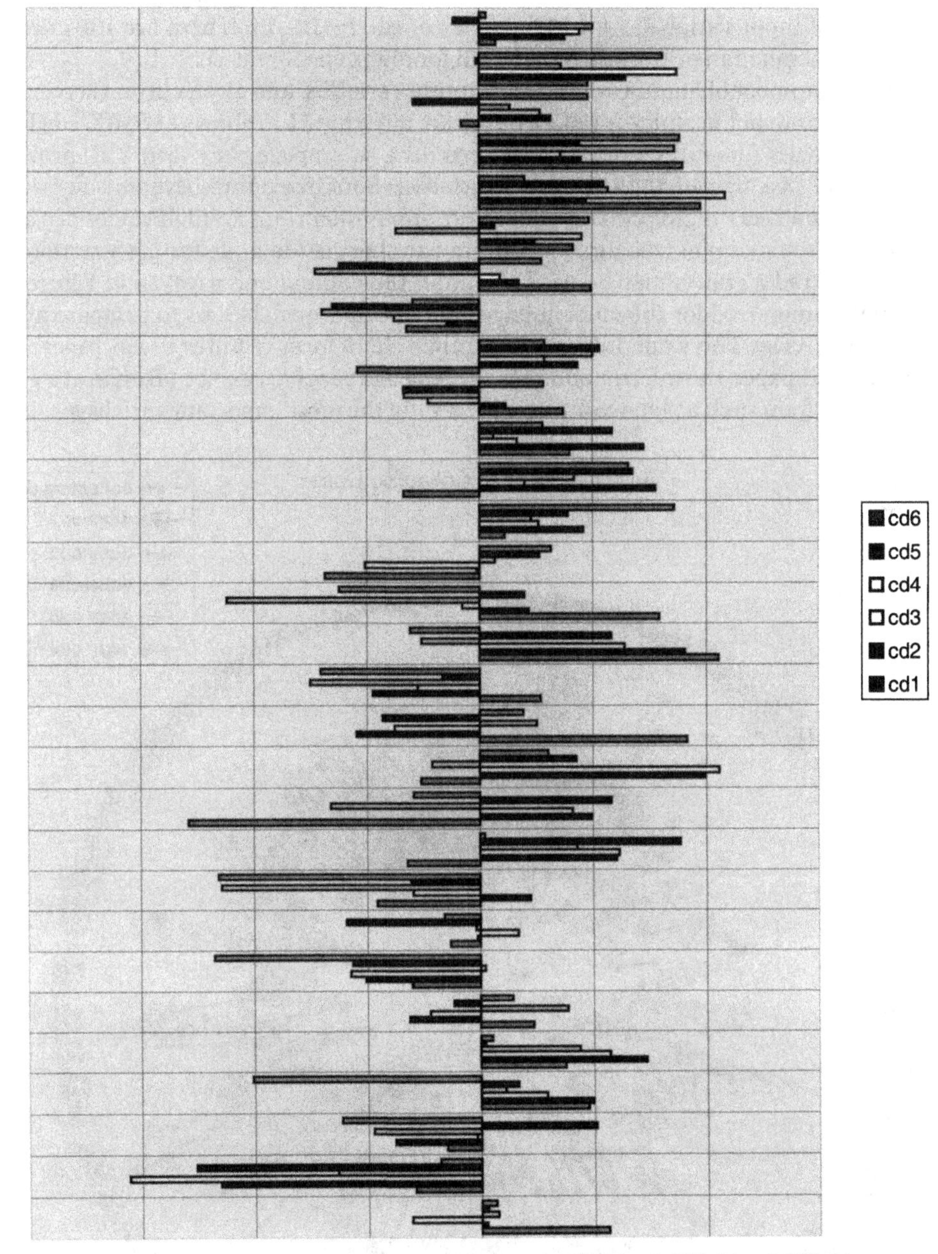

Figure 11.16 Bar chart of the LPSM.

organs, cd_i, $i=1, \ldots, 6$. Moreover, we are not interested in all values of these collective doses; rather, we are interested in those variables that are important for high values of collective dose, for some organ. Hence, the LPSM(X, cd_i) introduced above is applied to measure the sensitivity of input variable X for high values of cd_i, $i=1, \ldots, 6$. There are 161 uncertain variables that might in principle be important for high collective dose.

With this number of imput variables and output variables, and given current screen sizes, cobweb plots are not useful. An EXCEL bar chart in Figure 11.16 shows LPSM(X, cd_i) for the first 30 variables (the full set of variables requires a seven page bar chart). This provides a good way of picking out the important variables. Bars extending beyond, say, -0.2 or beyond 0.2 indicate that the corresponding variable is important for high values of a collective dose to some organ. Of course the entire bar chart is too large for presentation here. Instead, the radar charts may be used to put all the data on one page, as in Figure 11.17. Although compressed for the current page size, this figure enables us to compare all variables in one view. The same information in bar chart form requires seven pages. When printed on A3 paper, the information in the seven page bar chart can be taken in at a glance. In principle, from such a plot, we are able to identify the most important variables.

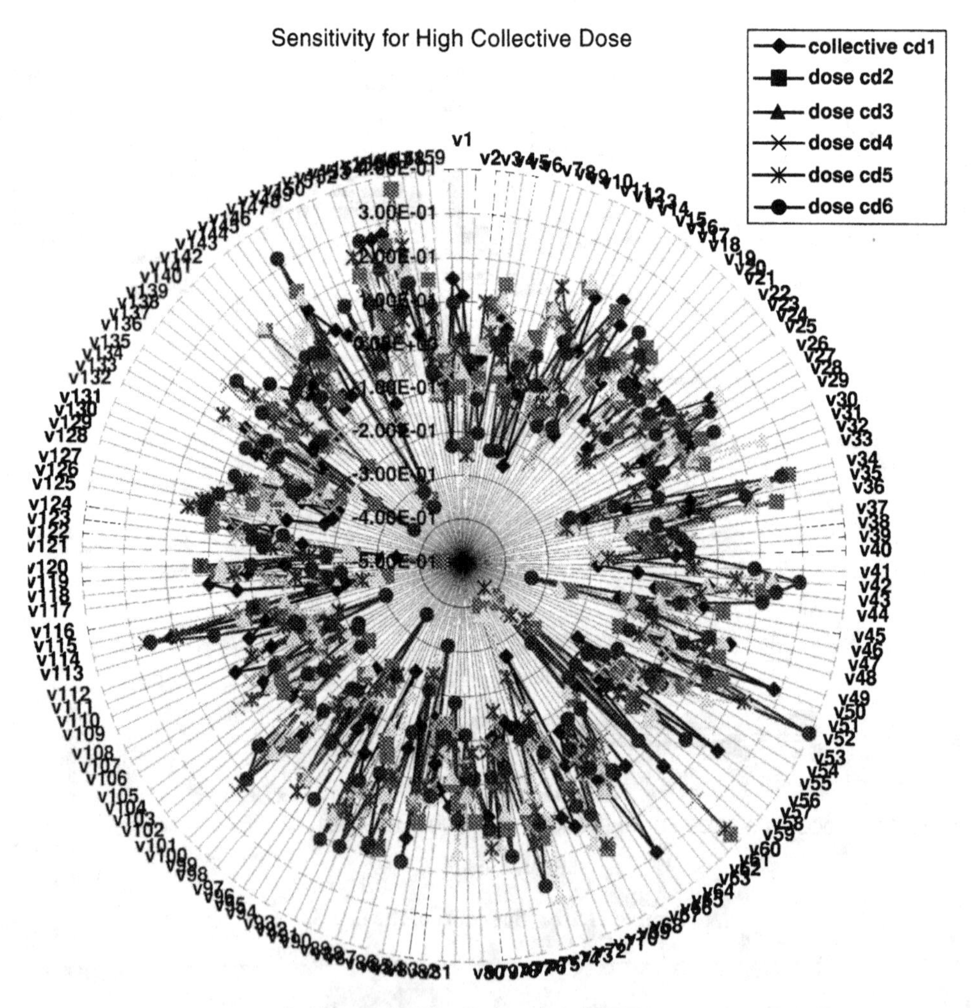

Figure 11.17 Radar plot of LPSM.

11.10 SCATTERPLOTS FOR STEERING OF DIRECTIONAL SAMPLES: UPLIFTING AND PIPING

An important use of graphical methods is for steering advanced Monte Carlo sampling routines. One such routine is directional sampling. When applicable, it can significantly improve the estimation of low probabilities. This section describes the use of scatterplots for steering a directional sampling routine. The data come from a study of dike failure due to uplifting and piping. For details, see Cooke and van Noortwijk (1999), and van Noortwijk *et al.* (1999).

A dike fails owing to uplifting and piping when water tunnels under the dike erode the back face. A reliability function Z is defined for uplifting and piping such that failure occurs when $Z = 0$. Failure due to uplifting and piping is a low-probability event. It requires very high local water levels, and many other structural variables are important as well. Local water levels are determined by the Rhine discharge and the North Sea water level. Z is a function of all these variables.

Suppose we hold all other variables at their nominal values and consider the uncertainty in Rhine discharge Q and the North Sea level S. For given values of Q and S, we can determine whether or not piping and uplifting occur. The probability of occurrence of uplifting and piping is thus determined by the joint distribution of Q and S.

In a full Monte Carlo exercise, we should have to sample very many times from the joint distribution of (Q, S) to build up a reliable estimate of the probability of uplifting and piping, since these are very rare events. In some cases, we can transform (Q, S), to independent exponential variables. If we then consider these variables in polar coordinates (ϕr), then

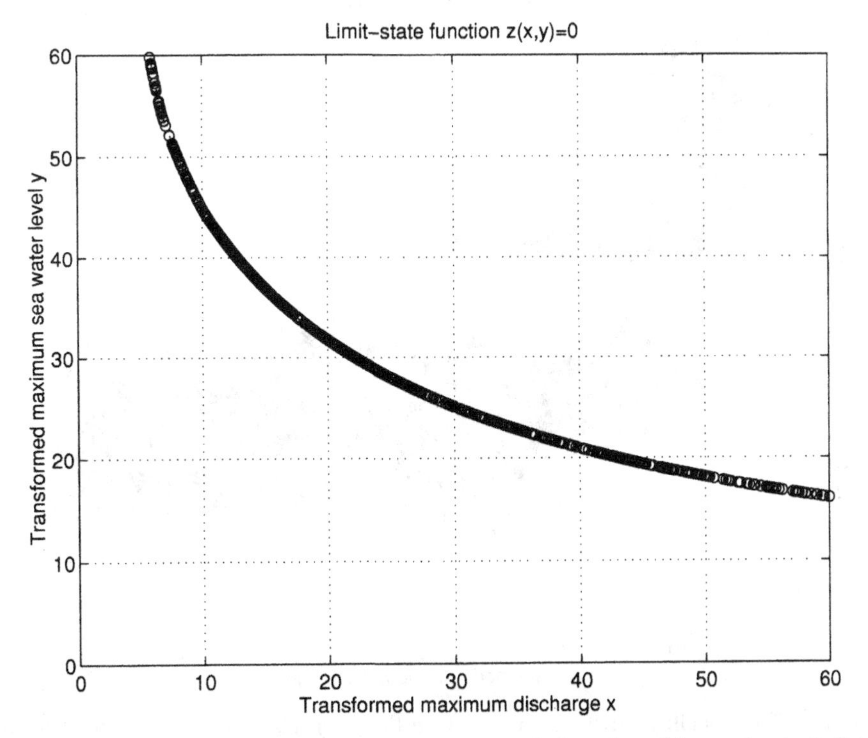

Figure 11.18 The reliability function $z(x, y) = 0$ in the (x,y) plane for a dike section including 1000 samples of $(r^*(\phi)\sin\phi, r^*(\phi)\cos\phi)$, where $r^*(\phi)$ is the zero of the reliability function, when only the inherent uncertainties in the river discharge and the sea water level are taken into account.

the following property holds: for a given sampled value ϕ, we can compute the probability of the set of rs for which uplifting and piping occur. The idea is then that we sample a direction ϕ and then compute the probability of uplifting and piping in direction ϕ (see Ditlevsen and Madsen, 1996, Chapter 9). This gives us a relatively inexpensive way of estimating the probability of uplifting and piping, *when other variables are held at their nominal values.* The question is how good is such an approximation?

To answer this question, graphical methods are brought into play. We first plot the failure hypersurface $Z = 0$ in the transformed (Q, S) plane (Figure 11.18), where the failure region is the upper right part of the figure. This is simply a scatterplot, conditional on nominal values of structural variables and conditional on $Z = 0$. Thus, in Figure 11.18, only the 'inherent' uncertainties in Q and S are taken into account. If uncertainties in other variables were taken into account, this might cause failure to happen at other values (Q, S) then those shown in Figure 11.18. These new samples might render a failure probability estimate based on Figure 11.18 unreliable.

To check this idea, we draw a sample of 10 000 from the uncertainty distributions of the structural variables, and for each sample, we sample a direction and compute the probability of failure in the given direction as before. When we look at these in the transformed (Q, S) plane, we get the cloud of points shown in Figure 11.19. We may think of these points as projections onto the (Q, S) plane of points on the failure hypersurface in the high-dimensional space containing all uncertain variables. Alternatively, the cloud of points indicates how much the line in Figure 11.18 may vary as the other structural variables are sampled from their uncertainty distributions.

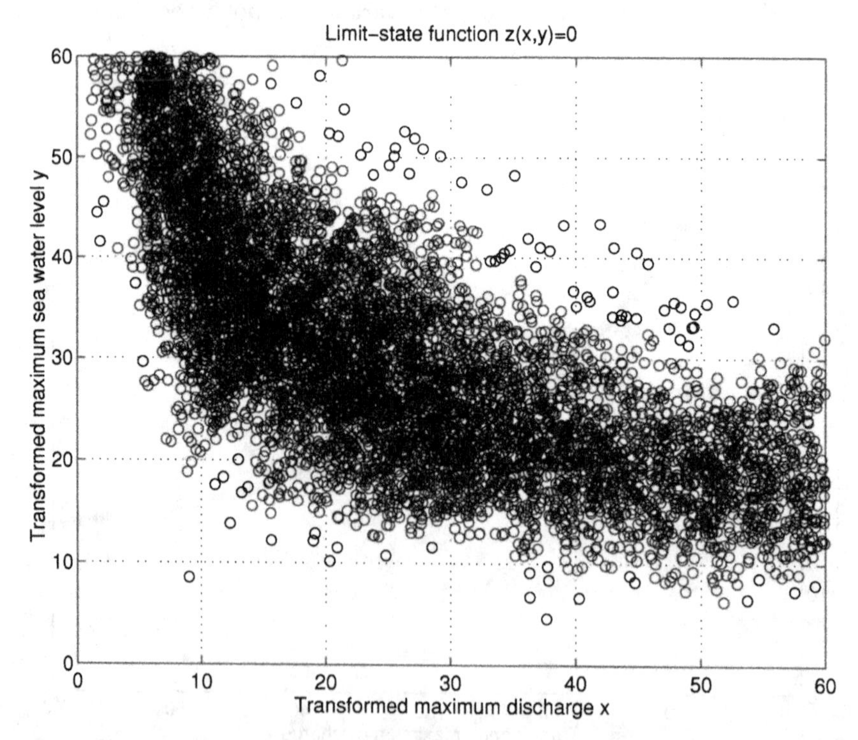

Figure 11.19 The reliability function $z(x, y) = 0$ in the (x, y) plane for a dike section including 10 000 sanokes if $(r^*(\phi) \sin \phi, r^*(\phi) \cos, \phi)$, where $r^*(\phi)$ is the zero of the reliability function, when all uncertainties (including the inherent uncertainties of the critical height over the length of the dike section) are taken into account.

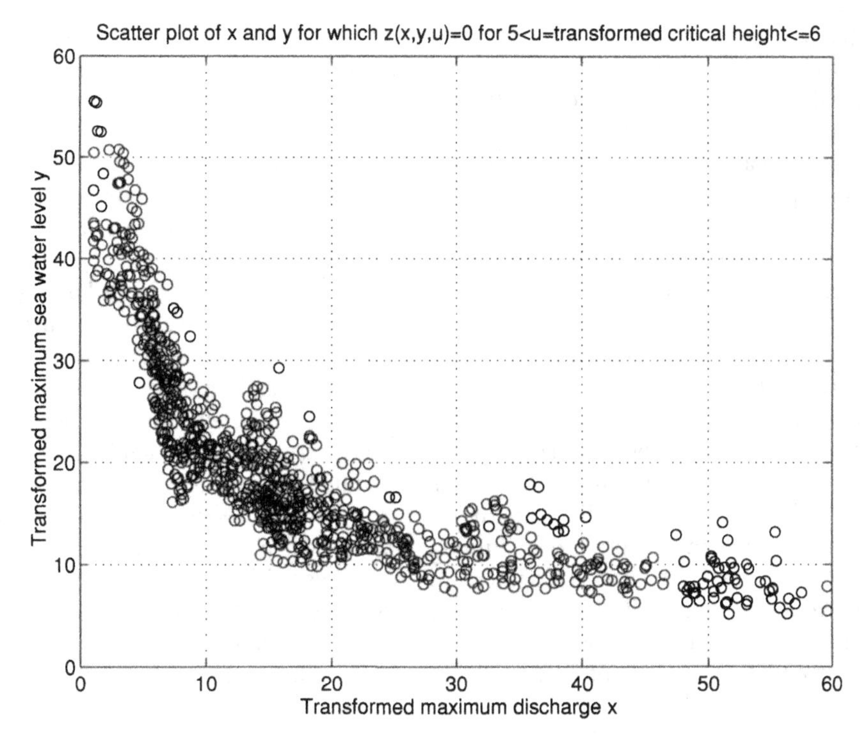

Figure 11.20 The reliability function $z(x, y, u) = 0$ in the (x,y) plane for a dike section including 947 samples of $(x, y, u) = (r^*(\phi, \psi) \sin\phi \sin\psi, r^*(\phi, \psi) \cos\phi \sin\psi, r^*(\phi, \psi) \cos\psi)$, where $r^*(\phi, \psi)$ is the zero of the reliability function and $5 < u \leqslant 6$, when all uncertainties are taken into account.

The dispersion of points in Figure 11.19 indicates that the failure probability is greatly influenced by the other structural variables. Reliable estimates of the failure probability using the two-dimensional directional sampling method sketched above still require about 1 000 000 samples. For this reason, the directional sampling is extended to three dimensions, the third dimension being critical height U. U is a function of several structural variables, and reflects the water level that the dike can withstand.

In order to calculate the probability of failure due to uplifting and piping, we transform (S, Q, U) to standard independent exponential variables, introduce polar coordinates (r, ϕ, ψ), and proceed as before *mutatis mutandis*.

Figure 11.20 is similar to Figure 11.19, except that the critical height U is conditioned to lie in the range between 5 and 6. Comparing Figures 11.19 and 11.20, we surmise that fixing U removes most of the dispersion in Figure 11.19 (of course, this must be verified for other values of U). When more ranges for U are given in one figure, this method is called the 'coplot graph' (see Cleveland, 1993, Chapter 4), but these coplots do not reproduce well in black and white, and so are not shown here. Using three-dimensional directional sampling—on the basis of the polar coordinates of the sea water level, the river discharge, and the critical height in the event of uplifting and piping—sample sizes of about 20 000 give satisfactory results.

11.11 CONCLUSIONS

This chapter has discussed a number of graphical methods; tornado graphs, radar plots, generalized reachable sets, scatterplots, matrix scatterplots, overlay scatterplots, and cobweb plots. These tools have different strengths and weaknesses, and lend themselves to different problems. The results deduced from the graphical analysis are backed up by the more formal SA tools described in other chapters of this book. Together, they provide powerful tools for exploring model behavior.

For presenting a large number of functional relationships, the radar plots of Figure 11.17 are the most powerful technique. For studying arbitrary stochastic relations between two variables, scatterplots are the most familiar technique, and therefore require no explanations. However, extensions of scatterplots to multivariate data, such as overlay scatterplots and matrix scatterplots, do require explanation, and do not always give the full picture. For multivariate problems with, say, less than 15 variables, cobweb plots give a full picture. Conditionalization can be used to discover relationships that are not reflected in global sensitivity measures. Cobweb plots are more complex and less familiar than other techniques, and their use therefore requires explanation. For problems of higher dimensionality, there are, as yet, no generic graphical tools. It is a question of 'flying by the seat of our pants', and finding subsets of variables of lower dimension that can be studied with the above techniques.

The examples discussed here illustrate the use of graphical methods in finding important parameters and steering Monte Carlo simulations. Another aspect of graphical methods that has not been the focus of attention here, but which was alluded to in the first sentence of the introduction, is equally important, and so deserves mention in closing. Graphical tools facilitate communication with decision makers, users, and stakeholders. All of the tools discussed here are used not only for analysis but also for communicating results.

III

Applications

12

Practical Experience in Applying Sensitivity and Uncertainty Analysis

E. Marian Scott

University of Glasgow, UK

Andrea Saltelli and Tine Sørensen

European Commission, Joint Research Centre, Ispra, Italy

12.1 INTRODUCTION

In Chapter 1, we have discussed and argued for the acceptance of SA as a key part of the modelling process and have highlighted only some of the many activities to which SA makes a contribution. In this chapter, we consider in more detail the development of 'good modelling practice' guidelines based around the use of SA before introducing a series of short application chapters showing SA in action. Interested readers are also directed towards the proceedings of the series of Sensitivity Analysis of Model Output (SAMO) conferences (SAMO, 1995, 1998) for further applications.

12.1.1 SA and its Relationship to Other Modelling Activities

Returning to SA for natural systems, we begin with Rosen's formalization of the scientific process (Rosen, 1991) introduced in Chapter 1 (Figure 12.1). We would like to identify constituent elements of the modelling process in the diagram, and try to associate them with a use of SA. The aim is to derive a map of SA in the context of the modelling process.

Sensitivity Analysis. Edited by A. Saltelli *et al.*

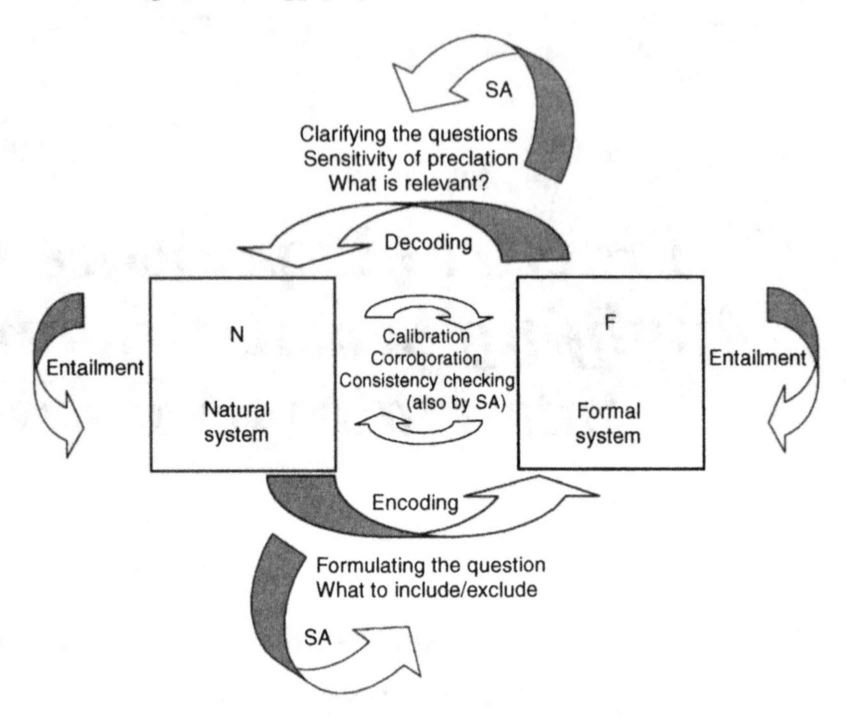

Figure 12.1 Rosen's formalization of the scientific process.

Following Rosen, the process of modelling is the establishment of a relation between the causal entailment in a natural system, N, and the syntactic entailment in a formal one, F. The relation is established by the opposite processes of encoding and decoding. The problem with the process is that while both natural causality and formal syntax are entailment systems (the elements of the systems are logically concatenated), the rules for encoding and decoding are—to say the least—rather open ended (they are *creative acts*, involving *art*, in Rosen's terminology). This leaves the modeller with a considerable freedom, while at the same time introducing in the system a painful degree of arbitrariness. This is why the use of models in the context of the scientific method is different from that of straight physical laws, amenable to clear-cut yes/no scrutiny against evidence. When a model is used to advocate a practice or to sustain a statement, it is more like to play the role of generic evidence in a trial, whose weight must be ultimately established by a jury.

This is what renders crucial the task of corroborating a model, a task of which SA can be part. Not only must a model be shown not to contradict the evidence, but it must do so when all driving forces relevant to the problem have been incorporated in a way that is plausible to the jury. In this sense, SA does more than providing a generic quality assurance to the model. It helps to demonstrate the worthiness of the model itself.

In science, it is often said, questions matter more than answers. Even in the formulation of a model, in its encoding/decoding processes, the questions being addressed to the system must be scrutinized carefully, and the formal structure possibly updated iteratively until it proves capable of providing an answer given the question. In this respect, SA can be helpful in indicating how much you can expect from the answer given the current model structure, and, more, it may also indicate which parts of the current model structure could be targeted for further development.

12.1.2 Good Modelling Practice

Rosen's formulation has made clear the stages of the modelling process and the modelling art.

In the creative modelling process, we are striving to improve the quality of the model, perhaps by reducing complexity, or by designing experiments to reduce uncertainty, and ultimately amending the model and repeating the testing process. SA plays a crucial role in this refinement process. Through an improved understanding of the interplay between the model factors and the relative contributions they make to the overall uncertainties, we can target specific factors for more detailed study to reduce uncertainty. We can identify factors that over the whole design space are non-influential in determining the outcome and that can perhaps be combined or omitted.

12.2 THE MODELLING PROCESS

The modelling process is often iterative, with five main stages being revisited over and over:

- formulation of the problem (phrasing the question), which impacts on the entire process of (encoding, decoding);
- conceptualization (encoding proper) (several alternative conceptualizations may be proposed);
- estimation and calibration (part of the formal system analysis);
- corroboration (an overall assessment and testing of the process, i.e. of the encoding/ decoding);
- analysis (answering the question(s)). This is the final decoding for prognostic models, whose quality and reliability depend upon the four previous aspects.

Of course, a different taxonomy can be proposed (Banks, 1998), but we propose to structure subsequent discussion around these stages.

Each stage may depend on different information sources; expert judgement may be used to formulate and conceptualize the model, and on-site (or site-specific) experiments can often provide information useful in parameter estimation and model corroboration. At each stage, however, decisions concerning structure and input must be made, and at each stage, arbitrariness uncertainties will be introduced that must be quantified in the corroboration and analysis stages.

12.2.1 Formulation

The first stage in the modelling exercise is to consider why a model is required. This is when the encoding/decoding activity first enters the modelling process. The model must be fit for the purpose for which it will be used, and so we must clearly define our objectives: 'Why model?' Answers could include:

- to estimate trends, rates and to study the dynamics of the system;
- to synthesize different (and perhaps) conflicting data sets;
- to describe and summarize;
- to test hypotheses;
- to provide a decision-making tool;
- to explore connections and relationships.

While first and foremost models are themselves tools to describe, explain, and explore observed phenomena, they may only be a part of a much larger management process. In many cases, they are used as decision aids, and to answer 'what if' questions. Thus, an important model use is the prediction of behaviors under new conditions, or simply predictions of future performance. It is clear that the model must be fit for the purpose for which it was created, and different purposes may require different levels of complexity. The formulation stage also identifies the response variables that will be used in the corroboration and analysis stages.

In the problem formulation stage, we must match the property of the formal system, F, with the objectives of the modelling process. SA should allow us to verify if such a matching is feasible and if it is informative. In turn, this helps (or forces) the analyst to define the objective of the modelling, and to select the *key statements and variables* that the analysis should either disprove or corroborate.

12.2.2 Conceptualization

The conceptualization stage includes visualisation of the structure, and identification and selection of the physical processes and parameters, which will be included in N. Obviously, the definition of the purpose for the model drives the selection of processes, and the conceptualization and scientific knowledge also determines the connections, which are included. Fundamental, then, is the knowledge that many conceptual structures are possible and may be supported by the same scientific evidence.

Part of the conceptualization process also concerns the evaluation of data resources, scientific knowledge, and the synthesis of information of many different types and sources. It is important to realize that the modelling process may overlook even fundamental aspects of N and its entailment structure, and that the model adequacy does not depend on the 'truth' of F versus N, but only upon the consistency of F with N given the question to be answered.

The choice of the most appropriate physical model is made through a process of selection and discrimination in which the 'bad' models are eliminated by comparison with experimental evidence. Before a given experiment of whatever complexity is implemented, in order to discriminate among competing physical models, it is necessary that the models exhibit non-overlapping bands of possible predictions, with input data uncertainty being taken into account. In other words, the possible ranges of outcomes for a set of candidate models should be investigated before the experiments are performed, each time taking the input parameters uncertainty into account.

12.2.3 Estimation and Calibration

This phase encompasses *model estimation*, *parameter estimation*, and *calibration*. The sources of information for parameter estimation can be many. In some situations, there will be sufficient experimental data available to estimate all the parameters. However, it should be borne in mind that later new data will be required for use in the model corroboration. More typically, only limited data will be available, which may in itself require to be interpolated to allow parameter estimation. Information can be supplemented by expert opinion, review of the literature, or comparison with analogues. Uncertainty in the data must also be incorporated into our overall analysis and corroboration.

12.2.4 Model Corroboration

Often in the context of a modelling exercise, the terms *verification* and *validation* are used. In Oreskes' terminology (see Oreskes *et al.*, 1994), *verification* means the establishment of truth, while *validation* refers to the establishment of legitimacy. Oreskes criticizes the use of both terms. Her main argument against the act of *verifying a model* is that natural systems are never closed, and thus absolute verification, in the sense of asserting the 'truth' of a model, is impossible. A more correct epistemological approach to the process of model building would rather be to use the term *corroboration*. A piece of evidence is said to *corroborate* a model when it does not contradict it. Oreskes and co-workers argue that the use of models should be limited to a restricted number of tasks: corroboration of theories, falsification of other models, and answering 'what if' questions, i.e. sensitivity analysis.

Model corroboration can only be attempted once the purpose of the model has been defined. The following is a classical example (Caswell, 1976) of two different corroboration contexts:

- corroboration of an 'explicative' model, to identify/understand fundamental mechanisms underlying the behaviors of the system (e.g. the pathways for sulfate formation in the atmosphere);
- corroboration of a model to be used for future predictions of future behaviors, or behaviors under conditions not previously encountered (e.g. the prediction of the total climate forcing by the year 2010).

We should like to conclude this subsection on model calibration and corroboration with a reminder of the so called 'optimism principle' (Chatfield, 1993). This states that a model always agrees better with the particular set of data used for its calibration (Carrera *et al.*, 1993) than with another independent set.

12.2.5 Analysis

This subsection is perhaps only relevant to prognostic modelling, i.e. to those simulation models aimed to query the future, make predictions, possibly expanding/contracting time. In fact, when models are used in a diagnostic mode, their use ends with the corroboration, i.e. when one among the proposed structures/processes for the system is shown to be more plausible. Analysis here indicates the final 'decoding' act — the one that cannot form part of the corroboration process as pointing to a region where there are no data. As discussed in Section 12.2.1, SA is essential in analytic prediction. Here we need to know in a transparent way the limits of our predictive power, and what is the limiting factor. This point is taken up at length in the final chapter of this volume.

12.2.6 The Modelling Process and the Role of SA / UA

The quality and reliability of the model is often difficult to judge, since they are multi-attribute, but the formal SA/UA process encourages the modeller to explore and gain greater understanding of the constructed model by considering:

- the magnitude of the uncertainty associated with the model responses (UA);

- the main factors that contribute to the uncertainties, including structural and model, parameter, and data uncertainties (SA);
- whether the uncertainty estimate is acceptable given the model purpose.

Factors affecting model quality include how the problem was specified, the formulation of the conceptual model (were important processes omitted, was a process inadequately described, are the model assumptions satisfied, is the model credible?), formulation of the computational model (errors in software and codes, do the model predictions seem reasonable?) and estimation of parameter values (do the predictions seem reasonable?), and finally how uncertain the results are. UA and SA are very important in the model evaluation phase, since the modeller must communicate confidence in the model properties and understanding of the sources of uncertainty to the decision-maker, who must deal with them within the decision-making process.

12.3 THE CHAPTERS AHEAD

In the series of eight chapters that follow this brief introduction, the abstract discussion above is developed into more concrete examples, showing the power of SA/UA in a wide (although not exhaustive) variety of application areas. Each chapter emphasizes the power of specific SA techniques (as presented in the methodological chapters) in understanding the sources of uncertainty and their quantification.

Chapter 13 (Scenario and Parametric Sensitivity and Uncertainty Analyses in Nuclear Waste Disposal Risk Assessment: The Case of GESAMAC). The authors use SA in the process of model audit, studying the scenario and parametric uncertainty in nuclear waste disposal risk assessment. They formulate the problem in a Bayesian context, and, after performing the SA, partition the total variance in max dose into two components, between scenarios and within scenarios, the second of which represents the component of uncertainty arising from lack of perfect knowledge of the scenario-specific parameters. As a result of the audit, the authors were able to estimate that the percentage of variance arising from scenario uncertainty was about 30% of the total variance for all the scenario distribution assumptions. In this way, SA helped to show how precise the answers expected from the model could be. The uncertainty audit shows us not to pretend to be able to predict quantitatively, but to rather reckon on the plausibility of different scenarios and consequences.

Chapter 14 (Sensitivity Analysis for Signal Extraction in Economic Time Series). The authors of this chapter discuss the estimation of unobserved components in a time series and present an SA approach to answering the question of how sensitive the unobserved components are to model and then parameter choice within the given model and to isolating and ranking the sources of uncertainty. Methods of analysis used include Bayesian techniques and importance measures (Chapter 8), and again SA/UA is used to explore the effect of different model assumptions and to direct the model choice.

Chapter 15 (A Dataless Precalibration Analysis in Solid State Physics). In this chapter, the authors introduce an example from physics, which shows the power of SA to investigate the usefulness of experimental measurements. Using a global SA approach involving first-order and total effects (Chapter 8), they explore which of a series of physical constants are indeterminate and which can be determined. In this way, the SA is used to design an experiment, whose value has been assured before its actual implementation. Thus SA is used as an important experimental design tool.

Chapter 16 (Application of First-Order (FORM) and Second-Order (SORM). Reliability Methods: Analysis and Interpretation of Sensitivity Measures Related to Groundwater Pressure Decreases and Resulting Ground Subsidence). The author makes use of the FORM and SORM reliability measures (Chapter 7) to study sensitivity measures for ground subsidence in an engineering context. The outcome variable is the probability of a failure (i.e. subsidence exceeding a fixed target), and the goal of the analysis is to identify those model parameters whose uncertainty is important in determining the probabilistic outcome and which could then be targeted for further experimental refinement.

Chapter 17 (One-at-a-Time and Mini-Global Analyses for Characterizing Model Sensitivity in the Nonlinear Ozone Predictions from the US EPA Regional Acid Deposition Model (RADM)). This chapter applies SA to a large, complex Eulerian air quality model. The authors use both one-at-a-time and global techniques for a restricted set of model inputs (Chapters 4 and 8). They also consider two scenarios of emissions (including an emissions control case). The model used is three-dimensional, and the authors conclude that a single sensitivity measure is inadequate for characterizing such a model. In this example, through the combination of different sensitivity runs, the authors were able to conclude that there was no evidence of large dependences in the model, and that the OAT and global analyses are strongly complementary, arguing that both should be used in characterizing the behavior of nonlinear models. This application shows the need to be flexible in the use of different SA strategies, specifically when dealing with complex models.

Chapter 18 (Comparing Different SA Methods on a Chemical Reaction Model). The authors compare several approaches of SA applied to three generations of the same model. Local SA and one-at-a-time methods were used first to conduct a screening exercise (Chapters 4 and 5); then, in the presence of possible interactions, the authors also computed first-order and total sensitivity indices (Chapter 8) to allow a comparison of the different approaches. The general advice prepared is that the Morris method (Chapter 4) is to be preferred to OAT methods, and that global, quantitative SA is preferred where finite parameter variations are involved. When the number of input factors is too high to afford such a global analysis, the authors recommend that a two-stage approach can be taken. The first stage involves a screening process, which should be global, and which can then be followed by a global quantitative analysis of the factors identified in the first stage.

Chapter 19 (An Application of SA to Fish Population Dynamics). The authors discuss an ecological model used to explore the dynamics of fish ecosystems, particularly the collapse and regeneration of fish species. The model used is a complex stage-based one with over 100 factors, and in the first phase of the work, the Morris screening technique (Chapter 4) was used. The results showed that environmental influences were of different importance for the different species and that the early stages of development were important in all species; thus, the SA was able to identify factors that required further investigation.

Chapter 20 (Global SA: A Quality Assurance Tool in Environmental Policy Modelling). In this final applications chapter, the authors deal with a policy problem: 'how to dispose of solid waste' and explore an *incineration*-versus-*landfill* option for solid waste using different (alternative) sets of indicators. They explore how sensitive the decision made is to different indicator systems. Extended FAST (Chapter 8) is used to quantitatively rank the groups of factors according to their influence on the output uncertainty. The indicator system and data all contribute to the final uncertainties, and, using the FAST method, the authors were able to partition the uncertainties. This use of SA can be labelled as 'helping to phrase the right question'.

12.4 CONCLUSIONS

This chapter has briefly introduced the modelling process and the links to SA to all stages of this process, before discussing the variety of applications presented by the different authors. Some themes are repeated in more than one chapter and can be briefly summarized. All of the applications deal with complex models, in both structure and number of parameters. The authors offer valuable practical experience in the use of screening techniques to identify important factors. Uncertainty concerning model structure and its effects on predictions are also topics of frequent discussion, with UA/SA being used to quantify this often-ignored source of uncertainty. Finally, the use of SA as a tool to target factors that make significant contributions to the overall uncertainty and that therefore could be the subject of further experiment is also a common theme. The following chapters show that SA/UA is an essential ingredient in the modeller's tool box.

13

Scenario and Parametric Sensitivity and Uncertainty Analyses in Nuclear Waste Disposal Risk Assessment: The Case of GESAMAC

David Draper

University of Bath, UK

Andrea Saltelli and Stefano Tarantola

European Commission, Joint Research Centre, Ispra, Italy

Pedro Prado

CIEMAT, Instituto de Medio Ambiente, Madrid, Spain

13.1 INTRODUCTION: THE GESAMAC PROJECT

Nuclear fission, as an energy source, has been employed in the United States and Europe for more than 40 years (Balogh, 1991), and yet the problem of safe disposal of radioactive waste arising as a by-product of power generation is still under study. Deep geological disposal—in which radioactive materials (e.g., spent fuel rods) are encapsulated and placed in a facility far below ground—is still the most actively investigated option, although in the 1970s disposal of nuclear waste in deep sea sediments was considered (Bishop and Hollister, 1974) and some objections to underground storage persist today (Keeney and von Winterfeldt, 1994; Schrader-Frechette, 1994). It is fair to say that even after decades of research, the physico-chemical behavior of deep geological disposal systems over

Sensitivity Analysis. Edited by A. Saltelli *et al.*

geological time scales (hundreds or thousands of years) is far from known with certainty (Pereira, 1989; Draper *et al.*, 1999).

From 1996 to 1999, with partners at the Physics Department at Stockholm University, we were involved in a project for the European Commission, GESAMAC (GEosphere modeling, geosphere Sensitivity Analysis, Model uncertainty in geosphere modeling, Advanced Computing in stochastic geosphere simulation: see http://www.ciemat.es/sweb/gesamac/), whose principal aim was to make progress in capturing all relevant sources of uncertainty when predicting what would happen if deep geosphere disposal barriers were compromised in the future by processes such as *geological faulting, human intrusion,* and/or *climatic change.* We used sensitivity analysis (SA) and uncertainty analysis (UA) in predicting the radiological dose for humans on the Earth's surface as a function of time, how far the disposal facility is underground, and other factors likely to be strongly related to dose. Complex computer simulation modelling features prominently in our work.

13.1.1 The System Model

The system model on which our predictions are based consists of a hypothetical underground radioactive waste disposal system represented by three coupled submodels: the *near field* (the radiological source term), a *far field* (the geosphere), and the *biosphere* (the region on, above, or near the Earth's surface in which people live and work). The first submodel—the near field (the underground repository itself)—does not include any consideration of spatial structure or chemical complexities. This submodel assumes an initial containment time for the waste materials (only radioactive decay is considered), followed by a constant leaching rate of the radioactive inventory present at the time containment fails. The third GESAMAC submodel—the biosphere—is very simple and assumes that the radionuclides leaving the geosphere enter a stream of water from which a human population obtains drinking water, so that the dose received depends on the ratio of the drinking water consumption to the stream flow rate. This is clearly not a real, site-specific safety study, but a simplified framework in order to illustrate SA and UA methodology.

The main focus of our computer simulations deals with the second submodel, the geosphere. GESAMAC has released for public use a program called GTMCHEM, which uses Monte Carlo methods to simulate the transport of radionuclides by groundwater through geologic formations, represented by a one-dimensional column of porous material (consisting of one or more layers) whose properties can change along the pathway and in which different chemical reactions (homogeneous or heterogeneous) can take place. An earlier version of the code, GTM-1 (Prado, 1992; Saltelli *et al.*, 1989; Prado *et al.*, 1991), was extensively tested through its inclusion in several versions of the LISA code and in the international PSACOIN benchmark exercises (NEA PSAG User's Group, 1989, 1990). The only chemical process considered in GTM-1 was adsorption by linear isotherm.

GTMCHEM expands on this by incorporating a number of other chemical phenomena (Eguilior and Prado, 1998), including equilibrium complexation in solution, homogeneous first-order chemical kinetics in solution, slow reversible adsorption, and a sink associated with filtration or biodegradation. The program works by solving the equation

$$\frac{\partial C_i}{\partial t} = -V\frac{\partial C_i}{\partial X} + D\frac{\partial^2 C_i}{\partial X^2} + \mathrm{SoSi}, \qquad (13.1)$$

where C represents concentration (mols/m^3), t is time (yr), X is the space coordinate (m), V is the groundwater velocity (m/yr), D is the hydrodynamic dispersion (m^2/yr), and SoSi is the

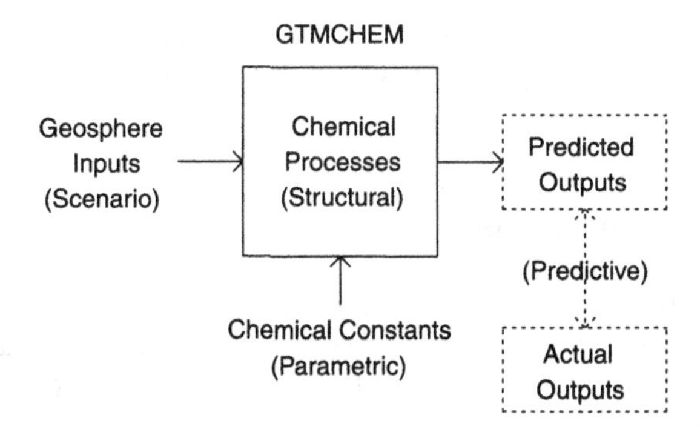

Figure 13.1 Schematic illustration of the four sources of uncertainty in GESAMAC.

source/sink term in which the chemical reactions are included. GTMCHEM produces the peak radiological fluxes at the end of each geosphere layer and the peak dose for each nuclide, as well as the associated time of the peak. The program may also be run with a fixed set of time points at which the fluxes and/or doses through space may be computed and stored.

13.1.2 The Uncertainty Framework

We have identified six ingredients, together constituting four sources of uncertainty, in the GESAMAC context, as follows (Figure 13.1; also see Draper, 1997).

1. *Past data D*, if available, would consist of readings on radiological dose under laboratory conditions relevant to those likely to be experienced in the geosphere or biosphere. (Fortunately for humankind, but unfortunately for the creation of *a predictive accuracy feedback loop* in our modeling—which would allow us to assess the most plausible structural and parametric possibilities—there have been no accidents to date of the type whose probabilities we are assessing.)
2. *Future observables* y^* consist of radiological dose values at given locations L, t years from now, as L and t vary over interesting ranges.
3. *Scenarios* $\mathcal{X}$ detail different sets of likely geosphere conditions at locations L and times t, as a result of human intrusion, faulting, and/or climate. The GESAMAC project found it useful to decompose scenario uncertainty hierarchically into two parts:
   - *macro-scenarios*, consisting of high-level statements of future geosphere conditions relevant to dose, such as climatic change; and
   - *micro-scenarios*; which are low-level characterizations of how the high-level scenarios (e.g., forces of climatic change such as erosion and deposition) would unfold chemically.
4. *Structural possibilities* $\mathcal{S}$ include different combinations of chemical processes (e.g., sorption, equilibrium, and matrix diffusion) and different sets of partial differential equations (PDEs, such as (13.1)) to model them.
5. *Parametric* uncertainty arises because the precise values of some of the relevant physical constants appearing in the PDEs are unknown. Note that parameters may be specific not only to structure but also to scenario (e.g., an early ice-age climatic scenario would have certain chemical constants driving it, whereas a worst-case geologic fracture scenario would be governed by different constants).

6. *Predictive* uncertainty is as speculative (at present) as past data in this project, and might be based on things such as discrepancies between actual and predicted lab results, extrapolated to field conditions.

13.1.3 Uncertainty Calculations

With the six ingredients above, the goal in uncertainty propagation is to produce two types of predictive distributions: *scenario-specific* and *aggregate*. It seems to us that the only hope of doing this in a way that captures all relevant sources of uncertainty is a fully Bayesian analysis, as in Draper (1995).

In the Bayesian approach, past data D (if any) are known; future observable outcome(s) y^* are unknown, and to be predicted; and we must regard the sets $\mathcal{X}$ and $\mathcal{S}$ of possible scenarios and structures as known (best results are naturally obtained if these sets are as close to fully exhaustive as possible). Then the *scenario-specific* predictive distribution $p(y^*|\mathcal{S}, x, D)$ for y^* given D, $\mathcal{S}$, and a particular scenario x is given by

$$p(y^*|\mathcal{S}, x, D) = \int_S \int_\Theta p(y^*|\theta_S, S, x) p(\theta_S|S, x, D) p(S\,|x, D)\, d\theta_s\, dS, \qquad (13.2)$$

and the *aggregate* predictive distribution $p(y^*|\mathcal{S}, \mathcal{X}, D)$ for y^* given D, $\mathcal{S}$, and $\mathcal{X}$ is

$$p(y^*|\mathcal{S}, \mathcal{X}, D) = \int_{\mathcal{X}} p(y^*|\mathcal{S}, x, D) p(x\,|\,D)\, dx. \qquad (13.3)$$

Here $p(y^*\,|\,\theta_S, S, x)$ is the *conditional predictive distribution* for y^* given specific choices for parameters, structure and scenario; and $p(\theta_S\,|\,S, x, D), p(S|\,x, D)$, and $p(x\,|\,D)$ are *posterior distributions* for the parameters, structure, and scenario (respectively) given the past data. Each of these posterior distributions depends on *prior distributions* in the usual Bayesian way; e.g., the posterior $p(S\,|\,x, D)$ for structure given the data and a particular scenario x is a multiplicative function of the prior $p(S\,|\,x)$ on structure and the likelihood $p(D\,|\,S, x)$ for the data given structure,

$$p(S\,|\,x, D) = c\, p(S\,|\,x) p(D\,|\,S, x), \qquad (13.4)$$

where c is a normalizing constant.

13.2 RESULTS FOR THE RADIONUCLIDE CHAIN

We have used Monte Carlo methods to approximate the intergrals in Equations (13.2) and (13.3). For instance, the following steps can be performed to simulate in a way that fleshes out all four sources of uncertainty in the previous section:

- First draw a scenario at random according to an appropriate probability distribution $p(x)$. In practice, this distribution typically includes a number of rare possibilities, so that it is much more efficient to stratify, over-sample on the low-probability scenarios, and reweight the findings at the end than to draw scenarios at random according to $p(x)$.
- Then select one or more structural choices (e.g., chemical processes and/or PDEs to implement them) internal to GTMCHEM according to a second probability distribution $p(S|x)$ specific to the chosen scenario.

- Parameters (chemical constants) specific to the chosen structure(s) may then be chosen according to a further set of appropriate probability distributions $p(\theta \mid S, x)$, yielding one or more GTMCHEM outputs, e.g., predicted maximum dose and dose values at location(s) L and time(s) t.
- Then these outputs may be compared with actual outcomes (if available) to estimate the likely size of GTMCHEM's prediction errors. Making a final series of draws (one for each value of L and t) from probability distributions to incorporate predictive uncertainty, which would then be added to GTMCHEM's predicted values, would complete one iteration of the Monte Carlo.

Repeating this simulation many times, making histograms or density traces of the resulting outputs, amounts to approximating the desired predictive distributions by Monte Carlo integration.

Since we have no past data D available on actual underground accidents (and we also have not made use of laboratory data in our work so far), there is no updating in the results given here from, e.g., a prior distribution $p(x)$ on scenarios to the corresponding posterior distribution $p(x \mid D)$. If we had past data available to support such updating, the Monte Carlo integration would become more complicated; Markov chain Monte Carlo (MCMC, see e.g., Gilks *et al.*, 1996) would then become a natural simulation-based alternative.

To obtain the results reported here:

- We focused on the scenario and parametric inputs to GTMCHEM of greatest interest in the standard reference test case in the nuclear safety community, the *PSACOIN Level E Intercomparison* (NEA PSAG User's Group, 1989). The Level E test case tracks the one-dimensional migration of four radionuclides — iodine (^{129}I) and a chain consisting of neptunium (^{237}Np), uranium (^{233}U), and thorium (^{229}Th) — through two geosphere layers characterized by different hydrogeological properties (sorption, hydrodynamic dispersion, and groundwater velocity). Here we present radiologic dose results for the chain and for all four radionuclides together; see Draper *et al.* (1999) for parallel results for iodine.
- GESAMAC developed a new test case called *Level E/G* (Prado *et al.*, 1998) by modifying the reference scenario in five different ways, to create a total of six scenarios:

1. Reference (REF): Level E.
2. A fast pathway (FP) to the geosphere, corresponding to a geological fault passing directly through the containment chamber, or to the reduction of the geosphere pathway by erosion of the upper layer, or to the bypassing of the second layer through human activities such as digging. This scenario thus represents a substantial decrease in radionuclide travel time through a large reduction in the geosphere path length.
3. An additional geosphere layer (AG), the opposite situation from the previous scenario. This case arises, for instance, from a retreating glacier leaving behind another barrier layer between the repository and the biosphere, or when a geological event creates an alternative pathway that is longer than that in the reference case.
4. Glacial advance (GA), related to the AG scenario but arising from an advancing rather than retreating glacier.
5. Human disposal errors (HDE), corresponding to deficiencies in the construction of the repository and/or in waste disposal operations leading to premature failure of the near-field barriers.
6. Environmentally induced changes (EIC), arising from human activities or geological events that are indirectly responsible for the modification of the disposal system

conditions, such as the drilling of a pumping well or mining tunnel at a dangerously small distance from the containment chamber.

The level E/G test case was arrived at by creating a total of nine micro-scenarios — three in each of the categories *geological changes, climatic evolution*, and *human activities* — and merging similar micro-scenarios into the five non-reference scenarios listed above.

- We focused in the work presented here only on the scenario and parametric components of overall uncertainty.

13.2.1 Radionuclide Migration Calculations and the Parallel Monte Carlo Driver

The Monte Carlo simulations of iodine migration were made by coupling the transport code GTMCHEM to a program written during the GESAMAC project called the Parallel Monte Carlo Driver (PMCD). This program, which is also publicly available, is a software package for risk assessments developed for a parallel computing environment. In developing this package, the aim was to offer to potential users of the code a high-performance computing tool that is user-friendly. The package is implemented using the Message Passing Interface (MPI), and the code presently runs on an IBM-SP2 parallel supercomputer.

Monte Carlo simulations are an ideal application for parallel processing in that every simulation can be made independently of the others. The GESAMAC Monte Carlo driver uses the single-program multiple data paradigm (SPMD) together with a master–slave approach. In brief, PMCD generates an input data matrix (master node), the rows of which are input parameters to the GTMCHEM code coupled to it. These parameters are sampled from probability distributions as indicated above. Each slave node uses information from the input matrix to perform a set of GTMCHEM migration calculations. Whenever a slave node completes them, it sends a message to the master, which in turn sends back more work to that slave.

The GESAMAC team performed 1000 runs of GTMCHEM for each scenario. Because the set of input parameters specified by the Level E/G test case for the chain of nuclides results in relatively fast runs, the strength of PMCD is not fully appreciated in this study. The run times and number of nodes varied as a function of the scenario being simulated: e.g., for 1000 simulations the number of nodes varied between 4 and 25 and the run times between 10 and 96 minutes. In the simulations, the inputs to GTMCHEM were random draws from uniform or log-uniform distributions, with minima and maxima given in the last two columns of Table 13.1 for the REF scenario for the chain. Full details on the Level E/G test case are available in Prado *et al.* (1998).

13.2.2 A Regression-Based Sensitivity Analysis

In this section, we present results for maximum dose of the nuclide chain, obtained by summing max dose across neptunium, uranium, and thorium (thus we are working not with the max over time of the sum across nuclides, $\max_t[N(t) + U(t) + TH(t)]$, but with the sum of the max doses over time, $\max_t[N(t)] + \max_t[U(t)] + \max_t[TH(t)]$, which will in general be somewhat larger). The max dose distribution was highly positively skewed for all scenarios, motivating a logarithmic transform prior to variance-based SA. For the EIC and FP scenarios, log max dose was close to Gaussian, but was negatively skewed in the AG case and appeared to have a two-component mixture form with the REF, GA, and HDE data (see

Table 13.1 Example of parametric inputs to GTMCHEM in the simulation study for the radionuclide chain: Reference scenario.

			Raw-scale	
Variable	**Meaning**	**Distribution**	**Min**	**Max**
CONTIM	No-leakage containment time	Uniform	100	1000
RLEACH	Leach rate after containment failure	Log uniform	10^{-6}	10^{-5}
VREAL1	Geosphere water travel velocity in layer 1	Log uniform	10^{-3}	10^{-1}
XPATH1	Geosphere layer 1 length	Uniform	100	500
RET11	Retardation coefficient 1 in layer 1	Uniform	300	3000
RET21	Retardation coefficient 2 in layer 1	Uniform	30	300
RET31	Retardation coefficient 3 in layer 1	Uniform	300	3000
VREAL2	Geosphere water travel velocity in layer 2	Log uniform	10^{-2}	10^{-1}
XPATH2	Geosphere layer 2 length	Uniform	50	200
RET12	Retardation coefficient 1 in layer 2	Uniform	300	3000
RET22	Retardation coefficient 2 in layer 2	Uniform	30	300
RET32	Retardation coefficient 3 in layer 2	Uniform	300	3000
STREAM	Stream flow rate	Log uniform	10^5	10^7

Figure 13.2 for the GA case). We do not pursue a mixture analysis here, focusing instead on discovering how much insight simple regression-based methods can provide.

We regressed a standardized version of log max dose on standardized versions of scenario-specific inputs, expressed on the raw or log scales as appropriate to produce approximate uniform distributions. All inputs were approximately pairwise-uncorrelated. Regressions were less well-behaved statistically (as measured, e.g., by residual plots) than in the case of ^{129}I, on which we have previously reported elsewhere. There was substantial nonlinearity in some scenarios, much (but not all) of which was captured by including quadratic and two-way interaction terms (we explored only quadratics and interactions among the variables with large main effects). As may be seen from the R^2 values in Table 13.2, the percentages of 'unexplained' variance even after the inclusion of interactions and quadratic terms ranged from 0.9% to 17.3%. Figure 13.3 illustrates the nonlinearity by plotting the standardized residuals against the predicted values for the HDE data; the dotted curve is a nonparametric smoother, showing that structure remains even after second-order effects of the important predictors are accounted for.

Table 13.3 gives an example of the regression results, in this case for the reference scenario. When the regression is performed with standardized versions of the outcome y and all of the predictors x_j, the squares of the standardized coefficients may be used as a measure of the variation in y (on the variance scale) 'explained' by each of the x_j, provided the predictors are (close to) independent. In this case, we have dealt with sample correlations of small size between the x_j by averaging the squared standardized coefficients over all possible orderings in which the variables could be entered sequentially into the regression equation. From Table 13.3, VREAL1 and STREAM are evidently the high-impact inputs for the reference scenario.

Table 13.4 summarizes the important variables for each scenario, by retaining only those inputs that 'explain' roughly 5% or more of the variance in log max dose. It is evident that, apart from occasional modest influence from other variables, the two inputs having to do with water travel velocity play the biggest role in the predicted variations in max dose of the nuclide chain, and in opposite directions: VREAL1 and max dose are positively related (large values of VREAL1 lead to faster travel times through the geosphere to the biosphere),

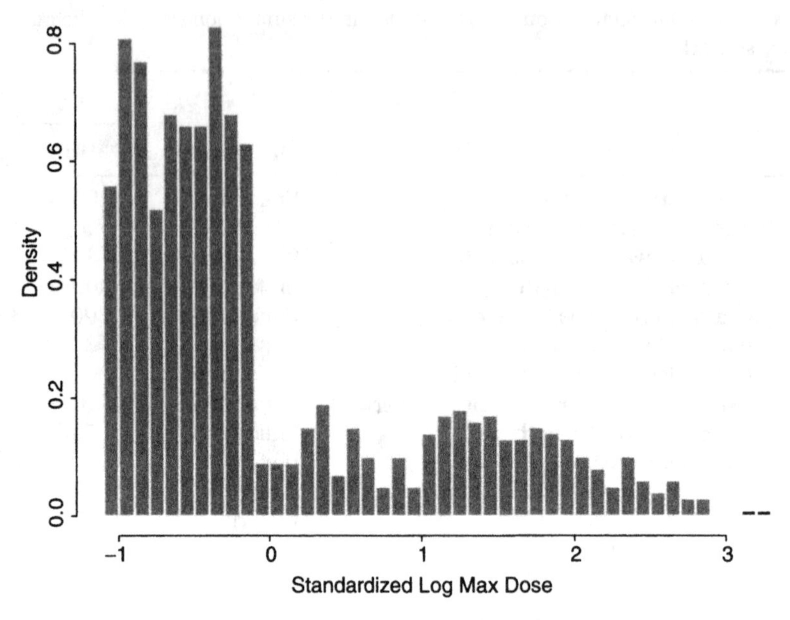

Figure 13.2 Histogram of standardized log max dose for the GA scenario.

Table 13.2 Scenario-specific output summaries. Note how much higher the dose is for the FP scenario.

Scenario	Number of geosphere layers	Maximum dose Min	Maximum dose Max	R^2 from: Main effects	R^2 from: Main effects + quadratics + interactions	Number of model inputs
REF	2	1.33×10^{-11}	9.18×10^{-6}	0.770	0.827	13
FP	1	1.59×10^{-5}	8.96×10^{-2}	0.967	0.991	17
AG	3	2.95×10^{-11}	2.12×10^{-4}	0.832	0.943	36
GA	2	2.66×10^{-12}	7.03×10^{-8}	0.685	0.884	25
HDE	2	1.35×10^{-11}	3.06×10^{-5}	0.847	0.923	28
EIC	2	2.73×10^{-11}	6.24×10^{-5}	0.864	0.907	28

whereas it is *small* values of STREAM that lead to large radiologic doses (arising from less dilution of the fluxes coming from the geosphere to the biosphere).

13.2.3 A Model Uncertainty Audit

How much of the overall uncertainty about maximum dose is attributable to scenario uncertainty, and how much to parametric uncertainty? To answer this question, following Draper (1995), we performed a *model uncertainty audit*, in which we partitioned the total variance in max dose into two components, between scenarios and within scenarios, the second of which represents the component of uncertainty arising from lack of perfect knowledge of the scenario-specific parameters. The relevant calculations are based on the double-expectation theorem (see e.g., Feller, 1971): with y as max dose, and scenario i

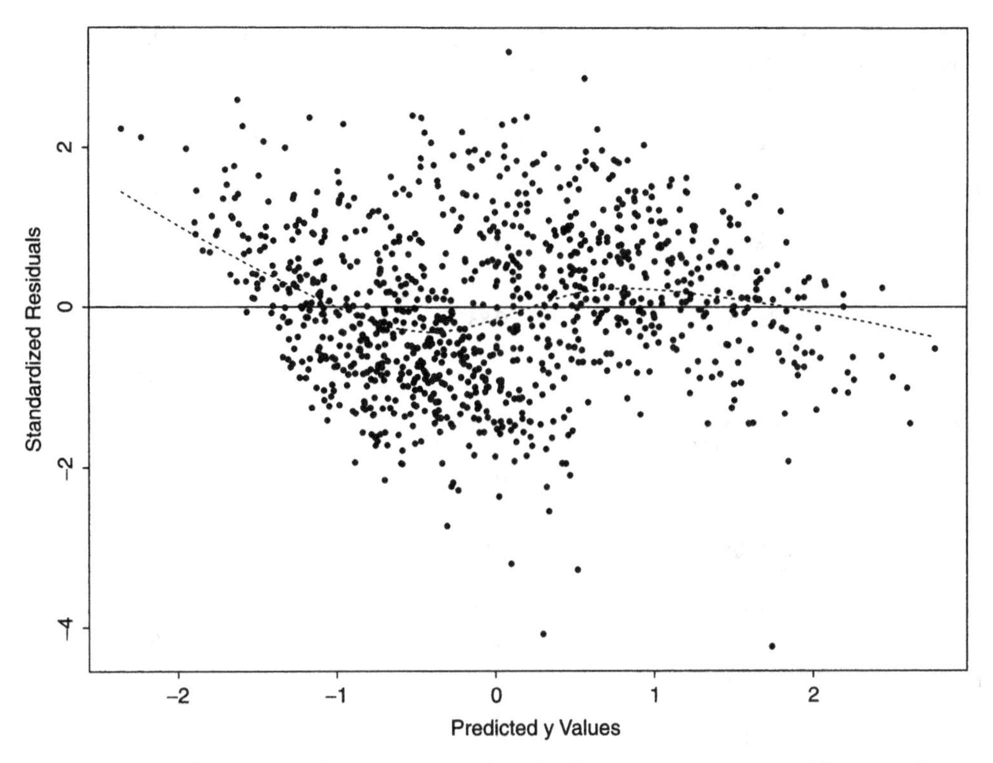

Figure 13.3 Standardized residuals versus predicted values in the regression of log max dose on main effects, interactions, and quadratic terms; HDE scenario. The dotted line is a robust smoother.

Table 13.3 Regression results for the REF scenario. L at the beginning of a variable name means that the variable entered the regression on the log scale. VR1=LVREAL1, VR2 = LVREAL2, XP1 = XPATH1; X1*X2 = interaction, X1∗X1=quadratic. Only terms whose standardized coefficient exceeded 0.1 in absolute value are included. The standard errors of the standardized coefficients were all roughly 0.012.

Variable	Standardized coefficient	Variance in log max dose 'explained' by variable
LVREAL1	0.622	0.387 ←
LSTREAM	− 0.457	0.209 ←
XPATH1	− 0.228	0.052
LVREAL2	0.221	0.049
RET11	− 0.191	0.037
VR1∗VR2	0.184	0.034
VR1∗VR1	0.170	0.029
VR1∗XP1	− 0.127	0.016
XPATH2	− 0.116	0.013
'Error'	—	0.173

Table 13.4 Summary of the important variables (those that 'explain' 5% or more of the variance in log max dose), by scenario. Naming conventions are as in Table 13.3.

Scenario	Variable	Standardized coefficient	Variance in log max dose 'explained' by main effect
REF	LVREAL1	0.613	0.376
	LSTREAM	− 0.443	0.196
	XPATH1	− 0.217	0.047
	LVREAL2	0.216	0.047
FP	LSTREAM	− 0.837	0.701
	LVREAL1	0.379	0.144
	RET1	− 0.254	0.065
AG	LVREAL1	0.743	0.553
	LSTREAM	− 0.384	0.147
	XPATH1	− 0.252	0.063
	VR1*VR1	− 0.230	0.053
GA	LVREAL1	0.678	0.460
	VR1*VR1	0.300	0.090
	LSTREAM	− 0.289	0.083
	RET11	− 0.254	0.064
HDE	LVREAL1	0.762	0.581
	LSTREAM	− 0.373	0.139
	XPATH1	− 0.246	0.061
EIC	LVREAL1	0.713	0.509
	LSTREAM	− 0.376	0.142
	LVREAL2	0.322	0.104
	XPATH1	− 0.220	0.048

occurring with probability p_i and leading to estimated mean and standard deviation (SD) of y of $\hat{\mu}_i$ and $\hat{\sigma}_i$, respectively (across the 1000 simulation replications),

$$\begin{aligned} \hat{V}(y) &= V_S[\hat{E}(y\,|\,S)] + E_S[\hat{V}(y\,|\,S)] = \hat{\sigma}^2 \\ &= \sum_{i=1}^{k} p_i(\hat{\mu}_i - \hat{\mu})^2 + \sum_{i=1}^{k} p_i\hat{\sigma}_i^2 \\ &= \begin{pmatrix} \text{between-} \\ \text{scenario} \\ \text{variance} \end{pmatrix} + \begin{pmatrix} \text{within-} \\ \text{scenario} \\ \text{variance} \end{pmatrix}, \end{aligned} \tag{13.5}$$

where

$$\hat{E}(y) = E_S[\hat{E}(y\,|\,S)] = \sum_{i=1}^{k} p_i\hat{\mu}_i = \hat{\mu}. \tag{13.6}$$

Table 13.5 presents the scenario-specific mean and SD estimates, together with three possible vectors of scenario probabilities. We obtained the first of these vectors by expert elicitation of the relative plausibility of the nine micro-scenarios described earlier in this section, and created the other two, for the purpose of sensitivity analysis, by doubling and halving the non-reference-scenario probabilities in the first vector. Table 13.6 then applies

Table 13.5 Scenario-specific estimated means and standard deviations of max dose, together with three possible sets of scenario probabilities.

	Estimated		Scenario probabilities (p_i)		
Scenario	**Mean ($\hat{\mu}_i$)**	**SD ($\hat{\sigma}_i$)**	**1**	**2**	**3**
REF	1.21×10^{-7}	6.25×10^{-7}	0.90	0.80	0.95
FP	6.54×10^{-3}	1.06×10^{-2}	0.0225	0.045	0.01125
AG	8.94×10^{-6}	1.91×10^{-5}	0.0125	0.025	0.00625
GA	1.20×10^{-9}	4.72×10^{-9}	0.0125	0.025	0.00625
HDE	3.10×10^{-7}	1.53×10^{-6}	0.02	0.04	0.01
EIC	1.07×10^{-6}	4.46×10^{-6}	0.0325	0.065	0.01625

Table 13.6 Sensitivity analysis of results as a function of scenario probabilities.

	Scenario probabilities		
Summary	**1**	**2**	**3**
Overall mean max dose $\hat{\mu}$	1.47×10^{-4}	2.95×10^{-4}	7.38×10^{-5}
Overall SD $\hat{\sigma}$	1.86×10^{-3}	2.63×10^{-3}	1.32×10^{-3}
Overall variance $\hat{\sigma}^2$	3.47×10^{-6}	6.89×10^{-6}	1.74×10^{-6}
Between-scenario variance	9.41×10^{-7}	1.84×10^{-6}	4.76×10^{-7}
Within-scenario variance	2.53×10^{-6}	5.06×10^{-6}	1.26×10^{-6}
% of variance between scenarois	27.1	26.7	27.3
$\hat{\mu}/\hat{\mu}_{REF}$	1218	2436	610
$\hat{\sigma}/\hat{\sigma}_{REF}$	2980	4201	2110

Equations (13.5) and (13.6) using each of the three scenario probability vectors. It may be seen that the percentage of variance arising from scenario uncertainty is quite stable across the three specifications of scenario probabilities, at about 27% of the total variance. Table 13.6 says that the mean maximum dose of the nuclide chain is 600–2400 times larger, when scenario uncertainty is acknowledged, than its value under the reference scenario, and the uncertainty about max dose on the SD scale is 2000–4000 times larger.

Figure 13.4 presents scenario-specific estimated predictive distributions for log maximum dose, and also plots the composite predictive distribution with scenario probability vectors 1–3. The fast pathway (FP) and glacial advance (GA) scenarios lead to max dose values that are noticeably higher and lower than the other four scenarios, respectively. Principally because of this, the composite distribution is considerably heavier-tailed than lognormal, in particular including a small but significant contribution of very high doses from scenario FP.

13.2.4 Sensitivity Analysis via Projection–Pursuit Regression

As Table 13.2 indicates, simple regression models with all relevant variables on the log scale — even models that include quadratics and interaction terms among predictors with large main effects — are inadequate to 'explain' all of the variance of max dose arising from the radionuclide chain. This suggests either (i) that other variables also play a subtle role in determining log max dose or (ii) that the predictors already included have even more highly

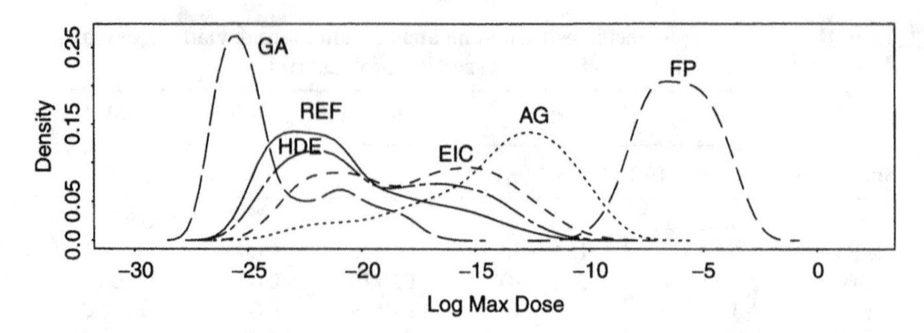

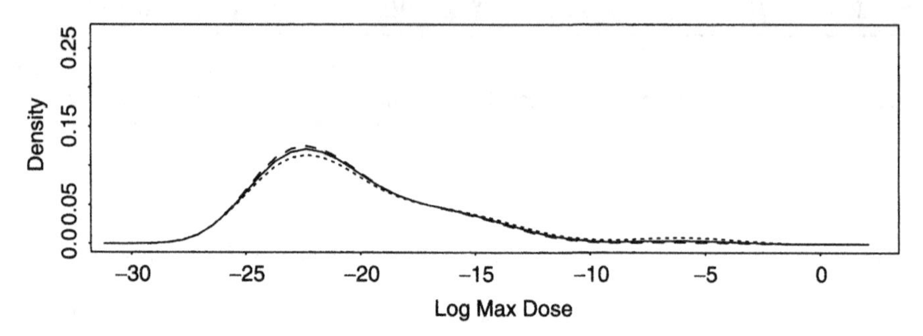

Figure 13.4 Scenario-specific predictive distributions for maximum dose (top panel), and composite predictive distribution with scenario probabilities 1–3 (bottom panel). The scenario probability vector again has little effect.

nonlinear relationships with the outcome of interest. A nonparametric regression technique developed in the 1980s, *projection pursuit regression* (**ppreg**; Friedman and Stuetzle, 1981), can shed some light on these questions.

Given a data set of (univariate) outcomes y_i and vectors $\mathbf{x}_i = (x_{i1}, \ldots, x_{ip})$ of explanatory variables (for i $= 1, \ldots, n$), the idea behind **ppreg** is to generalize the usual regression model

$$y_i = \mu_y + \sum_{j=1}^{p} \gamma_j (x_{ij} - \bar{x}_j) + e_i, \tag{13.7}$$

in which the e_i are IID with mean 0 and variance σ^2, by replacing the linear manner in which the x_j enter the prediction process by arbitrary nonlinear functions of the x_j determined nonparametrically:

$$y_i = \mu_y + \sum_{m=1}^{M} \beta_m \phi_m (\mathbf{a}_m^{\mathrm{T}} \mathbf{x}_i) + e_i. \tag{13.8}$$

Here the $\mathbf{a}_m$ are unit-length direction vectors onto which the predictors $\mathbf{x}$ are projected, and the ϕ_m have been standardized to have mean 0 and variance 1:

$$E[\phi_m(\mathbf{a}_m^{\mathrm{T}} \mathbf{x})] = 0, \qquad E[\phi_m^2(\mathbf{a}_m^{\mathrm{T}} \mathbf{x})] = 1, \qquad m = 1, \ldots, M. \tag{13.9}$$

In practice, the β_m, ϕ_m, and $\mathbf{a}_m$ are typically estimated by minimizing mean-squared error, with the ϕ functions determined by locally weighted regression smoothing (Cleveland, 1979) as in the **ppreg** function in the statistics package **S+** (MathSoft, 1998). The extra flexibility of the model (13.8) permits discovery of interaction and highly nonlinear relationships

Table 13.7 Estimated direction vectors $\hat{\mathbf{a}}_1$ and $\hat{\mathbf{a}}_2$ in the **ppreg** analysis of the REF scenario data with $M = 2$. L at the beginning of a variable name means that the variable entered the analysis on the log scale.

Variable	$\hat{a}_1$	$\hat{a}_2$
LVREAL1	0.759	0.777
LSTREAM	– 0.400	– 0.334
LVREAL2	0.290	– 0.316
XPATH1	– 0.278	0.246
RETC11	– 0.223	0.275
XPATH2	– 0.139	0.140
RETC12	– 0.102	0.115
RETC21	– 0.103	0.075
RETC22	– 0.093	0.077
RETC32	– 0.039	0.040
LRLEACH	0.034	– 0.032
RETC31	– 0.022	0.048
CONTIM	0.017	– 0.008

between y and the x_j, at least in principle. The user must choose M through a compromise between parsimony, interpretability, and explanatory power.

As an illustration of this method, we fitted **ppreg** models to the log max dose data under the REF and HDE scenarios. Table 13.7 presents the estimated direction vectors $\hat{\mathbf{a}}_m$ with the REF data in an analysis with $M = 2$. The predictors in the table have been sorted from largest to smallest in the size of their smallest $\hat{\mathbf{a}}$ values, because variables with small weights in both components of the direction vectors cannot play a large role in determining the outcome. One feature of the **ppreg** method is that simpler models with smaller M values are nested within those with larger M, so that the $\hat{\mathbf{a}}_1$ column in this table also provides the estimated direction vector with M set to 1. The estimated residual variance with $M = 1$ was $\hat{\sigma}^2 = 0.159$, about the same as that from the linear regression model summarized earlier in Table 13.3, and the top six variables in the $\hat{\mathbf{a}}_1$ column match the important predictors from the previous parametric model, providing some confirmation of the SA for the REF scenario presented earlier. The left panel of Figure 13.5 plots the estimated nonparametric regression function $\hat{\phi}_1(\mathbf{z}_1)$ against $\mathbf{z}_1 = \hat{\mathbf{a}}_1^T \mathbf{x}$ in the $M = 1$ case; this is approximately quadratic, verifying the earlier result that interactions and squared terms involving the predictors with large linear effects are needed.

However, if one wishes to 'explain' more of the output variance, functions more complicated than quadratics are required, as the right-hand panel in Figure 13.5 demonstrates. The middle and right panels of this figure plot the $\hat{\phi}_m$ functions from the $M = 2$ **ppreg** solution, and it is evident that something like a cubic function of $\hat{\mathbf{a}}_2^T \mathbf{x}$ would be required to go beyond the $M = 1$ solution. The corresponding estimated weights $\hat{\beta}_m$ with $M = 2$ were $(0.976, 0.275)$—indicating that significant weight is attached to the 'cubic' component—and the estimated residual variance was $\hat{\sigma}^2 = 0.094$, about half of its value from the linear regression model summarized in Table 13.3.[1] This indicates that, even when only two

[1] With these data, there is a diminishing return from including more than $M = 2$ components: the residual variance is still 0.054 even with $M = 9$, and only drops to 0.01 when $M \geqslant 20$.

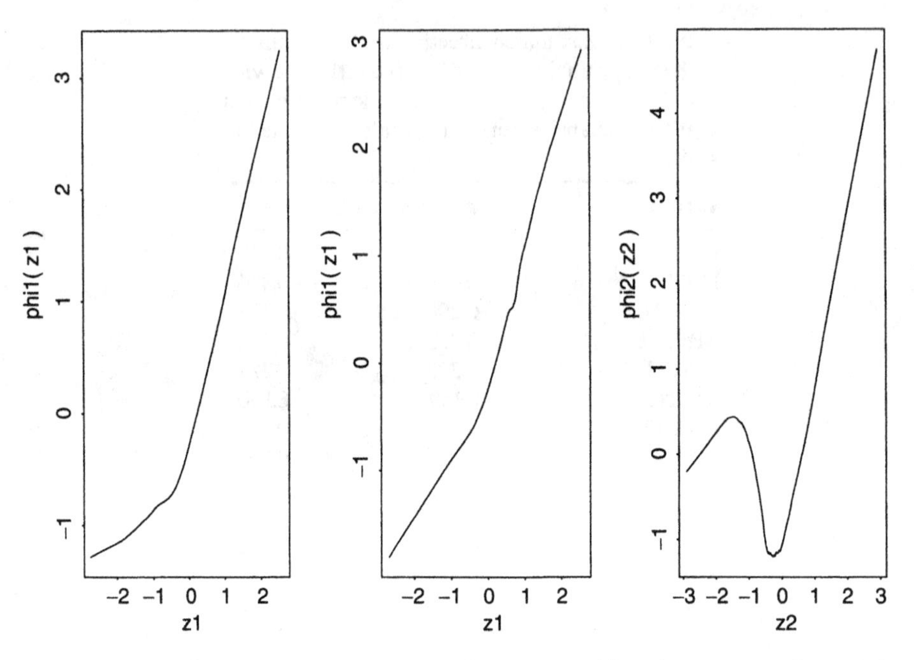

Figure 13.5 Estimated nonparametric regression functions $\phi_m(\mathbf{z})$ in the **ppreg** analysis of the REF scenario. The left-hand panel plots ϕ_1 with $M = 1$; the next two panels plot ϕ_1 and ϕ_2 for $M = 2$.

nonlinear components are used, considerably more signal can be extracted via **ppreg** than with ordinary regression methods.

With the HDE scenario a different message emerged. In an attempt to understand the nonlinearity evident in Figure 13.3, we fit **ppreg** models of varying complexity to the HDE data. To obtain a residual variance no more than half of the value from linear regression (0.077), $M \geqslant 4$ was needed; for example, $\hat{\sigma}^2$ with $M = 4$ was 0.037. An examination of the $\hat{\mathbf{a}}_m$ values to identify important predictors yielded a surprise: in addition to the eight variables already spotted with linear regression, four new variables had $\hat{\mathbf{a}}_m$ coefficients of at least 0.2. Because this scenario had 28 inputs, it was infeasible in the linear regression modeling to include all possible interaction and quadratic terms; we had contented ourselves with examining only interactions and quadratics among the variables with large linear effects. In this case, **ppreg** was able to identify four additional variables for further study.

The main drawback of **ppreg** for SA seems to be that re-interpreting its nonparametric models in parametric terms may not be easy. For example, while the middle panel of Figure 13.5 is approximately linear, a more careful attempt to fit the right-hand panel than that based on a cubic would involve the ratio of two polynomials of at least order 2, and the resulting function of the original variables is very complex.

13.2.5 Extended FAST: An Alternative for Nonlinear Sensitivity Analysis

Here we describe the results of a sensitivity analysis for total annual dose, arrived at by summing across all four nuclides — iodine plus the chain — monitored in the REF scenario. An alternative approach to the log-transformation regression method illustrated above involves performing the SA on the raw-scale variables, where relationships are considerably more nonlinear than with the log-scale data.

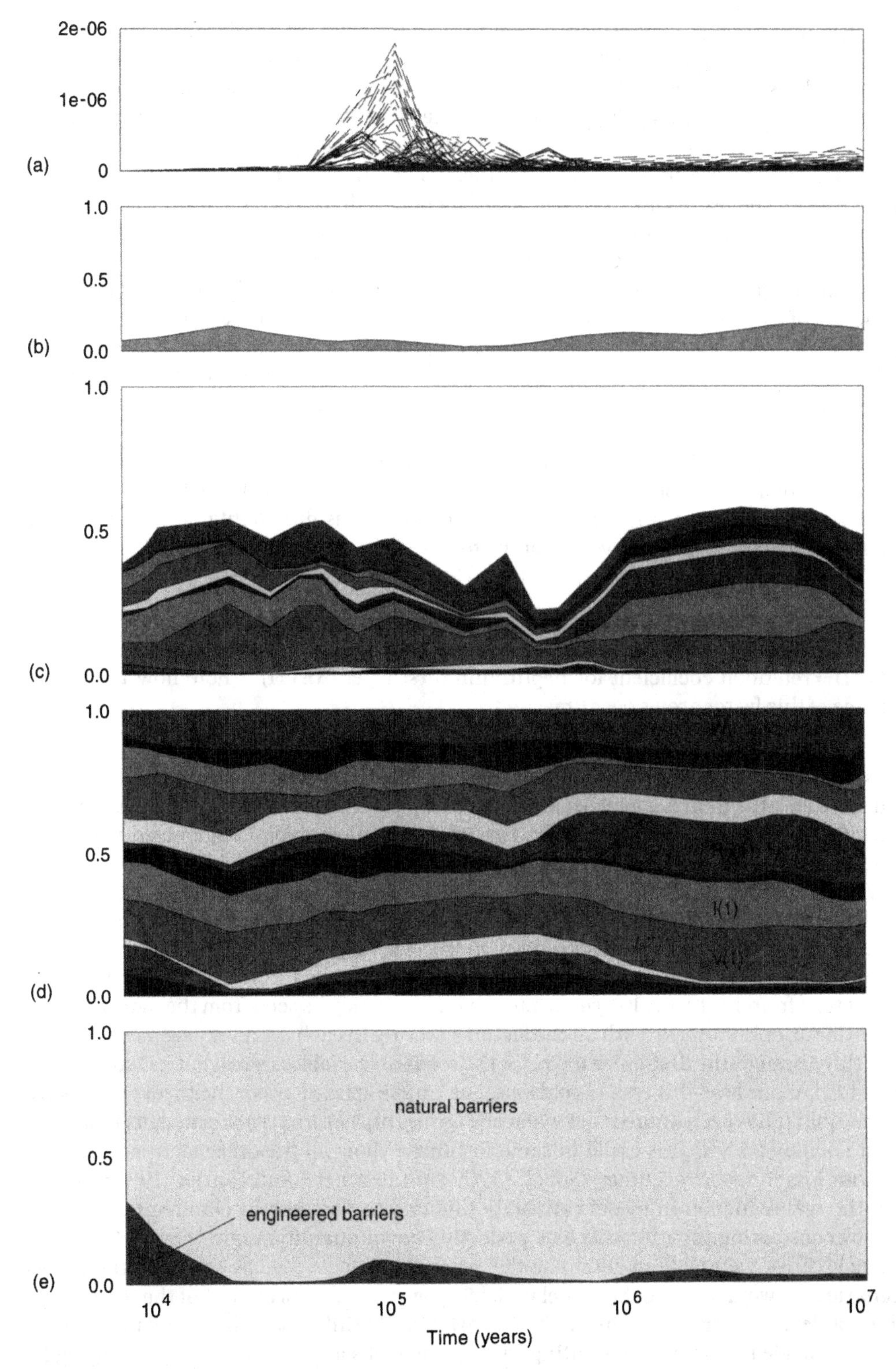

Figure 13.6 Extended FAST results from the REF scenario. The output variable is the total (iodine + chain) annual dose.

To deal with such situations, we have employed the *extended FAST* (Chapter 8) because of its ability to evaluate total effect indices for any uncertain factor involved in the model under analysis. We argue that each factor has to be described by a pair of indices—first-order and total—and that this kind of representation allows an exhaustive and computationally inexpensive characterization of the system under analysis.

An illustration of the first-order and total sensitivity indices is given in Figure 13.6 for the REF scenario. The results are expressed as a function of time, from 10^4 to 10^7 years into the future. We chose a sample size of 257 to run the model. The curves displayed in panel (a) of the figure are the result of 3084 model evaluations. We used these evaluations both for estimating R^2 (b) and for computing first-order (c) and total normalized indices (d) for all 12 factors of the underlying model ($257 \times 12 = 3084$). A much more restricted set of model outputs ($257 \times 2 = 514$) is sufficient to estimate the total normalized indices for engineered and natural barriers (e). Indeed, once the sample size is fixed, the computational effort is proportional to the number of factors or subgroups considered in the analysis.

Panel (b) of Figure 13.6 shows that the underlying model is strongly nonlinear, given that R^2 is always below 0.2. A cumulative plot of first-order indices is given in (c). The model under investigation is not additive, because the shaded region is below 0.6 everywhere; in other words, more than 40% of the output uncertainty is due to interactions occurring among the factors. A cumulative plot of the total indices for the 12 factors is given in panel (d). The most important factors can readily be identified:

- $v(1)$ = water velocity in the geosphere's first layer (VREAL1 in Table 13.1);
- $l(1)$ = length of the first geosphere layer (XPATH1);
- $Rc(1)$ = retention coefficient for neptunium (first layer; RET11)—note how the importance of this factor grows over time; and
- W = stream flow rate (STREAM).

These results agree broadly with those in Table 13.3 for max dose from the chain.

In panel (e), the total normalized indices are displayed for the factors being grouped into two subsets (natural and engineered barriers). The modest role of engineered barriers is highlighted, in agreement with previous risk assessment studies.

With an eye to uncertainty analysis, we have also made FAST computations for the annual radiological dose due to the chain across all six scenarios in the Level E/G test case, by estimating the first-order and total sensitivity indices for the scenario indicator variable via a set of 5763 model evaluations. The results, displayed in Figure 13.7, are expressed as a function of time from 10^3 to 4×10^7 years into the future. It may be seen from the figure that the scenario variable interacts with all the factors entering in each and every scenario.

In this example, the first-order index for the scenario variable is small but its total effect is close to 1! At one level this result is obvious, but it has implications for the theory of sensitivity analysis. It has been argued that when one (group of) factor(s) is important, its first-order effect should be high. This could in some instances allow all the other factors to be fixed without loss of model accuracy (Sobol', 1990). On the other hand, Jansen (1996b) argues that the real reduction in model variability that one can achieve by eliminating a factor i (i.e., by considering the ith factor as a perfectly known quantity) is given by its S_{Ti}. In our example, if we were able to eliminate the scenario variable (i.e., by selecting the proper scenario), we would reduce the model variability by around 95% at most of the time points. This is a clear measure of how much the scenario variable influences the output uncertainty.

We conclude that in problem settings where one seeks a group of factors accounting for most of the variance so that the others can be fixed, one should focus on the total effect of the target group, and not on its first-order effect.

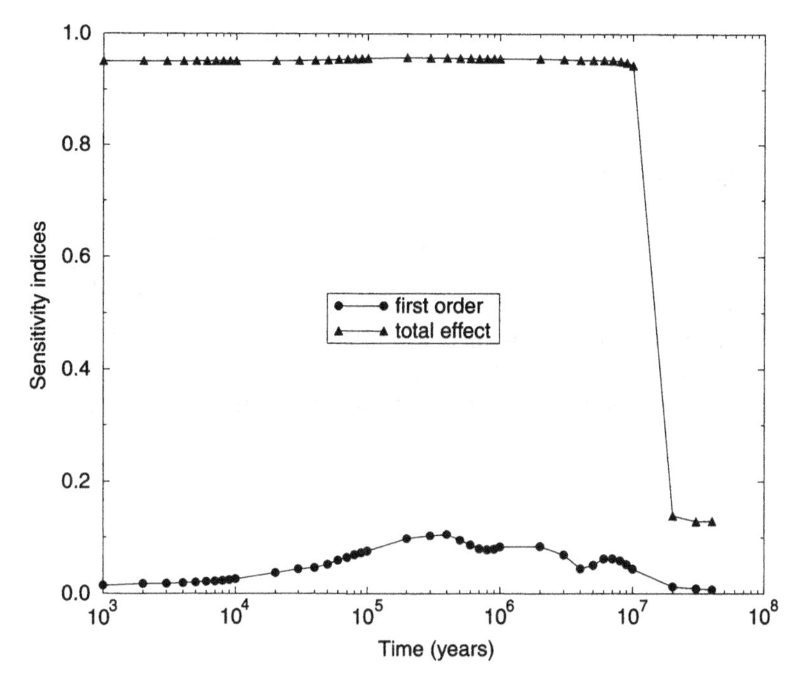

Figure 13.7 Results from the Level E/G test case when the output variable is the annual radiological dose due to the chain. First-order and total sensitivity indices are given for the scenario variable.

13.3 DISCUSSION

We draw the following conclusions relevant to sensitivity and uncertainty analysis from the GESAMAC study:

- Bayesian concepts and methods are the most flexible base on which to build detailed SA and UA calculations. The uncertainty framework in Section 13.1.2 is of potential interest in a wide variety of risk analysis settings.
- There is no need to treat scenario uncertainty differently from other sources of uncertainty in probabilistic risk assessment modeling. When scenario uncertainty is realistically assessed and propagated along with parametric uncertainty, the component of predictive variance associated with scenario uncertainty may well be 50% or more of the total across the other components. It has been common in the past to ignore this component of uncertainty or to treat it qualitatively; we believe that only through a quantitative approach like the one presented here can the full extent of the relevant uncertainties be appreciated.
- Parallel computing is a particularly effective way to make use of existing computational resources in Monte Carlo simulation. Other workers in nuclear waste disposal risk assessment may find our program GTMCHEM of use, and our parallel Monte Carlo driver (PMCD) may be useful in a much broader range of simulation studies.
- Regression-based sensitivity analyses often go some distance toward achieving an understanding of the mapping from inputs to outputs in complex simulation environments, but nonlinearities may well be present that are difficult to quantify with linear regression methods. Projection–pursuit regression deserves further study as a SA technique. The extended FAST technique is effective in partitioning variability across inputs in a

way that accounts fully for nonlinearities. Total effect indices are important to compute along with first-order effects.

ACKNOWLEDGMENTS

This work was partially funded under contract FI4W–CT95–0017 (the GESAMAC project) of the European Commission through its R + D program on 'Nuclear Fission Safety' (1994–1998). We are grateful to our GESAMAC partners Sonsoles Eguilior (CIEMAT, Madrid), Antonio Pereira and Bruno Mendes (Department of Physics, University of Stockholm), and Ryan Cheal (University of Bath, UK) for a fruitful and enjoyable collaboration, and to Henning von Maravic of the European Commission for helpful comments, discussion, and references.

14

Sensitivity Analysis for Signal Extraction in Economic Time Series

Christophe Planas

European Commission, Joint Research Centre, Ispra, Italy

Raoul Depoutot

European Commission, Eurostat, Luxembourg

14.1 INTRODUCTION

14.1.1 Overview

A common practice in official statistics and applied macroeconomics is to focus the analysis on some unobserved movements such as trend and seasonally adjusted series. Several methods are available for trend extraction and seasonal adjustment. With model-based methods (see e.g. Bell and Hillmer, 1984), the estimation of the unobserved components is carried out after selection of a model and maximum likelihood (ML) estimation of the parameters. When several models describe the data dynamics in a statistically correct way, an information criterion is typically used to select a single specification. Given the data, the chosen model and the parameter estimates, the unobserved components are computed by signal extraction via either Wiener–Kolmogorov or Kalman filtering.

Such unobserved component estimators can be seen as conditional on a model specification and on the estimated parameter values. In practice, however, time series models are only descriptions of data, so different models may describe the data equally well. A first problem for an official statistician who needs to produce trends and seasonally adjusted series is thus: within the class of the acceptable models, how sensitive are the unobserved components estimators to the model choice? Furthermore, for a given model specification,

Sensitivity Analysis. Edited by A. Saltelli *et al.*

the collection of new observations may slightly modify the ML parameter estimates. Consequently, the unobserved components estimators would show some instability due to the model coefficients update. A second problem is thus: for a given model, how sensitive are the unobserved components estimators to changes in model parameters?

In this chapter, our intention is to assess the uncertainty in the estimators of the components due to the uncertainty of model specification and of model parameters, and to apply sensitivity analysis techniques in order to isolate the sources of uncertainty that are the most influential for characterizing the trend, seasonal and short-term movements in time series. Some related works can be found in Carlin and Dempster (1989), mainly about parameter uncertainty in fractionally integrated auto-regressive moving average (ARMA) models, and in Bell and Otto (1993), about the joint effect of parameter and some type of data uncertainties. Parameter and model uncertainty have been considered in a forecasting context by Draper (1995). Our study can be distinguished from these works since it addresses the problem of model and parameter uncertainty in the unobserved component models framework, and as it applies the sensitivity analysis techniques presented in this book.

It is worth mentioning that other sources of uncertainty exist in the model-based signal extraction framework. Drawing up an exhaustive list of uncertainty sources, Bell (1989) also included: (i) sampling and non-sampling errors in data; (ii) decomposition uncertainty; and (iii) signal extraction error. We choose to ignore them on the basis of the following considerations. Firstly, we suppose that the series at hand does not result from a survey, so it will be free of sampling errors. Secondly, the effect of decomposition uncertainty on the unobserved component estimators can be assessed analytically (see Maravall and Planas, 1999). We shall use the same identification assumption throughout the analysis, namely the so-called *canonical* decomposition (see Box *et al.*, 1978a; Pierce, 1978). Thirdly, we view signal extraction error as a consequence of the other uncertainties. This is because we believe that it is completely determined by model, parameters, decomposition choice, and the sampling/non-sampling data error structure if present.

Focusing on trend and seasonal and irregular movements in an economic time series, we develop an in-depth analysis of the effects of parameter uncertainty. Next, for a given range of models, we discuss how sensitivity to model specification can be evaluated. The analysis puts together three main tools: the model-based signal extraction techniques (see Burman, 1980); a Bayesian framework for characterizing the distribution of model parameters and for assigning probability to time series models (see Kass and Raftery, 1995); and a variance-based sensitivity analysis technique (Saltelli *et al.*, 1993).

14.1.2 General Procedure

The implementation of uncertainty and sensitivity analysis tools in the time series analysis framework needs three main ingredients: an observed series, a parametric model that provides the analyst with a description of the uncertainties, and a functional of interest. By parametric model is intended a set of parameters associated with a model specification. In this study, we shall focus on regression with time series errors, with the stochastic part decomposed in unobserved movements such as trend, seasonal and irregular components. Information on the parameter uncertainty is obtained by combining, via Bayes' theorem, the likelihood function with a prior density for the parameters. This yields a posterior distribution that characterizes the parameter uncertainty (see Box and Tiao, 1973). For considering model uncertainty, a set of parametric models is needed so that a model averaging can be performed (see Draper, 1995). A probability is assigned to every model in the set of models under study using the Bayes factor (see Kass and Raftery, 1995). Finally, sensitivity analysis requires explicit statement of the objective of the analysis; we shall design a functional especially for addressing the issues of model assessment with respect to unobserved movements and of robustness of unobserved component estimators—see Section 14.2.

With these elements available, the effect of uncertainties on the functional of interest can be characterized. The overall variance is a first summary. We shall see that simple plots of the functional values with respect to each parameter provide relevant information, as discussed in Young (1999). Sensitivity analysis enables us to quantify how much of the overall variance is related to every uncertainty source. We shall make use of importance measures (see Saltelli *et al.*, 1993). An example developed in Sections 14.3 and 14.4 will illustrate the benefit obtained from considering sensitivity analysis techniques in modelling economic time series: mainly, sensitivity analysis gives some information about how accurately a model or a set of models defines a target of interest. It brings an insight into model assessment, and it enables an analyst to isolate the model with best properties with respect to the objective of the study. We shall see that sensitivity analysis can be seen as another way of studying model parsimony, complementary to information criteria such as the Bayesian information criterion (BIC) (see Hannan, 1980).

14.2 GENERAL FRAMEWORK

14.2.1 Model Specification

Given an observed $Y \equiv \{Y_t\}, t = 1, \ldots, T$, we consider a regression model with time series errors (see Fuller, 1996) specified as

$$Y_t = X_t'\beta + y_t, \tag{14.1}$$

where X_t is a vector of r deterministic regressors and $\beta = (\beta_1, \ldots, \beta_r)'$ a vector of coefficients. The regressors can represent some outlier patterns such as level shifts or additive outliers and some calendar effects. In official statistics, a great deal of attention is paid to the trading days rhythm, to the effect of Easter, and to the length of months. The stochastic term y_t is supposed to be well described by an ARIMA model of the type

$$\phi(B)y_t = \theta(B)a_t, \tag{14.2}$$

where a_t is a white noise, which is assumed normally distributed with variance V_a, B is the lag operator, $\phi(B) = \Phi(B)\delta(B)$, $\delta(B)$ is a differencing operator, and $\Phi(B)$ and $\theta(B)$ are finite polynomials, of respective orders p and q, that satisfy the stationarity and invertibility conditions. We shall use $\alpha = \{\phi_1, \ldots, \phi_p, \theta_1, \ldots, \theta_q, \beta_1, \ldots, \beta_r\}$ to denote a set of parameters of the polynomials $\Phi(B)$ and $\theta(B)$ and of the coefficients of the regressors. The stochastic term y_t is assumed to be made up of orthogonal unobserved components according to

$$y_t = s_t + p_t + u_t, \tag{14.3}$$

where s_t is the seasonal component, p_t the trend, and u_t a white-noise irregular component with variance V_u. The stochastic components s_t and p_t are also described by ARIMA models as

$$\begin{aligned} \phi_s(B)s_t &= \theta_s(B)a_{st}, \\ \phi_p(B)p_t &= \theta_p(B)a_{pt} \end{aligned} \tag{14.4}$$

where $\phi_l(B)$ and $\theta_l(B)$, for $l \equiv p, s$, denote finite polynomials having all roots on or outside the unit circle. The variables a_{st} and a_{pt} are independent white noise with variances V_s and V_p, respectively. The polynomials $\phi_s(B)$ and $\phi_p(B)$ are prime, while the moving average (MA) polynomials $\theta(B)$ and $\theta_p(B)$ have no unit roots in common. We identify a unique decomposition (14.3), (14.4) by imposing the canonical requirement on the models for the trend and seasonal components (see Hillmer and Tiao, 1982).

Given a plausible model (14.1), (14.2) and a decomposition (14.3), (14.4), the unobserved quantities can be estimated as (see Burman, 1980)

$$\hat{p}_t = E\left[p_t|Y\right], \qquad \hat{s}_t = E\left[s_t|Y\right], \qquad \hat{u}_t = E\left[u_t|Y\right]. \tag{14.5}$$

Finally, the regressors in (14.1) are allocated to the unobserved components estimators by matching the regressor type with the properties of the components; for example, calendar effects are assigned to the seasonal stochastic component, while a shift in level would be assigned to the trend component.

Equations (14.1), (14.4) represent the classical reduced-form approach to unobserved component analysis with deterministic exogeneous regressors, as thoroughly discussed for example in Bell and Hillmer (1984) and Maravall (1996). This model is routinely specified, estimated and decomposed using the software TRAMO-SEATS (see Gómez and Maravall, 1996). In the unobserved component models literature, it is usually considered that the overall model specification is unique and that the model parameters are estimated without errors. Since the final component estimators depend on the model chosen, say M, and on the ML estimates of the parameters, $\hat{\alpha}$, we shall rewrite (14.5) explicitly as

$$\hat{p}_t = E\left[p_t|Y, M, \hat{\alpha}\right], \qquad \hat{s}_t = E\left[s_t|Y, M, \hat{\alpha}\right], \qquad \hat{u}_t = E\left[u_t|Y, M, \hat{\alpha}\right].$$

14.2.2 Description of Uncertainties

Bayesian analysis offers the most natural framework for describing uncertainty. The joint posterior density of the parameters α and V_a is given by

$$\text{pr}(\alpha', V_a|Y, M) \propto \text{pr}(Y|M, \alpha, V_a)\,\text{pr}(\alpha, V_a|M)$$

where $\text{pr}(Y|M, \alpha, V_a)$ is the likelihood function associated with the model (14.1), (14.2) and $\text{pr}(\alpha, V_a|M)$ the prior distribution on the model parameters. Let y_D represent the series obtained by the differencing operation $\delta(B)y_t = \delta(B)(Y_t - X_t'\beta)$, with covariance matrix $\Sigma_{y_D} = V[y_D]$; then the likelihood function of the model (14.1), (14.2) is given by

$$\text{pr}(Y|M, \alpha, V_a) = \text{pr}(y_D\,|M, \alpha, V_a) \propto |\Sigma_{y_D}|^{-1/2}\exp\left(-\tfrac{1}{2}y_D'\Sigma_{y_D}^{-1}y_D\right).$$

Given a set of regressors, the Kalman recursions with suitable initial conditions offer the most convenient tool for evaluating the likelihood function (see e.g. Gómez and Maravall, 1994).

Regarding the prior distribution, we shall follow Bell and Otto (1993) and use a non-informative prior on the parameters α and V_a, namely $\text{pr}(\alpha, V_a|M) \propto 1/V_a$, over the parameter space where the ARMA parameters lie inside the stationarity and invertibility regions. This choice ensures that the posterior mode and the ML values coincide (see Box and Tiao, 1973). Integrating over V_a, the marginal distribution of α, $\text{pr}(\alpha|Y, M)$, can be obtained. We shall use an approximation to $\text{pr}(\alpha|Y, M)$ given by the normal distribution with mean $\hat{\alpha}$ and variance given by $\hat{\sum}$, the negative inverse Hessian of the log-likelihood evaluated at the ML parameters (for asymptotic properties of that approximation, see Box and Jenkins, 1970, pp. 252–258). It must be kept in mind that this distribution is truncated for stationarity and invertibility conditions. Although more elaborate sampling techniques could be used, we choose that approximation in order to simplify the computations so as to concentrate our efforts on the implementation of sensivity analysis tools.

Our treatment of model uncertainty will involve Bayesian model averaging. The problem of assigning a posterior probability to a model, $\text{pr}(M|Y)$, has recently received much attention in the statistical literature; see Kass and Raftery (1995) for an exhaustive survey. In short, if $M_1, \ldots, M_K$ are K models considered,

$$\text{pr}(M_k|Y) = \frac{\text{pr}(Y|M_k)\,\text{pr}(M_k)}{\sum_{i=1}^{K} \text{pr}(Y|M_i)\,\text{pr}(M_i)} \tag{14.6}$$

where $\text{pr}(M_k)$ is a prior probability on model k, and

$$\text{pr}(Y|M_k) = \int \text{pr}(Y|\alpha, V_a, M_k)\,\text{pr}(\alpha, V_a|M_k)\,d\alpha\,dV_a, \tag{14.7}$$

$\text{pr}(Y|\alpha, V_a, M_k)$ being the likelihood function. In practice, however, only models yielding residuals that pass all the usual diagnostic checks can be considered. This requirement constrains the range of plausible models, so in applications we are led to consider a limited number of possibilities. None of these models are true, since they only describe in a satisfying way the series' properties. In that context, it is difficult to anticipate that, for a given series, a model will be more likely. We shall thus be 'neutral' and impose the same prior probability on every model, $\text{pr}(M_k) = \text{pr}(M_i)$, $k \neq i$. Thus Equation (14.6) reduces to

$$\text{pr}(M_k|Y) = \frac{\text{pr}(Y|M_k)}{\sum_{i=1}^{K} \text{pr}(Y|M_i)}. \tag{14.8}$$

Several approaches are available for evaluating the integral in (14.7); a review can be found in Kass and Raftery (1995). We shall follow Draper (1995), who considers the large-sample approximation

$$\ln \text{pr}(Y|M_k) = \tfrac{1}{2} n_k \ln 2\pi - \tfrac{1}{2} n_k \ln T + \ln \text{pr}(Y|\hat{\alpha}, \hat{V}_a, M_k) + O(1), \tag{14.9}$$

where n_k is the number of parameters associated with model M_k. Notice that this last expression can be written as

$$\ln \text{pr}(Y|M_k) = \tfrac{1}{2} n_k \ln 2\pi - \tfrac{1}{2}\text{BIC}(M_k) + O(1), \tag{14.10}$$

where $\text{BIC}(M_k)$ is the Bayesian information criterion associated with model M_k (see Hannan, 1980). Hence the BIC of every fitted model provides a simple way through (14.10) to assign a probability to every model. The accuracy of this approximation is discussed for example in Kass and Wasserman (1995).

Finally, we need to measure the departure from the targets of interest. Most of the uncertainty analysis implemented in the time series literature has been devoted to the effect of atypical observations on parameter estimates and forecasts. In particular, departures from normal distribution of the innovations have received a lot of interest. Because we consider other types of uncertainty and other types of targets, we cannot make direct use of standard influence measures as defined in Martin and Yohai (1986) and in Peña (1990). Instead, we shall focus on the mean square deviation of the extracted component around the estimator related to the ML parameter values, defined as

$$\begin{aligned} f_p(\alpha, M) &= \frac{1}{T}\sum_{t=1}^{T}(E[p_t|Y, M, \alpha] - E[p_t|Y, M, \hat{\alpha}])^2, \\ f_s(\alpha, M) &= \frac{1}{T}\sum_{t=1}^{T}(E[s_t|Y, M, \alpha] - E[s_t|Y, M, \hat{\alpha}])^2, \\ f_u(\alpha, M) &= \frac{1}{T}\sum_{t=1}^{T}(E[u_t|Y, M, \alpha] - E[u_t|Y, M, \hat{\alpha}])^2. \end{aligned} \tag{14.11}$$

It is tempting to give (14.11) a 'robustness' interpretation. Trivially, low values reflect robust behavior of the component estimators with respect to the uncertainty source, while large values are associated with a poorly defined unobserved movement. Although some attention has been devoted to the problem of robustness in seasonal adjustment (see Riani, 1998), as far as we know, no formal definition of robustness in the unobserved component models framework is available in the statistical literature.

14.3 MODEL ASSESSMENT AND PARAMETER UNCERTAINTY

We develop the analysis for the case of an economic time series. We are interested in assessing the knowledge of the series' unobserved structure implied by the model selected. Given a model specification, we investigate the behavior of the mean sample deviations of the unobserved components estimates along the distribution of the model parameters. This will provide information about how tightly every acceptable model defines the unobserved movements.

The monthly series of Italian Production of Soft Drinks (IPSD, kindly provided by EUROSTAT), is made up of 107 monthly observations between January 1985 and November 1993. Three models correctly describing the data are detailed below. The ML estimates are reported, with their standard deviations given in parentheses. Also, we denote by Δ and Δ_{12} the differencing operations $1 - B$ and $1 - B^{12}$.

Model 1 $(2, 1, 1)(0, 1, 1)_{12}$, $\alpha = \{\phi_1, \phi_2, \theta_1, \theta_{12}\}$:

$$\begin{aligned} &(1 + \phi_1 B + \phi_2 B^2)\Delta\Delta_{12} y_t = (1 + \theta_1 B)(1 + \theta_{12} B^{12})a_t, \\ &\hat{\phi}_1 = -0.079(0.130), \quad \hat{\phi}_2 = 0.297(0.116), \\ &\hat{\theta}_1 = -0.733(0.106), \quad \theta_{12} = -0.606(0.117). \end{aligned} \tag{14.12}$$

Model 2 $(0, 1, 1)(0, 1, 1)_{12}$ plus two regressors for trading days, x_{1t}, and length of month effects, x_{2t}; $\alpha = \{\beta_1, \beta_2, \theta_1, \theta_{12}\}$:

$$\begin{aligned} y_t &= Y_t - \beta_1 x_{1t} - \beta_2 x_{2t}, \\ \Delta\Delta_{12} y_t &= (1 + \theta_{12} B^{12})a_t, \\ \hat{\theta}_1 &= -0.763(0.073), \quad \hat{\theta}_{12} = -0.658(0.120), \\ \hat{\beta}_1 &= 0.004(0.002), \quad \hat{\beta}_2 = 0.059(0.044). \end{aligned} \tag{14.13}$$

Table 14.1 Diagnostics.

	Model 1	Model 2	Model 3
$\hat{V}_a$	0.0695	0.0697	0.0706
Q_{24}	28.16	29.61	26.18
Q_{s2}	0.21	0.25	0.92
N	1.49	2.25	2.22
BIC	– 213.83	– 212.22	– 207.51

The regressor x_{1t} is constructed as the number of working days minus the number of Saturdays and Sunday times $\frac{5}{2}$ (see Gómez and Maravall, 1996).

Model 3 $(3,1,0)(0,1,1)_{12}$, $\alpha = \{\phi_1, \phi_2, \phi_3, \theta_{12}\}$:

$$
\begin{aligned}
&(1 + \phi_1 B + \phi_2 B^2 + \phi_3 B^3)\Delta\Delta_{12} y_t = (1 + \theta_{12} B^{12}) a_t,\\
&\hat{\phi}_1 = 0.506(0.105), \quad \hat{\phi}_2 = 0.527(0.010),\\
&\hat{\phi}_3 = 0.202(0.107), \quad \hat{\theta}_{12} = -0.733(0.134).
\end{aligned}
\tag{14.14}
$$

The three models are not nested. Table 14.1 reports several diagnostic checks on the residuals; namely, the innovations variance, the Ljung–Box statistics on the first 24 autocorrelations (Q_{24}), the Box–Pierce statistics on the first two seasonal lags (Q_{s2}), skewness S and kurtosis K aggregated to form the Bowman–Shenton test for normality $N = S^2 + K^2$ (see Harvey, 1989, pp. 258–260), and the BIC.

The tests considered suggest that the three models give a satisfactory description of the IPSD series. The BIC favors Model 1 against Model 2, and Model 2 against Model 3, but it does not tell us much about how well these models characterize the structure of the series. Yet the three specifications imply some differences in the modelling of the components. For Models 1 and 3, the AR polynomials have roots at frequencies close to the seasonal harmonics, so the AR polynomial is associated with the seasonal movements. So Models 1 and 3 imply a decomposition of the type

$$
\begin{aligned}
\Delta^2 p_t &= \theta_p(B) a_{pt},\\
\Phi(B)(1 + B + \cdots + B^{11}) s_t &= \theta_s(B) a_{st}.
\end{aligned}
$$

When decomposing the series under Model 2, the regressors x_{1t} and x_{2t} that describe some calendar effects are assigned to the seasonal component, so the decomposition is

$$
\begin{aligned}
\Delta^2 p_t &= \theta_p(B) a_{pt},\\
s_t &= s'_t + \beta_1 x_{1t} + \beta_2 x_{2t},\\
(1 + B + \cdots + B^{11}) s'_t &= \theta_s(B) a_{st}.
\end{aligned}
$$

The irregular term remains a white noise in the three cases. With these three ways of describing the data dynamics, the obvious problem for a practitioner who wishes to produce a trend or a seasonally adjusted series is which model to choose. The first question we consider is thus: given the parameter uncertainty related to every model, how well does each model characterize the series structure?

For every model, we simulate R values for the model parameters α according to $\alpha \sim N(\hat{\alpha}, \hat{\sum})$. The estimated covariance matrix $\hat{\sum}$ is given by the software used, TRAMO-SEATS for instance. For every simulated value $\alpha^{(j)}, j = 1, \ldots, R$, the observed series is decomposed into unobserved components, which are estimated by signal extraction. It is

Table 14.2 Experimental results ($\times 10^{-2}$).

	Model 1	Model 2	Model 3
$\bar{f}_p(\alpha)$	1.4950	0.3032	2.1974
$V[f_p(\alpha)]$	0.0757	0.0014	0.1032
$\bar{f}_s(\alpha)$	5.0351	1.7052	14.9918
$V[f_s(\alpha)]$	0.4900	0.0208	5.2156
$\bar{f}_u(\alpha)$	4.0802	1.5453	13.7823
$V[f_u(\alpha)]$	0.3332	0.0146	4.3882

then straightforward to compute the sample mean-square deviations given in (14.11) for every unobserved component estimator, $f_p(\alpha^{(j)}), f_s(\alpha^{(j)}), f_u(\alpha^{(j)}), j = 1, \ldots, R$. Table 14.2 below shows the means over 1000 replications of the sample mean-square deviations. It can be seen that Model 2 entails the least variation around the unobserved movements: the sample deviations around the estimators derived from ML parameters values have a much lower mean with respect to the other two models, and these deviations occur with a lower variance. Table 14.2 also shows that for the three models the trend component is more stable than seasonal and irregular movements.

A graphical analysis of the results of the simulation exercise can provide further information. Figures 14.1–14.3 show the plots of $f_{\cdot}(\alpha^{(j)})$ against the simulated parameters, $\phi_1^{(j)}, \phi_2^{(j)}, \phi_3^{(j)}, \theta_1^{(j)}, \theta_{12}^{(j)}, \beta_1^{(j)}, \beta_2^{(j)}$, according to their involvement in each model. The width of the horizontal axis is related to the uncertainty around each parameter. For the trend component, the scale of the plots related to Model 2 had to be reduced given the drop in the magnitude of the sample deviations. This drop is likely due to the incorporation of two regressors.

Figures 14.1–14.3 contain qualitative information about the relevance of each parameter in respect of the knowledge of the series' unobserved structure. When a parameter does not accurately define an unobserved movement, the corresponding plot fails to show a clearly

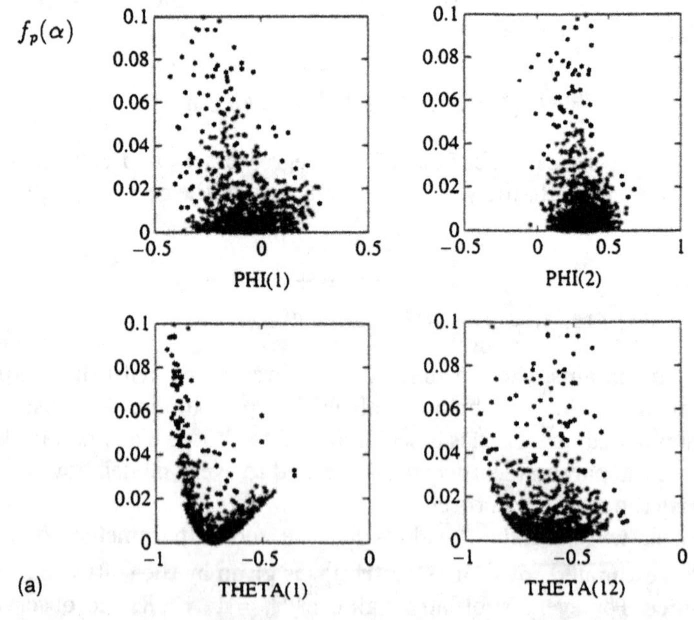

Figure 14.1 Trend: (a) Model 1; (b) Model 2; (c) Model 3.

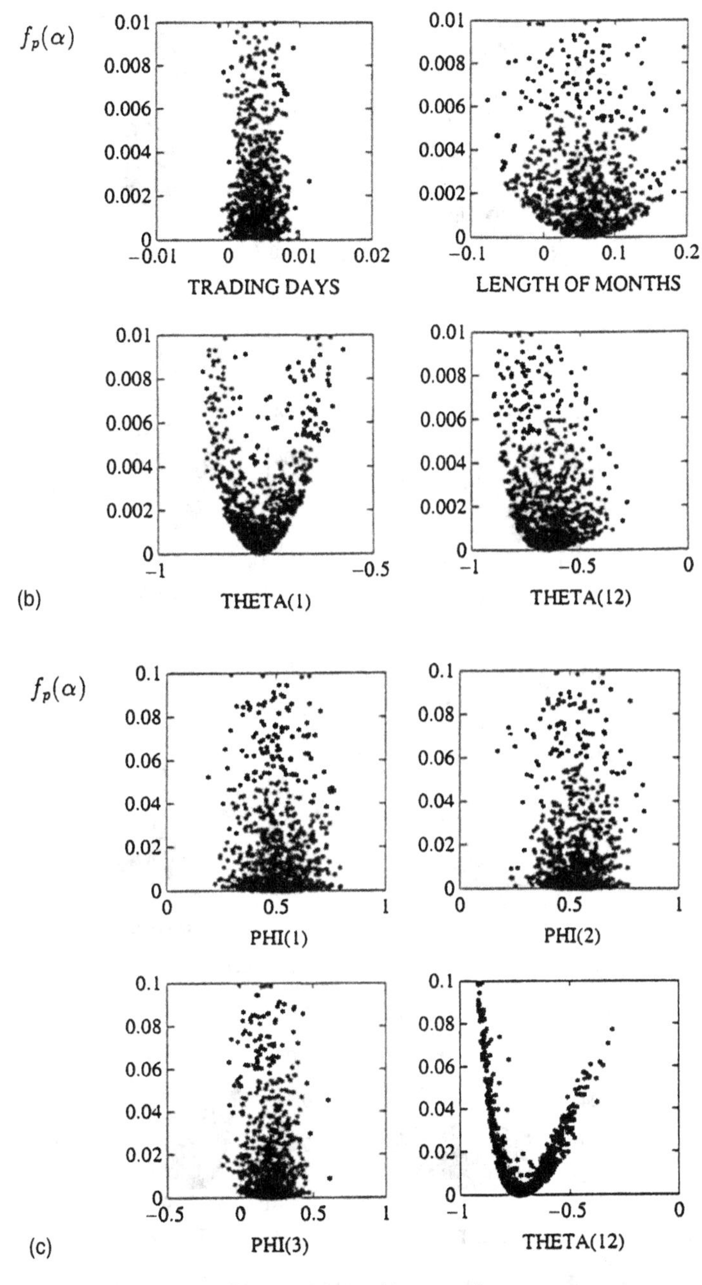

Figure 14.1 *(Continued)*

defined minimum: it is a case where many combinations of parameters lead to the same estimator, indicating some redundancy in the model specification. In contrast, when a parameter narrowly identifies the unobserved component, it yields a constant response whatever the values of the other parameters. This case fits with the concept of model parsimony (see Young, 1999).

It can be seen from Figures 14.1 (a–c) that the AR parameters nearly do not participate in the description of the series' long-term movements: for given values of the AR parameters,

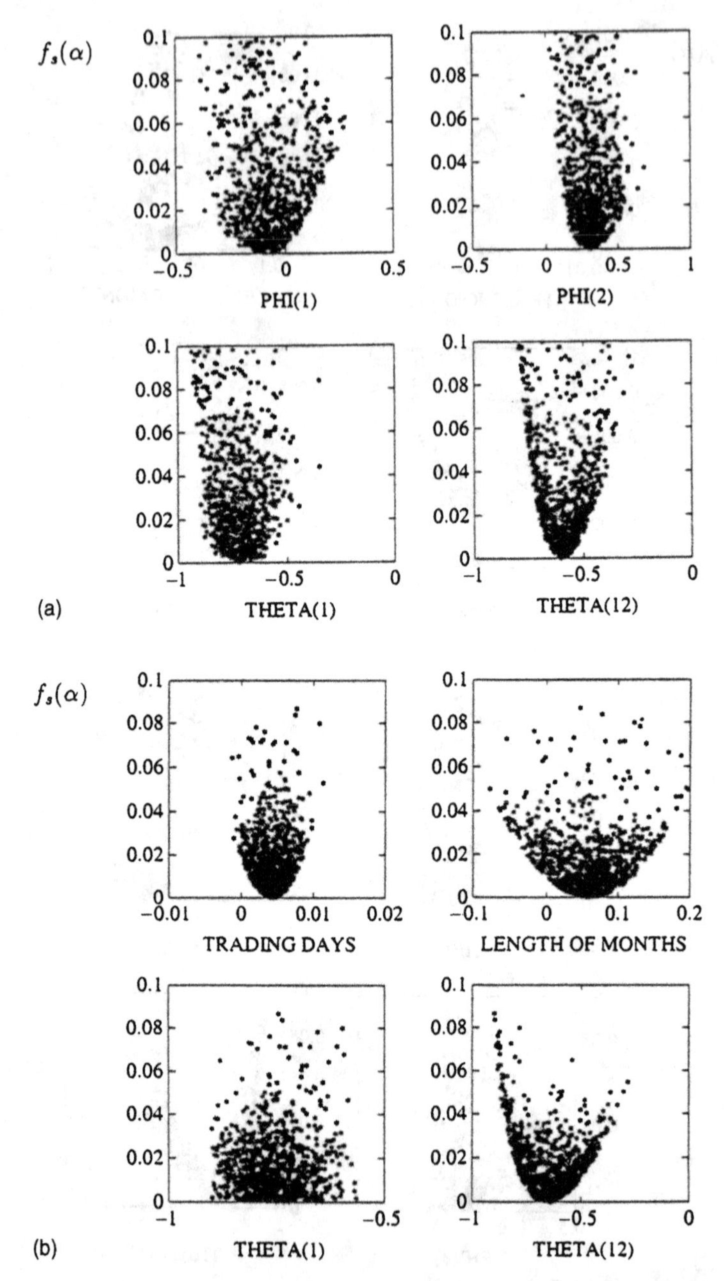

Figure 14.2 Seasonal: (a) Model 1; (b) Model 2; (c) Model 3.

large deviations of trend estimators from the estimator related to the ML parameters are observed. This is a consequence of the allocation of the AR polynomial roots to the seasonal component. On the other hand, the MA parameters define the long-term movements in a very tight way. Figures 14.2(a–c) show that Models 1 and 2 globally define the seasonal movements of the series quite narrowly. The parameter θ_{12} is remarkably influential in the definition of the seasonal movements. Short-term movements as displayed in Figures 14.3(a–c) seem well enough defined in Models 1 and 2, but not in Model 3, where only θ_{12} is influential.

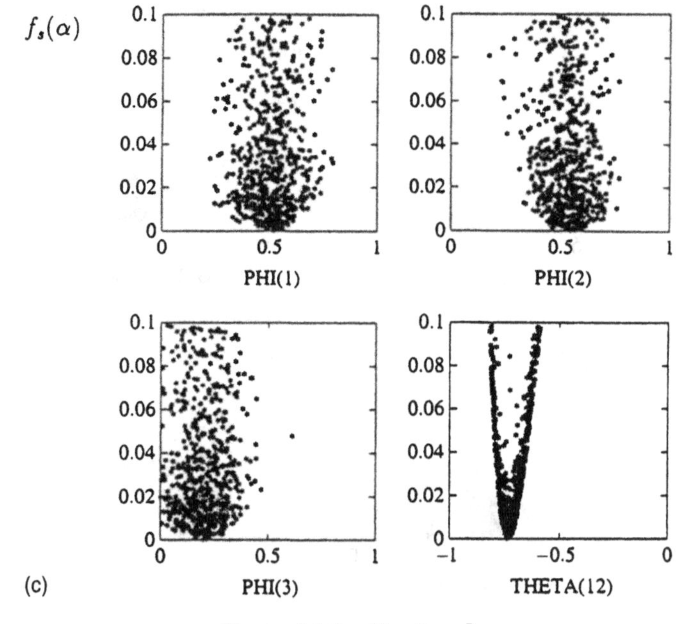

Figure 14.2 *(Continued)*

This graphical analysis suggests that if the goal of the analysis is the trend then models 1 and 3 are overparametrized because of the AR parameters. If the target is the seasonal movements then Model 3 is over-parameterised. As we now show, sensitivity analysis offers a quantitative complement to this graphical procedure for model assessment and analysis of structure definition.

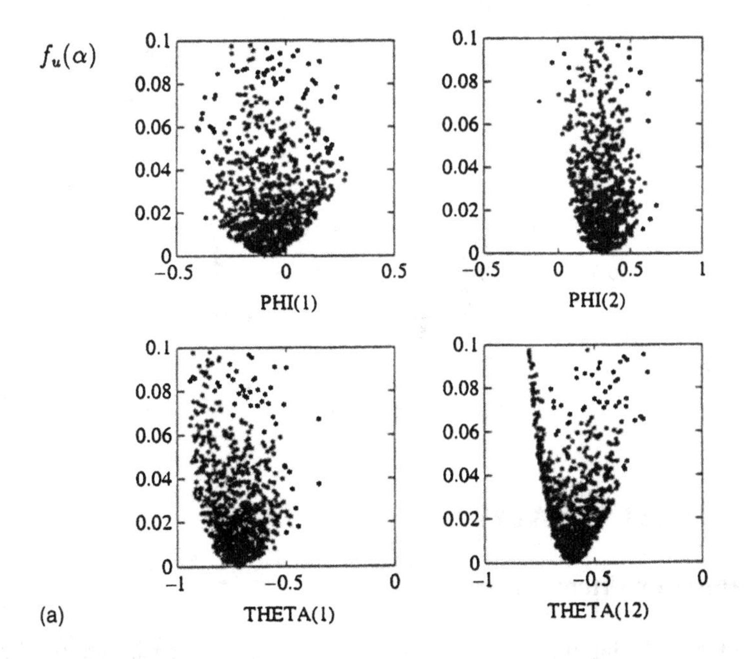

Figure 14.3 Irregular: (a) Model 1; (b) Model 2; (c) Model 3.

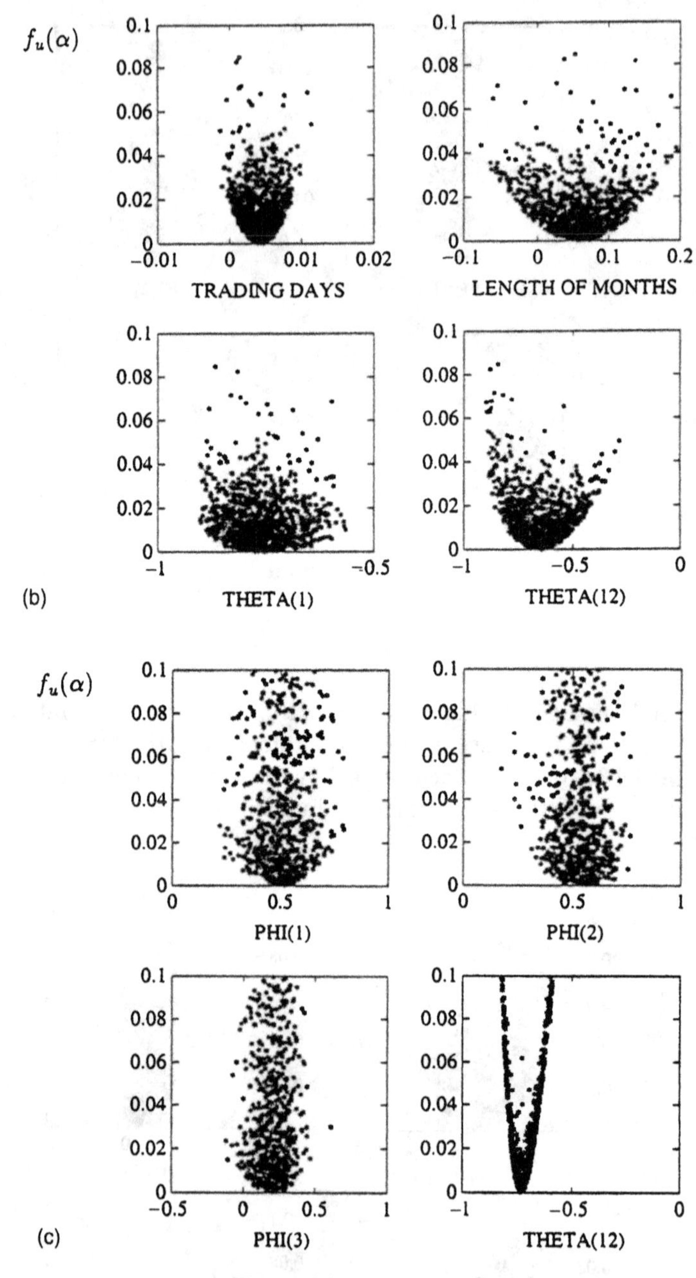

Figure 14.3 *(Continued)*

14.4 SENSITIVITY ANALYSIS

14.4.1 Parameter Uncertainty

How much of the deviations around $\hat{p}_t, \hat{s}_t$, and $\hat{u}_t$, can be attributed to the different parameters? Sensitivity analysis enables us to answer this question by focusing on the *importance measure* (IM) (see Saltelli *et al.*, 1993). Let α be written as $\alpha = (\alpha_i, \alpha_{-i})$, where α_i is a single parameter and α_{-i} embodies all the other elements. Making explicit conditioning arguments related to our context, the importance measure for parameter α_i is defined as

$$\mathrm{IM}_{\alpha_i}(f_l(\alpha)) = \frac{VE[f_l(\alpha)|Y, M_k, \alpha_i]}{V[f_l(\alpha)|Y, M_k]} \tag{14.15}$$

where $l \equiv p, s, u$. If the function of interest only depends on the parameter α_i while the other parameters are irrelevant then $E[f_l(\alpha)|\alpha_i] \simeq f_l(\alpha_i)$, and only α_i will contribute to variations in the function of interest: e.g. $\mathrm{IM}_{\alpha_i} \simeq 1$. Conversely if α_i is irrelevant, different values for α_i will not change $E[f_l(\alpha)|\alpha_i]$, so $\mathrm{IM}_{\alpha_i} \simeq 0$. In summary, the importance measure gives the contribution of every parameter to the variance of the objective function.

When the input parameters are orthogonal, the importance measure turns out to be equivalent to the Sobol' first-order sensitivity indices (see Archer *et al.*, 1997). In this case, the sum of the importance measures computed for all the parameters is bounded below the overall variance (see Sobol', 1993). With correlated parameters, the IM only gives the relative importance of every parameter because the Sobol' result has not yet been extended. Notice that the obvious possibility to orthogonalize the input parameters is not satisfactory in our case because the results would become difficult to interpret.

The computation of (14.15) is somewhat expensive. In our exercise, we generate R realizations of α_i from the normal approximation to the joint posterior $\alpha^{(j)} \sim p(\alpha|Y, M)$, and using these R values we simulate R values for the α_{-i} using the conditional distribution $\alpha_{-i} \sim p(\alpha_{-i}|Y, M, \alpha_i^{(j)})$, which is approximated as $N(E[\alpha_{-i}|\alpha_i^{(j)}], V[\alpha_{-i}|\alpha_i^{(j)}])$. This gives one of the main reason for using a normal approximation to the posterior distribution: simulating from a conditional normal distribution is straightforward. Notice that for orthogonal designs, an ingenious algorithm discussed in Saltelli *et al.* (1993) allows the importance measure to be evaluated with $2 \times R$ simulations instead of $R \times R$.

Consider for example Figure 14.1 (a). For an important contribution of the parameter θ_1 in characterizing the trend, a low variance of $f_p(\alpha)$ when θ_1 is fixed is expected. If we consider the uncertainty range for θ_1, we want a low mean value $E_{\theta_1} V[f_p(\alpha)|\theta_1]$. Standard results about variance decomposition give

$$\frac{E_{\theta_1} V[f_p(\alpha)|\theta_1]}{V[f_p(\alpha)]} = 1 - \mathrm{IM}_{\theta_1} \tag{14.16}$$

Hence, parameters that contribute mostly to the description of a pattern are associated with high importance measures. Conversely, low importance measure values point to irrelevant parameters with respect to the specific pattern. Table 14.3 displays the results for the three models fitted to the series IPSD.

Sensitivity analysis reviews in a very simple way the information contained in Figures 14.1–14.3. It confirms that for all three models, the MA parameters determine the pattern of the component estimates: θ_1 mostly drives the extracted trend, while θ_{12} is mostly important for seasonal and short-term fluctuations. In agreement with the graphical analysis, it tells us that the AR parameters are irrelevant, and that the external regressors of Model 2 determine in an important way the seasonal and short-term movements of the series. It is interesting to note that, in the time series decomposition framework, classical robustness concepts do not apply straightforwardly. Consider, for example, the case where the target is constant whatever the value a parameter takes in an uncertainty range: in this case, the parameter poorly defines the target, suggesting some over-parametrization. Instead, what is required for a good model definition is low posterior variance of the target and high

Table 14.3 Importance measures.

	$f_p(\alpha)$	$f_s(\alpha)$	$f_u(\alpha)$
Model 1			
ϕ_1	0.085	0.052	0.017
ϕ_2	0.042	0.068	0.028
θ_1	0.668	0.162	0.095
θ_{12}	0.039	0.822	0.906
Model 2			
β_1	0.001	0.143	0.193
β_2	0.021	0.274	0.256
θ_1	0.760	0.012	0.048
θ_{12}	0.256	0.609	0.463
Model 3			
ϕ_1	0.006	0.004	0.005
ϕ_2	0.011	0.005	0.005
ϕ_3	0.012	0.002	0.003
θ_{12}	0.914	0.992	0.995

sensitivity to the parameter values. This is illustrated by Model 2, where the uncertainty around the parameters implies less variation around the unobserved component estimates related to the ML parameters, and these variations are tightly associated with every parameter. Notice that this model was not favored by the BIC.

14.4.2 Combining Model and Parameter Uncertainty

The practitioner would be led to use Model 2 to decompose the time series. However, is it the case that choosing a different model would lead to very different trend and seasonally adjusted series? Clearly, a high model dependence of the extracted component is an undesirable feature. The problem is to evaluate the sensitivity of the unobserved component estimates relative to the model choice.

For simplicity, we write $f_l \equiv f_l(\alpha(M))$. For a given set of models $\mathcal{M} = \{M_1, \ldots, M_K\}$, a variance decomposition leads to (see Draper, 1995):

$$V[f_l | Y, \mathcal{M}] = V_M E[f_l | Y, M_k] + E_M V[f_l | Y, M_k]. \tag{14.17}$$

In the expression (14.17), the first term on the right-hand side gives the variance between models of the mean of the sample deviations, and the second term gives the mean over the models of the variances due to the parameter uncertainty. For every model, the mean and variance of the sample deviations are given in Table 14.2. Notice that the second term on the right-hand side can be decomposed as in (14.16), so, concentrating on a single parameter α_1, we have

$$\begin{aligned} V[f_l | Y, \mathcal{M}] = {} & V_M E[f_l | Y, M_k] + E_M V_{\alpha_1} E[f_l | Y, M_k, \alpha_1(M_k)] \\ & + E_{\mathcal{M}} E_{\alpha_1} V[f_l | Y, M_k, \alpha_1(M_k)]. \end{aligned}$$

The evaluation of (14.17) mainly follows from the previous computations. Let us denote $\pi_i \equiv p(M_i | Y)$, given by (14.8), and evaluated using (14.10), and $\bar{f}_{l,M_k} = E_\alpha[f_l | Y, M_k]$. Then, over the different models,

$$\bar{f}_l = E[f_l | Y] = \sum_{k=1}^{M} \pi_k \bar{f}_{l,M_k},$$

$$V[f_l | Y] = \sum_{k=1}^{K} \pi_k \frac{1}{R} \sum_{i=1}^{R} (f_l - \bar{f}_l)^2 \tag{14.18}$$

$$= \sum_{k=1}^{M} \pi_k (g_{l,M_k} - 2\bar{f}_{l,M_k}\bar{f}_l + \bar{f}_l)$$

where $g_{l_{M_k}} = R^{-1} \sum_{i=1}^{R} f_l(\alpha(M_k))^2$. The first term on the right hand side in (14.17) is evaluated as

$$V_M E[f_l | Y, M_k] = \sum_{k=1}^{K} \pi_k (\bar{f}_{l,M_k} - \bar{f}_l)^2 \tag{14.19}$$

Combining (14.18) and (14.19) yields an importance measure for model choice:

$$\text{IM}_{\mathcal{M}}(f_l) = \frac{V_M E[f_l | Y, M_k]}{V[f_l | Y, \mathcal{M}]}. \tag{14.20}$$

Applying these results to our case study, we get $\pi_1 = 0.67, \pi_2 = 0.30$, and $\pi_3 = 0.03$: the Bayes factor does not give much weight to Model 3. Then, from (14.20), we have

$$\text{IM}_{\mathcal{M}}(f_p) = 0.075,$$
$$\text{IM}_{\mathcal{M}}(f_s) = 0.152,$$
$$\text{IM}_{\mathcal{M}}(f_u) = 0.161.$$

Hence, for the IPSD series, the effect of model uncertainty is mostly relevant regarding seasonal and the short-term movements of the series, while the series trend is the component which is described in the most 'robust' way over the three models.

14.4.3 About the Normal Approximation

Before concluding, we briefly discuss the use of the normal approximation to the posterior distribution. In the Bayesian framework, several techniques are available for simulating from a posterior distribution (see Geweke, 1996; Gilks *et al.*, 1996). The main problem faced in our case is the number of replications needed to evaluate the importance measure (14.15), namely $R \times R$. We tried three algorithms: acceptance–rejection sampling, importance sampling, and Monte Carlo Markov chain with the Metropolis algorithm. A rejection rate of 98% made the first unsuited. Notice that in a similar context, Bell and Otto (1993) obtained a rejection rate of the same order when implementing the acceptance sampling scheme. Importance sampling (see Geweke, 1996) leads to computation of $R \times R$ weights, since an importance function is needed for computing the conditional expectation in (14.15), and another one for computing the variance of that conditional expectation. The computation of many weights considerably lowers the execution time. Furthermore, it does not provide a sample from the posterior distribution against which the normal approximation could be compared. The Metropolis algorithm (see Gilks *et al.*, 1996) was more convenient for our

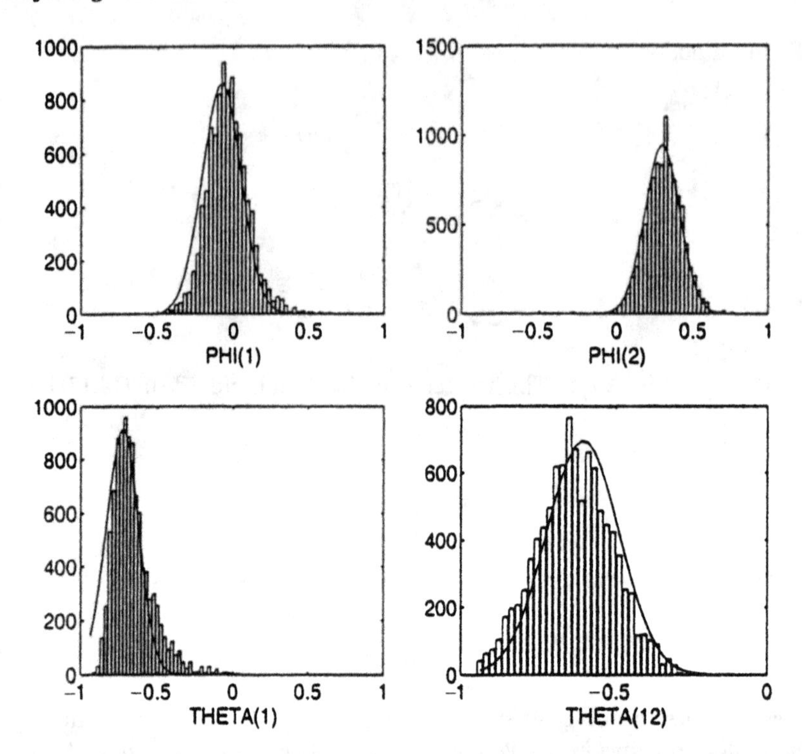

Figure 14.4 Parameter distribution in Model 1: normal distribution (curve) versus histogram of posterior samples.

case. Yet, the need to discard B first simulations actually leads to a total of $R \times B$ discarded steps, which is more costly. We use this approach, however, to compare the normal distribution with a sample of 2000 simulated α from the exact posterior. For the parameters ϕ_1, ϕ_2, θ_1, and θ_{12} in Model 1, Figure 14.4 displays the histogram of the output of the Metropolis algorithm, while the continuous line shows the normal distribution with mean $\hat{\alpha}$ and covariance matrix $\hat{\sum}$. The other two models essentially yielded similar results. Although not very accurate in the tails of the θ_1 distribution, it can be seen that the normal distribution is a reasonable approximation if emphasis is put on computing time.

14.5 CONCLUSIONS

We believe that within the vast battery of tools available in time series analysis, there is room for sensitivity analysis. The classical steps in handling time series, broadly known as Box and Jenkins (1970) methodology, are model specification, estimation, and diagnostic checking. Although there have been many developments in time series analysis, for example around the model specification procedure with structural modelling (see Harvey 1989), also in the range of specifications available (see e.g. Tong, 1990), that overall approach remains an unquestionable paradigm. The implementation of uncertainty and sensitivity analysis comes in a second stage, once the target of the analysis has been explicitly specified and when the statistical validity of the model considered has been checked.

We have shown that uncertainty and sensitivity analysis can provide complementary information about how tightly the objective of the analysis is described given the uncertainties faced. The contribution of every parameter of a given model to the definition of the target can be assessed, and which parameters have the most significant impact on the objective of the analysis can be determined. For a good model definition, a low posterior variance of the target and high sensitivity to the parameter values is required. Overparametrizations can be easily detected. Sensitivity analysis thus brings some insights into the analysis of model parsimony. Finally, sensitivity analysis enables us to quantify the importance of the model selection for characterizing a target of interest. Some possible model-dependent features of the target can be highlighted; as a preferred feature, a low discrepancy between the output of different models seems reasonable. Overall, sensitivity analysis can be seen as a further tool for model assessment.

For our case study, the results show that only a relatively low proportion of the uncertainty around unobserved components depends on the model used: the uncertainty about parameters had more important consequences than uncertainty about models. The effect of model uncertainty was mostly relevant regarding seasonal and short-term movements of the series, while the trend was the component described in the most stable way over the three models. The sensitivity analysis techniques pointed to Model 2 as the model that leads to a better definition of the series structure. This result does not agree with the BIC, which favored Model 1 instead. It shows that sensitivity analysis complements other classical tools, shedding light about the accuracy of model definition with respect to the target of interest.

15

A Dataless Precalibration Analysis in Solid State Physics

Stefano Tarantola

European Commission, Joint Research Centre, Ispra, Italy

Rosanna Pastorelli, Marco G. Beghi and Carlo E. Bottani

INFM, Politecnico di Milano, Italy

15.1 INTRODUCTION

Sensitivity Analysis (SA) can be used in the study of the stability of an inverse problem, in that whenever a computational model is employed to simulate an underlying physical process, it is of interest to ascertain whether experimental measurements, with their related uncertainties, allow the extraction, or estimation, of some parameters embedded into the simulation model (Saltelli and Scott, 1997).

Local sensitivity analysis approaches have been used in chemical kinetics to extract kinetic constants of complex systems from measured yield rates of components (Turanyi, 1990a; Rabitz, 1989). Variational methods have been employed in hydrogeology to identify uncertain parameters and geometry (Smidts and Devooght, 1997).

In this work, we propose the use of a global sensitivity analysis approach, based on the method of Sobol' (1990b) (see also Chapter 8), aimed at investigating the stability properties of an inverse problem in the field of solid state physics.

The inverse problem consists of deriving the elastic properties (i.e. the two factors Young's modulus E and Poisson's ratio ν) of a thin supported film from a set of measurements of surface acoustic wave velocities via surface Brillouin scattering (Lee *et al.*, 1988; Karanikas *et al.*, 1989; Carlotti *et al.*, 1991; Hua Xia *et al.*, 1992). The procedure followed to determine E and ν (see Section 15.2) is based on least-squares minimization of a target function, which is customarily implemented via a set of model simulations and a set of experimental measures. Since the analyst does not know *a priori* which elastic constants can be determined,

Sensitivity Analysis. Edited by A. Saltelli *et al.*

and to what degree of precision, he has to perform the measurements. He may discover, after the experiment, that the process is unstable for that specific case and, for instance, that E and ν cannot be estimated to the desired precision. SA is used here to overcome this situation. We shall show that SA can identify which elastic constants can be determined, before carrying out the measurements.

It has to be noted, however, that SA is not a tool for solving the inverse problem, but only to assess its stability. E and ν are determined by a least-squares procedure, where experimental data are obviously required.

Another pair of factors—the mass density ρ and the thickness t of the film—usually enter as constants in the procedure to determine E and ν. Actually, ρ and t are known to a finite level of precision. Therefore, in the present study, the whole set of factors (E, ν, ρ, t) is treated as uncertain. Their ranges are taken over physically meaningful domains (three representative cases are shown in Section 15.2), and the sensitivity results are discussed in Section 15.3.

Model predictions are assumed to perfectly mimic the reality of the underlying system. This assumption is not as strong as it looks: the model is built on the basis of primary mechanisms acting at microscopic level, and is therefore adequate to the task, as demonstrated by experience. In the SA performed hereinafter, we only use model predictions; missing experimental data have been replaced by simulated measurements. In view of what is explained above, the use of model predictions is plausible.

As we shall show in the next sections, SA is extremely helpful in the design of the experiment. Firstly, it actually yields a yardstick to quantify the stability of E and ν without requiring any measurements to be performed. Such information would allow the analyst to save expensive measurements whenever instability for E and ν is found. Secondly, sensitivity indices allow the analyst to tune the tolerance bounds for ρ and t—supplied by the provider of the film, according to the desired level of precision for E and ν. An example is shown in this study, where the uncertainty bounds for ρ are decreased to investigate how much the stability for ν improves.

15.2 DATA ANALYSIS TECHNIQUE

A thin film (e.g. amorphous carbon) is deposited on a substrate (typically silicon, Si). We are interested in estimating the elastic properties of the film, assuming its mass density ρ and thickness t to be known at a given confidence level, whereas the substrate properties are fully known.

Surface Brillouin scattering allows the elastic properties of the film to be derived by measurements of acoustic wave velocities at a number of incidence angels θ^i. A simulation model (Hardouin Duparc *et al.*, 1984) provides the velocities of acoustic waves v_c^i (c stands for 'computed') at the angles θ^i as a function of (E, ν, ρ, t). The inverse problem, i.e. the derivation of E and ν, is tackled by computing

$$\mathrm{SS}(E, \nu, \rho^*, t^*) = \sum_i \left[\frac{v_c^i(E, \nu, \rho^*, t^*) - v_e^i}{\sigma_e^i} \right]^2 \tag{15.1}$$

on a rectangular mesh over the (E, ν) plane, encompassing the domain of interest. ρ^* and t^* are nominal values of ρ and t, and v_e^i are experimental velocities with corresponding variances σ_e^i. The minimum of the SS function over (E, ν) (and the region around the minimum) identifies the pair $(\bar{E}, \bar{\nu})$ (and any related confidence level).

Table 15.1 Factors' ranges of variation for the three cases analyzed by SA.

	Case (a): fast on slow	Case (b): transition	Case (c): slow on fast
E (GPa)	300–800	400–550	1–800
ν	0.01–0.37	0.01–0.38	0.01–0.38
ρ (g/cm^3)	2.8–3.2	5.8–6.2	17.8–18.2
t (nm)	67–73	67–73	67–73

As shown above, in the least-squares method for deriving E and ν, the impact of uncertainties in ρ and t is not taken into account. In other words, the simulation model evaluates $v_c^i(E, \nu, \rho^*, t^*)$ at $\rho = \rho^*$ and $t = t^*$. In the SA performed in our study, we shall compute $\mathrm{SS}(E, \nu, \rho, t)$ by allowing the four factors to vary.

In this work, three physically representative cases have been studied, and SA has been applied to each of them. The ranges for the factors, assumed uniformly distributed, are summarized in Table 15.1.

These cases have been chosen since they cover all the possible behaviors of the film with respect to a Si substrate:

(a) a fast film on a slow substrate (fast on slow) is studied; this case is representative, for instance, of a tetrahedral amorphous carbon film;
(b) a transition region from 'fast on slow' to 'slow on fast' is investigated; this case is representative, for instance, of a nanocrystalline TiN film;
(c) a 'slow on fast' region is explored; this case is representative, for instance, of an Au isotropic film.

In all three cases, film thicknesses are taken in the representative range 67–73 nm, since the behavior of the film with respect to the substrate is independent of t.

15.3 SENSITIVITY ANALYSIS

The SA is performed by evaluating both the first-order and the total effect indices for each factor using the method proposed by Sobol' (1990b). A sample size of 512 is chosen for estimating each index. The total cost of the analysis is $(2k + 2) \times 512 = 5120$ model executions, as explained in Chapter 8. This sample size is large enough to guarantee that estimates of sensitivity are sufficiently precise and, at the same time, small enough to allow the analysis to be conducted in a reasonable time. The method of Sobol' has not been designed to operate when the input factors show a correlation structure or when, although uncorrelated, some constraints are imposed over the region where they are defined. The factors (E, ν, ρ, t) are uncorrelated and cover a 4D rectangular domain. Hence, the method of Sobol' is suitable for the task.

Let us denote the jth sample point by $(E_j, \nu_j, t_j, \rho_j)$ and the corresponding model output at the angle θ^i by $v_{c_j}^i$.

The target function used in the SA is

$$\mathrm{SS}_j(E, \nu, \rho, t) = \sum_i [v_c^i(E, \nu, \rho, t) - v_{c_j}^i(E_j, \nu_j, \rho_j, t_j)]^2, \tag{15.2}$$

Table 15.2 Average values $\overline{\widehat{S}(i)}$ and $\overline{\widehat{TS}(i)}$ computed over the 128 estimates of $S(i)$ and $TS(i)$ plus/minus their standard deviations, for the four factors and the three cases.

	Case (a): fast on slow		**Case (b): transition**		**Case (c): slow on fast**	
	$\overline{\hat{S}(\cdot)} \pm \sigma_{\hat{S}}$	$\overline{\widehat{TS}(\cdot)} \pm \sigma_{\widehat{TS}}$	$\overline{\hat{S}(\cdot)} \pm \sigma_{\hat{S}}$	$\overline{\widehat{TS}(\cdot)} \pm \sigma_{\widehat{TS}}$	$\overline{\hat{S}(\cdot)} \pm \sigma_{\hat{S}}$	$\overline{\widehat{TS}(\cdot)} \pm \sigma_{\widehat{TS}}$
E	0.89 ±0.05	0.96 ±0.02	0.70 ±0.10	0.88 ±0.03	0.99 ±0.01	0.99 ±0.01
ν	0.00 ±0.00	0.00 ±0.00	0.01 ±0.01	0.01 ±0.01	0.00 ±0.00	0.01 ±0.00
ρ	0.03 ±0.01	0.09 ±0.04	0.10 ±0.04	0.28 ±0.09	0.00 ±0.00	0.00 ±0.00
t	0.01 ±0.00	0.01 ±0.01	0.00 ±0.00	0.01 ±0.01	0.00 ±0.00	0.00 ±0.00

where, in place of the experimental velocity v_e^i, one generic model-predicted value $v_{c_j}^i$ can be used, by selecting it from the set of 5120 available predictions (σ_e^i, the precision of the measurement at a given angle, is cancelled in the analysis by assuming that its value does not depend on i).

The first-order sensitivity indices $S_j(E), S_j(\nu), S_j(\rho), S_j(t)$ and the total sensitivity indices $TS_j(E), TS_j(\nu), TS_j(\rho), TS_j(t)$ are then computed using the method of Sobol'.

The SA can be repeated by referring to a different point $\bar{j}$, corresponding to a new sample point $(E_{\bar{j}}, \nu_{\bar{j}}, \rho_{\bar{j}}, t_{\bar{j}})$. Various analyses can be repeated for sets of $\bar{j}$ point at no extra computational cost, since this depends only on the model's computational cost, while the effort of estimating the sensitivity indices given the set of model outputs is negligible. We have actually performed 128 sensitivity analyses by selecting the sample points $(E_{\bar{j}}, \nu_{\bar{j}}, \rho_{\bar{j}}, t_{\bar{j}})$ to uniformly cover the domain of the input factors. With this approach, the robustness of the sensitivity estimates can be investigated and their reliability assessed on points $(E_{\bar{j}}, \nu_{\bar{j}}, \rho_{\bar{j}}, t_{\bar{j}})$ that might correspond to actual elastic properties.

The average values $\overline{\hat{S}(i)}$ and $\overline{\widehat{TS}(i)}$ over the 128 estimates together with their standard deviations, are given in Table 15.2 for all the factors and cases.

15.4 DISCUSSION

In all cases, the $\overline{\widehat{TS}(E)}$ are very high in comparison with those for the other factors. The sensitivity analysis indicates that in these domains only E can be derived at a good level of stability. Furthermore, the $\overline{\widehat{TS}(\nu)}$ are zero, meaning that the factor ν is completely indeterminate. This indicates that experimental measurements are worth performing only if they aim at determining E.

As can be noted from the small values of $\sigma_{\widehat{TS}(E)}$, in all cases the 128 estimates of $\widehat{TS}(E)$ are robust to the choice of the reference points $(E_{\bar{j}}, \nu_{\bar{j}}, \rho_{\bar{j}}, t_{\bar{j}})$, which might be representative of actual specimen properties. This means that the (high) level of stability for E is very weakly influenced by the film characteristics.

It can be noted that in case (b), $\overline{\widehat{TS}(\rho)} = 0.28$, showing a relatively high influence of factor ρ on the SS target function, while the effect of t is negligible. A reduction of the uncertainty bounds for ρ could increase the stability in E and ν, and SA would indeed help to quantify this improvement. Therefore, we have performed an SA by halving the range for ρ, all the other factors varying as in case (b). This requires another set of model evaluations to be performed, but is a minor effort in comparison to that of performing a real experiment. The results show indeed that the estimate of E is more stable because $\overline{\widehat{TS}(E)}$ increases ot 0.95, but the indeterminacy on ν is not resolved $(\widehat{\mathrm{TS}(\nu)} = 0.02)$, i.e. it is not related to the degree of precision for ρ. The overall influence of ρ is obviously decreased to $\widehat{\mathrm{TS}}(\rho) = 0.09$, owing to the reduction in its range, while $\widehat{\mathrm{TS}}(t)$ remains unchanged.

We conclude that SA, used iteratively, can help the analyst to search for the best compromise between the desired precision for E and ν, and the precision to which ρ and t have to be known. In this case, our numerical (and hence costless) analysis shows that for realistic values of the design parameters only E can be successfully calibrated.

15.5 CONCLUSIONS

The least-squares method employed to solve the inverse problem provides a picture of the level of indeterminacy for the elastic properties. On the other hand, global sensitivity analysis offers a yardstick to quantify the stability of these physical quantities through a set of sensitivity indices.

An interesting result from an SA is that quantitative information on the stability of the procedure can be obtained essentially on the basis of computational modelling, i.e. no laboratory measurements are required *a priori*. SA indices could indicate, for instance, that the procedure is unstable. In this case, the experimental measurement is not worthwhile, thus saving extensive laboratory time and costs. In the opposite case, after demonstrating a sufficiently good degree of stability in the procedure, the same set of model evaluations used in the SA can be employed, together with the experimental measurements, to estimate the elastic parameters.

These cases show indeed that SA is a valuable tool for designing the experiment. SA indices supply an *a priori* evaluation of the stability, and they can help establish what is the allowable level of uncertainty in mass density and thickness that guarantees the desired level of stability in the results.

The best information would be to identify the maximum uncertainty in ρ and t that allows the analyst to estimate E and ν within a desired confidence interval. Sensitivity indices only link qualitatively to the confidence levels for $\bar{E}$ and $\bar{\nu}$. For instance, we know that the higher $\widehat{\mathrm{TS}}(E)$, the more precise is the confidence level for $\bar{E}$. A quantitative appreciation could, however, be obtained by appropriate calibrations.

ACKNOWLEDGMENTS

M. G. Beghi, C. E. Bottani, and R. Pastorelli acknowledge financial support from Progetto Finalizzato 'Materiali Speciali per Tecnologie Avanzate II' of Consiglio Nazionale delle Ricerche.

16

Application of First-Order (FORM) and Second-Order (SORM) Reliability Methods: Analysis and Interpretation of Sensitivity Measures Related to Groundwater Pressure Decreases and Resulting Ground Subsidence

Jeff D. Cawlfield

University of Missouri at Rolla, USA

16.1 INTRODUCTION

This chapter focuses on an application of reliability methods to an engineering problem involving the prediction of subsidence caused by groundwater withdrawals. In the example, uncertainty exists concerning the pumping rate, the transmissivity, the compressibility of the aquifer from which water is being pumped, and the compressibility of the clay layer overlying the aquifer. The overall goal is to evaluate the probability that the total subsidence will exceed some specified tolerance—perhaps the tolerable subsidence is dictated by building code, or perhaps it is limited by design considerations such as potential flooding, foundation or structural integrity, or utilities tolerance for differential settlement. Although the

Sensitivity Analysis. Edited by A. Saltelli *et al.*

example is hypothetical, it is a generic application with realistic values for the mean and variance for the uncertain variables of a typical sand aquifer overlain by a clay confining layer.

Subsidence due to many causes, not just groundwater withdrawals, is a worldwide problem. One of the more thorough compendiums of material related to subsidence worldwide is provided by Poland (1984). Another good reference on subsidence problems worldwide is Borcher (1998).

16.2 FORM SENSITIVITY INFORMATION

The first-order reliability algorithm (FORM) yields a wealth of sensitivity information, both directly and indirectly with only a little additional mathematical processing. The focus in this example is on the utility of the sensitivity information for engineering analysis and interpretation. The sensitivity measures of particular interest are the so-called gamma sensitivities, which, as described in Chapter 7, provide a measure of the sensitivity of the probabilistic outcome to equally likely changes in the uncertain variables. The gamma sensitivity measures, therefore, incorporate both deterministic and probabilistic information (e.g., distribution types) to yield valuable information about which of the uncertain variables, with their particular statistical characteristics, are most important to the outcome.

The gamma sensitivity measures, as defined by Der Kiureghian and Liu (1986), are obtained by evaluating the partial derivatives of the reliability index β with respect to each uncertain variable (with the variable transformed into uncorrelated standard variates). The gamma sensitivity measures are given by

$$\gamma = \frac{\boldsymbol{\Gamma}^{\mathrm{T}}\boldsymbol{\alpha}^{*}}{|\boldsymbol{\Gamma}^{\mathrm{T}}\boldsymbol{\alpha}^{*}|}. \tag{16.1}$$

The $\boldsymbol{\alpha}^{*}$ are the direction cosines at the design point (i.e., the point of maximum likelihood of 'failure'), which incorporate the partial derivatives as mentioned previously, and the matrix $\boldsymbol{\Gamma}$ is a rotation matrix that is used to transform the original uncertain variables into standard *uncorrelated* parameters. Thus, the gamma sensitivity measures are true probabilistic sensitivity measures—they measure how important each uncertain variable is to the reliability index (that is, the probability of 'failure') with respect to *equally likely* changes in each variable. All relevant information is incorporated into these sensitivity measures: the variance, correlation, and distribution type (through the matrix $\boldsymbol{\Gamma}$), and the instantaneous change in β with respect to each uncertain variable after transformation into standard variate space (through the $\boldsymbol{\alpha}$ values).

The computer code used in this example, CALREL (Liu *et al.*, 1989) also provides sensitivity information with respect to the distribution parameters for each uncertain variable. Thus, the FORM analysis provides information about how the probabilistic outcome is likely to be influenced if, for example, the mean value or variance for a particular uncertain variable is changed. This information can help guide the engineer or scientist in terms of developing laboratory or field sampling programs to obtain additional information about the variables impacting a problem.

Finally, CALREL provides sensitivity information with respect to deterministic parameters in the problem formulation that might be of interest. In this example, three parameters are evaluated for their affect on the outcome: the target subsidence, the thickness of the clay layer, and the thickness of the sand aquifer.

16.3 CASE STUDIES

16.3.1 Groundwater Pressure Declines: Governing Equations and Uncertain Parameters

In the example under consideration here, a sandy aquifer is being pumped via a fully penetrating well, and the fluid pressure is reduced as a result. The aquifer is assumed to be confined by an overlying clay layer. The decrease in groundwater pressure is typically expressed as a decrease in fluid head (drawdown), and this response is evaluated using the Theis equation or Jacob linearization of the Theis equation as described in any basic text on groundwater (e.g., Fetter, 1994; Domenico and Schwartz, 1998). The Jacob approximation gives drawdown (the decrease in fluid head) as follows:

$$s = \Delta h = \frac{2.3Q}{4\pi T} \log \frac{2.25Tt}{r^2 S}, \tag{16.2}$$

where s is the total drawdown (change in fluid head, Δh), Q is the pumping rate, r is the radius from the pumping well to the point of interest, T is the transmissivity of the aquifer, S is the storativity of the aquifer, and t is the elapsed time of pumping. For a confined aquifer, $S = S_s m = \rho g \beta_p m$, where S_s is the storage coefficient of the aquifer (neglecting the very small compressibility of the water), m is the original thickness of the aquifer, ρ is the water density, g is the acceleation due to gravity, and β_p is the compressibility of the aquifer pore space (in these general introductory equations, the compressibility of the pore space is given as β_p; later in the examples, specific compressibilities for sand or clay will be designated with appropriate additional subscripts). The Jacob approximation is valid when t is sufficiently long and/or r is sufficiently small. We shall assume that our interest is in the maximum fluid head decrease near the well bore, and for this problem we will assume that $r = 10$ cm and $t = 365$ days. The small value of r is used in order to calculate the head decrease at the edge of the well bore. Thus, we are evaluating the maximum head decrease expected at a given pumping rate over the course of one year of pumping. Since r is very small and t is relatively large in this case, the Jacob approximation is quite accurate.

In the example considered here, we assume that the radius r and the time t are known deterministic variables, and the pumping rate Q, the aquifer transmissivity T, and the aquifer pore compressibility β_p are uncertain variables.

16.3.2 Subsidence: Governing Equations and Uncertain Parameters

Domenico and Schwartz (1998, pp. 165–169) provide a concise overview of the equations governing elastic compaction and inelastic consolidation resulting from head (pressure) decreases in groundwater systems. The equations are fairly straightforward, and depend on the total pore fluid pressure decrease, the original thickness of the soil or rock, and the compressibilities of the soil or rock. For an elastic aquifer, such as a sand or gravel, with pore compressibility of β_p (in units of inverse pressure), the total compaction is given by

$$\Delta m = m\beta_p \,\Delta P \tag{16.3}$$

where ΔP is the change in pore fluid pressure and m is the original thickness of the soil or rock unit. In this example, we assume that the original thickness is known, but the compressibility is an uncertain variable. The pressure change is calculated using the previously mentioned Jacob approximation for drawdown (note that the equation is itself a function of uncertain variables Q, T, and β_p), $\Delta P = \rho g \Delta h$:

$$q = S_s H \left[\frac{\Delta h_1 + \Delta h_2}{2} \right] \tag{16.4}$$

For an inelastic confining layer, such as clay, the total consolidation is equivalent to the volume of water drained from a unit area of the layer, and is given as follows: where q is the value of consolidation in units of length, H is the original thickness of the clay, and Δh_i is the decrease in pore fluid head in layer i either above or below the clay. S_s is the storage coefficient of the clay, which incorporates the pore compressibility as follows (assuming compressibility of water is negligible):

$$S_s = \rho g \beta_p. \tag{16.5}$$

If the head is changing only in one layer (say in layer 1 below the clay), then $\Delta h_2 = 0$ in layer 2 above the clay. This will be the case in the example considered here. As before, we assume that the original thickness H is known, but the pore compressibility β_p is an uncertain variable. The head change in the aquifer is calculated using the previous Jacob approximation to the Theis solution for well drawdown (which also is a function of uncertain variables).

16.3.3 Example Problem Geometry and Boundary Conditions

The example considered here is a single elastic aquifer, which is being pumped at a rate Q, which produces a reduction in pore fluid head. This pore fluid head reduction causes compaction of the aquifer and consolidation of a single overlying clay layer. The aquifer is 100 m in original thickness m, and the clay layer is 50 m in original thickness H. For the purposes of the example, we designate the aquifer compressibility as β_{Ap} (in this example, we have a sand aquifer) and the clay compressibility as β_{Cp}. As initial values for the first part of the analysis, the mean and standard deviation for the uncertain variables are as given in Table 16.1.

Note that at the beginning of the analysis, we assume that the coefficient of variation is 0.10, which is at the low range of typical real-world values for uncertainty for transmissivity and compressibilities. We shall start with this low uncertainty, then consider what happens to the analysis and the sensitivity measures as the uncertainty increases. Some field studies have reported uncertainties for transmissivity as high as standard deviation equal to mean value (that is, coefficient of variation equal to 1.0 or higher).

As mentioned previously, we analyze here the reduction in head immediately near the wellbore, which gives a reasonable value for the absolute maximum subsidence that might be expected. In order to evaluate more precisely the subsidence over an area of concern, the

Table 16.1 Mean and standard deviations of uncertain variables.

	Mean	**Standard deviation**
Pumping rate Q	5×10^{-3} m^3/s	5×10^{-4}
Aquifer transmissivity T	1×10^{-4} m^2/s	1×10^{-5}
Aquifer compressibility β_{Ap}	1×10^{-9} m^2/N	1×10^{-10}
Clay compressibility β_{Cp}	1×10^{-7} m^2/N	1×10^{-8}

analysis would need to account for the average head reduction over various subdomains, and calculate the expected subsidence at different areas surrounding the well. Typically, subsidence will be greatest immediately overlying and near the pumping well, and will decrease radially away from the well. Also, we assume that the entire total amount of elastic compaction and inelastic consolidation in the two layers is transmitted perfectly to the surface. In reality, there is some reduction of the total subsidence, since overlying layers will often spread the effect over a larger area and 'bridge' over some of the subsidence occurring in the underlying layers.

16.3.4 CALREL Computer Code: Problem Assumptions and Program Input

CALREL is a computer program designed for structural reliability analysis, focused on computing probabilities of failure for components of a structural system or for series reliability problems. See the software appendix to this volume for further details about CALREL and other reliability programs. For the work described in this chapter, subroutines were compiled and linked using Microsoft Fortran Version 5.1.

CALREL is utilized to estimate FORM and SORM probabilities of failure associated with ground subsidence. CALREL also is used to run Monte Carlo simulations of the same problem, so that probabilities of failure for FORM, SORM, and Monte Carlo results may be compared. CALREL computes first-order sensitivity information in various formats; in this example, the program is used to compute sensitivity measures with respect to distribution parameters (such as means and standard deviations) and with respect to limit-state parameters (uncertain and deterministic variables in the performance function that is used to define failure).

For the example under consideration, component failure is defined in terms of subsidence exceeding a specified threshold level. The ground surface subsidence is calculated as the sum of the elastic compression occurring in the aquifer due to groundwater pressure declines, and the inelastic consolidation occurring in the overlying clay confining layer due to drainage into the aquifer in response to the pressure declines. The groundwater pressure decline is determined in this example by using an analytical solution that depends on the pumping rate, transmissivity, storativity, and radial distance from the pumping well. These equations must be coupled and then the resulting subsidence calculated in a subroutine that CALREL calls each time it requires an evaluation of the performance function. The performance function takes on a simple form as follows:

$$g(\mathbf{X}) = M - \Delta m, \tag{16.6}$$

where M is the threshold subsidence (i.e., that subsidence which is the limit of tolerance before failure occurs) and Δm is the actual (or estimated) subsidence calculated for the case under consideration. When Δm becomes greater than or equal to M, failure is defined to have occurred. The performance function and all relevant equations are written in subroutines that are compiled and linked to the main CALREL program.

As explained in Chapter 7, the objective is to find a design point at which the probability of failure can be estimated for this performance function. Any of the variables in the equations utilized to calculate Δm may be specified as uncertain variables $\mathbf{X}$ with specific distributions, distribution parameters (means and variances), and correlation coefficients (if correlation exists—in the example considered here, all variables are assumed uncorrelated).

During the FORM evaluations, CALREL calculates partial derivatives, which are used in the optimization algorithms for determining a design point and, which, ultimately, may be utilized to evaluate sensitivity measures for the FORM process.

16.4 RESULTS AND DISCUSSION

Although the example considered here is hypothetical, it is based on typical problems where subsidence due to groundwater pumping may be a factor in the design of some structures, or in the planning for flood control or transportation facilities. We shall consider four specific cases and examine the overall probabilistic outcome, with particular emphasis on the sensitivity measures as evaluated by FORM via the CALREL computer program.

16.4.1 All Uncertain Variables with Equal Uncertainty and Lognormal PDF

As a base case for initial study, we examine the probability that subsidence will equal or exceed a target value, given that the uncertain variables have mean values as specified previously, and coefficient of variation equal to 0.10 (i.e., the standard deviation for each variable is 0.10 of the mean value). We assume that each variable is described by a lognormal distribution (PDF), and that there is no correlation between the variables. How realistic are these assumptions?

The mean values are typical for geologic deposits consisting primarily of sand (in the case of the aquifer) or clay (in the case of the overlying layer). Transmissivity data have been examined in some detail over the past 20–30 years from a wide variety of geologic media, and most researchers have concluded that the lognormal distribution is most appropriate for describing such data. Data regarding the distribution for compressibilities are not so easily found, but logic dictates that the normal distribution is inappropriate because compressibilities can only take on positive values. The pumping rate Q is probably best described by a normal or a lognormal, because the uncertainty in Q is mostly due to measurement error and natural variation about the mean. However, the lognormal may be most appropriate again, because Q cannot be negative.

To begin the analysis, we analyze the subsidence obtained if we perform a strictly deterministic analysis and use the mean values as constants for the uncertain variables. In this example, the total subsidence is about 9.98 m. This value will give us a starting point for selecting target values of subsidence in the probabilistic model using FORM, because we would expect about a 50% chance of exceeding 9.98 m given the mean values. So we shall look at a suite of target values, and evaluate the probability of exceeding those target values, as well as interpreting the sensitivity information that we obtain. We are primarily interested in target values of about 10 m and higher, because our interest as engineers and designers is evaluating the probability of realizing high values of subsidence, which will have the greatest negative impact on the design or plan.

Table 16.2 summarizes results of FORM, SORM, and Monte Carlo (MC) simulation for probability of failure (i.e., exceeding the selected target values of subsidence) given the input variables and statistical characteristics as discussed previously (refer to Table 16.1).

As expected, the probability of exceeding the target value decreases as higher target values are evaluated. In essence, we are obtaining the cumulative distribution function; that

Table 16.2 Probability of exceeding target subsidence.

Target subsidence (m)	FORM	SORM	MC
10	0.4933	0.4934	0.5020
11	0.2814	0.2815	0.2905
12	0.1375	0.1375	0.1343
13	0.0589	0.0589	0.0550
14	0.0227	0.0227	0.0219
15	0.0080	0.0080	0.0079

is, the probability of failure versus target subsidence values. CALREL can be run for a suite of target values between, for example, 8.0 m and 20.0 m to obtain a reasonable approximation to the complete cumulative distribution function. The SORM results are almost identical to the FORM results, and both FORM and SORM are very, very close to the Monte Carlo simulation probabilities. In extremely nonlinear problems, or when extreme probabilities are evaluated for problems with highly non-normal distributions, the FORM results may not be as accurate as SORM or Monte Carlo simulation. Although the probability results are interesting and provide necessary information for evaluating risk for a design, our purpose here is to look at applications related to the sensitivity information and how it may be interpreted.

Consider first the gamma sensitivity results. The gamma sensitivity information does not change appreciably as the target values were increased, so consider target subsidence 12 m as typical of all the cases with gamma sensitivity results summarized in Table 16.3.

The gamma sensitivity information clearly indicates that the compressibility of the sand layer is not an important uncertain variable in this problem. In fact, even if the variance for the compressibility of the sand is increased to a relatively large value, the sensitivity for this variable does not increase significantly. Here is the first clear indication of how we can use such sensitivity information as engineers or scientists—since the uncertainty of this one variable is not important to the probabilistic outcome, we simplify the analysis by making that variable deterministic! CALREL has a convenient option for making any of the uncertain variables a deterministic variable during a specific simulation; as expected, when the sand compressibility is changed to a deterministic variable, the probability outcome and the sensitivity information are not influenced. It should be noted that most groundwater scientists and engineers would not intuitively have expected the sand compressibility to be a variable that would have such low influence on the probabilistic outcome. True, subsidence is primarily a function of the consolidation of clay layers, and even a deterministic analysis will show the compaction of the sand layer is relatively small and insignificant; however, the sand compressibility is also present as a variable in the Jacob approximation for calculating drawdown. It is not obvious, without carrying out the FORM sensitivity analysis, that the sand compressibility can be treated as deterministic without great impact on the results.

Table 16.3 Gamma sensitivity results: Case 1.

Uncertain variable	Gamma sensitivity
Transmissivity T	0.5592
Aquifer (sand) compressibility β_{Ap}	0.0245
Clay compressibility β_{Cp}	0.5838
Pumping rate Q	0.5881

The gamma sensitivity measures seem to indicate that the compressibility of the clay, the transmissivity T, and the pumping rate Q, are about equally important to the outcome in this case. Recall that each of the variables has relatively equivalent uncertainty as expressed by the coefficient of variation. It is expected that if the uncertainty of any of these variables is increased, the gamma sensitivity would reflect such increase, and we shall explore such results shortly.

Another important sensitivity measure is the sensitivity of the measures with respect to the deterministic parameters specified in the performance function. For this example, we specify three deterministic parameters of particular interest: the target subsidence, the thickness of the sand, and the thickness of the clay layer. The sensitivity measures are calculated as sensitivity of the reliability index with respect to each parameter, or sensitivity of the probability of failure with respect to each parameter. The results are shown in Table 16.4.

These values indicate how sensitive the reliability index β is to changes in the deterministic parameters. The reliability index is a measure of how likely it is that the value of the performance function will be 0 or negative. The reliability index and probability of failure are directly connected in a very nonlinear way through the failure approximation: $p_f \approx \Phi(-\beta)$, where $\Phi(\cdot)$ is the standard normal CDF function. We conclude that the probabilistic outcome is very sensitive to changes in the target subsidence, somewhat sensitive to the thickness of the clay layer, and not sensitive to the thickness of the sand layer. It is no surprise that the reliability and probability results are very sensitive to the choice of a target subsidence number; after all, we have already seen that the cumulative distribution function related to subsidence is directly a function of this target subsidence. It is very useful to note that the clay thickness is much more important than the sand thickness as a deterministic variable influencing subsidence, and therefore a field investigation program would be focused on the clay more than on the sand aquifer. Perhaps this result would be predicted by those familiar with subsidence problems, but the potential use of such sensitivity information in more complicated problems with perhaps many more deterministic variables should be obvious.

16.4.2 Effect of Assuming Normal PDF

A question that frequently arises in engineering applications such as that considered here is the question of what statistical distribution is most appropriate to describe each of the uncertain variables. In Case 1, we assumed that a lognormal distribution is a reasonable choice for the variables considered in this problem, because the variables were all limited to positive value. Additionally, at least for the transmissivity and compressibility variables, there has been some previous research and data analysis that indicates the lognormal is most appropriate. What happens if we assign a normal distribution to the variables? This should have some effect on the probability of failure and also the sensitivity information.

Table 16.4 Sensitivity of reliability index, and sensitivity of probability of failure, to deterministic parameters in the analysis.

Deterministic parameter	Sensitivity of reliability index	Sensitivity of probability of failure
Target subsidence	0.4913	0.1080
Sand layer thickness	0.0025	0.0005
Clay layer thickness	0.1170	0.0257

Table 16.5 Gamma sensitivity results: Case 2.

Uncertain variable	Gamma sensitivity
Transmissivity T	0.6075
Aquifer (sand) compressibility β_{Ap}	0.0250
Clay compressibility β_{Cp}	0.5595
Pumping rate Q	0.5634

All deterministic and statistical input is identical to Case 1 except that the variables are assigned normal marginal distributions. Consider the results for a target subsidence of 12 m as typical. The probability of exceeding 12 m is calculated as 0.1424, which is slightly higher than when the variables were all considered lognormal. The gamma sensitivity measures have changed slightly, as can be seen from Table 16.5.

Notice that the transmissivity is now the most important variable (as indicated by the gamma sensitivity), whereas it had slightly lower gamma sensitivity value than the compressibility of the clay and the pumping rate Q when all variables were assumed lognormal.

Although the effect of assigning normal versus lognormal distributions is somewhat small in the cases investigated here, the choice of distribution may have much more influence when extreme-probability events are being analyzed (because we would be evaluating probabilities in the tail of the distribution).

16.4.3 Effect of Increasing Uncertainty on Selected Variables

Often the uncertainty on the variables affecting groundwater flow will be relatively high, because of measurement uncertainties, inherent variability of the natural environment, and even human inaccuracies in testing procedures and measurements. How does increased uncertainty for one or more of the variables influence the gamma sensitivity measures?

If the uncertainty for transmissivity T is increased by a factor of 5, the coefficient of variation for the variable becomes 0.50—a relatively high uncertainty. Consider the case where the target subsidence is 12.0 m. All other input is identical to Case 1. The probability of failure increases dramatically to 0.4335, which is to be expected given that the uncertainty on a very important variable in the problem has been increased dramatically. The sensitivity information changes as well, as can be seen from Table 16.6.

The relative importance of the transmissivity has increased because the uncertainty on that variable has increased. The compressibility of the clay and pumping rate are still important, as reflected by the gamma sensitivity values, but their relative importance has decreased. If the uncertainties on those two variables are increased such that their

Table 16.6 Gamma sensitivity results: Case 3.

Uncertain variable	Gamma sensitivity
Transmissivity T	0.9544
Aquifer (sand) compressibility β_{Ap}	0.0088
Clay compressibility β_{Cp}	0.2103
Pumping rate Q	0.2118

Table 16.7 Gamma sensitivity results: Case 4

Uncertain variable	Gamma sensitivity
Transmissivity T	0.6887
Aquifer (sand) compressibility β_{Ap}	0.0304
Clay compressibility β_{Cp}	0.7192
Pumping rate Q	0.0869

coefficients of variation are also 0.5, the relative magnitude of the gamma sensitivity measures return to about the same form as for the original case.

16.4.4 Large Uncertainty for Transmissivity and Compressibility

Often in engineering practice, the analyst must begin a project with little data and great uncertainty. For the example subsidence case considered here, it would be prudent to start a real-world analysis by considering the physical parameters such as transmissivity and compressibility as very uncertain. Coefficients of variation of 1.0 or greater are not uncommon for such geologic parameters as transmissivity; therefore, we shall use that value for the transmissivity and the compressibilities. The pumping rate Q can be controlled fairly well in most cases, so a coefficient of variation of 0.10 is perhaps appropriate for that variable.

For comparison, consider the case where the target subsidence is 12.0 m. The probability of subsidence exceeding 12.0 m is now 0.437, which is only slightly higher than the previous Case 3, where only the transmissivity uncertainty had been increased. The gamma sensitivity measures now have the values shown in Table 16.7.

This result indicates that, in general, a FORM probability analysis of the subsidence problem considered in this example is most sensitive to uncertainty associated with the transmissivity of the sand and the compressibility of the clay layer. This is not to imply that the pumping rate or the compressibility of the sand are not important in any way—they are very important deterministic parameters that must be included in the analysis. We emphasize that the utility of the gamma sensitivity information is in the identification of those variables whose *uncertainty* is very important to the probabilistic outcome.

16.5 CONCLUSIONS

The reliability algorithms such as FORM and SORM are extremely useful tools for engineers and scientists with respect to evaluating sensitivity and for calculating probabilities related to hazards such as subsidence. In this chapter, we have examined an application to a hypothetical case, but with realistic values for aquifer and clay unit characteristics, where the maximum subsidence over a one-year period is of concern. The gamma sensitivity measures are shown to be very helpful in identifying those parameters whose uncertainty is most important to the probabilistic outcome. The FORM analysis also can be used to provide sensitivity measures related to deterministic parameters in the design formulation. Using a program such as CALREL, the engineer or scientist may investigate a wide range of statistical input assumptions such as means, variance, correlation, and distribution type. Often there will be little actual statistical information available at the beginning of a project, and a reliability approach such as that discussed in this chapter can provide insight to guide further field and laboratory investigations. In the example considered in this chapter, the FORM sensitivity measures tell us that the important uncertain variables are the aquifer transmissivity T and the clay compressibility. The thickness of the clay unit was identified as an important deterministic variable. Given these results, the project manager could direct efforts to reduce the uncertainty on T and clay compressibility and to obtain a good definition on the thickness of the clay unit.

17

One-at-a-Time and Mini-Global Analyses for Characterizing Model Sensitivity in the Nonlinear Ozone Predictions from the US EPA Regional Acid Deposition Model (RADM)

Robin L. Dennis, Jeff R. Arnold and Gail S. Tonnesen

NOAA, USA

17.1 INTRODUCTION

The air quality system of the urban and regional troposphere is a complex network of multiple competing interactions within and among chemical and physical atmospheric processes. These processes include emissions of primary pollutants and of secondary pollutant precursors, heterogeneous and homogeneous chemical production and loss, meteorological transport by advection and diffusion, and wet and dry deposition. This system, particularly as related to ozone (O_3), is known to be nonlinear with major autocatalytic cycles (Seinfeld *et al.*, 1998). Numerical photochemical air quality models (AQMs) such as the US Environmental Protection Agency's (USEPA) Regional Acid Deposition Model (RADM) (Chang *et al.*, 1987, 1990) represent the chemical and physical dynamics of the polluted troposphere by describing mathematically the species conservation equation using a set of coupled nonlinear partial differential equations (Russell and Dennis, 2000).

AQMs such as RADM have been developed for the purposes of better understanding tropospheric dynamics and assisting in planning for pollutant management. Both

Sensitivity Analysis. Edited by A. Saltelli *et al.*

purposes rely on model predictions of future O_3 concentrations ($[O_3]$) that might result from possible reductions in emissions of nitrogen oxides ($NO_x = NO + NO_2$) and/or volatile organic compounds (VOC), the two precursors necessary for O_3 build-up in the troposphere. Sensitivity analysis is an important tool for understanding the behavior of these complex process models, and, more specifically, for characterizing the role that input and parameter uncertainties can exercise singly and together in various combinations. These uncertainties increase the risk of errors in model guidance when the models are used to assist with policy-making for emissions controls. The scale of this air quality policy-making is quite large, so the cost of bad guidance can be very high: it is estimated that the annual cost of compliance with regulations for managing urban and regional O_3 in the USA exceeds US$1 billion (USEPA, 1997b). Because of the large economic and social costs of decisions affecting O_3 control, we wish to avoid potential errors where possible by using AQMs to provide a realistic simulation of future conditions and an accurate appraisal of the type and the amount of emissions control necessary to meet mandated air quality goals. Thus, evaluating the sensitivity of AQM predictions to uncertainties and dissaggregating the sensitivities into their sources helps make the operation of these physically based process models more transparent, and so may increase confidence in their use (Helton and Burmaster, 1996; Saltelli and Scott, 1997; see also Chapters 21 and 22).

Given these important science and policy issues, the AQM community has been acutely interested in understanding the effects of uncertainty perturbations on model predictions of O_3, and that interest is increasing as newer models and better computational resources are developed and applied. Sensitivity analysis on model resultants has traditionally been a part of both AQM development and performance evaluation as an aid to understanding the effects of input uncertainties on model resultant concentrations. However, the best use of sensitivity analysis in developing this understanding is not always obvious owing to anticipated dependencies in the models that derive from their strongly nonlinear chemical mechanisms. Hence, key methodological issues for applying sensitivity analysis to AQMs concern how it is that different sensitivity procedures and measures can lead to different interpretations of model behavior, and how to determine which procedures and measures are best suited for describing the models' sensitivity so that work can be directed to the most significant tractable uncertainty.

The chief regulatory use of AQMs is to predict for the photochemical regime in a particular region or urban area the potential change in $[O_3]$ resulting from a given change in either one or both of the emissions precursors VOC and NO_x. We define this quantity as the control response $\delta O_3/\delta_{\mathrm{Emissions}}$ where $\delta_{\mathrm{Emissions}}$ is often referred to as the 'control strategy', i.e., the type and amount of precursor control required to reduce O_3 to acceptable levels. The model's most important sensitivity, correspondingly, is how this control response is changed by uncertain input and parameter perturbations, and not simply how the resultant $[O_3]$ might be changed by uncertainties in a base case simulation without controls. We define that latter quantity as ΔO_3, or the total change in $[O_3]$ due to the uncertainty perturbation without controls.

However, producing model simulations with the range of input uncertainties necessary to fully characterize the $\delta O_3/\delta_{\mathrm{Emissions}}$ response for several uncertainties and levels of emissions control has been computationally expensive with these large models, and can greatly complicate the interpretation of results. For these reasons, AQM sensitivity analyses have been focused on the first step, characterizing ΔO_3, and have used the working assumption that $\delta O_3/\delta_{\mathrm{Emissions}}$ can be correctly inferred from that ΔO_3. We tested this assumption in previous sensitivity work (Dennis *et al.*, 1999) with a version of RADM that we instrumented to produce and save for post-processing and analysis several specific process characteristics (details of the analysis techniques are given in Tonnesen and Jeffries, 1994; Jeffries and Tonnesen, 1994; Tonnesen and Dennis, 2000). That sensitivity work showed that analyzing only the model resultants (e.g., changes in state variable time series or final concentrations of predicted species such as O_3) due to input perturbations gives an incomplete or distorted depiction of model behavior for the control strategy response. Briefly put, this is because model resultants such as $[O_3]$ are nonunique solutions to the equations representing the photochemical system's complex dynamics. That is to say, the complex interaction among the model's multiple chemical pathways as concentrations change over time, and between the chemistry and the meteorology represented in the model, provides any number of different ways to produce the same resultant species concentration. We concluded that it cannot be safely assumed that a measure of total change in a resultant concentration such as ΔO_3 from a perturbed input variable properly reflects the change in the photochemical system state that is of primary interest, $\delta O_3/\delta_{\mathrm{Emissions}}$, but rather that the $\delta O_3/\delta_{\mathrm{Emissions}}$ sensitivity must be evaluated directly.

Evaluating the $\delta O_3/\delta_{\mathrm{Emissions}}$ sensitivity directly, however, will require careful thought about the very important methodological issue of the degree to which individual sources of uncertainty in the model might be correctly added together to produce an overall model uncertainty in the $\delta O_3/\delta_{\mathrm{Emissions}}$ response. That is to say, we wish to know the strength of the anticipated dependencies in the model's response, and how best to characterize them with different sensitivity measures.

Confronting this methodological issue yields at least two interesting aspects of a model's sensitivity response that can be tested with a carefully designed analysis. One aspect is the common presumption in the AQM community that perturbations from uncertain inputs will be manifested as smaller sensitivities in the $\delta O_3/\delta_{\mathrm{Emissions}}$ response than in the ΔO_3 one. This belief about the full three-dimensional models seems to have been engendered and supported by calculations made using only the chemical mechanisms from AQMs in simple box models that suggest that the magnitude of $\delta O_3/\delta_{\mathrm{Emissions}}$ in the box model for chemistry will always be smaller than the magnitude of ΔO_3 for the same uncertainty (Gao *et al.*, 1995, 1996). But because the $\delta O_3/\delta_{\mathrm{Emissions}}$ response has not been systematically explored in a sensitivity series using a full AQM, this presumption has remained untested. A second aspect for testing concerns whether the sources of uncertainty for the $\delta O_3/\delta_{\mathrm{Emissions}}$ response are the same as those for ΔO_3. Knowing whether the sources are the same is important for establishing confidence in the model's guidance by better characterizing the model's relevant response for use in predicting O_3 change for a specified control strategy.

This chapter reports recent results from our work testing these aspects of AQM sensitivity using one model and a small set of uncertainty perturbations. These results provide data about the model's sensitivity response in the effort to better inform selection of the sensitivity methods best suited for developing an understanding of the model's behavior. In this way, the results address the key methodological issue that we defined just above. In addition, our research effort is aimed at exploring the behavior of AQMs generally and building a better understanding for using them in environmental policy-making at the fairly fine scale of the model's O_3 response to these uncertainties for these cases. We believe our results to be robust for this model and these sensitivities; however, because of the size and complexity of AQMs (requiring extensive computational resources) and the variation in response from point to point on the system's response surface (requiring that different physical sites and different days be treated separately and not concatenated), we cannot address with these results the larger-scale question of the overall uncertainty in model predictions from uncertain inputs and parameters.

17.2 METHODS AND EXPERIMENTAL DESIGN

Analyzing sources of uncertainty in the model results and testing the model system's degree of nonlinearity require that different sorts of sensitivity analysis be performed so that the effects of each perturbation and the effects of all changes simulated together can be assessed separately. Hence, we have used a mixed methodological approach wherein we simulate the uncertainties one-at-a-time (OAT) and then again together in various combinations including all-together in a global sensitivity simulation. Moreover, all these simulations are run multiple times, once for a base case series without controls, and then again for the emissions control case series.

Our experimental scheme is roughly as follows:

1. Simulated for the base case series, i.e., without projected emissions reductions in the NO_x and VOC precursors:
   - a best-estimate run with all model inputs unperturbed in RADM's standard configuration;
   - three model inputs perturbed separately, in OAT sensitivity runs;
   - three model inputs perturbed together in mini-global (MG) sensitivity runs.
2. Simulated for the control case series, i.e., with spatially uniform reductions in NO_x and VOC in three combinations: a 50% NO_x reduction series, a 50% VOC reduction series, and a series with 50% reductions in both NO_x and VOC:
   - a best-estimate run with all model inputs unperturbed in RADM's standard configuration;
   - three model inputs perturbed separately, in OAT sensitivity runs;
   - three model inputs perturbed together in mini-global (MG) sensitivity runs.

For the analysis presented here we evaluated RADM's ΔO_3 response through examination of the base case series and the model's $\delta O_3/\delta_{\text{Emissions}}$ response by differencing results from the base case and control case series. We describe the reasons for this approach and the basis for our experimental design below.

17.2.1 One-at-a-Time and Global Analysis Methods

Sensitivity analysis restricted to a ΔO_3 response to selected perturbations in individual input uncertainties simulated OAT has long been an officially recommended part of USEPA evaluations of AQMs used in regional and urban policy-making (USEPA, 1991). More recently, larger-scale sensitivity studies using combinations of input uncertainties varied together in a global analysis are being conducted as new model formulations and more computational power make it possible to use even Monte Carlo-type methods on some uncertainties associated with AQMs (Hanna *et al.*, 1998). Global analysis of combined input uncertainties such as the Monte Carlo variance is thought to be a more appropriate method of sensitivity analysis than OAT for complex models such as photochemical AQMs where system nonlinearities may be important (Saltelli and Hjorth, 1995). This is because a large-scale global or even a mini-global (MG) analysis where perturbations in a small number of uncertainties are simulated together can incorporate system feedbacks and interactions among inputs or driving functions in the model, leading to dependences and nonlinear system response, while an OAT analysis cannot. However, such nonlinear system responses can complicate the interpretation of results from global or MG sensitivity analyses by obscuring important changes in a model's internal processes. For this reason, the interpretation of global and MG sensitivity results for these models will be conditional on the particular parameter set and uncertainty range evaluated in the sensitivity simulations, and most likely cannot be extended to an understanding of the sources of the changes revealed by the sensitivity results. Conversely, where the dependences and degree of nonlinearity are small, it can be possible to learn the sources for change in the system by performing a directed series of OAT sensitivity simulations with the model and analyzing those results in the light of results from the global or MG simulation. With either an OAT or a global analysis, however, the inputs to be perturbed should be chosen to provide the maximum number of effectively different variations within the uncertainty ranges appropriate for the model and the particular sensitivity study.

We note a tendency in some applications to treat OAT and global analysis as exclusive or competing methods of evaluating model sensitivities. However, we believe that they are strongly complementary and have employed them together in a mixed approach to better learn how RADM represents the change of state of particular photochemical systems in response to selected uncertainty perturbations. Using both OAT and MG sensitivity simulations in a mixed approach is, in fact, the only means we know to evaluate dependences in the model by testing whether the effects of perturbations in individual uncertain inputs are roughly additive for ΔO_3, for $\delta O_3/\delta_{Emissions}$, or for both, or whether the known nonlinearities in the photochemical system dominate the combined response in the MG. As such, we believe that it is a useful approach for exploring the methodological issue in sensitivity testing that we set out above. Furthermore, our results indicate to us that a mixed approach is obligatory in all sensitivity studies for understanding the model's behavior and correctly describing its sensitivity response.

17.2.2 The O_3 Response Surface and VOC–NO_x–OH Photochemistry

Since the relevant sensitivity for photochemical AQMs is not ΔO_3 but rather the change in the photochemical system's state as precursors and products change, tracking that change requires knowing the initial location, and the degree and direction of the system's change on its response surface. An example O_3 response surface is shown in Figure 17.1. The shape of the O_3 response surface is in fact determined by nonlinearities in O_3 production $P(O_3)$ which changes over time as the NO_x and VOC levels change throughout the day and from one location to another. This is the nonlinearity that leads to the expectation in the AQM community of strong dependences between input uncertainties that may alter $P(O_3)$ and, hence, change the relevant sensitivities, too. The O_3 response will also be modified by perturbations in the model other than in NO_x and VOC emission levels, such as by changes in parameters of the meteorology driver or the chemical mechanism. Uncertainties of these sorts may, in fact, alter the spacing and shape of the response surface contour lines, affecting both the ΔO_3 and the $\delta O_3/\delta_{Emissions}$ sensitivity responses.

Given the importance of the photochemical response surface for understanding our AQM sensitivity results, it will be useful to consider it in a little more detail. We provide some detail here, together with a very brief overview of the chemistry of O_3 formation in the troposphere that helps explain several important features of the O_3 response surface.

Figure 17.1 shows the response surface of peak $[O_3]$ to various levels of VOC and NO_x emissions from a simulation for Atlanta, GA using a trajectory model (a description of the model and its set-up can be found in Tonnesen and Dennis, 2000). The isopleth lines of the

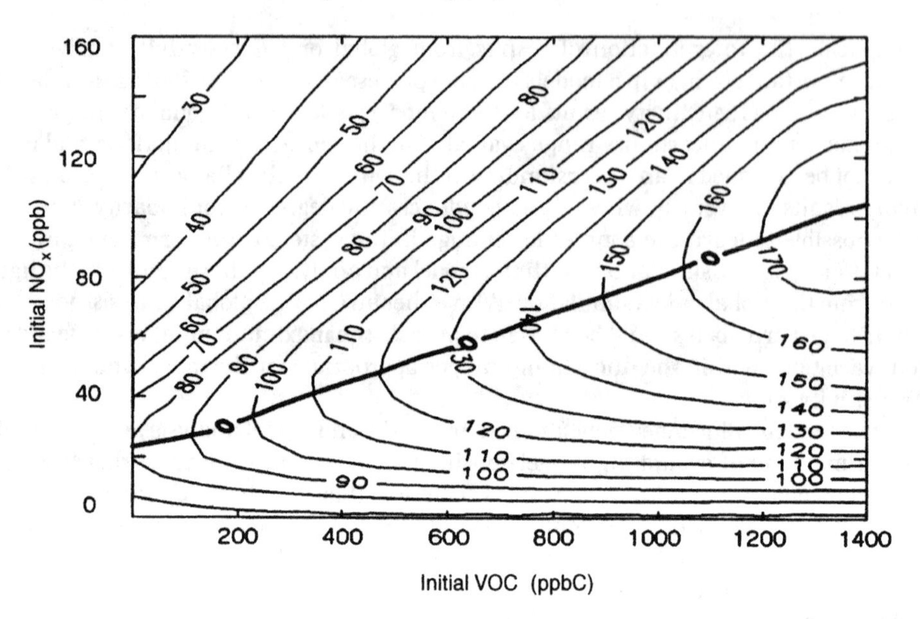

Figure 17.1 O_3 response surface from box model predictions with varying levels of NO_x and VOC precursors. Isopleths of peak $[O_3]$ (ppb) and the ridgeline of maximum $[O_3]$ across all combinations of the precursors are shown.

surface are derived by fitting contours to the peak model-predicted $[O_3]$ in multiple simulations using different VOC and NO_x emissions. Note that many different combinations of NO_x and VOC can produce the same $[O_3]$: for example, 20 parts per billion (ppb) NO_x and 400 ppb carbon VOC produce 100 ppb O_3 in this simulation, but 80 ppb NO_x and 450 ppb carbon VOC also produce 100 ppb O_3. The heavy line drawn across the ridges of the surface contour lines is the ridgeline of maximum $[O_3]$ and divides the response surface into two domains. In these two domains, $P(O_3)$ is limited in different ways: below the ridgeline, $P(O_3)$ is limited by NO_x availability, while above the ridgeline, radical availability from VOC limits $P(O_3)$.

The structure of the O_3 response surface as seen in Figure 17.1 depicts explicitly how $P(O_3)$ in the troposphere varies with VOC and NO_x concentrations. But, in addition to these two necessary precursors, there are a few key radical species that initiate and propagate $P(O_3)$ by attacking either VOC and NO_x preferentially, since concentrations of all species in the pollutant mix change throughout the day and from one site to another. A brief description of the general nature of this chemistry will likely aid in an understanding of our results, so we include that here.

$P(O_3)$ is initiated by creation of the hydroxyl radical (OH) through photolysis of O_3 or nitrous acid (HONO) as follows:

$$O_3 + h\nu \rightarrow O(^1D) + O_2 \quad \text{(R1)}$$

$$O(^1D) + H_2O \rightarrow 2\,OH \quad \text{(R2)}$$

$$HONO + h\nu \rightarrow OH + NO \quad \text{(R3)}$$

Other, multistep pathways such as the photolysis of formaldehyde (HCHO) and its subsequent reactions also exist for creating OH through intermediate compounds from species present in urban settings.

Once created, the OH radical can react with carbon monoxide (CO) or VOC to produce the peroxy radicals HO_2 and RO_2 as in (R4). At high [NO], peroxy radicals react with NO to split an oxygen–oxygen bond, thereby creating an odd oxygen (O_x) in the form of NO_2, as in (R5):

$$OH + CO + O_2 \rightarrow CO_2 + HO_2 \qquad (R4)$$

$$HO_2 + NO \rightarrow OH + NO_2 \qquad (R5)$$

Note that (R5) also recreates the original OH radical, which is then available for propagation through additional cycles. A pseudo-photostationary-state equilibrium exists between NO_x and O_3 such that at high [NO], the O_x remains in the form of NO_2 and at low [NO] the O_x may be converted to O_3 by

$$NO_2 + h\nu \rightarrow NO + (O^3P) \qquad (R6)$$

$$(O^3P) + O_2 \rightarrow O_3 \qquad (R7)$$

Note here that (R6) recreates the original NO, making it, too, available for propagation through additional cycles.

$P(O_3)$ may proceed by (R1)–(R7) in an autocatalytic cycle without terminating either OH or NO_x. At conditions with high $[NO_2]$, this photochemical cycle can be stopped as OH radicals preferentially attack NO_2 to produce relatively inert nitric acid (HNO_3) as in (R8):

$$OH + NO_2 \rightarrow HNO_3 \qquad (R8)$$

thereby effectively removing OH and NO_2 from the reaction system in one step. The O_3–NO_x equilibrium causes [NO] to decrease as $[O_3]$ increases or as NO_2 is terminated by (R8) and other reactions that convert NO_x to other nitrogen-containing compounds that no longer react on time scales relevant to the urban setting. At low levels of NO, the photochemical system will also be stopped when peroxy radicals preferentially self-terminate to produce peroxides, as (R9) depicts for HO_2:

$$HO_2 + HO_2 \rightarrow H_2O_2 \qquad (R9)$$

instead of recreating the OH by (R5). These termination reactions end the propagation of O_x through the radical–NO_x cycles and shut-down local chemical $P(O_3)$.

With this basic understanding, the O_3 ridgeline on the response surface in Figure 17.1 can now be explained in terms of the VOC–NO_x–OH cycles as the region of maximum OH production and propagation. As such, this concentration ridgeline separates the response surface into two domains and marks an area of subtle changes in the response of $[O_3]$ to changed NO_x and VOC levels, i.e., the $\delta O_3/\delta_{\mathrm{Emissions}}$ response, a significant consideration for policy-makers planning emissions reductions of these precursors. Figure 17.1 shows that as NO_x and VOC levels are changed, the photochemical system will change its state and move over this response surface, often crossing the ridgeline into the other domain. This change of state as the system moves over the response surface has important implications for understanding and using an AQM, since the photochemical domains above and below the ridgeline are substantially different. Systems with a low ratio of VOC/NO_x emissions are found in the domain above the ridgeline where $P(O_3)$ is limited by the availability of radicals. Under conditions of high $[NO_x]$ in this radical-limited domain, NO_2 reacts with OH and terminates to HNO_3, as in (R8) above, removing both OH and NO_2, which limits $P(O_3)$ by reducing the production of OH. Furthermore, excess NO in this high-NO_x region titrates O_3 back to NO_2 thereby reducing O_3 photolysis as a source of new OH for additional cycling. For photochemical systems in these radical-limited conditions, the efficiency of O_3 production per NO_x terminated is low, and $P(O_3)$ is more responsive to reductions in VOC than in NO_x. In fact, as Figure 17.1 makes clear, reducing NO_x emissions without concomitant VOC reductions for photochemical systems in this domain will cause increases in $[O_3]$.

Conditions below the ridgeline with high VOC/NO_x emissions ratios, however, present a very different case. There, [NO] is relatively low, allowing the peroxy radicals to self-terminate as in (R10) above. This reduces the number of times that OH can be propagated and so lowers the efficiency of $P(O_x)$ per radical. Termination of NO_2 by OH as in (R8) is also reduced because the higher VOC/NO_2 ratio causes NO_2 to compete less effectively for the available OH radicals. Hence, in these cases, although the efficiency of $P(O_3)$ per NO_x terminated is high, less NO_x is available for reaction, resulting in lower $P(O_3)$ and a lower final $[O_3]$. In contrast to the radical-limited domain above the ridgeline, in this NO_x-limited domain, $P(O_3)$ is more responsive to reductions in emissions of NO_x than of VOC. However, although VOC emissions reductions are less effective in reducing $[O_3]$ for photochemical systems in this domain of the response surface, in no case will reductions in either precursor increase the $[O_3]$ there.

17.2.3 Description of RADM

We carried out our series of OAT and MG sensitivity experiments using our RADM version 2.6.2 both at the relatively low grid cell resolution of 80 km on a side, and at the higher resolution of 20 km on a side, one-way nested in the 80 km model. For each sensitivity run at both resolutions, the model simulated the atmospheric conditions over the eastern USA for the time period 19 July to 12 August 1988. RADM is a full three-dimensional AQM that incorporates emissions, advection, vertical eddy mixing, dry and wet deposition, land-use specific air-surface exchange, and complex cloud effects including vertical redistribution, aqueous chemistry, scavenging loss, and radiative effects. Anthropogenic emissions for the model are taken in part from measured 1988 emissions and in part from the USEPA 1991 emissions inventories adjusted back to 1988 levels through changed activity and population levels. Emissions of biogenics are taken from the second-generation Biogenic Emissions Inventory System (BEIS2). Photochemical production and loss is accomplished with the RADM2 chemical mechanism (Stockwell *et al.*, 1990), which has 162 gas-phase reactions involving 60 reactive gas-phase species, 41 of which are advected and tracked through the model in parts per million (ppm) volume mixing ratios. Dry deposition of species is computed with the resistance-in-series method. Clear-sky photolysis rates are calculated by the delta–Eddington method, and then are attenuated by the model-generated clouds. Meteorology in RADM is driven by the fifth-generation Penn State/NCAR primitive equation mesoscale meteorological model, MM5 (Grell *et al.*, 1994), with results interpreted by the RADM Meteorology–Chemistry Interface Program (MCIP) to derive the various surface and planetary boundary layer (PBL) parameters required by the AQM.

The 80 km version covers the region from east of central Texas in the USA to James Bay in Canada to the southern tip of Florida with 35×38 cells in the horizontal dimension; the 20 km domain covers the northeastern USA from Kentucky to Maine with 61×69 horizontal cells. Subgrid-scale turbulence in the PBL at both grid resolutions is described using the local eddy diffusivities K_z based on boundary-layer scaling theory (Byun, 1990, 1991). The simulations at each grid resolution were made with 21 vertical layers extending from the surface to the top of the free troposphere. Layer thickness increases with height above the surface from the first layer, which is nominally 35 m, so that there are 14 layers up 2400 m, the height of a typical daytime mixed layer. RADM has on the order of 50 hourly inputs, with time-steps in the chemistry on the order of minutes or less, and advective time-steps at 10 min in the 80 km version and 5 min in the 20 km version. Outcome variables are generally output each hour.

This is a large, complex, system of models that together make-up the physical–chemical process-based AQM referred to as 'RADM'. The RADM model system is computationally quite intensive: a single 36 h simulation at the 20 km resolution typically requires 10 h on a Cray Research T3D supercomputer using 16 processors. Executing the simulations that support the work reported here — the best-estimate and all uncertainty-perturbed simulations for the 20 km experiments in both the uncontrolled base case series and emissions reductions series — required in excess of 3300 CPU hours on that computer.

17.2.4 Design of the Uncertainty Simulations

Uncertainties in key elements of the emissions and meteorology inputs to AQMs are in the range of 50–100%, with some areas of emissions uncertainty even higher (Russell and Dennis, 2000). Uncertainties in the chemical mechanisms are generally thought to be somewhat lower (Russell and Dennis, 2000), but can vary through a range of 30% or more as new techniques are applied to re-measure reaction rate constants and yields. OAT sensitivity analyses have traditionally been used with full three-dimensional AQMs in efforts to describe the effects of these uncertainties individually on peak predicted $[O_3]$. As noted above, however, $\delta O_3/\delta_{Emissions}$ and not ΔO_3 is the important sensitivity for understanding and using these models, since ΔO_3 cannot denote change of the system state on the response surface, and is not the measure most relevant for actual use of the models in policy-making.

For the work reported here we designed a series of sensitivity experiments to characterize the degree of nonlinearity in RADM's sensitivity response and to explore the relation between the ΔO_3 and $\delta O_3/\delta_{Emissions}$ measures of that response. In these runs, three model inputs were perturbed to the limit of the known range of their uncertainty. Those inputs are a key meteorological parameter, (mixing height), the anthropogenic emissions inputs for VOC and NO_x, and the rate constant for one important photochemical reaction. We made the following changes in RADM inputs to reflect these three uncertainties: a 50% reduction ($z50$) and 50% increase ($z150$) in mixing height from the current best estimate; a 50% increase in NO_x and VOC emissions (E_{150}), and a 20% reduction in the rate constant for (R8), $OH + NO_2 \rightarrow HNO_3$ ($R8_{New}$). This reaction is considered of central importance in the photochemistry (Gao *et al.*, 1995, 1996), since it directly affects the propagation of OH and availability of NO_2 as described in Section 17.2.2, and its expression has recently been reassessed and revised downward (Donahue *et al.*, 1997).

We have run the model in its standard configuration with current best estimates for all inputs and parameters, and with the three selected inputs perturbed in nearly every permutation of uncertainties simulated OAT, in pairs, and in MG combinations of all three. In addition, we have run all these simulations for two series of sensitivity cases. The first series is the base case uncontrolled simulation, and the second series uses realistic control estimates of 50% future reductions in VOC and NO_x emissions as a means to test the representativeness of the $\delta O_3/\delta_{Emissions}$ measure. We reiterate that these VOC and NO_x emissions reductions in the control case are not sensitivity parameters for this series of experiments; rather, they provide the benchmark information about how the $z50$, E_{150}, and $R8_{New}$ perturbations shift the photochemical system on its response surface, and demonstrate the actual $\delta O_3/\delta_{Emissions}$ response used for our comparisons.

17.3 RESULTS

Although we produced RADM output at both the 80 km and 20 km resolutions for every hour on each of the 25 simulation days in all cells of the model domains, for brevity and clarity the results we report here are concentrated on five days, 28 July to 1 August 1988, and at only the higher, 20 km, grid cell resolution. To focus our presentation further, we selected four sites in a coherent air mass or photochemical regime that began in the urban core of New York City, and continued downwind northeasterly from there up along the eastern seaboard of the USA. The four sites we have used begin outside the urban New York City core: the first (NY1) is over New York's central Long Island, the second (NY2) is two cells east of there at the eastern edge of Long Island, the third (NY3) is north and east of NY2 across Long Island Sound over inland Connecticut, and the fourth (NY4) is two cells east of NY3 in the southwest corner of Rhode Island.

The sensitivity results reported here are drawn from our simulations using the E_{150} and $R8_{New}$ perturbations, but only using the reduced mixing height (z50) not the increase (z150), and only for the 50% NO_x control strategy (δE_{NO_x}) and not the VOC controls. We have further restricted our analysis here to comparing the model's standard configuration best estimate run against the $z50$, E_{150}, and $R8_{New}$ perturbations simulated OAT and treated separately, simulated OAT but summed together, and simulated together in one realization of a MG; and we have made these comparisons for both the uncontrolled base case series and the reduced-emissions control case series. Our intent in varying the three particular inputs we selected in this way was to increase $P(O_3)$ in the perturbed runs over the best-estimate result chiefly by altering the availability of NO_x to the photochemical system using a manageably small number of changes. This was done successfully. Figure 17.2 shows that each uncertainty simulated OAT increased peak predicted $[O_3]$ over the best-estimate result for nearly every day of the time series at NY2. At the three other sites in our analysis (not shown here), similar increases were evident. In fact, we chose these particular uncertain inputs and perturbation levels for this relatively large and unidirectional effect on O_3, anticipating that the contribution from each uncertainty would thus be clearly expressed in the MG case, allowing us to more easily evaluate any dependences or deviation from additivity in the MG simulation.

All results using the RADM peak $[O_3]$ predictions that we describe in the text and show in the figures are made with a three-hour average peak O_3 value, even though the model $[O_3]$ output is hourly. We calculated a three-hour time average for several reasons. First, the model's sensitivity response varies greatly over the diurnal cycle as the photochemistry begins, increases, peaks, and then wanes with diurnal insolation. To clearly delineate system behavior for this sensitivity work, we required a relatively short time interval during that part of the cycle when photochemical production was actively driving the system, and desired that that interval capture the system's maximum response. Work with these uncertainties simulated OAT showed that the maximum O_3 sensitivity response measured as change in the ppb mixing ratio occurs with the daily $[O_3]$ peak, as can be seen in Figure 17.2. Also, we elected to average over the highest three hours of daily $[O_3]$ rather than use the one-hour model prediction directly in order to make the model-predicted value slightly more robust for our comparisons. Furthermore, model predictions of peak O_3 values have traditionally been of interest to the AQM community, since USEPA ambient air quality standards are based on peak concentrations. The US standard for $[O_3]$, for example, has been a statistical interpretation of a one hour average, but may soon be changing in statistical form to an eight-hour daily average at a lower concentration (USEPA, 1997a). Given this very current policy interest in time-averaged peaks, we performed our analysis for an eight-hour

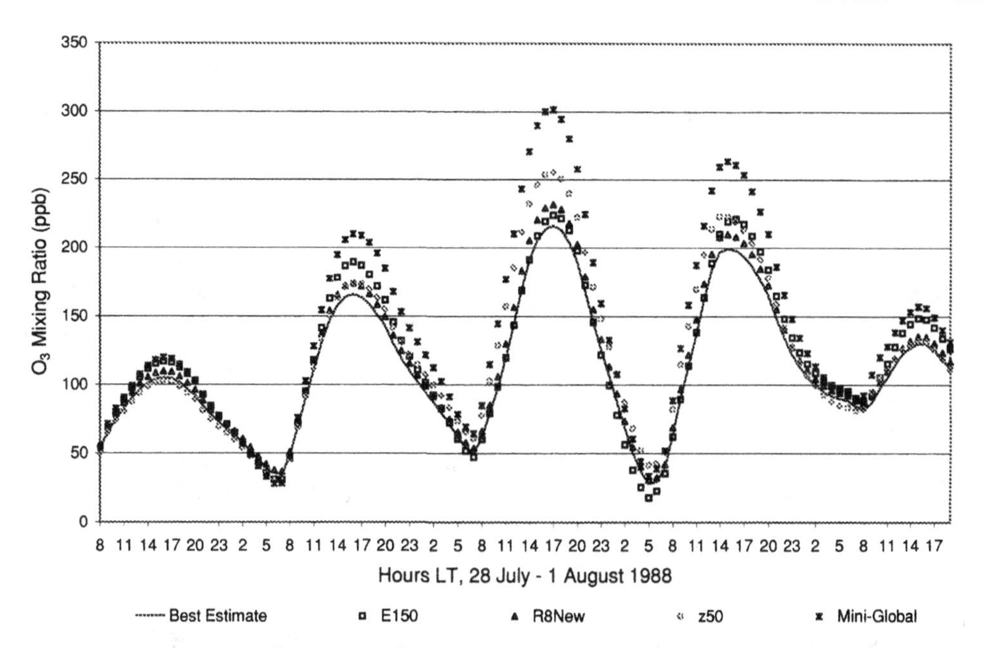

Figure 17.2 Time series of RADM-predicted hourly [O_3] (ppb) in the 20 km cell over eastern Long Island, NY (NY2) for all days (28 July to 1 August 1988) in the base case series. The RADM best-estimate simulation and effects of each uncertainty perturbation simulated OAT and together in the MG simulation are depicted.

average as well as for the three-hour average that we report here; however, because the eight-hour time period incorporates more diurnal variability in the overall sensitivity response, we have analyzed the three-hour peak first in an effort to establish a baseline for the maximum sensitivity of the system. Results for the eight-hour average will be included in future reports of our work with these uncertainties and sensitivity measures.

17.3.1 OAT and MG analyses for interpreting model behavior

Reliability of ΔO_3 OAT sensitivities

We evaluated the degree of dependency among the three uncertainties in the model's base case ΔO_3 sensitivity as the deviation from additivity when peak O_3 predictions from the MG perturbations are compared against predictions from the sum of the three OAT perturbations (OAT Sum). Deviations from a strict additivity of the OAT uncertainties would then be represented as large differences between the MG and OAT Sum predictions. Table 17.1 shows that for the uncontrolled base case series, the mean deviation from additivity using the three-hour averaged peak O_3 value for all four sites in the NY air mass over all five days is approximately 6 ppb or 8% with an absolute value deviation of approximately 6 ppb or 12%. The difference MG – OAT Sum is positive at all four sites for all five days with one exception, and most often the difference is less than 10 ppb. Using time series results for the uncertainties simulated OAT and MG at two sites, Figure 17.3 shows that this result is obtained irrespective of whether the effect of the uncertainty perturbation is positive or negative on the change in O_3. Closer evaluation of these results revealed a tendency for this deviation to increase with increasing model-predicted peak O_3. Interestingly, this tendency itself appears strongly nonlinear when represented in absolute units (ppb), and Figure 17.4 shows

Table 17.1 ΔO_3 from best estimate for the base case series.

Site	Day	Best estimate $[O_3]$ (ppb)	E_{150} (ppb)	$R8_{New}$ (ppb)	$z50$ (ppb)	MG (ppb)	OAT Sum (ppb)	MG – OAT Sum (ppb)	MG – OAT Sum (% of MG)
	28 July	113.5	4.2	7.3	−8.0	2.6	3.5	−0.9	−34
	29 July	105.1	4.8	10.3	12.6	32.0	27.7	4.3	14
NY1	30 July	184.4	−12.3	17.0	51.9	71.8	56.6	15.2	21
	31 July	179.0	1.0	10.9	25.8	43.6	37.7	5.9	14
	1 Aug	145.1	12.2	7.3	4.8	25.9	24.2	1.7	6
	28 July	104.4	11.7	4.7	−2.3	14.3	14.1	0.2	1
	29 July	164.4	23.4	8.0	8.5	43.7	39.8	3.9	9
NY2	30 July	214.2	7.3	15.3	39.0	84.6	61.6	23.0	27
	31 July	198.1	18.7	10.5	23.4	63.5	52.7	10.9	17
	1 Aug	128.9	18.0	5.1	1.9	26.5	25.0	1.5	6
	28 July	109.7	15.9	4.3	0.7	21.6	20.9	0.7	3
	29 July	140.4	22.6	5.8	10.5	42.5	38.9	3.6	8
NY3	30 July	174.4	25.3	8.4	31.5	75.8	65.2	10.7	14
	31 July	153.8	24.0	7.0	25.6	64.9	56.6	8.3	13
	1 Aug	119.0	18.3	4.3	4.0	28.0	26.5	1.5	5
	28 July	104.6	14.3	3.6	−0.7	17.9	17.3	0.6	4
	29 July	150.8	24.4	5.7	4.7	37.3	34.8	2.6	7
NY4	30 July	182.6	30.4	8.7	26.2	74.6	65.3	9.3	12
	31 July	193.3	31.3	8.7	22.7	71.4	62.7	8.6	12
	1 Aug	122.3	17.5	4.2	3.1	26.3	24.8	1.6	6

Summary statistics: all sites all days

Measure	Best estimate $[O_3]$ (ppb)	E_{150} (ppb)	$R8_{New}$ (ppb)	$z50$ (ppb)	MG (ppb)	OAT Sum (ppb)	MG – OAT Sum (ppb)	MG – OAT Sum (% of MG)
Mean	151.64	15.65	7.84	14.29	43.43	37.78	5.65	8.3
Median	150.45	17.72	7.29	9.49	39.93	36.22	3.75	8.7
Standard error	7.53	2.40	0.81	3.51	5.41	4.24	1.34	2.6
Standard deviation	33.68	10.72	3.61	15.69	24.17	18.96	5.99	11.8

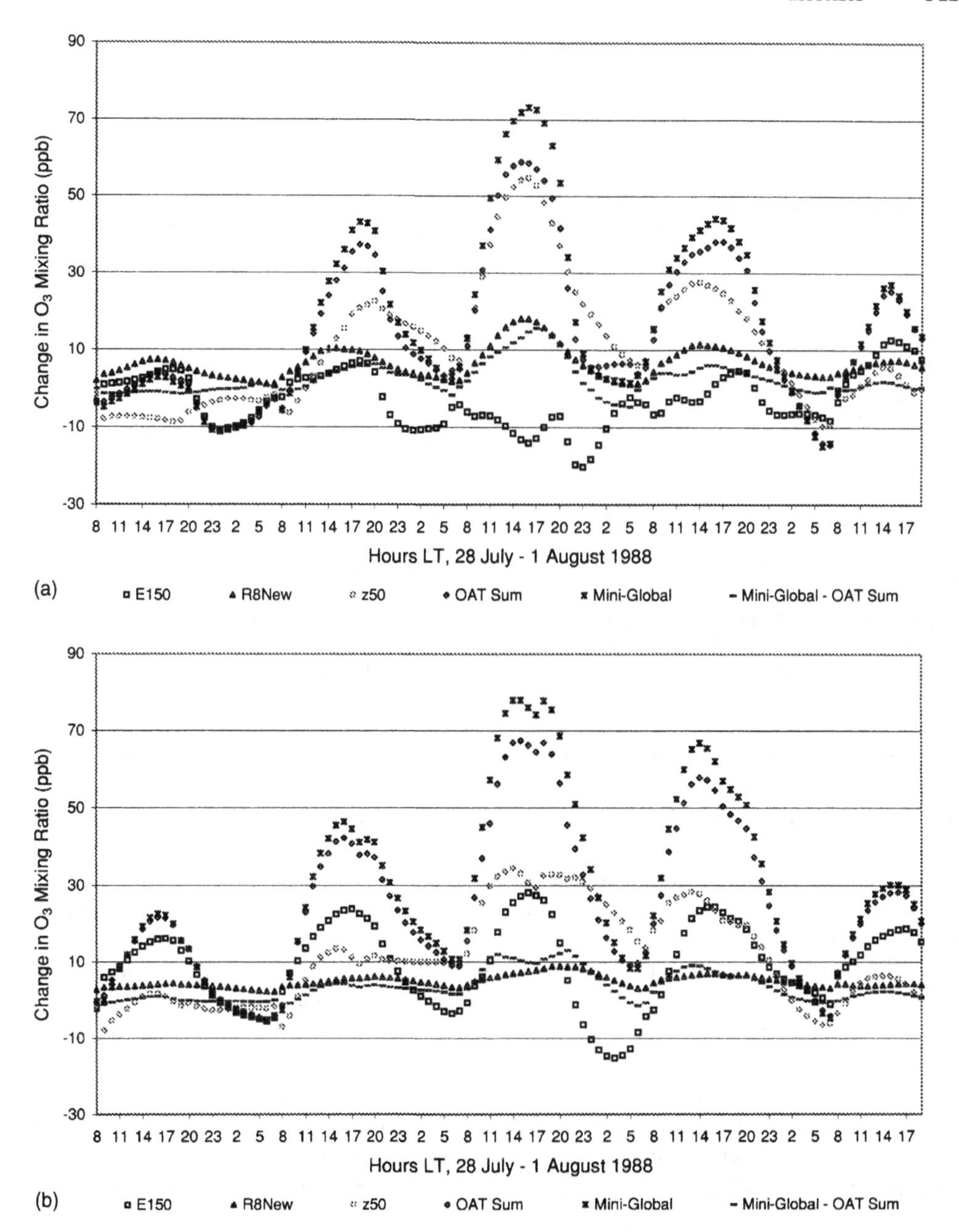

Figure 17.3 Time series of the changes in RADM-predicted hourly $[O_3]$ (ppb) in the 20 km cells (a) over central Long Island, (NY1) and (b) over south-central Connecticut (NY3) from the base case series due to each uncertainty perturbation simulated OAT, all perturbations simulated together in the MG, and the difference MG − OAT Sum.

proportionately larger deviations for higher base O_3 levels. But when the relation is expressed as the relative percent of the MG result (not shown here), it appears more nearly linear in base $[O_3]$.

Overall, the deviation from additivity in the base case series appears small, as Table 17.1 and Figure 17.3 illustrate, and seems explicable in terms of $P(O_3)$ changes in the coherent urban New York air mass that we analyzed. The low degree of nonlinearity that we observed

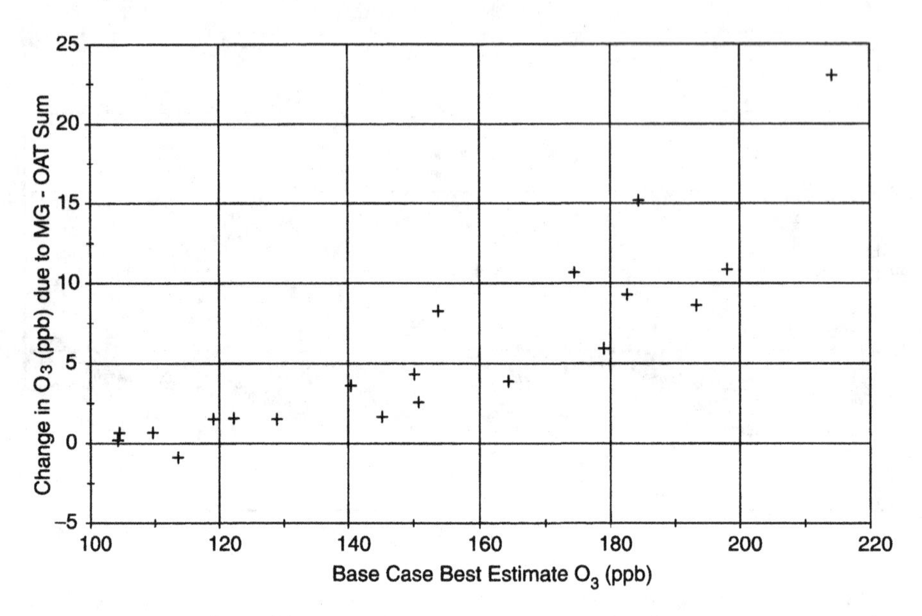

Figure 17.4 Deviation from additivity in the three-hour averaged peak RADM-predicted $[O_3]$ (ppb) in the base case series for all days and sites due to dependences among the three uncertainty perturbations evaluated as the difference MG − OAT Sum and displayed as a function of the base case best estimate $[O_3]$.

was unexpected, since the chemistry of the model is known to be highly nonlinear, and our uncertainty perturbations were specifically designed to affect $P(O_3)$ by altering NO_x availability in different ways. The low degree of nonlinearity is an advantageous result, however, since it means that the OAT sensitivities can be used to interpret the MG analysis of ΔO_3 to elucidate sources of the model sensitivities.

While the OAT ΔO_3 sensitivities combined in a near-linear way, Table 17.1 shows that the MG ΔO_3, with one exception, is always greater than the OAT Sum ΔO_3. We interpret this to mean that there are some dependences in the model represented in the MG but not captured by the OAT simulations. However, our previous work (Dennis *et al.*, 1999) showed, using the same OAT simulations, that the E_{150} perturbation changed the actual shape of the O_3 response surface, altering the spacing between isopleth lines, while the $R8_{New}$ and $z50$ uncertainties appeared only to shift the location of the photochemical system on that surface. Hence, we suppose that a similar response surface change-of-shape by one of these uncertainties—and most likely the emissions—may be responsible in part for the apparent enhancement in the MG ΔO_3, but we are unable to conclude definitively from these results.

$\delta O_3/\delta E_{NO_x}$ sensitivities

We evaluated the degree of dependence among the uncertainties in the control case sensitivity $\delta O_3/\delta E_{NO_x}$ in the same way as we did the base case ΔO_3. Table 17.2 shows for the control case series that the mean deviation from additivity using the averaged peak O_3 value for all four sites in the NY air mass over all five days is approximately 2ppb or 12%, with an absolute value deviation of approximately 3 ppb or 16%. We note that because the mean deviation from additivity for $\delta O_3/\delta E_{NO_x}$ is larger than that for ΔO_3, the $\delta O_3/\delta E_{NO_x}$ standard deviation is only about 35% larger than its mean, while the standard deviation for the ΔO_3 sensitivity is about 50% larger than its mean. Figure 17.5 illustrates the five-day time series

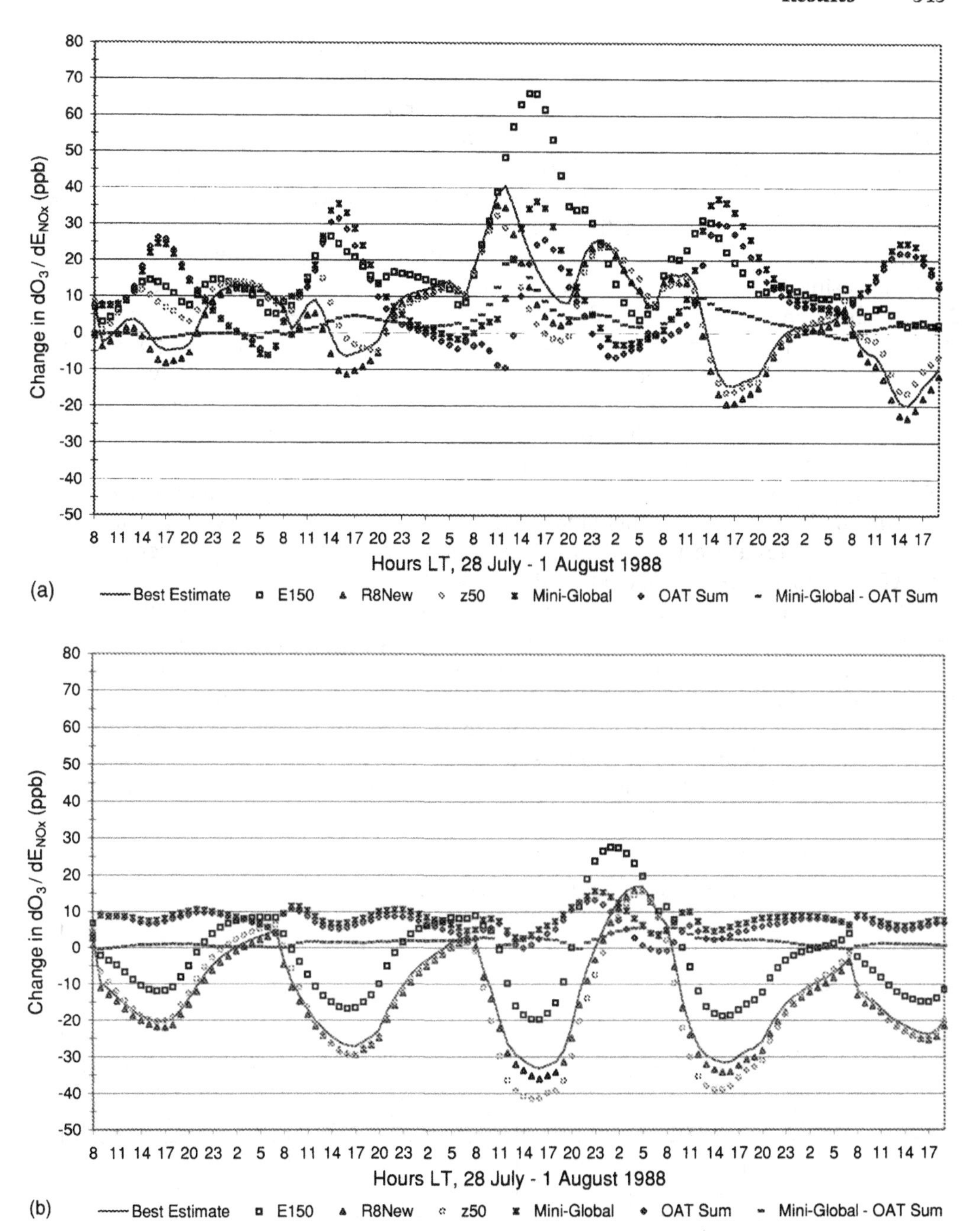

Figure 17.5 Time series of the changes in RADM-predicted dO_3/dE_{NO_x} (ppb) in the 20 km cells (a) over central Long Island (NY1) and (b) over south-central Connecticut (NY3) from the 50% NO_x control case series due to each uncertainty perturbation simulated OAT, all perturbations simulated together in the MG, and the difference MG – OAT Sum.

of the $\delta O_3/\delta E_{NO_x}$ sensitivity at two sites for each uncertainty OAT and MG. The difference MG − OAT Sum is mostly positive during the photochemically active period of peak O_3 for all days at all sites (NY2 and NY4 not shown here). Comparison of Figures 17.3 and 17.5 indicates that the deviations from additivity at NY1 and NY3 for both the ΔO_3 and the $\delta O_3/\delta E_{NO_x}$ sensitivities are nearly always positive and small during the photochemically active period of the days, irrespective of whether the results from individual uncertainty perturbations are positive or negative. This result is also obtained at sites NY2 and NY4 (not shown here). The deviation from additivity in the $\delta O_3/\delta E_{NO_x}$ sensitivity does seem to track increasing O_3, as Figure 17.6 shows, though the trend here is not so strong as it is for ΔO_3. Furthermore, deviation from additivity in $\delta O_3/\delta E_{NO_x}$ expressed as the relative percent of the MG result as in Figure 17.6 also clusters with a coarse division of the five episode days into high and low $[O_3]$ such that for the low-$[O_3]$ days 28 July, 29 July, and 1 August, the deviation in additivity in the sensitivity measure is mostly less than 15%, whereas for the higher-$[O_3]$ days 30 July and 31 July, the deviation is mostly greater than 25%. This clustering is not apparent in similar treatment of the ΔO_3 sensitivity.

The $\delta O_3/\delta E_{NO_x}$ sensitivity for uncertainties simulated MG is greater than it is for the OAT Sum on all day and site combinations except two: 28 July at NY1 and 30 July at NY2. While this indicates at least some dependences among the uncertainties not captured by the OAT simulations, we note again that this response is consistent with the response surface change-of-shape induced by an unusually large sensitivity to the E_{150} perturbation as described above. In the 30 July NY2 control case, for example, Table 17.2 shows that the $\delta O_3/\delta E_{NO_x}$ sensitivity to E_{150} OAT is approximately 44 ppb, substantially greater than the approximately 19 ppb mean response for all days and sites to E_{150} OAT. For the same day at NY1, the $\delta O_3/\delta E_{NO_x}$ sensitivity to E_{150} OAT is also large, approximately 47 ppb; but the sensitivity response to all three uncertainties simulated MG is 37% greater than the response to the uncertainties in the OAT Sum here, and not less than the OAT Sum as was the case at NY2.

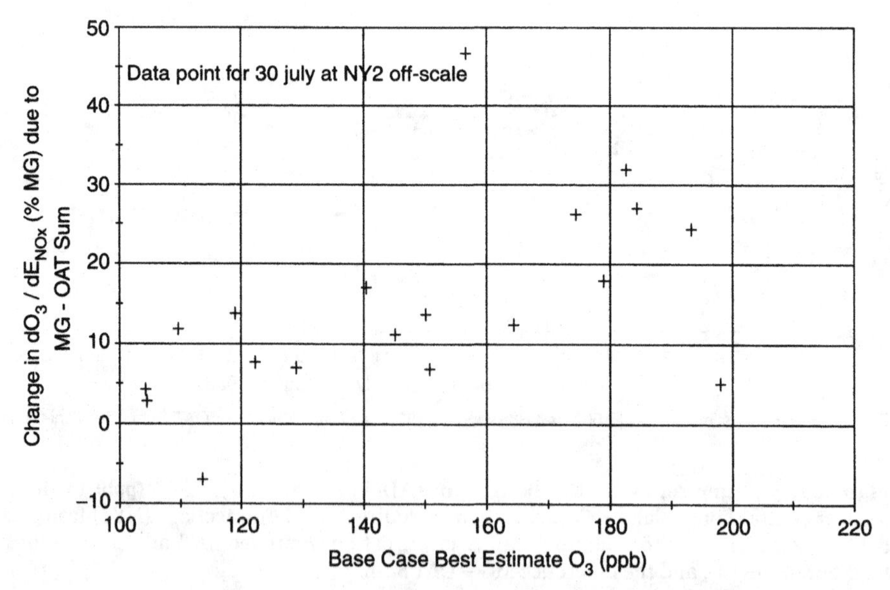

Figure 17.6 Deviation from additivity in the three hour averaged peak RADM-predicted dO_3/dE_{NO_x} in the 50% NO_x control case series for all days and sites due to dependencies among the three uncertainty perturbations evaluated as the difference MG − OAT Sum and displayed as a percentage of the MG value as a function of the base case best estimate $[O_3]$ (ppb).

Table 17.2 $\delta O_3/\delta E_{NO_x}$ from best estimate for the NO_x control case series.

Site	Day	E_{150} (ppb)	$R8_{New}$ (ppb)	$z50$ (ppb)	MG (ppb)	OAT Sum (ppb)	MG − OAT Sum (ppb)	MG − OAT Sum (% of MG)
	28 July	16.9	−3.5	11.8	23.5	25.2	−1.6	−7
	29 July	28.2	−5.0	4.7	32.4	27.9	4.4	14
NY1	30 July	46.5	−8.7	−13.6	33.3	24.3	9.0	27
	31 July	36.1	−5.1	−2.1	35.3	29.0	6.3	18
	1 Aug	21.4	−3.3	3.6	24.4	21.7	2.7	11
	28 July	10.4	−1.9	2.1	11.1	10.7	0.5	4
	29 July	17.0	−3.3	−0.9	14.7	12.9	1.8	12
NY2	30 July	44.3	−6.8	−15.5	16.6	22.1	−5.5	33
	31 July	30.8	−4.5	−8.4	18.9	18.0	1.0	5
	1 Aug	11.3	−2.0	−0.4	9.6	8.9	0.7	7
	28 July	8.7	−1.7	0.3	8.3	7.3	1.0	12
	29 July	11.1	−2.2	−2.0	8.3	6.9	1.4	17
NY3	30 July	16.6	−2.9	−8.0	7.6	5.6	2.0	26
	31 July	12.9	−2.7	−7.4	5.2	2.8	2.5	47
	1 Aug	9.0	−1.6	−1.0	7.4	6.4	1.0	14
	28 July	8.1	−1.4	0.2	7.2	7.0	0.2	3
	29 July	11.2	−2.2	−0.7	8.9	8.3	0.6	7
NY4	30 July	14.8	−3.3	−7.4	6.1	4.2	2.0	32
	31 July	16.4	−3.6	−7.3	7.3	5.5	1.8	24
	1 Aug	9.9	−1.6	−0.8	8.2	7.6	0.6	8

Summary statistics: all sites all days

Measure	E_{150} (ppb)	$R8_{New}$ (ppb)	$z50$ (ppb)	MG (ppb)	OAT Sum (ppb)	MG − OAT Sum (ppb)	MG − OAT Sum (% of MG)
Mean	19.08	−3.35	−2.63	14.72	13.10	1.62	12.4
Absolute value mean	19.08	3.35	4.90	14.72	13.10	2.33	15.8
Median	15.59	−3.10	−0.96	9.24	8.60	1.22	12.2
Standard error	2.65	0.42	1.43	2.21	1.96	0.64	3.6
Standard deviation	11.84	1.88	6.38	9.87	8.74	2.85	16.1

Mean response: each site all days

Site	E_{150} (ppb)	$R8_{New}$ (ppb)	$z50$ (ppb)	MG (ppb)	Best estimate (ppb)	Mean MG response (% of best estimate)
NY1	29.83	−5.13	0.90	29.78	−5.54	−538
NY2	22.77	−3.67	−4.62	14.18	−23.59	−60
NY3	11.64	−2.23	−3.63	7.35	−26.28	−28
NY4	12.08	−2.39	−3.18	7.55	−29.48	−26

How uncertainties combine

While the deviation from strict additivity is not zero for our sensitivity measures in either the base case or control case series, we believe that it remains sufficiently small in both ΔO_3 and $\delta O_3/\delta E_{NO_x}$ to enable us to usefully compare them, and to correctly interpret the probable sources of uncertainty in the MG results. The implication of these results is that, by-and-large, uncertainties combine or are propagated through the model without attenuation, even for the more derivative—mathematically speaking—compound measure of the model's behavior in the control case. For the uncertainties we have simulated here and with the day and site combinations we have examined, there is a slight enhancement in the model's response to the MG simulations, which we believe we understand. However, the perturbations used in our simulations tended to produce unidirectional effects on the system as we intended, but not all model uncertainties would be expected to do so or be uniformly distributed around the current best estimate. Our results to this point summarized in Tables 17.1 and 17.2 suggest that uncertainties producing opposite effects in either the ΔO_3 or $\delta O_3/\delta E_{NO_x}$ sensitivity when simulated OAT tend to cancel in a near-linear way when considered together. We note that our study has perturbed only a small set of uncertainties, and we might expect that for the complex AQMs, as the number of uncertainties considered is increased, they will increasingly cancel. There is in fact substantial, if indirect, evidence of this effect in the off-setting errors that have repeatedly been discovered in control applications of AQMs (Tesche and McNally, 1995).

17.3.2 Inability to Infer $\delta O_3/\delta E_{NO_x}$ from ΔO_3

We noted in Section 17.1 that sensitivity work with AQMs has generally focused on the base case ΔO_3 sensitivity, with the assumptions that the relevant control sensitivity $\delta O_3/\delta E_{NO_x}$ can be inferred from that, and that the $\delta O_3/\delta E_{NO_x}$ will in any case always be less than ΔO_3. We tested these assumptions by comparing the response of ΔO_3 with that of $\delta O_3/\delta E_{NO_x}$ to the three uncertainties simulated MG. Figure 17.7 shows that there is no apparent relation between the two sensitivity measures, and that the control response is not always less than the base case response. This suggests that separate analyses must be performed for each sensitivity to fully characterize model behavior.

17.3.3 Mixing Ratio Magnitudes and the Relative Effects of Simulated Uncertainties

ΔO_3 *sensitivities*

Table 17.1 shows that the mean ΔO_3 sensitivity (approximately 43 ppb) due to the uncertainties simulated MG is about 28% of the mean predicted base $[O_3]$ (approximately 151 ppb). However, Figure 17.8 illustrates a strong trend in the relative magnitude of the ΔO_3 sensitivity as a function of base case $[O_3]$ such that at O_3 values between 100 and 120 ppb, the ΔO_3 sensitivity is less than 20% of the base, whereas at O_3 values between 180 and 200 ppb, the sensitivity is nearer to 40%. This trend can be attributed in large part to the somewhat weaker tendencies of the ΔO_3 response to increase with increasing $[O_3]$ in each uncertainty simulated OAT: Table 17.1 shows this is most pronounced for the $z50$ perturbation. But the trend is also partly the result of the increased deviation from strict additivity in the MG with increasing $[O_3]$ already described.

The mean ΔO_3 responses to the E_{150} and $z50$ perturbations are quite similar in magnitude: approximately 16 ppb for the E_{150} and approximately 14 ppb for the $z50$, and together these two account for the majority (about 69%) of the MG response. Across sites on the same day or over time at the same site, however, the ppb ΔO_3 response can vary greatly for both of these uncertainties, and differently for one than the other. Table 17.1 shows that the standard

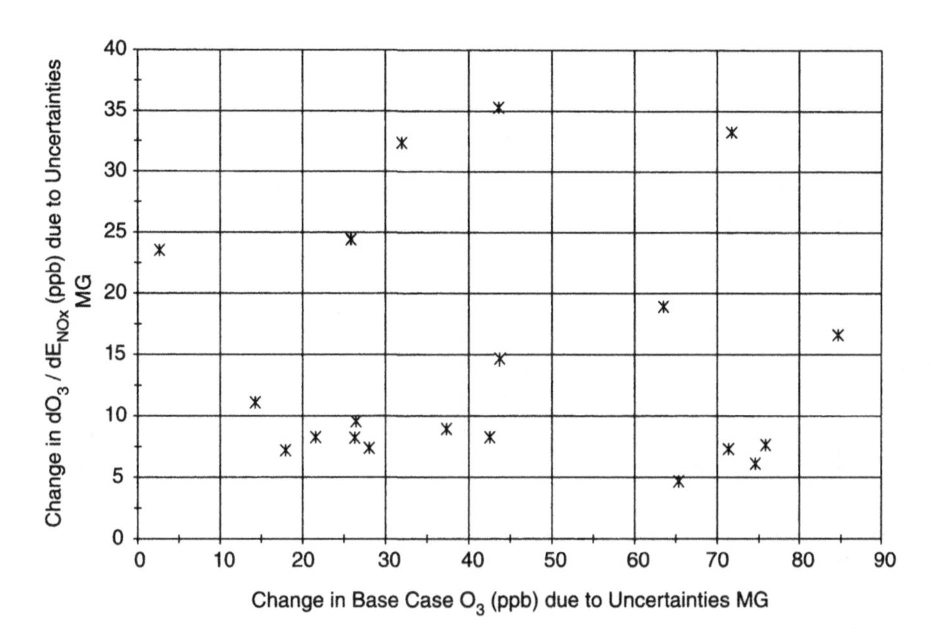

Figure 17.7 System response showing no correlation among uncertainty perturbations, simulated MG, when comparing the change in the RADM-predicted dO_3/dE_{NO_x} (ppb) in the 50% NO_x control case series with the change in the RADM-predicted $[O_3]$ (ppb) in the base case series for all days and sites. The three hour averaged peak O_3 value was used in both measures.

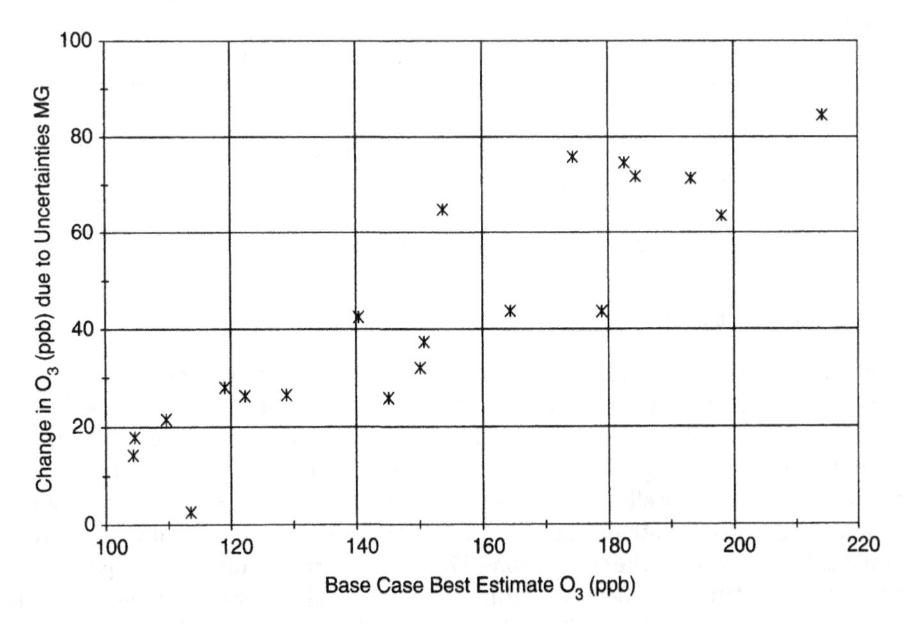

Figure 17.8 System sensitivity to uncertainty perturbations, simulated MG, as a function of the base case best estimate $[O_3]$ (ppb) for all days and sites and depicted as the change in the three hour averaged peak RADM-predicted $[O_3]$ (ppb) in the base case series.

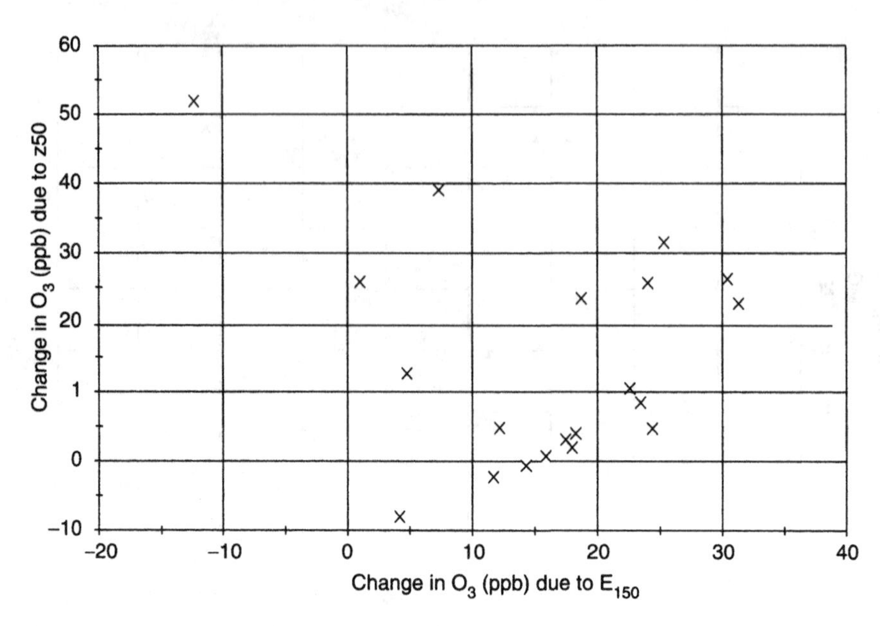

Figure 17.9 System response showing no correlation between the z50 and E_{150} uncertainty perturbations in the three hour averaged peak RADM-predicted $[O_3]$ (ppb).

deviation of the $z50$ response (approximately 16 ppb) is slightly greater than the $z50$ mean response, whereas the standard deviation of the E_{150} response (approximately 11 ppb) is substantially smaller than its mean. Moreover, the variability in the $z50$ response (approximately 240 ppb) is more than 50% larger than the variability in the E_{150} response (approximately 114 ppb). Figure 17.9 shows that the system response to these two uncertainties is strongly uncorrelated: the R^2 value is 0.04. This high variability across sites and days suggests that a Monte Carlo-type analysis would be unable to produce an accurate measure of the effects of these uncertainties on the model's ΔO_3 response unless constructed for a day-by-day analysis over several days at one site.

We note that Table 17.1 and Figure 17.3 indicate that there is little association of the ΔO_3 response for the uncertainties simulated MG across the four sites in our analysis, but that the variability is mostly associated with the day-to-day differences in meteorology at each site and their concomitant effects on the photochemistry.

$\delta O_3/\delta E_{NO_x}$ sensitivities

The three perturbations simulated OAT affected the system's predictions of $\delta O_3/\delta E_{NO_x}$ differently; i.e., the change in O_3 for a 50% reduction in NO_x was not the same for each uncertainty. In policy terms, the effectiveness of the 50% NO_x control in reducing O_3 was different in important ways for the three uncertainties. For example, the E_{150} perturbation simulated OAT decreased the effectiveness of the NO_x control for each site on every day, while the $z50$ and $R8_{New}$ perturbations simulated OAT generally, though not always, increased effectiveness. These trends can be seen in Table 17.3 by comparing values for the ppb change in O_3 due to NO_x control in the model best-estimate run, with values for the change in O_3 due to NO_x control in each uncertainty run. Table 17.3 shows that on all days at NY1, for example, the E_{150} run for the control case resulted in substantial O_3 increases over the best-estimate run for NO_x control: on 28 July O_3, was increased approximately 14 ppb by the E_{150} perturbation

Table 17.3 Change in $\delta O_3/\delta E_{NO_x}$ and best estimate response for the control case series.

Site	Day	E_{150} (ppb)	$R8_{New}$ (ppb)	$z50$ (ppb)	MG (ppb)	Best estimate $\delta O_3/\delta E_{NO_x}$ (ppb)
NY1	28 July	13.6	−6.8	8.5	20.2	−3.3
	29 July	22.5	−10.7	−1.0	26.7	−5.7
	30 July	60.3	5.1	0.3	47.1	13.8
	31 July	22.7	−18.5	−15.5	21.8	−13.5
	1 Aug	2.4	−22.4	−15.4	5.4	−19.0
NY2	28 July	−5.2	−17.6	−13.6	−4.6	−15.7
	29 July	−9.6	−29.9	−27.5	−12.0	−26.7
	30 July	26.8	−24.8	−33.6	−1.4	−18.0
	31 July	−3.0	−38.3	−42.2	−14.9	−33.8
	1 Aug	−12.5	−25.8	−24.2	−14.2	−23.8
NY3	28 July	−11.5	−21.8	−19.9	−11.9	−20.1
	29 July	−14.8	−28.1	−27.9	−17.6	−25.9
	30 July	−14.1	−33.7	−38.7	−23.1	−30.7
	31 July	−18.4	−34.0	−38.7	−26.1	−31.3
	1 Aug	−14.4	−25.0	−24.4	−16.0	−23.4
NY4	28 July	−11.4	−20.9	−19.3	−12.4	−19.5
	29 July	−17.6	−31.0	−29.5	−19.8	−28.8
	30 July	−20.8	−38.8	−42.9	−29.4	−35.5
	31 July	−24.5	−44.4	−48.2	−33.6	−40.9
	1 Aug	−13.8	−25.2	−24.5	−15.5	−23.7

Mean Response: all sites all days

	E_{150} (ppb)	$R8_{New}$ (ppb)	$z50$ (ppb)	MG (ppb)	Best estimate $\delta O_3/\delta E_{NO_x}$ (ppb)
Mean	−2.2	−24.6	−23.9	−6.6	−21.3

in the control case, whereas the control case best-estimate result was an O_3 decrease of 3.3 ppb. Recall from Figure 17.1 that O_3 can be increased even if NO_x emissions are decreased because of nonlinearities in the chemistry that define the O_3 response surface and the position of a particular photochemical system on it relative to the O_3 ridgeline. While the $R8_{New}$ perturbation always resulted in more effective NO_x control of O_3 at all sites every day, the $z50$ change at NY1 resulted in smaller O_3 decreases or larger increases—less-effective control—each day except 31 July. However, Table 17.3 also shows that the $z50$ change resulted in more effective NO_x control on all days after the first day, 28 July, at NY2, NY3, and NY4.

The dominance of the E_{150} perturbation simulated OAT is also shown in means and totals calculated for the absolute value of the changes due to each perturbation for the $\delta O_3/\delta E_{NO_x}$ sensitivity. These values are shown in Table 17.2. About 70% of the mean absolute value change of approximately 27 ppb is accounted for by the mean absolute change of approximately 19 ppb due to E_{150}. Moreover, the magnitude of the mean E_{150} absolute change in $\delta O_3/\delta E_{NO_x}$ is very near but not precisely the same as the magnitude of the mean absolute value change of approximately 23 ppb that resulted from the NO_x control in the best-estimate run. The E_{150} perturbation increased both VOC and NO_x by 50%, while the NO_x control case series brought the NO_x emissions level down by 50%. But because the photochemical system is nonlinear in VOC and NO_x, the control case did not return the system to the same point on the O_3 response surface—hence the slight difference in these two values.

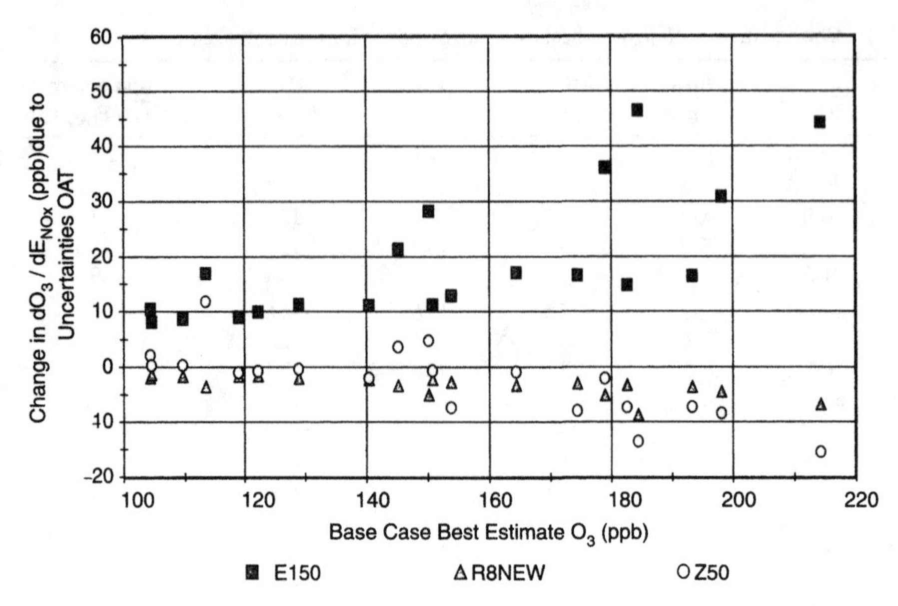

Figure 17.10 Trends in the effects of uncertaintyperturbations, simulated OAT, in dO_3/dE_{NO_x} (ppb) in the 50% NO_x control case series as a function of the base case best estimate $[O_3]$.

Table 17.2 also reveals trends in the relation of the ppb change in $\delta O_3/\delta E_{NO_x}$ to the $[O_3]$ in the best-estimate run when the uncertainties are simulated OAT, but not when they are simulated MG. This results from separate and different trends in best-estimate $[O_3]$ with each perturbation. Figure 17.10 shows that the trends in the $z50$ and the $R8_{New}$ perturbations simulated OAT are nonlinear, of similar magnitude, and negative, whereas the trend in the E_{150} perturbation is also nonlinear, but larger in absolute magnitude than the others, and positive. When these trends are combined in the MG simulation as shown in Figure 17.11, the OAT trends cancel owing to the near-additivity when the uncertainties are simulated MG, as described in Section 17.3.1.

Comparing all the daily values and the means and totals for the uncertainties simulated OAT and MG for the base case series and the control case in Tables 17.1 and 17.2, shows that the $\delta O_3/\delta E_{NO_x}$ sensitivity is distinctly different from the ΔO_3 in that the change of sign between the E_{150} perturbation on one hand and the $z50$ and $R8_{New}$ perturbations on the other is not reflected in the ΔO_3 response. Hence, the MG and OAT Sum totals for the ΔO_3 sensitivity are also affected. The mean change in $\delta O_3/\delta E_{NO_x}$ when the uncertainties are simulated MG (approximately 15 ppb) is substantially less than the mean ΔO_3 change for the MG runs (approximately 43 ppb). However, expressed as a percentage of the mean best-estimate $\delta O_3/\delta E_{NO_x}$ change (approximately 21 ppb), the mean relative change in $\delta O_3/\delta E_{NO_x}$ due to uncertainties simulated MG is about 69%, and is somewhat larger than the corresponding value calculated for the ΔO_3 sensitivity and the ΔO_3 best estimate, about 29%.

A strong trend is apparent in the ppb change in the $\delta O_3/\delta E_{NO_x}$ sensitivity to the uncertainties simulated MG across the four sites in our analysis. Table 17.2 shows that the mean change drops from approximately 30 ppb at NY1 to approximately 14 ppb at NY2 to approximately 7 ppb at NY3 and NY4. This trend follows the path of the air mass as it leaves the relatively more urban regime over Long Island (NY1 and NY2), passing north and east up to relatively more rural locations (NY3 and NY4). Furthermore, this pattern is consistent with our other results (Li *et al.*, 1998) simulating 20 summer days of 1988 using the 80 km resolution RADM. This trend is accentuated when the change in the $\delta O_3/\delta E_{NO_x}$ sensitivity

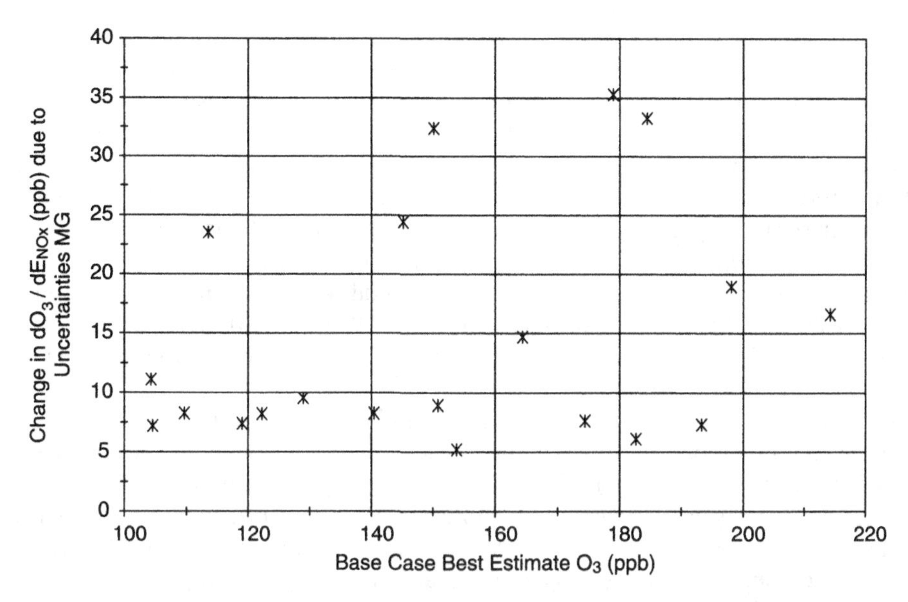

Figure 17.11 Lack of trend in the effects of uncertainty perturbations, simulated MG, in dO_3/dE_{NO_x} (ppb) in the 50% NO_x control case series as a function of the base case best estimate $[O_3]$

due to the uncertainties simulated MG is measured relative to the $\delta O_3/\delta E_{NO_x}$ in the best-estimate run. Table 17.2 shows that the mean change in the sensitivity response is about 540% of the best-estimate response at NY1; however, this value is inflated owing to the very small size (approximately 6 ppb) of the best-estimate sensitivity response for the photochemical system at these conditions. The high response of about 540% at NY1 falls sharply to about 60% at NY2 and to about 27% at NY3 and NY4. Given the results already described, it is not surprising that this trend in response to uncertainties simulated MG across sites is dominated by the E_{150} perturbation, but the effects of the $z50$ and $R8_{New}$ perturbations are not negligible. When the mean $\delta O_3/\delta E_{NO_x}$ responses due to the $z50$ and $R8_{New}$ perturbations are summed, their combined influence expressed relative to the mean $\delta O_3/\delta E_{NO_x}$ in the best-estimate run is about 76% at NY1, about 35% at NY2, and about 21% at NY3 and NY4. The mean value of these two perturbations across all sites for all days (approximately 6 ppb) is about 28% of the mean $\delta O_3/\delta E_{NO_x}$ response in the best-estimate run (approximately 21 ppb), whereas the mean of all three perturbations simulated MG (approximately 15 ppb) is about 69% of the mean best-estimate. As above, the $\delta O_3/\delta E_{NO_x}$ sensitivity response here is distinctly different from the ΔO_3 response: corresponding means and sums calculated for the ΔO_3 sensitivity show that the $z50$ and $R8_{New}$ summed response (approximately 23 ppb) is about 15% of the mean $[O_3]$ in the base case best-estimate run.

17.4 DISCUSSION AND CONCLUSIONS

Our results show that the three input uncertainties we considered have a moderate effect (20–40%) on the change in $[O_3]$ in the base case series, but can have a moderate to large effect (20% to >100%) on the change in $[O_3]$ predicted for 50% NO_x control. The magnitude of this change for the $\delta O_3/\delta E_{NO_x}$ sensitivity is substantial, and was entirely unanticipated. However, these results are illustrative only, and we cannot make definitive conclusions about the model's overall sensitivity from them.

Our results do show that changes in the $\delta O_3/\delta E_{NO_x}$ response for the different uncertainties are strongly dependent on the position of the photochemical system on the response surface, and, at least for the E_{150} perturbation, on the shape of that response surface. This means that the sensitivity results will also be a function of the model grid-size resolution, of the geographic location of the site relative to major urban emissions, and of the characteristic of the photochemical system as either NO_x or radical-limited as depicted in one domain or the other on the O_3 response surface. We conclude that it cannot be assumed that different urban areas will produce the same $\delta O_3/\delta E_{NO_x}$ response unless it can be established that they occupy the same position on the response surface — and we know of no means at the present time to establish that reliably except by brute force. We also note that results from different models may differ in unexpected ways unless the models can be configured very similarly.

Variability in the magnitude of the model's ΔO_3 and $\delta O_3/\delta E_{NO_x}$ responses over time and across space greatly complicates how a summary sensitivity statistic should be constructed. It is apparent from the results reported here for our relatively small number of sites and days that a single value of a sensitivity measure, similar to those produced in box model sensitivity studies, will be inadequate to fully characterize a three-dimensional model. It is likely that the ΔO_3 response will vary greatly across space, tracking changing [O_3]. Because [O_3] is influenced by both meteorology local to a site and transport of precursors and products from upwind sites, the local ΔO_3 sensitivity will most likely vary day-to-day on both a percent and absolute ppb basis. This conclusion is supported for our sites and cases, as Table 17.1 shows. In contrast to the ΔO_3 sensitivity, the $\delta O_3/\delta E_{NO_x}$ response is likely to be most influenced by variations in local $P(O_3)$ due to changing precursor concentrations around major urban areas. Our previous work with a large number of sites at the RADM 80 km resolution suggests that different urban areas can demonstrate very different $\delta O_3/\delta E_{NO_x}$ sensitivities.

Given these results, we conclude that it is unlikely that a single all-inclusive metric can be constructed that correctly and usefully relates the model's ΔO_3 sensitivity to [O_3]. We cannot conclude at this time whether the sensitivity response could be a vector associated with O_3 magnitude or should more properly be a matrix that includes location. However, with the variation that we have seen in our study, we suggest that individual metrics will need to be developed to properly reflect differences in the $\delta O_3/\delta E_{NO_x}$ response for urban and rural sites to capture the variation across space and pollutant mixes. We conclude from our study of a New York air mass with a coherent photochemical production potential that some measures or indicators of a photochemical system's degree of NO_x or radical limitation, i.e., its position on the response surface relative to the ridgeline, should be evaluated for possible incorporation into such metrics.

Our results suggest that uncertainties can be propagated through the model and combine without attenuation. We cannot conclude from these results whether the state variable response to uncertainties is ultimately bounded by some value, nor can we say whether the uncertainty in these inputs and parameters leads to bias rather than imprecision in model predictions. We recognize that these are important questions, however, and intend to pursue them in future sensitivity work with RADM and other air quality models.

Concerning sensitivity analysis methods, we conclude that OAT and global analysis should not be viewed as exclusive or competitive, but rather as strongly complementary. Our mixed approach using OAT and a MG sensitivity analysis in this work demonstrated the importance of establishing some basic characteristics of the system, and of developing an understanding of any possible system dependences. The exploratory runs that we made with uncertainties simulated OAT were crucial for interpreting the MG results of both of our sensitivity measures. We further conclude that it may in fact be necessary to split a global analysis for instances where an uncertainty such as the E_{150} perturbation and a response such as the $\delta O_3/\delta E_{NO_x}$ sensitivity measure here are perhaps too-closely linked. Additional OAT analysis should help with determining this. Moreover, an understanding of the departure from additivity in the MG would not have been possible without OAT simulations. Thus, we conclude that some OAT analysis should precede any full-scale global analysis, especially where the subject model is nonlinear such as RADM.

Although our results point to surprisingly small deviations from additivity in both the ΔO_3 and $\delta O_3/\delta E_{NO_x}$ sensitivities, we believe that it is still possible that dependences exist in the photochemistry that were not resolved at the 20 km grid scale but that would affect the sensitivity response at finer resolutions. If such dependences are present, we would expect to detect them at the much finer grid resolutions becoming available with the next generation of air quality models (Dennis *et al.*, 1996). Previous work with RADM has, in fact, demonstrated the effects of grid-size resolution on $[O_3]$ and transport (Jang *et al.*, 1995), and we have noted some extreme model behavior in the results reported here at one site over several days. Dependences emerging at resolutions finer than 20 km will require careful experimental design, and more detailed interpretation of results, most likely including analysis of calculations of internal model processes and rates in addition to use of changes in state variable concentrations.

The system that we studied has a strong potential for spatial and temporal gradients in sensitivity response, especially for its $\delta O_3/\delta E_{NO_x}$ sensitivity. Results from the OAT simulations helped us interpret the MG responses, and suggest to us that a global analysis with these systems should be constructed for separate days and sites, or, at least, that the days and sites should be carefully characterized with OAT analyses before the design of the global study is determined. We realize that this may complicate a global analysis and increase its computational demand to ensure that sufficient realizations of the model are run to provide robust statistical precision in the results and more accurate inferences for building an understanding of the model; however, the system will be much better characterized and comprehensible if the analysis follows this line.

ACKNOWLEDGMENTS

Support for J. R. A. and G. S. T. was provided by the NOAA Atmospheric Sciences Modeling Division, and was administered by the Visiting Scientist Program of the University Corporation for Atmospheric Research.

Note that this paper has not been reviewed by the US EPA, and so does not represent Agency policy.

18

Comparing Different Sensitivity Analysis Methods on a Chemical Reactions Model

Francesca Campolongo and Andrea Saltelli

European Commission, Joint Research Centre, Ispra, Italy

18.1 INTRODUCTION

This chapter serves two main purposes: to demonstrate a classical application of SA, namely as a tool to elucidate mechanisms in complex chemical kinetics reaction schemes, and, at the same time, to contrast some of the methods presented in Part II of this volume, on Methodology. Further, we introduce here an example of a hybrid, or multistep, approach, where different methods are used in sequence.

In this respect, the present chapter can be considered as an extension of Section 2.10 on 'When to use what'. Both there and here, our suggestions are not to be read as prescriptions, but are derived from our experience as modellers and practitioners.

18.2 LOCAL OAT APPROACHES: THE EOAT AND THE DERIVATIVE-BASED DESIGNS

In Chapter 4, we have described the OAT approach to SA. In an OAT experiment, SA is performed by changing 'one factor at a time', and exploring what the model does with the new datum. Several examples of application of the OAT approach can be found in the literature. It is perhaps the first intuitive method used by most modellers to probe their models. In particular, the most widely used method is what we call 'elementary OAT', or EOAT. In this type of analysis, a baseline value is selected for each of the input factors. Then, factors are moved away from the baseline one at a time, while all other factors are kept constant at the

Sensitivity Analysis. Edited by A. Saltelli *et al.*

baseline; the baseline is not changed throughout the analysis. While this approach is easy to implement, computationally inexpensive, and useful to provide a glimpse of the model behavior, it has limited descriptive power. Any conclusion drawn on the relation between the output considered and the individual factor being varied is only legitimate around the baseline case. If the model considered is nonlinear then the overall effect of a given factor could well escape or deceive the analyst when EOAT is used. This may easily happen because in some corner or edge of the input-factor space a different pattern of sensitivity exists, or because factors interact with each other.

An alternative to the EOAT is the use of the derivative-based approach, where the effect of a given input x on the output y is assumed to be proportional to the derivative $\partial y/\partial x$. Derivative-based sensitivity analysis methods have been used extensively in chemistry in a variety of applications, such as solution of inverse problems—for example, computing kinetic constants from measured flow rates in a batch or flow reactor (see Turanyi, 1990a for a review) or relating variables at the molecular scale to those at the macroscopic one (Rabitz, 1989). Derivative-based methods, used to solve inverse problems, have proven their worth. Nevertheless, we contend that neither an EOAT approach nor a derivative-based SA should be used to rank the impact of different uncertain (or variable) input factors in determining the variations of the output under examination, unless the model is known to be linear or the range of variation is small. Some other methods like that of Morris (Chapter 4) or the SRC (Chapter 6) should be used instead (see also Saltelli, 1999).

In the following, we try to substantiate the above claim by comparing different SA methods on a real, moderately nonlinear model. The model under study is a chemical kinetics model (KIM) of the tropospheric oxidation pathways of dimethyl sulfide (DMS), a sulfur-bearing compound of interest in climatic studies. A description of its main features (taken from Campolongo *et al.*, 1999a) is given in the next section.

18.3 THE KIM MODEL

KIM stands for Kinetic Model for OH-initiated oxidation of DMS (CH_3SCH_3), and incorporates a description of the tropospheric reaction pathways for the formation of sulfur-containing molecules, such as sulfur dioxide (SO_2) and methanesulfonic acid (MSA, CH_3SO_3H), from DMS. KIM is zero-dimensional and includes multiphase (droplets–air) transport and chemistry. The KIM model is relevant to climate change studies, because of the important contribution of DMS emissions to the formation of climatically active atmospheric aerosols and in particular, the hypothesized feedback mechanism linking the biogenic sulfur cycle to the greenhouse effect (Charlson *et al.*, 1987, 1992).

The oxidation can be enhanced by the presence of water droplets in the troposphere, and provides an aqueous pathway for the formation of sulfur-containing molecules.

The process of building the KIM model involved tackling uncertain mechanisms and reaction rates. Certain mechanistic aspects of the chemistry of DMS were so poorly understood that the knowledge of the various reactions and of their relative weights was very imprecise. Thus, the reaction scheme adopted in the model was affected by large uncertainty (structural uncertainty). Furthermore, given the large uncertainties in the parameter values governing DMS oxidation kinetics, the error bars associated with the rate constants involved were large (factor uncertainty).

All the above reasons, together with the scarcity of observed data for a full model calibration, led to the implementation of a model-building process where uncertainty and sensitivity analysis played a central role. The KIM model was run within a Monte Carlo (MC) driver, capable of propagating the uncertainty in the input parameters onto the output variables. The MC analysis of Saltelli and Hjorth (1995) allowed the quantification of (i) the uncertainty in model prediction and (ii) the relative importance of each input parameter in determining such an uncertainty. The study was based on KIM-I, a purely gas-phase chemistry version without droplets.

One of the main limitations of studies done with KIM-I was that they neglected the temperature effects on the DMS-oxidation process. Remedio *et al.* (1994) extended the KIM-I model to include latitude dependence, producing a second version of the model, KIM-II. The MC analysis was then performed again on KIM-II, and a latitudinal analysis, emphasizing the possible regional differences on the main oxidation pathways of DMS and on the relative amounts of end-products formed, was carried out. Results of the analysis agreed generally with those found in a bibliography by Koga and Tanaka (1993). However, the conclusions of this second study were still conditional upon the model and data assumptions underlying the experiment. Among these assumptions, the non-inclusion of the heterogeneous chemistry (aqueous phase) and dry deposition, by far the largest sink for SO_2 molecules, was the most severe.

In 1997, a third version of the KIM model (KIM-III) was produced (Campolongo *et al.*, 1999a). In KIM-III, the heterogeneous chemistry is dealt with, and a first attempt is made to include some elements of cloud processing in the model. Liquid-phase chemistry occurs inside the water droplets in the troposphere, and involves transfer to the droplet, chemical reactions inside the droplet, and the sink terms for the droplet (e.g. wash-out). The assumptions and the choices made in KIM-III are not discussed here because they are beyond the scope of this chapter. A detailed discussion can be found in Campolongo *et al.* (1999a).

The homogeneous chemistry scheme of KIM-III is adopted with minor modifications from Saltelli and Hjorth (1995), and is illustrated in Figure 18.1(a). The hypothesized heterogeneous chemistry reactions are outlined in Figure 18.1(b). The chemical reactions considered in KIM-III, their input data, and their sources of information, are given in Campolongo *et al.* (1999a).

The KIM-III model solves a system of approximately 40 ordinary differential equations involving approximately 60 species. Uncertain input factors include initial concentrations, kinetic constants, and thermodynamic parameters that enter into the definition of the kinetic constants. Uncertainties in the heterogeneous oxidation of DMS and of its intermediates in the liquid phase are even more severe than for the homogeneous chemistry mechanism.

As discussed by Ayers *et al.* (1996) the concentration ratio in marine aerosol between MSA and non-sea-salt sulfates (nss-SO_4^{2-}, including SO_2 and H_2SO_4), i.e.

$$\alpha = \mathrm{MSA}/(\mathrm{SO_2} + \mathrm{H_2SO_4}),$$

seems to offer the best opportunity for comparing observed data with those predicted by models, in particular, for considering the temperature dependences involved in the various branches of the oxidation processes. Further, the MSA/ nss-SO_4^{2-} ratio may be used for estimating the actual contribution of DMS to observed nss-SO_4^{2-} from measurements of MSA, if the dependences of the ratio on temperature and other ambient conditions are sufficiently well known (Savoie and Prospero, 1989; Saltzman *et al.*, 1996).

The temperature dependence of α was the focus of the study by Campolongo *et al.* (1999a). The values of α predicted by KIM-III were compared with field observations of ratios of MSA to non-sea-salt sulfate (Bates *et al.*, 1992).

Note that in the work by Campolongo *et al.* (1999a), the performance of the KIM-III model was tested twice. First, the model was run assigning to the kinetic constant k_{21} the value as

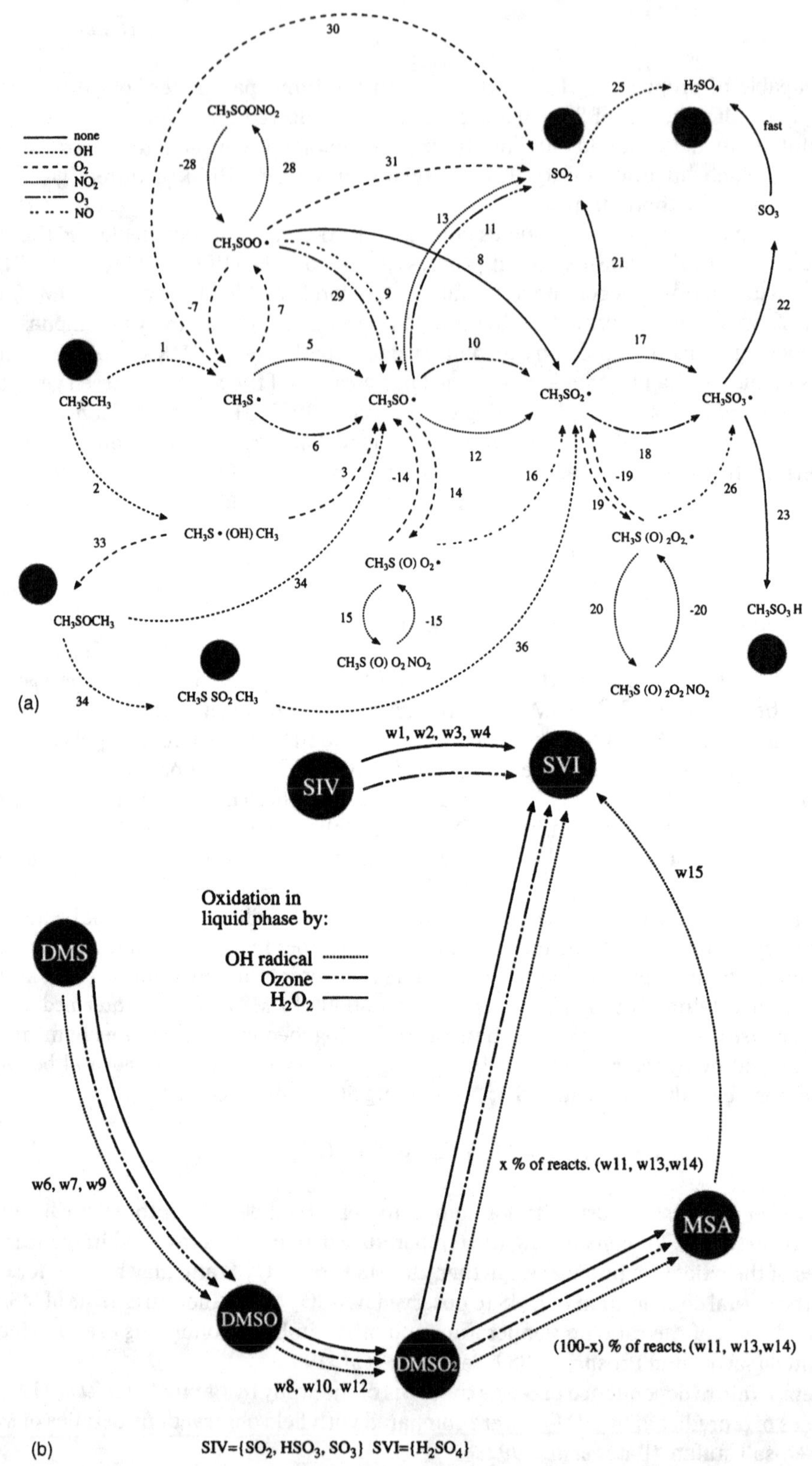

Figure 18.1 A scheme of the KIM model structure: (a) the homogeneous chemistry scheme; (b) the heterogeneous chemistry scheme.

reported in Ray *et al.* (1996). Then, a sensitivity analysis exercise (see Section 18.4) underscored a strong sensitivity of the model to the k_{21} value. Thus, given the strong discrepancies among the different estimates of k_{21} present in the literature (for a review, see Saltelli and Hjorth, 1995), Campolongo *et al.* (1999a) felt it appropriate to repeat the analysis, and explored the outcomes (both in terms of model structure and system behavior) derived from an alternative choice for k_{21}. The model was then run again, replacing the k_{21} value taken from Ray *et al.* (1996) with that given in Mellouki *et al.* (1988).

In the following, when referring to a SA experiment conducted on the KIM-III model, the value adopted for k_{21} will be specified.

18.4 COMPARING DIFFERENT SA APPROACHES ON KIM

In its final version (KIM-III), the model contains 68 uncertain input factors. It is important then to conduct a screening exercise to establish which are the most important factors in the model (i.e. those that mostly drive the variations in the model output).

To rank the input factors in order of importance, local SA based on the EOAT and the derivative-based methods mentioned in Section 18.2 are not appropriate. A screening method that belongs to the class of the OAT designs but that, differently from the EOAT and the derivative-based approaches described in Section 18.2, can be considered as global is the method proposed by Morris (Chapter 4). The Morris sensitivity measure is global in the sense that it is obtained by computing a number r of local measures at different points $\mathbf{x}_1, \ldots, \mathbf{x}_r$ of the input space, and then taking their average (thus losing the dependence on the specific point at which the measure was computed). In other words, the Morris method changes the baseline value at each step; that is, the method wanders in the whole space of input factors rather than oscillating around the baseline as in EOAT. Thus, the information on the model sensitivity produced by the Morris method is more general with respect to that produced by an elementary OAT (in the sense that the whole input-factor space is explored). Furthermore, with respect to other SA methods (such as the regression-based ones), the Morris method is computationally cheap (see Chapter 4).

The Morris method is applied to the KIM model (version KIM-III and using the value of k_{21} reported by Ray *et al.*, 1996), and its results are contrasted against local SA in Table 18.1. Derivatives of the output are computed by changing all inputs by 1% around the nominal

Table 18.1 Comparison of the Morris and derivative methods. Ranks in parentheses are not among the top five factors for one of the methods. Dots stand for factor undetected by the method (zero sensitivity). Reprinted from Saltelli, A., Sensitivity Analysis. Could better methods be used? *Journal of Geophysical Research*, **104**(D3), pp 3789–93, 1999.

Factor	Ranking from Morris	Ranking from $\partial\alpha/\partial x$
WATLIQ	1	1
QW7	2	...
DEHH202	3	(6)
RHLH202	4	(8)
RHLDMS	5	2
RHLO3	(6)	3
YOOHRAD	(8)	4
W11	(14)	5

value. Only those factors that are in the list of the top five for either method (Morris or derivative) are displayed. The factors are not described here, nor are their roles in the model, for which the reader is referred to the original article by Campolongo *et al.* (1999a).

The examples of application of the Morris method available in the literature up to now have all been based on the assumption of uniform distributions for the input factors. When such an assumption holds, the levels of the experiment are simply obtained by dividing the interval in which each factor varies into equal parts. In the KIM model, factors were assigned different ranges of variation and different distribution types, mostly based on the literature (uniform, log-uniform, normal, log-normal, . . .). In this case, a simple choice of levels would result in a loss of information, since it would neglect the statistical information contained in the distribution functions. The procedure adopted by Campolongo *et al.* (1999b) is the following: instead of sampling the input values directly in Ω, they first sampled in the space of the quantiles of the distributions, which is a k-dimensional hypercube (each quantile varies in [0,1]). Then, given a quantile value for a given input factor, the actual value taken by the factor was derived from its known statistical distribution.

As one can see from Table 18.1, the disagreement is already significant at the level of the second most important factor. One should recall here that Morris also computes 'derivatives', albeit over larger ranges, and averages them over the space of definition of the factors. As this space is much larger than the 1% variation allowed to the differential analysis, the results differ.

In the work by Campolongo *et al.* (1999b), the KIM model was further investigated in the same configuration by fixing the factors not identified by the screening. An MC analysis was then applied to just 20 variables (Table 18.2). Although hypothesis testing was used there, here we show the first five factors in the set of 20 that are identified by either the standard regression coefficients, their rank equivalent SRRC, or the $\partial\alpha/\partial x$.

The disagreement between derivatives and regression is pronounced. It is important to appreciate that even the SRC analysis is not fully successful, since, based on $R^2 = 0.74$, 26% of the variation of the output is not accounted for. Nevertheless, the analyst is informed of this. This was not the case with either the Morris or the local approach. Likewise, the value for the rank-based measures ($R^2 = 0.87$) flags the possible existence of some non-monotonicity, or interaction, likely involving WATLIQ, accounting for about 10% of the variance of α.

Table 18.2 Results of Monte Carlo Analysis. SRC denotes standardized regression coefficients and SRRC denotes standardized rank regression coefficients. Ranks in parentheses are not among the top five. Reprinted from Saltelli, A., Sensitivity Analysis. Could better methods be used? *Journal of Geophysical Research*, **104**(D3), pp 3789–93, 1999.

Factor	**Ranking from SRC ($R^2 = 0.74$)**	**Ranking from SRRC (R2 = 0.87)**	**Ranking from $\partial\alpha/\partial x$**
Q1	1	1	. . .
Q21	2	2	. . .
WATLIQ	(15)	3	1
QW7	5	4	. . .
RHLH202	4	(6)	(8)
RHLDMS	(12)	(10)	2
RHLO3	3	5	3
YOOHRAD	. . .	. . .	4
W11	. . .	. . .	5

This example has shown that even a zero-dimensional, purely chemical, model such as KIM is nonlinear enough to render the local approach unwarranted, especially since one cannot appreciate its error without contrasting it with another method. A global screening method that allows a scan of the whole input parameter space is certainly to be preferred. Furthermore, it is clear that the presence of interactions should not be overlooked.

On the basis of these considerations, Campolongo *et al.* (1999b) carried out further SA studies on the KIM-III model (see Section 18.5). The value chosen for k_{21} was that taken from Mellouki *et al.* (1988).

18.5 A QUANTITATIVE SENSITIVITY ANALYSIS

The problem of interactions may be tackled by computing the first-order as well as the total sensitivity indices for each variable (Chapter 8) following either the FAST or the 'Sobol' methodologies. If the two indices are different for a given factor, the factor is involved in interactions. As a drawback, these two methods require a large number of model evaluations when the number of input factors involved in the analysis is high (tens or hundreds).

Campolongo and Saltelli (1997) suggested that, when dealing with models containing a large number of parameters, a possible procedure that would match sensitivity information achieved with computational cost would be the one made by two consecutive experiments. A preliminary screening exercise could be conducted with the goal of identifying the parameter subset that controls most of the output variability with low computational effort. Then, the screening could be followed by a quantitative method applied to the subset of pre-selected inputs. For a successful experiment, both the exercises should be run with global techniques. This procedure can be effective, especially since often, among a large number of input parameters involved in a model, only a few have a significant effect on the model output. Furthermore, the approach has a significant advantage of providing quantitative sensitivity measures while controlling the computational cost of the experiment.

Campolongo *et al.* (1999b) applied this idea to the KIM-III model (in its version KIM-III with the k_{21} value taken from Mellouki *et al.*, 1988), and implemented a quantitative analysis, using the extended FAST method on a subset of input factors of the KIM model. Again, the Morris method was employed as the screening procedure to select the subset of most influential factors.

The Morris method was applied to the model with sample size $r = 10$ and number of levels $l = 4$ (see Chapter 4). The values for the four levels were derived from the statistical distribution function of each input factor as described above.

Results of the Morris screening exercise are given in Figure 18.2. The two Morris sensitivity measures μ and σ are plotted for the 68 input factors (only the 10 most important factors are named). Note that μ and σ provide two complementary measures of sensitivity; however, for this exercise, the two sets of the 10 most important factors as identified by μ and by σ are identical. The ranking of the 30 most important factors obtained according to μ is given in Table 18.3.

The 10 most important factors resulting from the Morris screening have been selected for further investigation. A quantitative appreciation of their influence on the output variable has been obtained by performing the extended FAST. The ranges and distributions adopted for the 10 factors are the same as in the screening exercise. The remaining factors have been assumed irrelevant and, therefore, fixed at their nominal values.

The sample size $N = 1026$ has been used for the calculation of each pair of indices (S_i, S_{T_i}), with $M = 4$, $\omega_i = 64$ and $N_r = 2$. The total cost of the analysis is

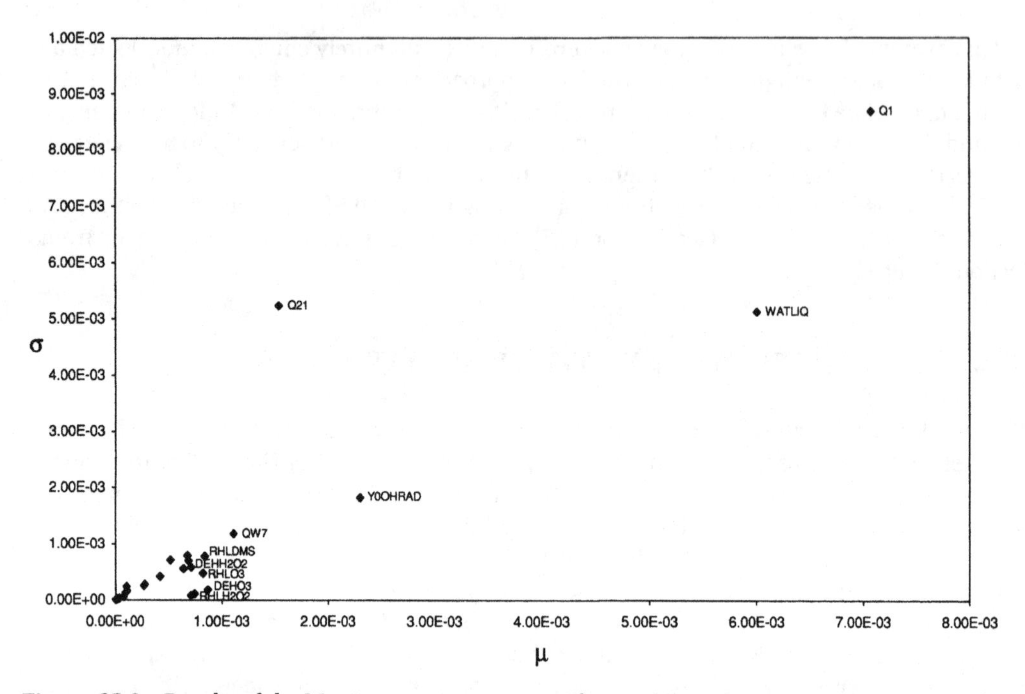

Figure 18.2 Results of the Morris screening exercise. The two Morris sensitivity measures μ and σ are plotted for the 68 input factors. Only the 10 most important factors are named. Reproduced from *Computer Physics Communications*, **117**, Tackling quantitatively large dimensionality problems, by Campolongo, F., Tarantola, S., and Saltelli, A., pp 75–85, 1999, with permission from Elsevier Science.

Table 18.3 Results of the Morris experiment on the KIM-III model. The 30 most important factors according to the SA measure μ are listed.

Factor	Rank	Factor	Rank
Q1	1	B7	16
WATLIQ	2	DEHDMS	17
YOOHRAD	3	R8	18
Q21	4	Q6	19
QW7	5	R30	20
DEHO3	6	B6	21
RHLDMS	7	W11	22
RHLO3	8	R29	23
RHLH202	9	B32	24
DEHH202	10	R19	25
R10	11	R14	26
R18	12	QM14	27
W15	13	R12	28
R31	14	QM19	29
Q7	15	R16	30

$C = N \times k = 10\,260$, since $k = 10$. The results are illustrated in Table 18.4 and Figure 18.3, where two pie charts, respectively for the S_i (a) and for the S_{T_i} (b), represent the percentage contributions of the 10 factors to the total output variance.

The execution of the classic FAST would have yielded the pie chart (a) only. According to Cukier's empirical formula (Cukier *et al.*, 1975), the computational cost of the classic FAST

Table 18.4 The FAST indices computed for the 10-factor model. The remaining 58 uncertain factors were kept fixed at their nominal values. Reproduced from *Computer Physical Communications*, **117**, Tackling quantitatively large dimensionality problems, by Campolongo, F., Tarantola, S., and Saltelli, A., pp 75–85, 1999, with permission from Elsevier Science.

Factor	First-order index (S_i)	Total index (S_{T_i})
WATLIQ	0.075 95	0.136 68
YOOHRAD	0.006 63	0.017 26
Q1	0.932 84	0.971 96
QW7	0.012 07	0.059 23
Q21	0.000 03	0.013 91
RHLO3	0.001 50	0.023 45
RHLH202	0.000 08	0.013 72
RHLDMS	0.002 01	0.018 98
DEHH202	0.000 07	0.011 86
DEHO3	0.000 13	0.017 52

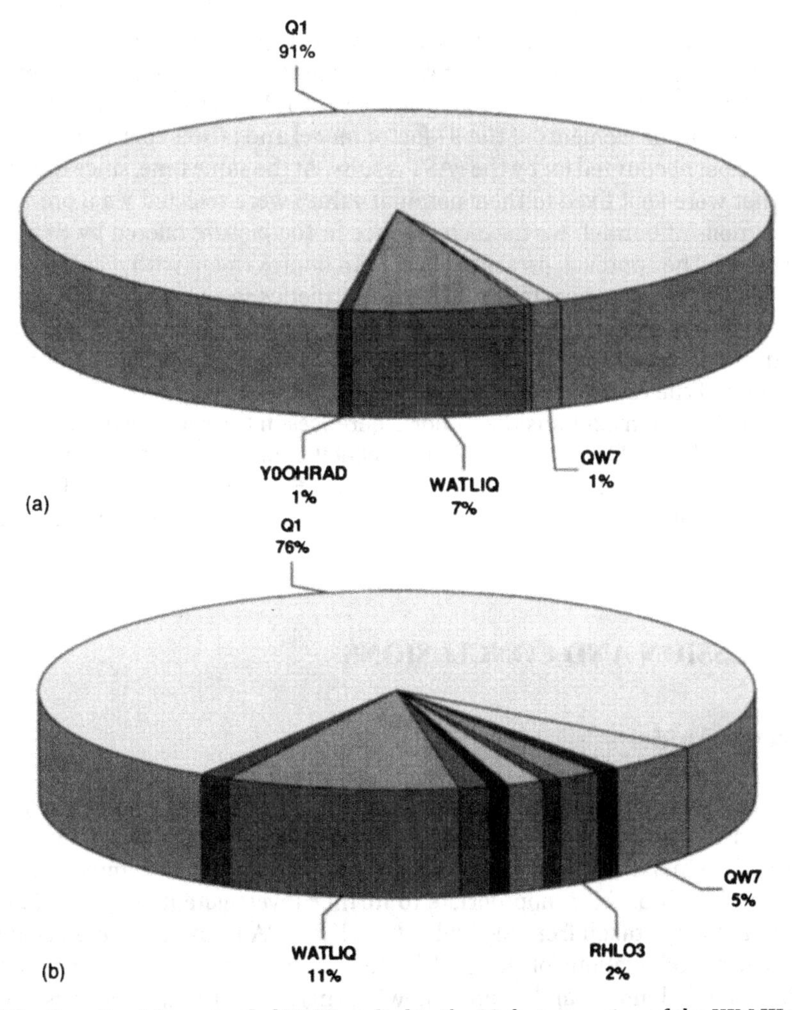

Figure 18.3 Results of the extended FAST applied to the 10-factor version of the KIM-III model. The two pie charts, respectively for the S_i (a) and for theS_{T_i} (b), represent the percentage contributions of the 10 factors to the total output variance. Reproduced from *Computer Physical Communications*, **117**, Tackling quantitatively large dimensionality problems, by Campolongo, F., Tarantola, S., and Saltelli, A., pp 75–85, 1999, with permission from Elsevier Science.

would have been $C = 806$. In this particular exercise, standard FAST would have likely sufficed, given that, within error, the model behaves additively (the sum of the S_i is 1 on the pie chart (a)). The pie chart (b) indicates non-negligible interactions, though these are likely to be an overestimate. Indeed $S_{Q_1} = 0.93$, while $S_{TQ_1} = 0.97$; once all the S_{T_i} are renormalized with the sum of the total indices, the influence of Q_1 ends up being reduced. The true relevance of interactions is likely to lie between the values given by the two charts. If present, interactions do not play a dominant role.

The pie chart (b) in Figure 18.3 yields useful quantitative information about the overall relative importance of the 10 factors. It shows that the factor Q_1, which is the quantile of k_1, is by far the most important among the 10 factors, accounting for approximately 76–90% of the output variance *D*. This can be explained based on the chemistry of the system (Campolongo *et al.*, 1999a). Furthermore, the same result was found by Campolongo *et al.* (1999b) when studying the sensitivity of the KIM-III model with the old value of k_{21}. The second most important factor is WATLIQ, explaining about 7–11% of *D*. The other seven factors included in the analysis account for 4–13% of the total output variance *D*.

The above results are valid and quantitative for the 10-factor model. When we extend them to the original (68-factor) model, the results can be taken as qualitative, since possible interactions between the elements of the 10-factor model and others that were kept fixed in this analysis are not accounted for by the FAST results. At the same time, since the 58 uncertain factors that were kept fixed to their nominal values were selected via a pre-screening experiment as non-influential, we have confidence in the picture offered by FAST also for the original model. The approach presented here is an improvement with respect to the work of Falls *et al.* (1979). These authors tackle a chemical kinetics system with 57 factors, similar to ours, but perform a pre-screening by arbitrarily varying all factors by 5% around their nominal value. FAST indices are then computed on a selected set of 17 factors. In this case, the extrapolation of the results to the original 57-factor model is arguable.

The same consideration explains the minor differences in the ranking of the importance of the factors provided by Morris and FAST. The reliability of the FAST results is conditional upon the choice of the subgroup of factors to which FAST is applied. As shown in this section, the choice of the factors to be used in FAST is a delicate task for the analyst to decide upon.

18.6 DISCUSSION AND CONCLUSIONS

18.6.1 On the Model

The SA conducted on the KIM model (version KIM-III) run with the k_{21} value from Ray *et al.* (1996) identified the subset of input factors that mostly control the model output, i.e. α. Among these, the kinetic constant k_{21} was indicated. Given the major importance of this constant, the investigator felt it appropriate to further investigate its value and ended up changing it to the value taken from Mellouki *et al.* (1988). SA is an iterative procedure: first its results suggest modifications of the model, then another SA experiment has to be conducted on the modified model, and so on. Following this line of thought, SA was performed again on the KIM-III model run with the k_{21} value from Mellouki *et al.* (1988). The results of this second set of experiments (Morris plus FAST on the 10-factor model) indicated which factors need to be more accurately estimated in order to improve the agreement between model predictions and observed data.

18.6.2 On the Practice of SA

A major conclusion of our set of experiments is that the use of EOAT or a local SA method to draw conclusions on the relative impact of uncertain or variable input factors on the prediction of a model should be avoided unless the variation of the factors is small. Even when the investigator does not desire to embark on a quantitative SA study, the Morris method, or even a plain Monte Carlo analysis at low sample size, should be preferred over elementary OAT. Similarly, the use of local methods should be confined to situations where they can perform effectively, as in the case of inverse problems or in the computation of large sets of sensitivity coefficients as a function of time.

More generally, we advocate the use of global, possibly quantitative, sensitivity analysis methods for all problem settings where finite parameter variations are involved.

Experience with environmental models (e.g., those involving mass transfer with chemical reaction) shows the following: (i) Different factors are affected by different ranges of variation / uncertainty. Different sensitivity patterns predominate in different regions of the space of the input, especially when models are nonlinear. (ii) The presence of non-negligible interactions (i.e. the effect of changing x_j and x_i is different from the sum of the individual effects) cannot be discounted; see Chapter 13 for an example. For these reasons, statements of the kind 'x_j is more important than x_i' using EOAT or local SA approaches cannot be sustained (e.g., they might be true, but thus cannot be inferred from the SA).

When the number of input factors involved in the model is too high to afford a computationally expensive quantitative analysis, a two-step procedure may be adopted. The final results of such a procedure are conditional upon the success of the first-phase screening exercise. The choice of screening method is therefore extremely delicate and important. The screening has to be a global method, and it is very important that the probability of an error of Type II, i.e. that of not identifying a factor that indeed is important, is low. Previous exercises and iterations of the present one seem to indicate that Morris does not make Type II errors (Campolongo *et al.*, 1999a,b). This is reasonable, since the influence of factors in models follows—according to our experience—a Pareto-like distribution, with few factors accounting for most of the variance, and most of the factors taking up the remainder. One has to build a model artificially in order to achieve a case where the influence of factors is more uniformly balanced.

For a model with about 70 factors (such as KIM) it is very likely that less than 10 have some sizeable influence, and in this case two of them clearly contribute more than 90% of the variance. In other words, although it is an arbitrary choice, by selecting the first 10 factors identified by the screening for the quantitative analysis we feel reasonably protected from an error of Type II.

We also want to underline one more time, that the results obtained for the 10-factor model must be carefully interpreted, and cannot be extended to the original version of the model, where 68 input factors are uncertain.

19

An Application of Sensitivity Analysis to Fish Population Dynamics

José-Manuel Zaldívar Comenges and Francesca Campolongo

European Commission, Joint Research Centre, Ispra, Italy

19.1 INTRODUCTION

The central problem of ecology concerns the relationships of individual organisms with their environment, the interactions and diversity of species, and the fluxes of energy and materials through ecosystems. Although ecological modelling has experienced great development in recent years, the identification of nonlinear demographic behaviour in ecological systems continues to provoke debate; see Dennis *et al.* (1995). This is mainly due to the difficulty of connecting models and data, i.e. there is a lack of explicit connections between theories and experiments. This is not surprising, since data collection from ecological systems requires careful sampling techniques and the level of noise contamination is high. The difficulty is heightened because ecosystems evolve over very long periods of time, and hence, to characterize their dynamic behavior, we assume that the system has reached a steady state and is not on a transient. This may not be the case for an important number of ecological systems. Furthermore, environmental fluctuations play a fundamental role in such systems, and hence it is really difficult to distinguish when a population change is due to nonlinear interaction between different species, e.g. competition and predation, or has been provoked by an environmental factor, e.g. temperature and solar intensity. Moreover, there is a lack of rigorous parametric sensitivity analysis in existing models that makes it difficult to assess which factors are important and with which degree of precision experimental data are needed.

Zaldívar *et al.* (1998) addressed the problems of modelling the dynamics of fish ecosystems using Lotka–Volterra nonlinear differential equations and stage-based discrete models.

Sensitivity Analysis. Edited by A. Saltelli *et al.*

Models were developed to mimic the data on scale deposition rates of small pelagic fish, namely sardine and anchovy, in different locations: the California current off western North America and the Benguela current off southwestern Africa. The simulation results showed that, although environmental fluctuations can explain the magnitude of observed variations in geological recordings and catch data of pelagic fishes, they cannot explain the low observed frequencies. This implies that relevant nonlinear biological mechanisms must be included when modelling their dynamics. From the comparison between the geological data and simulation results, it was concluded that environmental fluctuations seemed to be responsible for the higher frequencies (periods shorter than 150 years), whereas population-density-dependent interactions seemed to be responsible for low frequencies (periods longer than 450 years).

In this chapter, the discrete stage-based model developed by Zaldívar *et al.* (1998) is further investigated. In a stage-based model, the life cycle is defined in terms of size classes or development stages, rather than age classes (Caswell, 1989). It assumes that vital rates depend on body size and that growth is sufficiently plastic that individuals of the same age may differ appreciably in size. Despite the fact that the ecological structure of the model has been kept as simple as possible, in its final version the model contains over 100 biological and physical factors. A sensitivity analysis exercise is conducted in order to assess the relative importance of the various factors and physical processes involved. The method used is the screening method proposed by Morris (1991).

The guiding philosophy of the Morris experiment is to determine, within reasonable uncertainty, which input factors can be considered to have effects that are (a) negligible, (b) linear and additive, and (c) nonlinear or involved in interactions with other inputs. The method is based on a composite experiment based on individually randomised one-factor-at-a time (OAT) designs in the input factors (Chapter 4). For the results, data analysis is based on examination of changes in output that are unambiguously attributable to changes in individual inputs. *A priori* correlation among the input factors is assumed to be negligible. The method requires a number of model evaluations of the order of k, where k is the number of factors under examination.

19.2 GENERAL CHARACTERISTICS OF PELAGIC FISH ASSEMBLAGES ANALYZED

The term pelagic fish refers to fish that move freely in open seas, spend most of their time in the water column, and usually feed in the same realm. Pelagic fish can be divided into three categories: small pelagic fish (e.g. anchovies, sardines, and herring), mid-size pelagic fish (e.g. horse mackerel, mackerel and yellowtail), and large pelagic fish (e.g. tuna and swordfish). Practically all the differences in their behavior, i.e. food, feeding strategy, migrations, etc., depend on their size.

Most pelagic fish live in shoals that give them an advantage in swimming, searching for food, and avoiding predators. However, shoaling makes pelagic fish particularly vulnerable to modern fishing gear and intensive exploitation that often leads to overfishing. On the other hand, since pelagic stocks fluctuate naturally over long periods of time, it is difficult to determine the real cause of stock decline without long-term population and environmental studies. From data on long-term records, it can be concluded that fish abundance can decrease to levels similar to those we could call 'collapses' or 'failures' without fishing pressure, and, reciprocally, we could expect recoveries of fish stocks from very low levels of abundance due solely to natural factors. Obviously, fishing reinforces natural changes, and, on some occasions, will significantly affect natural trends. But it is very difficult to distinguish fishing mortality from natural mortality, if we consider the life cycle of the population as a whole. Hence, the sustainability of a given yield cannot be assured in most cases for these types of fish, given that fishing mortality is not the only factor regulating abundance.

The world's main concentrations of small pelagic fish coincide with the major upwelling regions. The general locations of the most highly productive upwelling regions are well known; they are the coastal waters of Japan, and four eastern boundary current systems: the California current off western North America, the Peru current off western South America, the Canary current off the Iberian Peninsula and north-western Africa, and the Benguela current off southwestern Africa. Bakun and Parrish (1982) showed that there was a group of a half-dozen species common to all regions: anchovy, sardine, horse mackerel, hake, Spanish mackerel, and bonito.

19.2.1 Fossil Records of Sardine and Anchovy Scales

From the above-mentioned assemblages of pelagic fish, geological data in the form of time series of fish-scale counts exist for three different regions:

- Pacific sardine and northern anchovy from the Santa Barbara basin (Soutar and Isaacs, 1969, 1974; Baumgartner *et al.*, 1992);
- DeVries and Pearcy (1982) constructed sequences of fossil scales from the slope sediments off Peru;
- Shackleton (1986, 1987) studied the Benguela margin off west Africa.

The study of fossil fish scales is based on the assumption that the number of fish scales preserved in the sediment is related to the size of the fish populations that produced them. This relationship depends on the number of scales from a fish of a particular species that arrive at the sea bed, and the degree to which the scales are preserved in the sediment during and after burial. The relationship between fossil scales and the fish populations that produced them need not necessarily be the same from species to species, since there are factors that are different, for example scale preservation (Shackleton, 1986). Furthermore, there are other related problems, including misidentification (i.e. confusion between the scales of the two species), exclusion of the small scales to avoid the above problem and reduce the error, etc.

Soutar and Isaacs (1969) developed time series of fish-scale counts for small pelagic species, including the Pacific sardine and northern anchovy; these series were based on the analysis of a piston core from the Santa Barbara Basin and extend back over nearly two millennia. After this, they constructed (Soutar and Isaacs, 1974) shorter series covering the 160 years from 1810 to 1970, which allowed them to compare and integrate the paleoecological record with direct estimates of population biomass. Baumgartner *et al.* (1992) constructed a more complete time series by integrating and adding to the data sets developed by Soutar and Isaacs and by providing a detailed analysis to determine the strength of the signal compared with the noise in scale-deposition rates. Furthermore, they also recalibrated the scale-deposition data using available population estimates for recent years and converted the scale-deposition series into units of biomass series. Sequences of fossil scales have since been constructed from the slope sediments off Peru (De Vries and Pearcy, 1982) and for the Benguela margin off west Africa (Shackleton, 1986, 1987). But the lack of well-developed, continuously varved records in these two last places has so far prevented the reconstruction of time series of equal quality to those from Santa Barbara Basin.

In this study, we have considered the data record provided by Baumgartner *et al.* (1992) and the longest data record provided by Shackleton (1987). Spectral analysis of the California cores (Baumgartner *et al.*, 1992) shows that at the higher frequencies (periods shorter than 150 years), sardines and anchovies both tend to vary over a period of approximately 60 years. In addition, the anchovy fluctuate at a period of 100 years. At lower frequencies, the anchovies appear to fluctuate with longer period (680 years) than the sardines (480 years). Furthermore, there is a weak positive correlation between the sardine and anchovy series, which is carried by the low-frequency component of the variances (periods of 150 years or longer). The scale-deposition record shows nine major recoveries (defined as an increase from less than one million to over four million metric tons of biomass) and subsequent collapses of the sardine population over 1700 years. The average time for recovery of the sardine is 30 years.

In the data from the Namibian core, Shackleton (1986, 1987) showed that there are a series of cycles with a period of about 25 years for pilchard and 14 years for anchovies. Within this general pattern, the pilchard and anchovy interact in two different ways: during some periods, it seems that there is the replacement of one species by the others (see Figure 19.1), i.e. when pilchard scales are at a maximum, anchovy scales are at a minimum; on the other hand, during some periods, they seem to be in phase, i.e. when pilchard scales are maximum, anchovy scales are also maximum (see Figure 19.2).

The general conclusion after analyzing the data is that large, natural fluctuations in the size of fish populations occurred long before fisheries perturbed the stocks, and hence we must understand the causes of these fluctuations before being able to assess the impact of human intervention or before being able to decide a coherent fisheries management strategy.

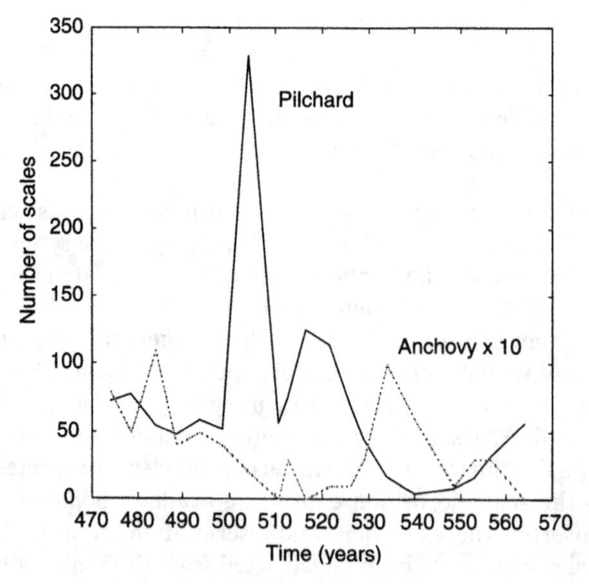

Figure 19.1 Distribution of anchovy and pilchard scales showing the periodic replacement pattern (data from Shackleton).

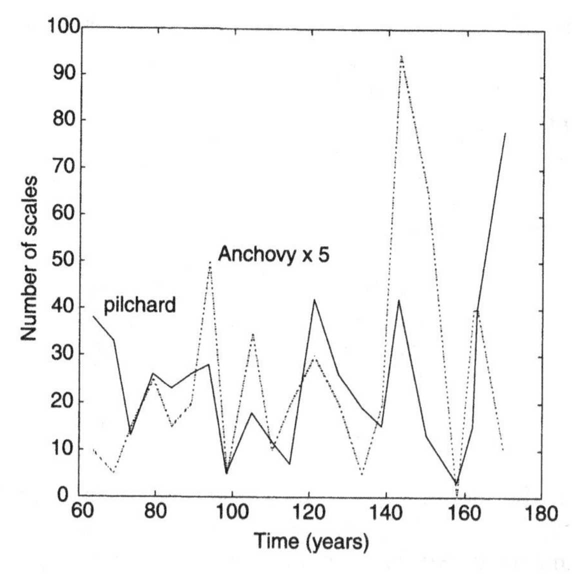

Figure 19.2 Distribution of anchovy and pilchard scales showing the in phase pattern (data from Shackleton).

19.3 A STAGE-BASED MODELLING APPROACH TO ANALYZE FLUCTUATIONS IN PELAGIC FISH POPULATIONS

Regimes of high and low abundance of pelagic fish have attracted considerable interest, especially the frequently observed alternation between sardines and anchovies. Various hypothesis regarding the initiation and maintenance of regimes of high or low abundance have been proposed:

- fluctuating environment;
- density-dependent reproduction rates;
- interspecific competition;
- predation/cannibalism.

In order to model the impact of all these effects on the population dynamics of pelagic fish, a discrete modelling approach, based on stage-specific matrix transitions (Caswell, 1989), has been used. The existence of demographically important differences among individuals is obvious in the case of pelagic fish that undergo several stages before maturity. The use of continuous ordinary differential equations would ignore population structure by treating all individuals as identical. Matrix population models (Caswell, 1989) integrate population dynamics and population structure particularly clearly, and they are really useful when the life cycle is described in terms of size classes or development stages, rather than age classes. There are fundamentally two types of approaches: the age-classified model and the stage-classified model. The first assumes that age-specific survival and fertility are sufficient to determine population dynamics. However, if the dependence of vital rates on body size and growth is sufficiently plastic that individuals of the same age may differ appreciably in size, then age will provide little information about the fate of an individual. In the case of fish, it seems more convenient to use a stage-based model, since some species must achieve a threshold size before beginning to reproduce, and once reproduction begins, reproductive output is strongly dependent on adult body size.

In the stage-based type of modelling, the matrix **A** describes the transformation of a population from time t to time $t+1$:

$$\mathbf{n}_{t+1} = \mathbf{A}\,\mathbf{n}_t, \tag{19.1}$$

where **A** has the following structure:

$$\mathbf{A} = \begin{bmatrix} P_1 & m_2 & m_3 & m_4 & \dots & m_{q-1} & m_q \\ G_1 & P_2 & 0 & 0 & \dots & 0 & 0 \\ 0 & G_2 & P_3 & 0 & \dots & 0 & 0 \\ \vdots & \vdots & \vdots & \vdots & & \vdots & \vdots \\ 0 & 0 & 0 & 0 & \dots & G_{q-1} & P_q \end{bmatrix}, \tag{19.2}$$

$\mathbf{n}_t$ is a vector describing the population at each stage at time t, P_i is the probability of surviving and staying in stage i, G_i is the probability of surviving and growing into the next stage, and m_i is the maternity per fish per unit time (day), $i = 1, 2\dots, q$. Both P_i and G_i are functions of the survival probability p_i and the growth probability γ_i (Caswell, 1989):

$$P_i = p_i(1-\gamma_i), \tag{19.3}$$

$$G_i = p_i\,\gamma_i, \tag{19.4}$$

where

$$p_i = e^{-z_i} \tag{19.5}$$

and

$$\gamma_i = \frac{(1-p_i)p_i^{d_i-1}}{1-p_i^{d_i}}, \tag{19.6}$$

where z_i is the daily instantaneous mortality rate (IMR) and d_i is the duration (days) within the ith stage.

19.3.1 Fluctuating Environment Models

Physical factors may influence biological systems directly or indirectly. Direct physical influences include changes in transport pathways in both the horizontal and vertical planes, alterations in physiological rates, and shifts in the spatial boundaries of acceptable habitat. Indirect pathways include secondary responses of production to influences on nutrient concentrations, food availability, the distribution of predators, growth rates, and reproductive capabilities.

To consider a model that incorporates environmental fluctuations, a stage-based model is employed. Fortunately, for the different classes and species of interest, the derivation of the model factor values can be found in the literature (see Smith *et al.*, 1992; Dickerson *et al.*, 1992; Butler *et al.*, 1993). The values of the different factors for the sardine used in this work are specified in Table 19.1. For the anchovy, we used the values from Butler *et al.* (1993) with nine stages, i.e. egg, yolk-sac larvae, early larvae, late larvae, early juvenile, late juvenile,

Table 19.1 Stage-specific life-history factor values of the Pacific sardine (Smith *et al.*, 1992).

	Size (mm)		Daily natural mortality			Duration (days)			Daily fecundity		
Stage	**Min.**	**Max.**	**Min.**	**Best**	**Max.**	**Min.**	**Best**	**Max.**	**Min**	**Best**	**Max.**
Egg			0.31	0.72	2.12	1.4	2.5	3.9	0	0	0
Yolk-sac larvae	Hatch	4	0.394	0.6698	0.971	1.4	3.1	3.9	0	0	0
Early larvae	4	10	0.1423	0.2417	0.3502	5	11	21	0	0	0
Late larvae	10	35	0.057	0.0964	0.139	20	35	50	0	0	0
Early juvenile	35	60	0.029	0.056	0.081	17	25	40	0	0	0
Juvenile I	60	85	0.0116	0.0197	0.0285	30	50	80	0	0	0
Juvenile II	85	110	0.0023	0.004	0.0058	80	110	146	0	0	0
Juvenile III	110	135	0.0016	0.0028	0.004	105	146	185	0	0	0
Juvenile IV	135	160	0.0012	0.0022	0.0032	110	170	220	0	0	0
Prerecruit	160	185	0.0006	0.0011	0.0015	110	175	220	0	80	161
Early adult	185	210	0.0006	0.0011	0.0015	190	381	570	286	389	489
Adult	210	235	0.0006	0.0011	0.0022	400	663	920	730	946	1114
Late adult	235	260	0.0006	0.0011	0.0022	1908	2773	3473	1064	1688	3123

prerecruit, early adult, and late adult (see Table 19.2). For chub mackerel, we used data from Dickerson *et al.* (1992) and also nine stages, i.e. egg, early larvae, late larvae, early juvenile, juvenile, late juvenile, prerecruit, early adult, and late adult (see Table 19.3).

The 'best' factors were chosen to produce a dominant eigenvalue λ_{max} (population growth rate) close or equal to 1 in the matrix **A**, i.e. the population is stationary. By measuring the variation of the dominant eigenvalue, Butler *et al.* (1993) concluded that natural variations in individual stage-specific mortality or duration result in changes from equilibrium to an annual increase or decrease of about threefold. This magnitude is sufficient to explain the large changes in population size observed in nature.

To study the dynamic behavior of such a system, we have simulated the sardine population dynamic behavior assuming a cyclic fecundity period of one year with values between the maximum and minimum in Table 19.1 and a random variation for the rest of the factors within the same limits. To avoid an exponential explosion or an extinction of the population, it was found necessary to correct the limiting values to give a random distribution centered around $\lambda_{max} = 1$. Figure 19.3 shows the fluctuations in the different population classes for

Table 19.2 Stage-specific life-history factor values of the Northern anchovy (Butler *et al.*, 1993).

	Size (mm)		Daily natural mortality			Duration (days)			Daily fecundity		
Stage	**Min.**	**Max.**	**Min.**	**Best**	**Max.**	**Min.**	**Best**	**Max.**	**Min**	**Best**	**Max.**
Egg			0.12	0.231	0.45	1.4	2.9	3.9	0	0	0
Yolk-sac larvae	Hatch	4	0.19	0.366	0.59	1.4	3.6	3.9	0	0	0
Early larvae	4	10	0.187	0.286	0.345	8	12	23	0	0	0
Late larvae	10	35	0.047	0.0719	0.087	35	45	71	0	0	0
Early juvenile	35	60	0.0009	0.02796	0.017	45	62	100	0	0	0
Late Juvenile	60	85	0.0029	0.0044	0.0053	60	80	138	0	0	0
Prerecruit	85	110	0.002	0.0031	0.0037	200	287	632	0	10.5	19.4
Early adult	110	135	0.0011	0.0021	0.0036	750	1000	1250	143.8	199.2	230.7
Late adult	135	160	0.0011	0.0021	0.0036	1000	1250	1500	284.2	448.4	529.0

Table 19.3 Stage-specific life-history factor values of the chub mackerel (Dickerson *et al.*, 1992)

	Size (mm)		Daily natural mortality			Duration (days)			Daily fecundity		
Stage	**Min.**	**Max.**	**Min.**	**Best**	**Max.**	**Min.**	**Best**	**Max.**	**Min**	**Best**	**Max.**
Egg			0.126	0.240	1.614	6.43	16	21.14	0	0	0
Early larvae	4	10	0.020	0.150	0.360	3.03	23	33.74	0	0	0
Late larvae	10	35	0.016	0.050	0.079	29.08	89	121.2	0	0	0
Early Juvenile	35	60	0.0009	0.020	0.055	20	64	144.5	0	0	0
Juvenile	60	110	0.0009	0.010	0.045	17	128	289	0	0	0
Late Juvenile	110	135	0.0016	0.027	0.045	45	64	144.5	0	0	0
Prerecruit	135	225	0.0005	0.00056	0.0018	190	346	570	288.2	411.7	452.9
Early adult	225	300	0.0005	0.00056	0.0018	400	663	920	691.7	988.14	1086.1
Late adult	300	340	0.0005	0.00056	0.0018	1908	2773	3473	1165.4	1664.9	1831.4

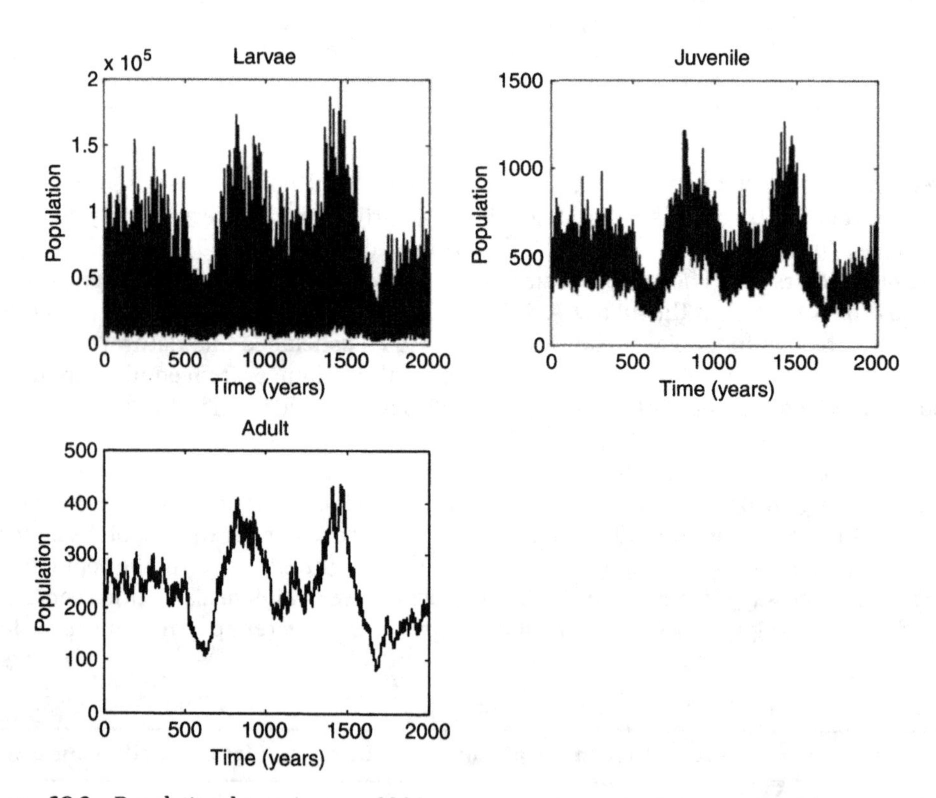

Figure 19.3 Population dynamics over 2000 years assuming cyclic (one-year) fertility and random factors for sardine.

the sardine (larvae = class 2 + 3 + 4, juvenile = class 5 + 6 + 7 + 8 + 9, and adult = class 10 + 11 + 13). As can be seen, even daily random environmental fluctuation can produce the observed fluctuations.

In a recent study (US GLOBEC Executive Summary) on the consequences for fish under a global warming scenario, the effects of different physical processes on different pelagic fish populations were assessed. The results showed that, for example, the effect of a change in sea temperature has different effects on sardine and anchovy, whereas stratification has the same effect. Hence, in-phase and out-of-phase (replacement) fluctuations can be explained by environmental effects. However, it is difficult to assign the different frequencies found in the scale deposition rates and to explain the correlation found between sardines and anchovy time series in the low-frequency component of the variance assuming only environmental effects.

19.3.2 Intraspecies Competition

Let $x(t)$ be the population of the species at time t; then the rate of change

$$\frac{dx}{dt} = \text{births} - \text{deaths} + \text{immigration} - \text{emigration} \tag{19.7}$$

is a conservation equation for the population. The form of the various terms on the right-hand side of Equation (19.7) necessitates modelling the situation that we are concerned with. The necessity of density dependence in this equation was stated by Haldane (1953), who noted that if $dx/dt \approx 0$ over a number of generations, some or all the variables must be functions of $x(t)$ unless the sum of births − deaths + immigration − emigration is always identically zero, which is unlikely. Thus a population persisting for many generations must be regulated by density dependence.

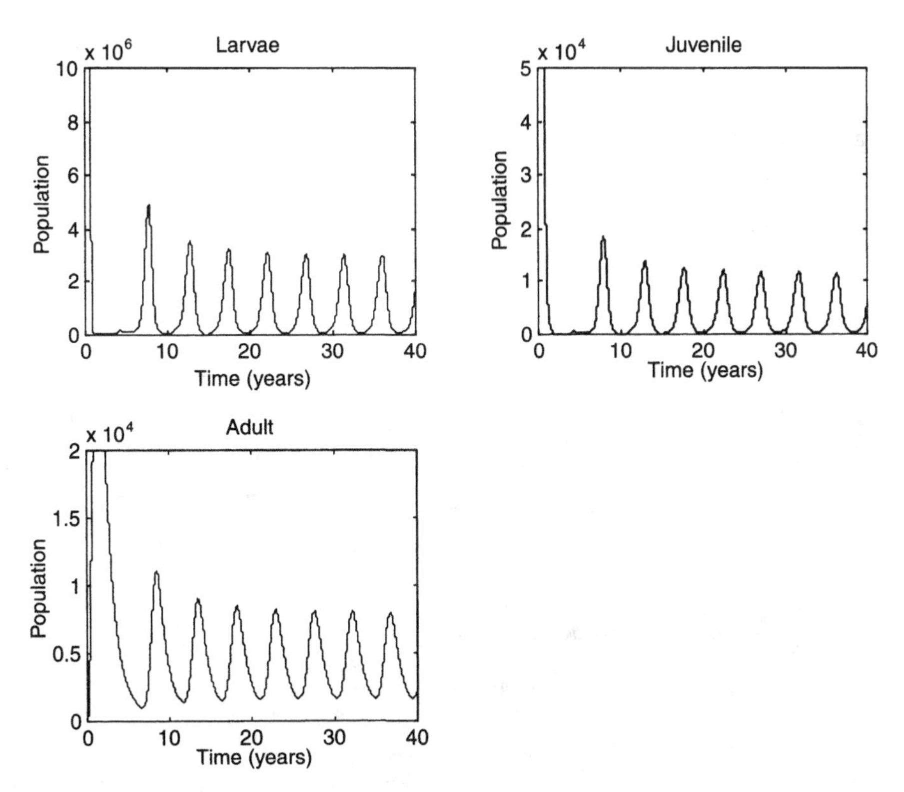

Figure 19.4 Population dynamics over 40 years assuming density-dependence fertility for anchovy (larvae = class 2 + 3 + 4, juvenile = class 5 + 6, adult = class 7 + 8 + 9).

In the case of a stage-based discrete system, the projection equation can be written as Equation (19.1):

$$\mathbf{n}_{t+1} = \mathbf{A}_{\mathbf{n}}\mathbf{n}_t \tag{19.8}$$

where the notation $\mathbf{A}_{\mathbf{n}}$ indicates that the matrix entries $a_{ij}(\mathbf{n})$ depend on the population vector $\mathbf{n}$. At each stage in the life cycle, the individual may utilize resources at different rates, and may thus differ in its contribution to 'density'.

In our case, we have chosen the discrete model for the anchovy (Butler *et al.*, 1993) and we have simulated density-dependent fecundity, using the Ricker equation, and density-independent survival rates in a similar way as Levin and Goodyear (1980) did in a study of the striped bass. In this case, the fecundity of the last three stages of anchovies, i.e. pre-recruit, early adult, and late adult, is given by

$$f_i = \alpha f_i(0)e^{-\beta N}, \tag{19.9}$$

where $f_i(0)$ is the fecundity given by the stage-specific life-history factors (Butler *et al.*, 1993), and α and β are two factors that measure the strength of the density dependence.

Figure 19.4 shows the simulation results. As can be seen, after a transitory phase, the oscillations become stable and the solution is a limit cycle. However, as observed for the continuous case, the time interval between fluctuations tends to be no longer than one generation, which does not correspond to the observed data on scale deposition rates.

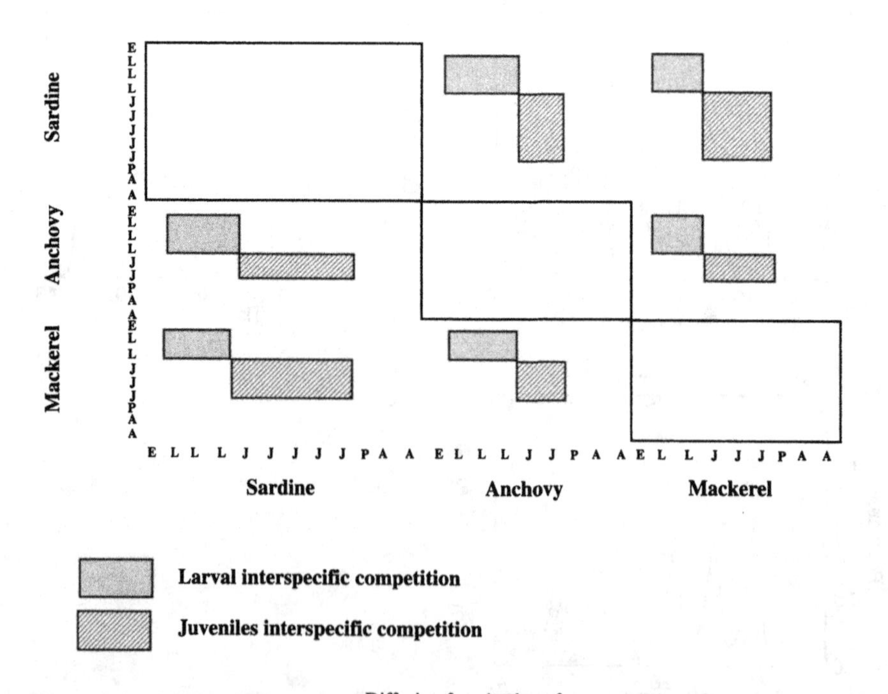

Figure 19.5 Schematic layout of the discrete version of the replacement model (E = egg, L = larvae, J = juvenile, P = prerecruit, A = adult).

19.3.3 Interspecific Competition

Four main types of interaction between species are recognized in ecology: protocooperation, mutualism or symbiosis; mutual competitive suppression, or competition for a common resource; predator–prey or parasite–host interactions; and interactions in which one species has either a positive or negative effect on the other, but is completely unaffected by the other species (Odum, 1971).

Zaldívar *et al.* (1998) considered, according to the ecosystem they were interested in modelling, different plausible alternatives and discussed the consequences as a function of the simulation results and the experimental data. Their study led to the conclusion that it is difficult to justify the use of continuous models to study population structure. For this reason, they developed a discrete version of the cyclic advantage relationship in competitive ability model (Matsuda *et al.*, 1992). The structure of such a model is shown in Figure 19.5. Each population is represented by its discrete matrix transition **A** (see Equation (19.2), factors in Table 19.1, and references). Furthermore, density-dependent competition between different life stages, i.e. larvae, juveniles, etc., is allowed by a discrete function similar to the continuous model (Matsuda *et al.*, 1992). However, in this case there is a matrix of interactions. The approach used, to identify the factor region in which the three populations coexist, is the same as in the continuous case (Zaldívar *et al.*, 1998).

Furthermore, the model also includes the diffusion to a refuge. This is modelled by doubling the dimension of the matrix by including the free systems plus a common term allowing the movement from one patch to the other. In this case diffusion is only permissible for the adult population.

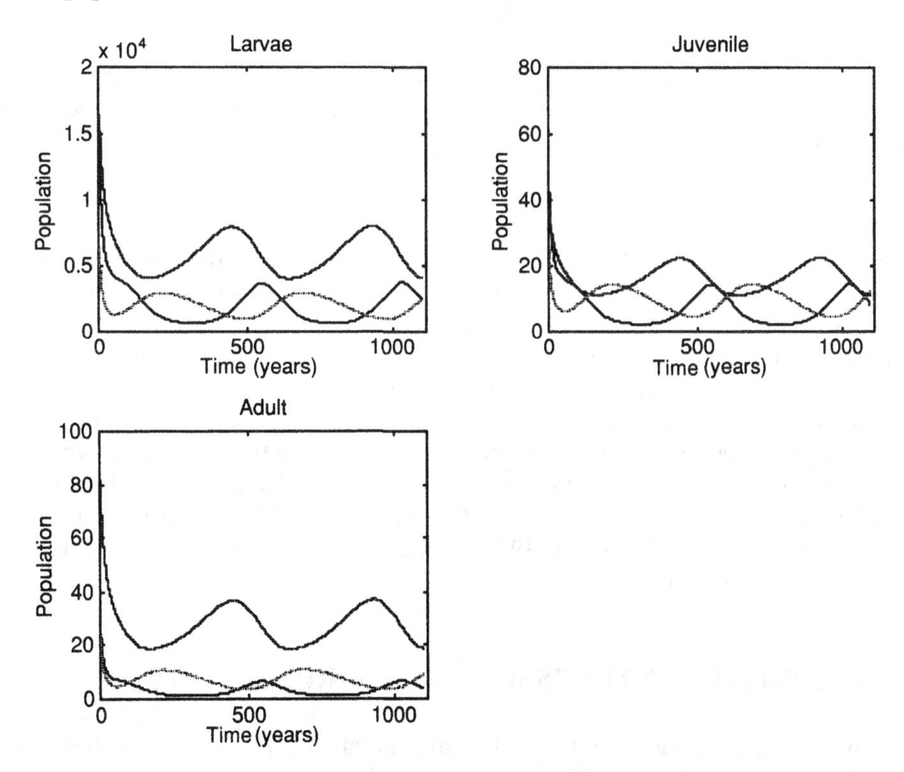

Figure 19.6 Simulation of the discrete version of the cyclic advantage relationship in the competing ability model in the common habitat. Competition between larvae and juveniles.

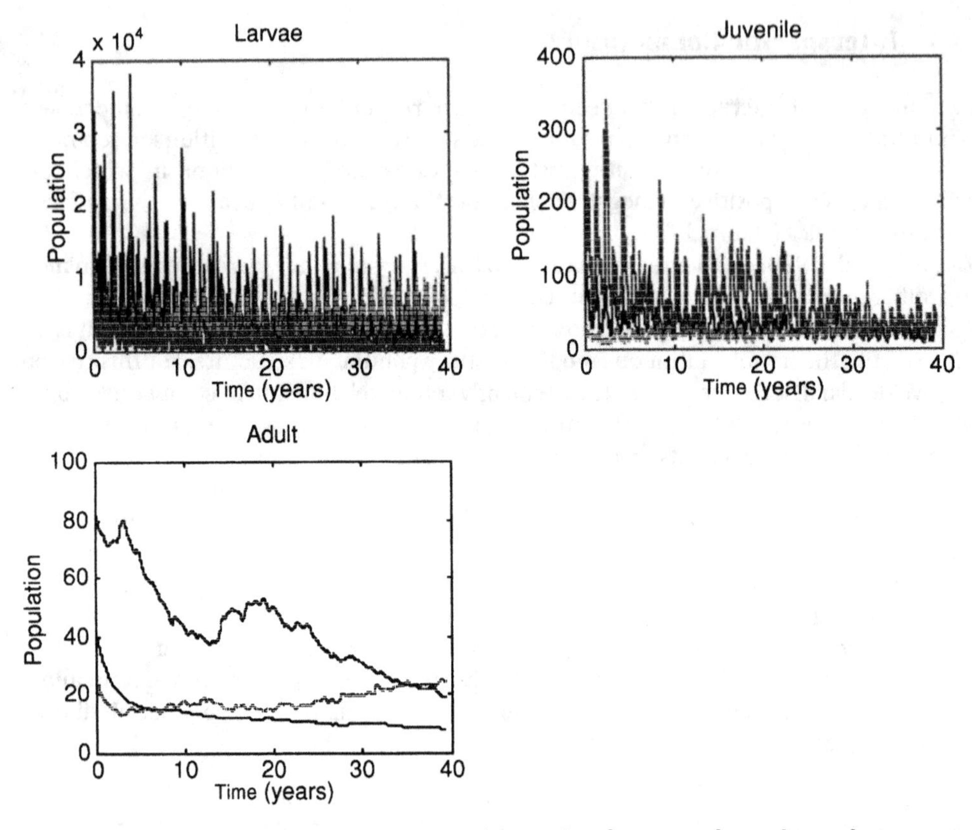

Figure 19.7 Simulation of the discrete version of the cyclic advantage relationship in the competing ability model in the common habitat. Competition between larvae and juveniles plus random environmental fluctuations.

Figure 19.6 shows the simulation of approximately 1000 years using the discrete model and allowing only competition between larvae and juveniles but not the adult population. As can be seen, periodic replacement induced by interspecific competition between larvae and juveniles of the three species induces replacement in the adult population, even though there are no specific competition terms between adults.

Moreover, it is important to notice the low frequencies generated by this type of interaction, which produces periods of approximately 500 years (the time step of our simulation is one day). Including competition between adults, the dynamic behavior is similar. However, the period between replacements increases (Zaldívar *et al.*, 1998). The inclusion of random environmental fluctuations (see Figure 19.7) introduces high-frequency fluctuations superimposed on the replacement pattern.

19.4 SENSITIVITY ANALYSIS BY THE MORRIS METHOD

In order to improve our understanding of the fish population dynamics, a sensitivity analysis (SA) exercise is conducted on the stage-based model described in the previous section (Figure 19.5). The first step in conducting such an analysis is to determine which are the quantities of interest. First of all, the investigator has to specify which is the model response (or model output) that he/she considers as most informative for the goal of his/her analysis. In this study, the chosen model output is λ_{max}, i. e. the dominant eigenvalue in the matrix **A** after one year's simulation time.

Then, the input factors to be included in the analysis have to be selected. In its final version, the fish population model described in Section 19.3.3 contains 103 input factors. All of them are included in the present analysis. Of those 103, 72 are factors that represent the daily natural mortality (Z), duration (D), and daily fecundity (F) of each of the three species under study (sardines (I), anchovies (J), and mackerels (K)). Their best values and ranges of variation are those given in Tables 19.1–19.3 (note that the daily fecundity factors of early development stages with min = max = 0 are not considered). To simplify notation, these 72 factors are denoted by two capital letters, the first indicating the type of factor (Z, D, or F), and the second indicating the species to which it is referring (I, J, or K). Numbers in parentheses denote the life stage: $i = 1, \ldots, 13$ for sardines, i.e. from egg to late adult following the life stages given in Table 19.1, and $i = 1, \ldots, 9$ for anchovies and mackerels, from egg to late adult following the life stages given respectively in Tables 19.2 and 19.3. For example, ZJ(3) denotes the mortality (Z) of anchovies (J) in the early-larvae stage. The remaining 31 inputs are factors involved in the migration and interspecific competition between larvae and juveniles. Lower-case letters denote these factors.

As mentioned several times in the present volume, the choice of SA method depends strongly on the number of input factors examined. The choice of the Morris method is in this case perfectly motivated, since the model contains over 100 input factors (Morris, 1991). A detailed description of the Morris method is given in Chapter 4.

19.4.1 Application of the Morris Method to the Fish Population Model

The Morris method has been applied to our fish population model with a sample size $r = 10$. Each of the 103 input factors is assumed to follow a uniform distribution among its extreme values, reported in Tables 19.1–19.3. In the design, each factor is varied across four levels ($l = 4$). A total number $N = 1040$ of model evaluations is performed ($N = r \times (k + 1)$, where k is the number of input factors).

19.4.2 Results and Discussion

The Morris method provides two sensitivity measures for each input factor: a measure μ of the 'overall' effect of an input, and a measure σ that is the sum of all the second- and higher-order effects in which the factor is involved (including curvatures and effects due to interaction with other factors). Results of the Morris screening exercise are given in Figure 19.8, where μ and σ are plotted for the 103 input factors. Labels indicating names of factors are reported only for the most important factors.

An immediate conclusion that can be drawn from an analysis of Figure 19.8 is that each input factor having a high value of the estimated mean μ also has a high value of the estimated standard deviation σ; all the points lie around the diagonal. This implies that none of the input factors that affect the output have a purely linear effect.

The main goal of a screening exercise is to rank the factors in order of importance, and to identify the subset of the most important. Of the two Morris measures μ and σ that provide different information about the effects of each factor on the model, μ, which estimates the

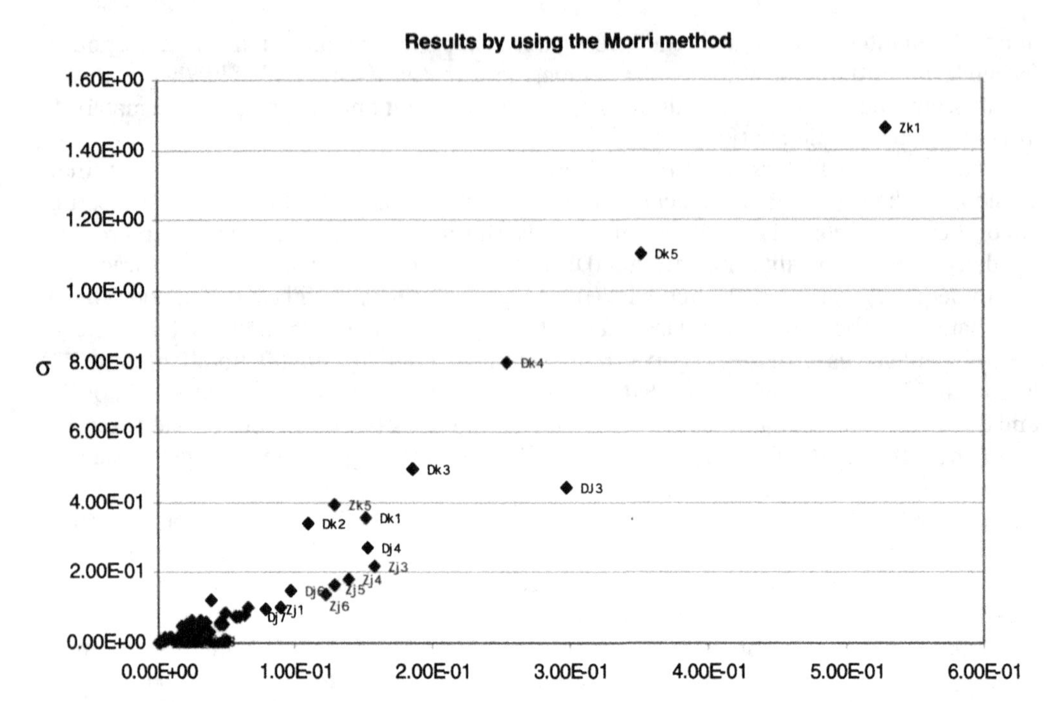

Figure 19.8 Results of the Morris screening exercise. The two Morris sensitivity measures μ and σ are plotted for the 103 input factors. Labels indicating names of factors are reported only for the most important factors.

'overall' effect, is certainly the most informative. Thus, factors are ranked in (decreasing) order of importance according to μ (see Table 19.4). However, in this case, the order of importance that would have been obtained by using σ is very similar to that reported in Table 19.4.

It is worth noting that also in two other application examples of the Morris method (Campolongo and Saltelli, 1997; Campolongo *et al.*, 1999b), the rankings provided respectively by μ and σ were almost identical. This is probably due to the fact that the majority of the models encountered in natural sciences are highly nonlinear. Thus, a factor that is important in the model (high μ) is usually also involved in curvature or interaction effects (high σ). On the other hand, a factor having important curvature or interaction effects (high σ) consequently has an 'overall' importance in the model (high μ).

In Figure 19.8 a first group of influential factors clearly separated from the others can be identified. These are (in decreasing order of importance) ZK(1), the daily natural mortality for mackerels at the egg stage; DK(5) and DK(4), the duration for mackerels respectively at the juvenile and early-juvenile stages; and DJ(3), the duration for anchovies at the early-larvae stage. Among these factors, DJ(3) is the one less involved in interaction and/or curvature effects; it does not lie exactly on the diagonal of the (μ, σ) plane but in the $\mu > \sigma$ zone.

A second group of factors, which seem to have a Euclidean distance from the origin substantially different from zero in the (μ, σ) plane, is that including (not in order of importance) DJ(4), DJ(6), and DJ(7), i.e. duration for anchovies at the late-larvae, late-juvenile, and pre-recruit stages; DK(1), DK(2), and DK(3), i.e. duration for mackerels at the egg and early- and late-larvae stages; ZJ(1), ZJ(3), ZJ(4), ZJ(5), and ZJ(6), i.e. mortality for anchovies at the egg,

Table 19.4 Results of SA by using the Morris method. Order of importance of factors. 103 factors are included in the analysis.

Factor	μ	Factor	μ	Factor	μ
ZK(1)	5.29×10^{-1}	FJ(8)	2.22×10^{-2}	ZK(8)	1.03×10^{-3}
DK(5)	3.51×10^{-1}	DK(6)	2.21×10^{-2}	aj31	1.01×10^{-3}
DJ(3)	2.98×10^{-1}	DI(2)	1.95×10^{-2}	DI(13)	7.67×10^{-4}
DK(4)	2.54×10^{-1}	aj22	1.88×10^{-2}	b1	7.00×10^{-4}
DK(3)	1.86×10^{-1}	DI(9)	1.86×10^{-2}	DJ(9)	6.07×10^{-4}
ZJ(3)	1.58×10^{-1}	FJ(7)	1.85×10^{-2}	aj13	5.13×10^{-4}
DJ(4)	1.54×10^{-1}	ZI(2)	1.80×10^{-2}	D2b	4.70×10^{-4}
DK(1)	1.52×10^{-1}	ZK(6)	1.74×10^{-2}	b2	4.53×10^{-4}
ZJ(4)	1.39×10^{-1}	FI(13)	1.67×10^{-2}	al22	3.67×10^{-4}
ZJ(5)	1.29×10^{-1}	DI(8)	1.02×10^{-2}	DK(9)	3.47×10^{-4}
ZK(5)	1.29×10^{-1}	ZI(12)	1.02×10^{-2}	al31	3.43×10^{-4}
ZJ(6)	1.23×10^{-1}	DI(7)	8.69×10^{-3}	aj21	3.33×10^{-4}
DK(2)	1.10×10^{-1}	DI(11)	7.47×10^{-3}	D1c	2.50×10^{-4}
DJ(6)	9.78×10^{-2}	ZI(6)	6.97×10^{-3}	al13	2.37×10^{-4}
ZJ(1)	8.94×10^{-2}	ZI(8)	6.60×10^{-3}	al32	1.90×10^{-4}
DJ(7)	7.82×10^{-2}	DI(12)	6.04×10^{-3}	aj32	1.67×10^{-4}
DJ(2)	6.53×10^{-2}	ZI(11)	5.44×10^{-3}	DK(7)	1.53×10^{-4}
ZJ(7)	6.36×10^{-2}	ZI(13)	5.37×10^{-3}	aj33	1.37×10^{-4}
DJ(1)	5.89×10^{-2}	DJ(8)	5.12×10^{-3}	D1b	1.37×10^{-4}
DI(3)	5.75×10^{-2}	FJ(9)	5.10×10^{-3}	FK(8)	1.33×10^{-4}
ZJ(8)	4.89×10^{-2}	aj11	4.50×10^{-3}	D2c	1.33×10^{-4}
ZI(1)	4.87×10^{-2}	DI(10)	4.00×10^{-3}	D2a	1.33×10^{-4}
ZI(4)	4.60×10^{-2}	ZJ(9)	3.68×10^{-3}	al11	1.33×10^{-4}
ZJ(2)	4.55×10^{-2}	ZI(7)	3.62×10^{-3}	al12	1.17×10^{-4}
ZK(7)	3.96×10^{-2}	FK(7)	3.61×10^{-3}	al33	1.10×10^{-4}
DI(5)	3.82×10^{-2}	b3	3.46×10^{-3}	DK(8)	1.03×10^{-4}
DJ(5)	3.57×10^{-2}	ZI(9)	3.27×10^{-3}	D3b	1.00×10^{-4}
DI(4)	3.42×10^{-2}	FI(12)	2.60×10^{-3}	al21	8.67×10^{-5}
ZK(2)	3.18×10^{-2}	FI(11)	2.37×10^{-3}	D1a	8.33×10^{-5}
ZI(3)	3.16×10^{-2}	ZI(10)	1.89×10^{-3}	D3a	7.67×10^{-5}
DI(6)	3.09×10^{-2}	ZK(9)	1.86×10^{-3}	D3c	6.67×10^{-5}
DI(1)	2.88×10^{-2}	aj23	1.30×10^{-3}	D1d	3.33×10^{-6}
ZK(4)	2.54×10^{-2}	FI(10)	1.09×10^{-3}	al23	3.33×10^{-6}
ZK(3)	2.42×10^{-2}	FK(9)	1.07×10^{-3}		
ZI(5)	2.22×10^{-2}	aj12	1.07×10^{-3}		

early- and late-larvae, and early- and late-juvenile stages; and ZK(5), i.e. mortality for mackerels at the juvenile stage. Then other factors follow immediately after: DJ(2), i.e. duration for anchovies at the yolk-sac larvae stage; and ZJ(7), i.e. mortality for anchovies at the prerecruit stage, and so on.

Apart from the four or five most important factors that are well apart from the others, there are no clearly separated clusters in Figure 19.8. The values of μ decrease smoothly, without discontinuity. Thus, it is very difficult to distinguish a group of important factors from a group of non-important ones. However, we know from experience that the influence of factors in models follows a Pareto-like distribution (see Chapter 4), with a few factors accounting for most of the variance, and most of the factors taking up the remainder. Thus, for a model with over 100 factors, it is very likely that less than 20 factors have some sizeable influence; the remaining 80 can be neglected.

An important conclusion that can be drawn from the results of the Morris exercise is that the input factors involved in the migration and interspecific competition between larvae and juveniles are not very important with respect to the others, using as output the dominant eigenvalue of the matrix $\mathbf{A_n}$. Even though these factors are responsible for the fluctuations observed in the simulations, none of these factors (denoted by lower-case letters) appear among the first 30 identified by Morris. This implies that this group of factors do not play a significant role in the magnitude of the fluctuations, and, hence, if our modelling approach is correct, it would be difficult to identify interspecific competition in real time series data since it will be masked by environmental fluctuations.

Daily fecundity (F) factors are also not very significant for any of the three species at any life stage. The first of these factors to appear in Table 19.4 is only rank = 36. Furthermore, none of the most important 20 factors are related to adult life stages. Daily natural mortality and duration factors play a more substantial role in the dynamics of the three populations. Our results are in agreement with the findings of Butler *et al.* (1993). They found by analyzing anchovies and sardines separately that the instantaneous mortality rates of the early- and late-larval stages of anchovy, ZJ(3) and ZJ(4), and the duration at the early-larval stage, DJ(3), had the greatest effect on population growth. Furthermore, as we have also observed, the sardine population appears to be less sensitive to changes in vital rates than the anchovy population by an order of magnitude. In fact, there is in Table 19.4 an absence of factors related to sardines amongst the most important ones: DI(3) is ranked in 20th position.

19.5 CONCLUSIONS

The anthropogenic pressure on the environment has increased. This will continue in the foreseeable future. This makes it necessary to minimize the consequences of anthropogenic influences on natural systems. The first step in this direction consists of learning how to estimate the character and size of the impact of these influences and to predict their consequences. On the other hand, assessing environmental influences and predicting their consequences are also closely related problems despite some significant differences. For these reasons, population ecologists have long been interested in two key topics: the first is the relative importance of intrinsic factors, such as self-regulation by density-dependent biological mechanisms, and extrinsic environmental variations in determining population dynamics; the second is nonlinearity in the process that generates these fluctuations.

Pelagic fish are known for their spectacular fishery collapses, for example in the cases of the Monterrey sardines (1940s) and the Peruvian anchovetas (1970s). Such collapses seem to be an inevitable consequence of inadequate understanding of the resources of the ecosystems. There is a need to improve resource management models based on understanding of qualitative state shifts and to improve our capability to recognize and predict state shifts. Zaldívar *et al.* (1998) developed different models and compared the simulation results with the patterns obtained from geological records. Their conclusions were that, although environmental fluctuations can explain the magnitude of observed variations in geological recordings and catch data of pelagic fish, they are difficult to link to the observed patterns. Further study is needed to link long-term environmental fluctuations with population changes, and biologically relevant nonlinear mechanisms must be included when modelling the dynamics of pelagic fish populations.

Among the several models examined by Zaldívar *et al.* (1998), the stage-based discrete model seems to be the one that best represents the patterns observed in the geological data. From a comparison between those data and discrete simulation results, it has been observed that environmental fluctuations seem to be responsible for the higher frequencies (periods shorter than 150 years) whereas density-dependent interactions seem to be responsible for low frequencies (periods longer than 400 years).

In this study, we attempted to improve our understanding of the population dynamics by performing a sensitivity analysis of the stage-based discrete model. The methodology followed was the one proposed by Morris (1991) and described in detail in Chapter 4. Results show that one of the reasons why interspecific competition is so difficult to identify from experimental data may be the small influence it has on the magnitude of the observed fluctuations, being masked by environmental effects. Furthermore, different species have different sensitivities to changes in the environment, and these differences can be greater than an order of magnitude. In our model, it seems that the order of species sensitivity is mackerel, anchovy, and sardine. Finally, the early stages of development seem to be the most important in all the species considered, so a better understanding of these stages would improve the modelling approach.

Following an idea originally introduced by Sobol' (1990b), the sensitivity of a model can be analyzed by dividing its input factors into groups according to their logical role in the model (see Chapter 20 for an example). The performance of a 'by groups' analysis on the fish population model would be interesting to confirm the order of sensitivity of the different species involved. Furthermore, such an analysis would have the advantage of reducing the computational cost of the experiment because it reduces the number of factors. This would allow the use of a quantitative method (i.e. a method capable of decomposing the total output variance into the percentages that each group is accounting for) that could not be employed here because of its high computational cost. On the basis of the results found in this study, we believe that the performance of a 'by groups' SA, using a quantitative method, should be considered for further studies.

ACKNOWLEDGMENTS

The authors gratefully acknowledge the Institute for Systems, Informatics and Safety of the European Commission, which has partially funded this work through an Exploratory Research entitled 'Use of Scale Deposits for Fish Population Modelling', Project 2, 1998.

20

Global Sensitivity Analysis: A Quality Assurance Tool in Environmental Policy Modelling

Stefano Tarantola

European Commission, Joint Research Centre, Ispra, Italy

Jochen Jesinghaus and Maila Puolamaa

European Commission, Eurostat, Luxembourg

20.1 INTRODUCTION

This chapter illustrates an application of sensitivity analysis (SA) in the context of quality assessment in environmental statistics, showing that SA is an indispensable tool for the management of uncertainties in models used for environmental policy and for the transparency of the models themselves.

The case study originates from Eurostat, the Statistical Office of the European Commission. We illustrate the role that global sensitivity analysis can play as a quality assurance tool for the optimization of the uncertainties on a worked example related to the use of environmental indicators for policy-making. In the first stage, uncertainty analysis (UA) will be employed to test the robustness of the underlying model, where the uncertain factors (namely input variables and model parameters) are explored over their range of variation. In this specific study, UA will help us investigate whether or not the underlying indicator-based model allows a given policy decision to be taken.

In the second stage, the extended FAST (see Chapter 8) will be used to reveal to what extent the uncertain model factors affect the model response. In this study, the factors

Sensitivity Analysis. Edited by A. Saltelli *et al.*

will be clustered according to different logical groups (see Table 20.1), and the extended FAST will allow us to rank quantitatively the groups of factors according to their influence on the output uncertainty (taking into account all possible interactions occurring between the groups).

The information provided by a global SA can indeed be used in decision-making to try to balance the influence of the various uncertain elements. The global SA procedure could be performed iteratively until the optimal situation is reached, which is ideally that corresponding to the case where the elements of the chain are more or less equally influential or otherwise optimal, once the cost of minimizing the various uncertainties is taken into account.

20.2 WHY SENSITIVITY ANALYSIS?

In the context of environmental modelling, model factors can be affected by large uncertainties. The extended FAST has been performed to identify those model factors that essentially drive the uncertainty on the decision to be made. SA can help the modeller in improving the process of building up highly aggregated information (e.g. pressure indices), i.e. in reducing the overall output uncertainty below reasonable levels.

The extended FAST will be used to apportion the output uncertainty to different subgroups of factors. Factors can be logically regrouped in different fashion, for example controllable versus uncontrollable sources of uncertainty, or uncertain data versus uncertain weights, etc.

'Controllable' sources of uncertainty are, for example, the errors due to lack of harmonization in official statistics. 'Uncontrollable' refers for example to stochastic processes or lack of consensus in the scientific community, or to uncertainty, for example on temporally distant impacts of climate change that would have to be included in the model.

The model that is considered in this study contains 196 uncertain factors. They have been categorized into 10 groups, and the total sensitivity indices have been estimated for each group. The groups of uncertain factors are described in the following sections. Associated with each factor is a probability distribution function and range, which are shown in Table 20.1.

In this work, we aim to explore the 'sensitive points' of the chain *data collection–data processing–calculation of indicators–calculation of indices–usage of indices for decision-making*, trying to give recommendations on where improvements to data quality and/or methodologies would yield the maximum added value for decision-makers using environmental indicators.

The present exercise focuses on a problem of environmental management. It has been assumed that two exclusive options for the disposal of solid waste are available to the policy-maker, who must choose the 'right' option between incineration and landfill. The model employed to assist the policy-maker supplies a pressure-on-decision index (PI) for each policy option. The index is proportional to the overall hazard to the environment of the corresponding option. The model output is defined as a suitable combination of the pressure-on-decision indices for the two options

$$Y = \log\left(\frac{\mathrm{PI}_{\mathrm{incineration}}}{\mathrm{PI}_{\mathrm{landfill}}}\right),$$

so that positive values for the output indicate that landfill is the preferred option, and vice versa.

Table 20.1 The 10 groups of uncertain factors: the total number of factors in the analysis is 196.

Groups and cardinality	Definition	p.d.f.	Range (values)
TU (1)	Territorial unit: two levels of aggregation for collecting data	Discrete uniform	0 ≡ national level 1 ≡ regional level
DATA (176)	Activity rates (120) and emission factors (37)	Uniform	Nominal value (NV) from CorinAir database Range: 2 QF × NV; Quality factors (QF): A = 5%, B = 10%, C = 20%, D = 50%, E = 90%
	National emissions (19)	Normal	$\sigma = 0.1 \times$ nominal value
E/F (1)	Approach for evaluating indicators: Eurostat or Finnish set	Discrete uniform	0 ≡ Eurostat list (Figure 20.1) 1 ≡ Finnish list
GWP (1)	Weights for greenhouse indicator (in the Finnish set): three time-horizons	Discrete uniform	0 ≡ GWP at 20 yr 1 ≡ GWP at 100 yr 2 ≡ GWP at 500 yr
W_E (11)	Weights for Eurostat indicators	Uniform	Nominal value from Scientific Advisory Groups Range = ±20% of nominal value
EEC (1)	Approach for evaluating environmental concerns	Discrete uniform	0 ≡ target values (Adriaanse, 1993) 1 ≡ expert judgement (Puolamaa *et al.* 1996)
ADR (1)	Type of target values according to Adriaanse (1993)	Discrete uniform	0 ≡ policy targets 1 ≡ sustainable targets
SUS (2)	Sustainable target for ozone depletion	Uniform	Range (0.01–0.10)
	Sustainable target for dispersion of toxics and eco-toxicological effect	Uniform	Range (0.01–0.02)
F/G (1)	Type of expert judgement: Finnish or German stakeholders	Discrete uniform	0 ≡ German stakeholders (RSD survey, see Figure 20.2 and Table 20.3) 1 ≡ Finnish stakeholders (see Table 20.2)
STH (1)	Preferences of Finnish stakeholders	Discrete uniform	i ≡ stakeholder i; $i = 1, \ldots, 8$

20.3 MODEL STRUCTURE AND UNCERTAINTIES

In the first stage, the waste management model evaluates a set of environmental pressure indicators, by aggregating atmospheric emissions through weighting coefficients (the intra-weights). These weights measure the harmfulness of a given pollutant to a specific environmental theme (e.g. greenhouse effect). Pressure indicators are defined on a set of selected environmental themes (see Figure 20.1 for the list of themes suggested by Eurostat).

Then, aggregating pressure indicators by means of inter-weights yields a PI. Inter-weights are weighting coefficients attached to the environmental themes, representing their relative importance in terms of the belief of a given stakeholder.

The model embodies a chain of uncertainties. Sources of uncertainty in data collection include roughly estimated statistics, technical coefficients (evaluated by expert judgement, estimated via *in situ* measurements, or derived from the scientific literature), and the choice of aggregation levels for gathering data, since these are available at different spatial resolutions (i.e. Nomenclature of Territorial Units for Statistics, NUTS). Uncertainties in indicator building include the selection of the environmental themes for inclusion in the aggregation model, and the choice among different intra-weights available for a specific environmental theme. In pressure indices building, both policy/sustainable targets (Adriaanse, 1993) and stakeholders' preferences (Puolamaa *et al.*, 1996) are affected by uncertainties.

Furthermore, different modules can be employed in some parts of the model; for example, the use of policy targets for a given year instead of sustainable targets as a design for inter-weights, or the use of Eurostat indicators rather than Finnish indicators, giving rise to 'structural' uncertainties.[1]

20.4 DATA AVAILABILITY

Atmospheric emissions (e.g. kilogram of CO_2 emitted per year), emission factors (e.g. kilogram of CO_2 emitted per metric ton of waste incinerated) and production rates (e.g. kilogram of municipal waste incinerated) are collected from the CorinAir data base, which is maintained by the European Environment Agency (EEA, 1996). All these emission data are available in a spatially disaggregated form, i.e. either at national level (NUTS0), or at a more detailed level (NUTS1). CorinAir data include nominal production rates and emission factors for a given type of waste, according to possible destinations (SNAP codes, i.e. Selected Nomenclature for Air Pollution). The pollutants considered in this study are SO_2,NO_x, NMVOC (non-methane volatile organic compounds), CH_4, CO, CO_2, N_2O, NH_3, dioxins, furans, plus the heavy metals As, Cd, Cr, Cu, Hg, Ni, Pb, Se, and Zn. CorinAir files often supply quality factors in a qualitative fashion, i.e. A (for best quality data) to E. When the quality is not reported, the value E is chosen. Uncertainty is then added to nominal values on the basis of the corresponding quality factor (see Table 20.1).

20.5 COMPUTING AIR EMISSIONS

Two types of waste—municipal and industrial—are considered in this study. Municipal production rates are obtained by collecting and aggregating incineration of municipal waste, open burning of agricultural waste, landfill, and compost production. Industrial production rates are obtained by aggregating incineration of industrial waste and landfill. Given that a vague idea exists about the amount of waste coming from municipalities that is placed in landfill, we assumed that 70% is municipal and the remaining 30% industrial.[2]

[1] We term 'structural' those uncertainties involving the question 'Was the right model selected?'.

[2] Very poor estimates of this fraction are currently available. It is noteworthy that the model has revealed the necessity to have this kind of distinction, which is another source of uncertainty (although not yet quantitatively included here, but planned to be accounted for in the future).

Emission factors are poor, and sometimes missing. They are often integrated with data suggested in the literature (EEA, 1996). Air emissions are finally evaluated by weighting production rates with the corresponding emission factors.

20.6 PRESSURE INDICATORS

Two alternative sets of indicators have been used: the set of Finnish indicators proposed in (Puolamaa *et al.*, 1996), and the list of Eurostat indicators (Eurostat, 1999).

20.6.1 Finnish Indicators

Three Finnish indicators have been employed: the greenhouse indicator, the acidification indicator, and the indicator of eco-toxicological effect.

The greenhouse indicator is calculated by considering the effects of carbon dioxide, methane and nitrogen oxide emissions (i.e. N_2O and NO_x). The acidification indicator refers to the atmospheric deposition of acidifying compounds, and is derived by taking into account the effect of sulfur dioxide, nitrogen oxide (NO_x), and ammonia emissions, which are the most acidifying compounds. The indicator of eco-toxicological effect includes heavy metals in the list above, and organic dibenzodioxins and -furans.

When evaluating the greenhouse indicator, the global warming potentials (GWPs) represent a source of uncertainty because three sets of GWPs have been proposed by the Intergovernmental Panel for Climate Change (Houghton *et al.*, 1996) according to three different time-horizons (see Table 20.1). Given that there is no scientific basis for choosing any one of them, this (subjective) uncertainty has been tackled in the exercise by considering the three sets of values as occurring with the same probability.

For the acidification indicator, the intra-weights for the various compounds are represented by acidification potentials, which are defined on the basis of the quantity of hydrogen ions that would be formed in complete atmospheric deposition of the gas in question. The values for the acidification potentials adopted in this study, expressed as kilograms of sulfur dioxide equivalents per kilogram of the compound considered, are taken from Puolamaa *et al.* (1996).

The intra-weights necessary to implement the eco-toxicological indicator are the eco-toxicity potentials (ETP). They give a measure of the persistence of toxic substances accumulated in the aquatic or terrestrial ecosystem. Weights for heavy metals have been gathered from ICI (1997).

20.6.2 Eurostat Indicators

In order to define the contents and structure of the pressure indices project, Eurostat organized surveys among natural scientists, the so-called Scientific Advisory Groups (SAG) (Jesinghaus, 1999). The SAG ranked the indicators within each environmental theme in order of importance. The Eurostat core list of environmental themes and the six most influential indicators for each theme are displayed in Figure 20.1 from left to right in order of decreasing importance (Eurostat, 1999). The rankings are expressed by means of intra-weights, which constitute the group W_E of uncertain factors (see Table 20.1). The intra-weight for a given indicator is defined as the percentage of SAG experts who selected that indicator for the core list.

Air pollution	Emissions of nitrogen oxides (NO_X)	Emissions of NMVOC	Emissions of sulfur dioxides (SO_2)	Emissions of particles	Consumption of gasoline and diesel oil by road veh's	Primary energy consumption
Climate change	Emissions of carbon dioxide (CO_2)	Emissions of methane (CH_4)	Emissions of nitrous oxide (N_2O)	Emissions of chloro-fluoro-carbons	Emissions of nitrogen oxides (NO_X)	Emissions of sulfur oxides (SO_X)
Loss of biodiversity	Protected area loss damage, and fragmentation	Wetland loss through drainage	Agriculture intensity: area used for intensive...	Fragmen-tation of forests and landscapes	Clearance of natural and semi-nat'l forests	Change in traditional land-use practice
Marine environment and Coastal zones	Eutrophi-cation	Overfishing	Develop-ment along shore	Discharges of heavy metals	Oil pollution at coast and at sea	Discharges of halogena-ted organic compounds
Ozone-layer depletion	Emissions of bromo-fluoro-carbons	Emissions of chloro-fluoro-carbons	Emissions of hydro-chlorofluoro-carbons	Emissions of CO_2	Emissions of NO_x	Emissions of methyl bromide (CH_3Br)
Resource depletion	Water consumption per capita	Use of energy per capita	Increase in territory occupied by urbanization	Nutrient balance of the soil	Electricity production from fossil fuels	Timber balance (new growth/ harvest)
Dispersion of toxic substances	Consumption of pesticides by agriculture	Emissions of persistent organic pollutants	Consumption of toxic chemicals	Index of heavy metal emissions to water	Index of heavy metal emissions to air	Emissions of radioactive material
Urban environmental problems	Energgy consumption	Non-recycled municipal waste	Non-treated wastewater	Share of private car transport	People endangered by noise emissions	Land use (change from natural to built-up area)
Waste	Waste landfilled	Waste incinerated	Hazardous waste	Municipal waste	Waste per product	Waste recycled/ material recovered
Water pollution and water resources	Nutrient use (nitrogen and phosphorus)	Ground-water abstraction	Pesticides used per hectare of agric'l area	Nitrogen used per hectare of agric'l area	Water treated/ water collected	Emissions of organic matter as BOD

Figure 20.1 Eurostat environmental pressure indicators (from Eurostat, 1999). In the first column, the 10 environmental themes are given. For each environmental theme, the six most important indicators, ranked from left to right, are reported along each row.

Four themes have been selected because of their relevance to the study: air pollution, climate change, ozone-layer depletion, and dispersion of toxic substances. The theme 'waste' itself is, of course, also relevant for the study. However, the SAG of waste experts has expressed its subjective preference, attributing a higher score to landfill than to incineration (Figure 20.1). Given that it would be double-counting to include the implicit valuation of the waste experts at this level, we have decided not to include the theme 'waste' in the analysis, but to use it as a reference for interpreting the results of the more detailed analysis performed here.

In the theme 'air pollution' the emissions of NO_x, NMVOC, and SO_2 are included in the model. Particulate emissions have not been included for lack of data.

Waste management activities affect the theme 'climate change' through all six indicators; however, emissions of chlorofluorocarbons (CFCs) are not included because of lack of data on emission factors.

In the theme 'ozone-layer depletion', emissions of carbon dioxide and of nitrogen oxides (NO_x) are included. Emissions of halons, CFCs and HCFCs are omitted because of lack of data on emission factors.

In the theme 'dispersion of toxic substances', emissions of persistent organic pollutants (dioxins and furans), to air (for incineration) and of heavy metals (to air for incineration, to groundwater for landfill) are considered.

20.7 INDICATORS OF TOTAL BURDEN

Indicators at national level describe the total burden of human activities in a given country. These indicators are needed for computing both target values (Section 20.8.1), and for indicator normalization (Section 20.8.2). In the present experiment, depending on the set of indicators that is used (either Finnish or Eurostat) in a specific run of the model, the corresponding set of indicators at national level has to be evaluated. Firstly, atmospheric emissions from all the economic sectors and for all the pollutants of interest are collected. Emissions of SO_2,NO_x,NMVOC,CH_4,CO,CO_2,N_2O,NH_3 and heavy metals are gathered from CorinAir at NUTS0, while dioxins and furans national emissions are provided through a survey conducted by the Landesumweltamt NRW (NRW State Environmental Agency, 1997). Finally, national emissions are aggregated to indicators of total burden as illustrated in Section 20.6.

20.8 PRESSURE-TO-DECISION INDICES

The inter-weights used to aggregate indicators to indices must be representative of the perceived impact of the problem covered by the themes, and should include both objective and subjective elements. An objective approach should be able to quantify the distance of the actual situation from a policy target[3], whereas a subjective approach should take into account experts' and public opinion. Here we present two alternative approaches: the first is based on objective criteria (Section 20.8.1), while the second encompasses subjective aspects (Section 20.8.2). The triggering factor [EEC] has been introduced to select between the two approaches, which have the same probability of occurrence in each model execution.

20.8.1 Target Values

In the objective approach, a target for each theme indicator is set, and the inter-weight is defined as the reciprocal of the target value (Adriaanse, 1993).

Targets can be formulated by relating either to policy targets, which are chosen for a certain reference year, or to sustainability levels, representing the targets proposed to reach an environmental sustainable development.

20.8.2 Expert Judgement

The subjective approach is based on expert judgement (Puolamaa *et al.*, 1996). The pressure indicators are first normalized using the total national burdens and then fed into the

[3]'Objective' in the sense of 'observable', which does not necessarily mean that a political decision for a specific target is scientifically and/or economically sound.

Table 20.2 Result of the analytical hierarchy process. Weighting of environmental problems by stakeholders.

Finnish indicators	**Greenhouse**	**Eco-toxicological**	**—**	**Acidification**
Eurostat indicators	**Climate Change**	**Dispersion of Toxics**	**Ozone-layer Depletion**	**Air Pollution**
Agriculture	15.6	11	14.1	6.5
Manufacture	14.2	10.6	14.2	7.3
Environmental journalists	15.5	10.7	17.1	4.6
Environmental NGOs	11.8	9.3	14	3.3
Administration	20	8.1	14.1	5.2
Traffic	13.8	11.2	8.7	6.7
Environmental scientists	22.3	8.7	10.5	5.1
Politicians	9.9	10.4	16.5	5.5
All groups	**15.4**	**10**	**13.7**	**5.5**

inter-weighting stage. Two sets of inter-weights are used in this study. A trigger factor [F/G] has been set up to randomly select one of the two sets of inter-weights.

In the first set, a group of Finnish stakeholders from eight different domains (Table 20.2) were selected and interviewed (Puolamaa *et al.*, 1996). The analytical hierarchy process (AHP) performed in Puolamaa *et al.* (1996) provides the weighting of the environmental problems by stakeholders.

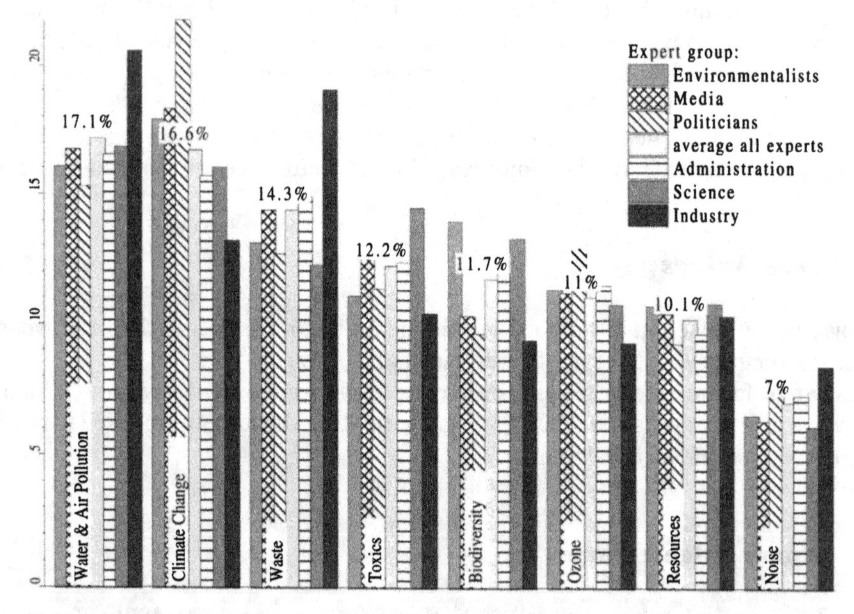

Figure 20.2 Overall importance of environmental themes (RSD, 1990). In the German study, over 600 experts had been asked how they perceive the overall importance of various environmental themes. 'Ordinary' water and air pollution had been sub-summed into one theme. To enable comparisons with the Finnish and Eurostat indicator sets, the result (17%) has been split artificially on the basis of the water/air relation used by Adriaanse for the Dutch pressure index (water pollution was not considered in our study, because of lack of data).

Table 20.3 Inter-weights obtained from the survey conducted by RSD (1990).

Finnish indicators	**Greenhouse**	**Eco-toxicological**	**—**	**Acidification**
Eurostat indicators	**Climate Change**	**Dispersion of Toxics**	**Ozone-layer Depletion**	**Air Pollution**
Average weighting factors	0.166	0.12	0.11	0.12

A subjective source of uncertainty is due to the fact that different weights, associated with the various domains (the first column of Table 20.2), are available for each environmental indicator. In a typical deterministic analysis, the average inter-weight (the last row in Table 20.2) would be employed. In the present study, we allow each domain to occur with the same probability in a given model execution, thus taking this source into account (factor [STH]).

The second set of inter-weights is obtained from a survey performed by the Research Unit for Societal Development, University of Mannheim in 1990 (RSD, 1990) where a group of German experts from different sectors (Figure 20.2) were asked to provide a weighting coefficient for various themes (Table 20.3).

20.9 RESULTS AND DISCUSSION

Three cases are illustrated in this section: they relate to Austria in the year 1994.

The first exercise is carried out on a preliminary version of the model where the indicators of eco-toxicological effect (ET) and of dispersion of toxics (DT) are not taken into account, and the factors [ADR], [SUS] and [F/G] are not sampled. The bimodal histogram displayed in Figure 20.3 represents the outcome of the uncertainty analysis. The left-hand region, where incineration is preferred, comprises approximately 40% of the total area, and this shows that the model does not allow a ranking of the two options. The pie chart in Figure 20.4 confirms that the choice of the indicator set has an overwhelming influence and reveals that the scientific community must work to find a consensus on the proper set. Here sensitivity analysis does not help to decide whether one should use the Finnish or the Eurostat indicators, but about the relative importance of the various types of uncertainty, including that arising from the choice of indicators.

In the second experiment, SA has been performed for the two cases separately, under the hypothesis that a consensus among the experts has been already established on the use of a given set of indicators.

Figure 20.5(a) illustrates the results when employing the Finnish set, whereas Figure 20.5(b) illustrates the use of the Eurostat set. In Figure 20.5(a) data account for 45% of the total output variance V (this is expected, because of the poor quality of waste data). The [GWP] factor shares 40% of V, thus indicating that more effort is required from the IPCC panel of experts in order to converge towards a more stable set of [GWP] values. The remaining 15% includes all the other factors.

In summary, 'data uncertainty' (the factor [Data]) is about as important as 'model uncertainty' (all the other factors), i.e. the uncertainty introduced by the analyst. This situation could lead to criticism, and would not be acceptable in most practical applications. If a reduction in output uncertainty were required in order to better support an eventual decision, the analyst would know how to prioritize his/her data acquisition effort.

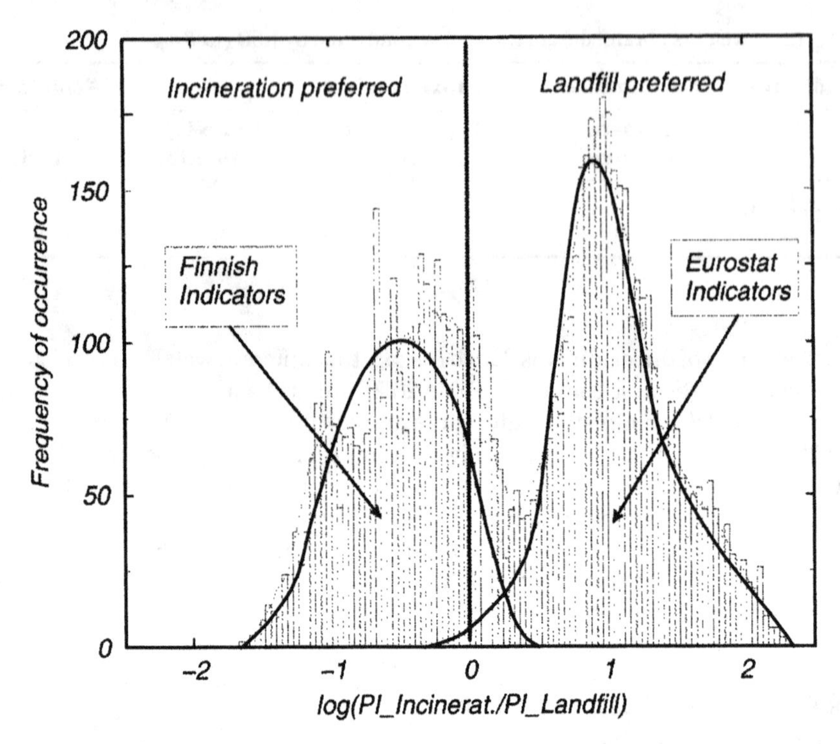

Figure 20.3 Results of UA for the exercise where the indicators of eco-toxicological effect (ET) and of dispersion of toxics (DT) are not included.

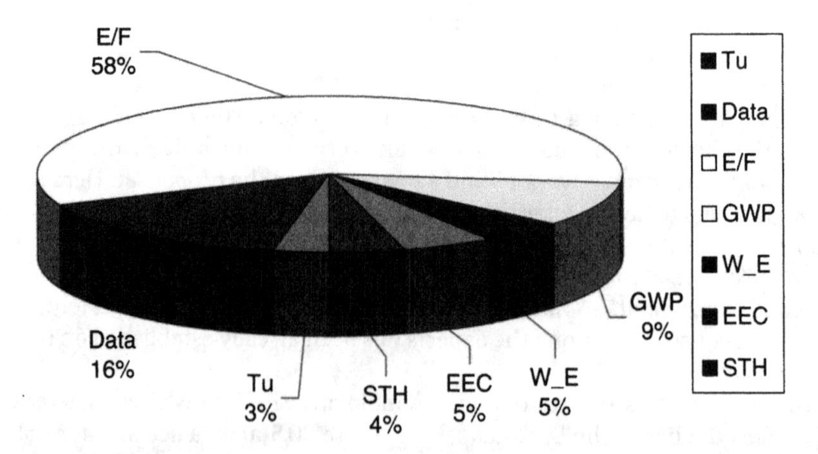

Figure 20.4 Results of SA for the first exercise. Both the Finnish and the Eurostat sets of indicators have been used (factor [E/F]). Uncertainty on indicators is much larger than that of the available data.

Figure 20.5(b) shows that the impact of 'data uncertainty' on the PI is very large (82%), underlining high model relevance and quality, the model being weakly influenced by subjective uncertainties.

In the third exercise, the eco-toxicological and dispersion indicators have been included in the model, and all groups of factors have been considered. Here a consensus on the policy

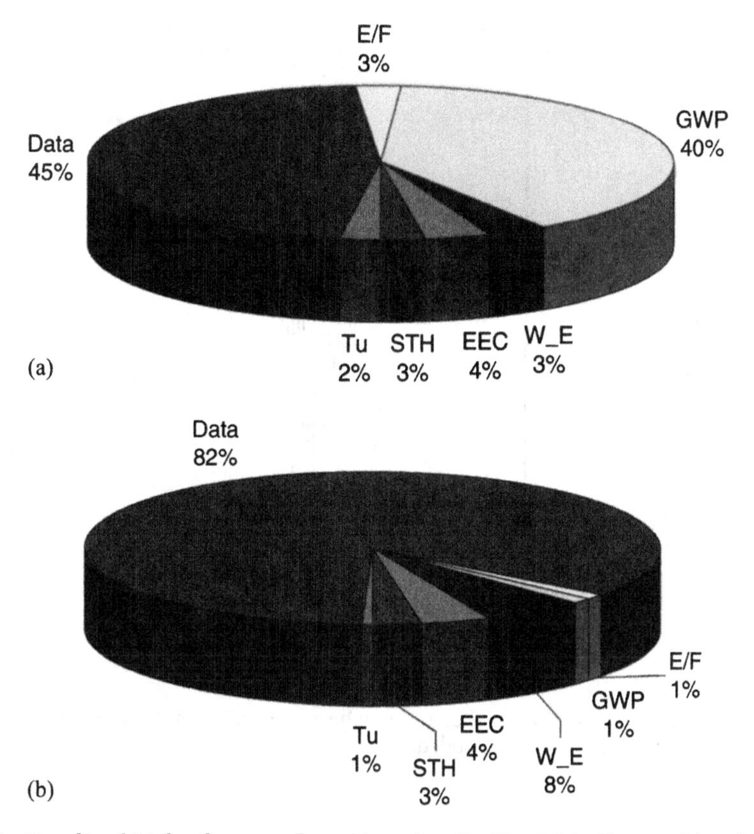

Figure 20.5 Results of SA for the second exercise using the Finnish indicators (a) and the Eurostat indicators (b).

has been reached (landfill is preferable) although the uncertainty in the resulting output is quite large (the histogram in Figure 20.6).

The sensitivity analysis for the third exercise (Figure 20.7) shows that the total impact of 'data uncertainty' on the output index is low (37%), the largest fraction being due to 'subjective' uncertainties. Addressing further investigation on [E/F] and [GWP] (15% each) will enhance model accuracy.

These results strengthen the role of SA as a tool for helping the analyst to test and improve the quality of the model and the transparency of the decisions. A reduction of the model output uncertainty below reasonable levels increases the likelihood for a certain environmental policy to attain the suitable scientific added value that may support its implementation.

In summary, the results reported above show the importance of including dioxin and furan emissions in the model. The design of data collection should try to strike a balance between accuracy and complete coverage. A few 'precise' indicators may please the eye of the statistician, but a more complete 'optimally inaccurate'[4] set better fits the requirements of decision-making processes and can yield more reliable support for the ultimate decision.

[4]The notion of 'optimally inaccurate indicators' was developed by John O'Connor, formerly of the World Bank (personal communication).

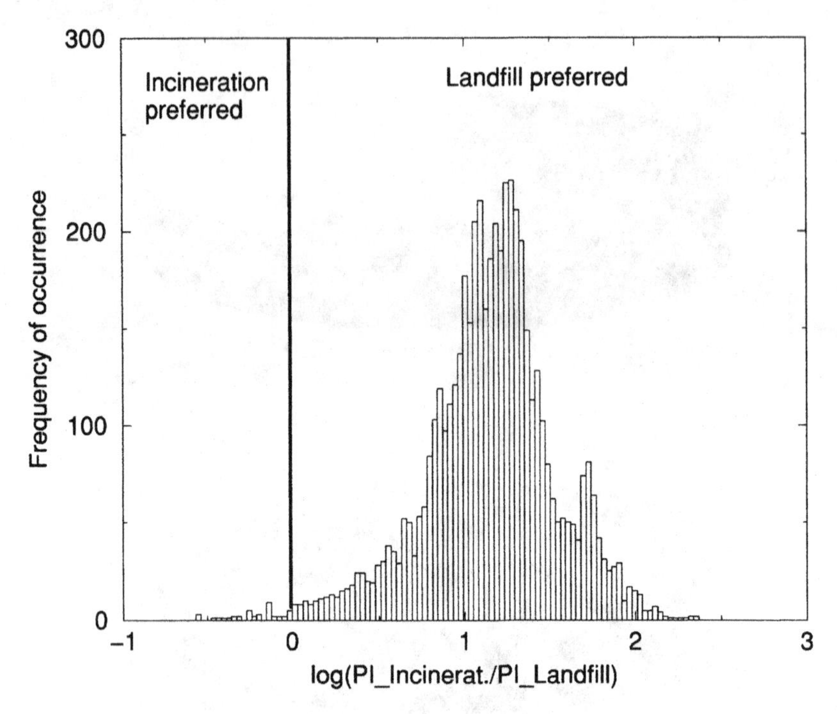

Figure 20.6 Results of UA for the third exercise, where the indicators of eco-toxicological effect (ET) and of dispersion of toxics (DT) are included.

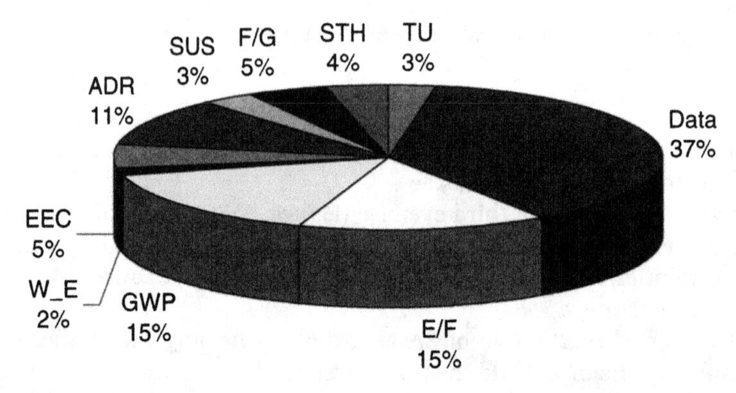

Figure 20.7 Results of SA for the third exercise, where the indicators of eco-toxicological effect (ET) and of dispersion of toxics (DT) are included.

Other indicators could be further included in the model: particulate matter emissions, which have a negative impact on incineration, are still not considered. However, this study is purely methodological. The focus is indeed on the novelty of the approaches used to tackle the problem of uncertainties in environmental policy. In standard approaches, single (nominal) values are given to uncertain model factors, resulting in outputs with no confidence levels attached. Such a situation can hardly be relevant to decision-making.

The model used in this work could be employed to examine the combination of other policy options. For instance, it could be used to study what happens if municipal waste is placed in landfill and industrial/hospital wastes incinerated, or the effects of incinerating 40% of the total waste production and placing the rest in landfill, whatever the composition of the waste.

A generalization of the model could allow the investigation of other interesting policy options, such as, for instance, the quantification of the consequences of banning the open burning of agricultural waste in rural regions, or the effect of using modern incineration plants for industrial waste instead of old ones without particle abatement systems. A more complex generalization could involve the inclusion of socio-economic aspects of environmental change, in the perspective of making policies for sustainable development.

ACKNOWLEDGMENTS

This work has been partially funded by the Statistical Office of the European Union (Eurostat), Contract 13592-1998-02 A1CA ISP LU, Lot 14, SUP COM 97. The authors gratefully acknowledge Tine Sørensen for her useful comments and suggestions.

IV

Conclusions

21

Assuring the Quality of Models Designed for Predictive Tasks

M. B. Beck

University of Georgia, Athens, USA

Jining Chen

TsingHua University, Beijing, PR China

21.1 INTRODUCTION

In 1992, an article with a rather arresting title was published in the journal *Advances in Water Resources*: 'Ground-water models cannot be validated' it proclaimed (Konikow and Bredehoeft, 1992). Who, one wondered, would be impudent enough to respond 'Yes, they can!'? Who would gainsay the weighty authorities—of Popper and Hawking—wheeled out in support of the paper's title? Thus far, it would seem, essentially no-one. For, phrased in these terms, it is hard to disagree with such a proclamation. Much the same sentiment, of switching the intent of the process from validation to invalidation, in this particular subject area, was expressed long ago by Young (1978). Perhaps the more interesting response is to ask the questions: 'How, then, shall we move forward; and what might be the role of sensitivity analysis in doing so?' Assembling a response to these questions is our purpose in the present chapter.

Validation (or invalidation) has never *not* been a problem (Caswell, 1976; Beck *et al.*, 1997). Konikow and Bredehoeft were therefore quite justified in drawing our attention back to it, yet again. For over three decades, mathematical models encoded in a computer program have been manipulated for the purposes of interpreting and predicting the behavior of environmental systems. Continuing expansion in the capacity of the digital computer has essentially removed all constraints on the number of equations to be solved and the degree of nastiness of their nonlinear properties. These models are a secondary knowledge base, however, fuelled by the ever-expanding primary knowledge base of atmospheric chemistry,

Sensitivity Analysis. Edited by A. Saltelli *et al.*

aquatic biology, ecology, and so forth. They are formally organized distillates of the raw material of relatively recently discovered 'facts'; 'facts', in fact, that may be rather qualitative, if not speculative, in character, especially in the fields of biology and ecology (Shrader-Frechette, 1995). Much of the rich distillate of interpreted empirical evidence in ecology (e.g. Porter, 1996; Porter *et al.*, 1996) is mightily hard to express in the strict, formal rigidities of mathematics. Or, more accurately, perhaps it is not so hard to write down some equations, but it is difficult to know what values should be assigned to the parameters (coefficients, constants) appearing in them in great profusion. Above all, if one looked for deliverance to the field of system identification—to the body of methods, that is, for reconciling a model with the *in situ* field data and hence extracting the sought-for parameter estimates—frustration would be the most likely outcome, at least for all but the lowest-order (small) models (Beck, 1987, 1994; Beven, 1989, 1993). No formal proof can be offered for this, the widespread lack of identifiability of anything but a small model. It is in general not possible to state how many, or which, of the parameters in a state-space model of the system's behavior can be evaluated precisely and unambiguously from a given record of observations, and especially so if the model is nonlinear (Beck *et al.*, 1990; Kleissen *et al.*, 1990). Yet the overwhelming empirical evidence is that a single input–output couple of observations will at best yield only a handful of well-identified estimates of the model's parameters, regardless of the length of the observed record (Jakeman and Hornbeger, 1993; Jakeman *et al.*, 1994; Stigter and Beck, 1994; Young and Beven, 1994).

In short, the ever growing stockpiles of raw material quarried from the frontiers of the primary (mono-disciplinary) sciences are increasingly less likely to receive rigorous, formal refinement through the devices of more extensive (multidisciplinary) mathematical models. Put another way, it is becoming increasingly difficult to calibrate the instruments of prediction (the models) we construct from these stocks of raw material. Irrespective of whether one considers the process of evaluating the success of this refinement as a matter of 'validation' or 'invalidation', grave doubt must be cast on the prospects for success in the classical sense of model validation, i.e., in obtaining a close match of the model's performance and observed behavior *together with* well-defined estimates of all the model's parameters. The former is often attainable without the latter, such that a substantial capacity for ambiguity, even absurdity, may lie latent in the supposedly validated model—although this might be of no material consequence for prediction and decision-making (Klepper *et al.*, 1991; Beck and Halfon, 1991). Alternatively, one might impose well-definedness on the parameters *a priori*, by composing the structure of the model in a maximally 'physically meaningful' manner, and then seek to match history without any further corruption (through model calibration) of this relatively pure articulation of theory. Such a model would almost certainly be of a very high order, as a consequence of having to reach down to the microscopic scale at which the laws of physics can safely be deemed to operate. But when the model fails to match macroscopically observed behavior, it will not be at all obvious to which defective, constituent, microscopic hypotheses this failure is due.

At least for those working in the middle ground, where there *are* data yet *few* incontrovertible prior theories, these things have largely begun to be accepted by the majority. The current fascination with the issue of uncertainty is a reflection of this acceptance (Beck, 1987; Funtowicz and Ravetz, 1990; Beven and Binley, 1992; Beven, 1996; van Asselt and Rotmans, 1996). Where there *is* some supposedly incontrovertible prior theory, confidence in the prospect of (in)validation might yet continue to flourish (Woods, 1999). But even in these fields, in the subjects of meteorology, oceanography, and subsurface water flow—underpinned, as they are, by geophysics—what we now see is a rise to the fore in the application of algorithms of data assimilation (Evensen, 1994; McLaughlin, 1995; Eigbe *et al.*,

1998). In other words, our outlook is coalescing around the view that the relatively sparse data can but be assimilated into the current theory, not employed to root out ruthlessly its inadequacies. We do indeed have a problem with validation.

In this chapter, we shall take as our point of departure a recent paper reviewing the nature of the problem (Beck *et al.*, 1997). What is striking is the intellectual impasse into which collectively, as advocates of the development and application of models, we seem to have been driving ourselves. The chapter recounts this tendency, but then suggests a possible path out of the impasse. Central to the opening up of this new avenue of research are three ingredients: first, the recall of some seminal ideas of Caswell (1976), i.e., that validation is essentially a task of design; second, a shift in the focus of our attention away from the need primarily, if not exclusively, to demonstrate a match of the model's output with past observations; and third, the commandeering of a specific form of so-called regionalized sensitivity analysis (Hornberger and Spear, 1980; Spear and Hornberger, 1980) and turning it to our present purposes of model (in)validation. This leads to the development of a novel, but admittedly tentative, statistic for gauging the quality, reliability, or relevance of a candidate model in fulfilling a given predictive task. How this measure might work in practice is then illustrated with a case study in predictive exposure assessment (Chen and Beck, 1998). Last, the chapter touches briefly on a quite different form of sensitivity analysis applicable to another facet of the problem of validation, one of empirical corroboration/refutation of what we shall call high-level conceptual descriptions of massively complicated systems (such as the insight of the ocean conveyor belt; Weaver, 1995). Our principal contribution, however, shares a very great deal in common with methods now being proposed for use in assessing the potential and origins of structural change in the future behavior of a system. It has therefore another closely related point of departure, in the area of forecasting environmental change (Beck, 1998), and a companion article setting out how a regionalized sensitivity analysis can likewise be applied to identification of those key parameters in a model on which the reachability of some feared (desired) pattern of future behavior may crucially turn (Beck *et al.*, 2000).

It cannot have escaped attention by now that the word (in)validation is itself part of the problem (Konikow and Bredehoeft, 1992; Beck *et al.*, 1997; Oreskes, 1998a). We shall try therefore to adhere to the term 'quality assurance' (Funtowicz and Ravetz, 1990), so that our discussion is concerned with establishing what evidence must be assembled in order to assure the quality of a model (relative to its performing a specified task) beyond reasonable doubt. But let us begin by recounting why the term 'validation' has become so troublesome in recent times.

21.2 THE IMPASSE

Put rather more expansively than in the bald title of their 1992 article, these are the concerns of Konikow and Bredhoeft (Bredehoeft and Konikow, 1993):

> To the general public, proclaiming that a ground-water model is *validated* carries with it an aura of correctness that we do not believe many of us who model would claim. We can place all the caveats we wish, but the public has its own understanding of what the word implies. Using the word *valid* with respect to models misleads the public; *verification* carries with it similar connotations as far as the public is concerned.
>
> Models are adjusted until an *adequate* match to some set of historical data is achieved. . . . Once an adequate match between historical data and model output is achieved, the model is commonly

> used to predict the response of the system into the future. . . . Usually care is taken to predict only for a time comparable to the period that was matched. In other words, if we matched to a 10-year history, we would make a 10-year prediction with some confidence.
>
> The ground-water community is attempting to place confidence bounds on predictions arising out of the uncertainty in parameter estimates. This is a welcome development; the single valued predictions of the past were overly simplistic. However, these confidence limits do not bound errors arising from the selection of a wrong conceptual model, or from problems arising with numerical solution algorithms.
>
> Much of the current impetus for model validation has originated in the nuclear waste community. This discussion would be purely academic were it not for the fact that we, as ground-water hydrologists, are dealing with problems of great public importance with high public visibility.

These remarks, made within the editorial columns of the journal *Ground Water*, have rekindled the debate, just as intended (see, e.g., McCombie and McKinley, 1993; Bair, 1994; and see also Oreskes *et al.*, 1994). To paraphrase, the four issues beyond which we would like to proceed are therefore as follows:

(i) Extrapolation over periods greater than the span of 'history matching' is not recommended, or at least should be viewed with little confidence when practised.
(ii) Quantifying the uncertainty (error) attaching to a model's prediction is highly desirable, but present methods for so doing are heavily circumscribed in being unable to account for the structural (conceptual) error in the model.
(iii) The word 'validation' has been used too loosely and associated too frequently with poor science and the sloppy implementation of evaluation procedures (see also McCombie and McKinley, 1993).
(iv) The public has a legitimate stake in being assured of the quality of models that are used to inform the decision-making process.

Since both the practical interests of the lay public and the philosophical positions of Popper and Hawking on epistemology are fused in the one 'problem', the potential difficulty of its clear and straightforward discussion can readily be appreciated.

In the present chapter, we shall address primarily just the first of the above points and, to a lesser extent, the second.[1] To all intents and purposes, the most widely accepted test of model validation has not changed over the years, not even recently. It is still as it was in the following definition, taken from Standard E 978-84 of the American Society of Testing and Materials on *Evaluating Environmental Fate Models of Chemicals* (ASTM, 1984):

> Comparison of model results with numerical data independently derived from experience or observations of the environment.

In this the word 'experience' should not pass unnoticed.

21.2.1 Behavior under Novel Conditions: The Paradox of (In)validation

Let us consider the nature of the task of making predictive exposure assessments. A novel substance with potentially beneficial attributes, but also some known adverse side-effects, is

[1] Structural error and structural change are matters dealt with in a companion paper, using an argument largely identical to that used herein (Beck *et al.*, 2000).

being manufactured. The need is to predict the possible consequences of releasing this chemical into new environments. Substantial costs, and substantial damages to the environment, may attach to the decisions regulating the release (or otherwise) of this substance. Because this is a novel substance, no field observations of the fate and effect of the substance are possible, *by definition.*

The problem of predictive exposure assessment cannot therefore rely on the availability of field observations for the purpose of history-matching. In principle, any attempt at prediction is an excursion beyond the span of time covered by 'history-matching'. We could hardly disagree with the principle of eschewing the temptation to make such excursions, as advocated by Konikow and Bredehoeft (1992), were it not for the fact that we build models *precisely* for the purpose of extrapolation beyond the confines of the record of the past, have always striven so to do, and especially so now in attempting to forecast environmental change (Beck, 1998; Oreskes, 1998a). The most urgent and challenging issues of management are so often those for which extrapolation beyond what has been observed is unavoidable. This is true not only in the grand issues of our time, but also in some of the more mundane areas of enquiry. In predicting the impact on a receiving water body of the sudden failure of a wastewater treatment system, for example, transient anaerobic conditions might thereby be induced, with thus an altered set of degradation pathways for the accidentally released organic contaminants. The historical observations, however, might cover merely predominantly aerobic conditions in the river and thus be of strictly limited value for the purposes of forecasting the fate of the contaminants (Reda, 1996; Reda and Beck, 1997).

There is a paradox then (Beck *et al.*, 1997). The greater the degree of extrapolation from past conditions, so the greater must be the reliance on a model as the instrument of prediction; hence, the greater is the desirability of being able to quantify the validity (or reliability) of the model, yet the greater is the degree of difficulty in doing just this. Should the use of models be put aside in these situations? We think not, for reasons stated more fully elsewhere (Beck *et al.*, 1997). In essence, our argument for a way out of the possible intellectual impasse of model validation takes its lead from Caswell's seminal contribution, in particular, the following (Caswell, 1976; p.317):

> Models of systems are systems themselves, the interacting components of which are mathematical variables and expressions. They are, however, man-made systems. . . . The construction of such artificial systems is a design problem, and the process of design is in essence a search for agreement between properties of the artificial system and a set of demands placed on it by the designer. It is impossible to evaluate the success or failure of a design attempt without specification of these demands, the task environment in which the artificial system is to operate. Validation of a model is precisely such an evaluation. . . .

21.2.2 Towards an Exit from the Impasse

Let us assume, as a point of departure, a fairly standard representation of the model in terms of the following state variable dynamics,

$$\frac{d\mathbf{x}(t)}{dt} = \mathbf{f}\{\mathbf{x}, \mathbf{u}, \boldsymbol{\alpha}, t\} + \boldsymbol{\xi}(t), \tag{21.1a}$$

with the observed outputs being defined as follows:

$$\mathbf{y}(t) = \mathbf{h}\{\mathbf{x}, \boldsymbol{\alpha}, t\} + \boldsymbol{\eta}(t), \tag{21.1b}$$

in which **f** and **h** are vectors of nonlinear functions, **u**, **x**, and **y** are the input, state, and output vectors, respectively, $\boldsymbol{\alpha}$ is a vector of model parameters, ξ and η are notional representations respectively of those attributes of behavior and output observations that are not to be included in the model in specific form, and t is continuous time. Should it be necessary, spatial variability of the system's state can be assumed to be accounted for by, for example, the use of several state variables of the same attribute of interest at several defined locations.

Acceptability of a model is conventionally gauged in two ways. First, and perhaps more subtly, it is assessed through a process of peer review of the primary, theoretical material and constituent hypotheses of which the model is composed. This is essentially a matter of making judgements about the quality of the *internal* properties of the model, in particular, of whether the functions **f** and **h** have been 'properly' expressed and 'realistic' values assigned to the parameters ($\boldsymbol{\alpha}$) appearing in them. Propriety and realism are here matters of where consensus and longevity of experience reside; they are not generally the subject of quantitative, statistical testing procedures.

Second, much more obviously, and almost universally, we think of validation as a function of matching observed behavior, as reflected in **y**, with estimated behavior, i.e., $\hat{\mathbf{y}}$. In other words, the quality of the model is assured in terms of its *external* performance, *if*, of course, some historical observations are available. We can all agree that the image should simply look something like the real thing. In the context of the foregoing model, $\hat{\mathbf{y}}$ will normally have been generated from substitution of the solution of Equation (21.1a) for **x**, given the observed values of the inputs (**u**), into the function $\mathbf{h}\{\cdot\}$ of Equation (21.1b), with ξ and η often referred to as noise processes, assumed to have been identically zero for all t. Acceptance or rejection of the model as being worthy of use in some way—typically as an instrument of prediction—will be based on more or less quantitative manipulations of **y**, $\hat{\mathbf{y}}$, and the divergence between them.

All the current quantitative tests of (in)validity are cast therefore in the *output* space of the model, as opposed to its *parameter* space, and acceptance/rejection of the model as a *whole* is implied, as opposed to acceptance of some of its *parts* and the rejection of others (Beck *et al.*, 1997). In the absence of a history to be matched, yet assuming the user of the model is able to specify a task or a kind of performance, against which the model is to be designed, our search is for a quantitative measure—cast in terms of the internal parameter space of the model—by which to gauge the appropriateness of its design composition. More specifically, we would like to be able to discriminate between whether one model is better suited to a given task than another, for example, in being able to forecast the effects of a high-end exposure of an organism to a toxic substance. Put another way, could we develop a test statistic capable of discriminating between those tasks for which the given model is well suited and those for which it is not?

Our purpose then is not to put aside the conventional assessments of the model's match with history, or the process of qualitative peer review, but rather to augment these with something else. However, we might well want to put aside the convention of requiring a model to generate nothing but predictions that aspire to being facts, which subsequent observation may reveal, in the event, as true or false (see also McCombie and McKinley, 1993). Instead, what we should seek is a judgement about the quality of a model as a tool that has been designed against a task specification. In this respect, a policy forecast can be just as 'objective' as a scientific prediction. Other products of design, such as computer programs and screwdrivers, are no less objective than scientific facts, just because they perform defined tasks rather than state intended truths. The quality of the model would then derive from the extent to which a design task is fulfilled, not from some measure of the model's distance from the (unknowable) truth—which is *not* to say that we do not hope there are very strong 'elements of truth' in the policy forecast and the model from which it is generated. We shall see that a form of sensitivity analysis may play a vital role in enabling development of the sought-for quantitative measure of a well designed, relevant model.

21.3 A REGIONALIZED SENSITIVITY ANALYSIS

We are strongly accustomed to the idea of performance being specified in terms of a time series of observations of the model's state (or output) variables. This is indispensable to history-matching. But it is also a rather restrictive outlook, as admirably demonstrated in the work of Hornberger, Spear and Young in the late 1970s (Young *et al.*, 1978; Hornberger and Spear, 1980; Spear and Hornberger, 1980). As their exploration of eutrophication in an estuarial inlet shows, it is possible to determine what might be the critical components of a model in discriminating between a match and a mismatch of *qualitatively* observed behavior, and under gross uncertainty. In order to present their method, and the reasons for its development, we must first reconstruct the contemporary knowledge base, hopes, and aspirations of those times, now some two decades ago.

21.3.1 Origins of the Method

In its simplest form the problem of eutrophication may be defined as follows. It is the artificial acceleration of the natural ageing of bodies of water as a result of an elevated rate of accumulation therein of nutrients (carbon-, nitrogen-, and phosphorus-bearing substances). This much was already understood by the late 1960s. What was intended for the beneficial enhancement of primary biological production on the land surface was instead being diverted into enhanced primary production in the aquatic environment. This was manifest in the high-frequency, high-amplitude perturbations of sudden bursts of growth and rapid collapse in the biomass of phytoplankton—microscopic organisms at the base of the food chain in an aquatic ecosystem. Given a ready supply of the principal nutrients, it was well known that under an appropriate combination of environmental factors (solar irradiance, temperature, concentrations of other trace nutrients, and so on), populations of the various species of phytoplankton would grow rapidly, with a succession of species becoming dominant over the annual cycle. The phytoplankton would be preyed upon by zooplankton, which in turn were themselves prey to the fish. Dead, decaying, and fecal matter from all of this activity was equally well known to be re-mineralized by a host of bacterial species—operating either in the water column or in the bottom sediments of the lake. Such re-mineralization had an influence over the concentration of oxygen dissolved in the water, and this, in its turn, exerted an influence over the degradation pathways operative in the overall process of microbially mediated re-mineralisation. Thus the cycling of elemental material fluxes was closed; thus the disturbances of excessive phytoplankton growth would be propagated around these cycles.

Such was the raw material of the primary science base of the time. It was rapidly being refined and encoded in mathematical form, typical of which was the model CLEANER (Park *et al.*, 1975). We presumed that as the order of the internal description of the system's behavior grew, i.e., the order of $[\mathbf{x}, \boldsymbol{\alpha}]$ in Equation (21.1), so would the order of access to its external description, i.e., the order of $[\mathbf{u}, \mathbf{y}]$ characterizing observed behavior. We expected the estimation of parameters to become 'automated', subordinated to the formal algorithms of constrained optimization instead of practised 'by hand'; the observed features of the system's behavior were expected, just as much, to become increasingly detached from our personal, qualitative, and arguably subjectively tainted, powers of observation (of sight, sound, and smell); and the time series of data on [**u**, **y**] were expected to become increasingly available and, eventually, sampled at an appropriately fast frequency for a sufficiently long period.

Experience, however, was failing to bear out these expectations, and rather consistently so. To their credit, Hornberger, Spear, and Young were the first to recognize the situation as such and deal with it effectively. Their approach has come to be known as a regionalized sensitivity analysis (RSA); it has been widely applied, yet little changed over the intervening years.

21.3.2 Essence of the Analysis

Two presumptions underpin the classical approach to model parameter estimation and, by implication, the matching of history. First, there will be some 'dots' through which the 'curve' of the model's output can be made to pass closely (Figure 21.1a). Second, there will be a uniquely defined combination of values for the model's parameters ($\boldsymbol{\alpha}$) attaching to this optimal match of estimated and observed behavior (the point O in Figure 21.1b). Neither plays

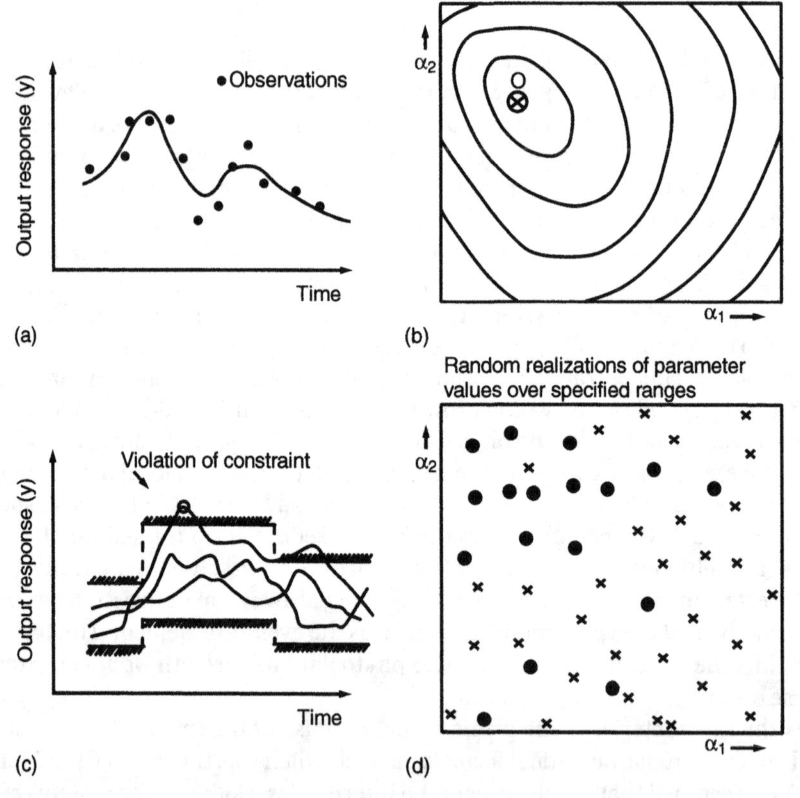

Figure 21.1 A comparison of (a and b) the classical approach to model parameter estimation and (c and d) regionalized sensitivity analysis: (a) fitting the model response to the data; (b) contours of the fitting-function surface in the parameter space; (c) specification of the constraints on acceptable model response; and (d) analysis of model parameter values (dots indicate values giving rise to acceptable behavior, and crosses indicate values giving rise to unacceptable behaviors).

any role whatsoever in the RSA. The problem statement remains as inelegant and uncertain as it truly is, but *not* without the prospect of making meaningful progress, thus:

- The trajectories of the time-series observations $\mathbf{y}(t_0), \mathbf{y}(t_1), \ldots, \mathbf{y}(t_N)$, against which the performance of the model would normally be evaluated, are replaced by a definition of (past) behavior (B) in terms of less detailed, more qualitative, possibly subjectively derived, constraints, such as those shown in Figure 21.1 (c).
- The conventional error-loss (objective) function for locating a uniquely best estimate ($\hat{\boldsymbol{\alpha}}$) of the parameter vector is replaced by a criterion for either accepting or rejecting a candidate parametrization $\boldsymbol{\alpha}^j$ as giving rise to the defined behavior (B) introduced in the foregoing (Figure 21.1d).

For example, in the original study of eutrophication in the Peel Inlet of western Australia (Hornberger and Spear, 1980; Spear and Hornberger, 1980), one item of the behavior definition (B) was chosen to constrain the estimated yearly peak biomass of the nuisance alga *Cladophora* to be greater than 1.5 times and less than 10.0 times its initial biomass at April 1 (defined as t_0), i.e.,

$$1.5x(t_0) \leqslant x_{\max}(t) \leqslant 10.0x(t_0) \qquad \text{for } t_0 \leqslant t \leqslant t_N \tag{21.2}$$

In addition, the ranges of permissible values from which the candidate model parameter vectors are to be drawn were specified as rectangular distributions with upper and lower bounds, i.e.,

$$\boldsymbol{\alpha}_l \leqslant \boldsymbol{\alpha}^j \leqslant \boldsymbol{\alpha}_u \tag{21.3}$$

The two types of inequalities (21.2) and (21.3) reflect the uncertainty of the empirical evidence and the uncertainty of the prior hypotheses, respectively. In the original study, the pattern of the input disturbances $\mathbf{u}(t)$ and the initial conditions $\mathbf{x}(t_0)$ were assumed to be known and not subject to uncertainty. Both could, of course, be treated as parts of the parameterization of the model, obviously so in the case of the latter, and possibly via the use of time-series models in the case of the former. This same principle could in fact be further extended to parameterizations of the processes $\boldsymbol{\xi}$ and $\boldsymbol{\eta}$ in Equation (21.1), although this remains an as yet unexplored option.

The RSA comprises two steps, the first being a combination of straightforward Monte Carlo simulation with a binary classification of the candidate model parametrizations. A sample vector $\boldsymbol{\alpha}^j$ is drawn at random from its parent distribution and substituted into the model to obtain a sample realization of the trajectory $\mathbf{x}(t)$, which is then assessed for its satisfaction, or otherwise, of the set of constraints defined in the form of the inequality (21.2). Repeated sampling of $\boldsymbol{\alpha}^j$, for a sufficiently large number of times, allows the derivation of an ensemble of parameter vectors giving rise to the behavior (B) and a complementary ensemble associated with not-the-behavior ($\bar{B}$). For this analysis, therefore, there is no meaningful interpretation of a degree of closeness to a uniquely best set of parameter estimates. Each sample vector $\boldsymbol{\alpha}^j$ giving rise to the behavior is equally as good or as probable as any other successful candidate parameterization of the model.[2]

Two sets of candidate parameterizations can therefore be distinguished: those m samples $\{\boldsymbol{\alpha}(B)\}$ that give the behavior and those n samples that do not, i.e., $\{\boldsymbol{\alpha}(\bar{B})\}$. In the second step of the RSA, the objective is to identify which among the hypotheses parametrized by $\boldsymbol{\alpha}$ are

[2] A number of subsequent variations on this theme can be found in Keesman and van Straten (1990), Beven and Binley (1992), and Klepper and Hendrix (1994).

the significant determinants of observed past behavior. 'Significance' is here indicated by the degree to which the central tendencies of the marginal and joint distributions of the (*a posteriori*) ensembles of the 'behavior-giving' parameter values $\{\boldsymbol{\alpha}(B)\}$ and their complement $\{\boldsymbol{\alpha}(\bar{B})\}$ are distinctly separated. For each constituent parameter α_i the maximum separation, distance d_{max}, of the respective cumulative distributions of $\{\alpha_i(B)\}$ and $\{\alpha_i(\bar{B})\}$ may be determined and the Kolmogorov–Smirnov statistic, $d_{m,n}$ then used to discriminate between significant and insignificant separations for a chosen level of confidence (Kendall and Stuart, 1961; Spear and Hornberger, 1980). Relatively large separation implies that assigning a particular value to the given parameter is key to discriminating whether the model does, or does not, generate the specified behavior. Relatively small separation of the two distributions implies that evaluation of the associated parameter is redundant in so discriminating the performances of the model. For these latter, it matters not, in effect, what value is given to the parameter; the giving or not giving of the observed behavior is more or less equally probable whatever value of the parameter is realized. Thus, for instance, the distinct clustering of parameter combinations that give the behavior, towards high values of α_2 and low values of α_1 in Figure 21.1(d), suggests that both parameters are key in the sense now defined. Indeed, one could also conclude that both of the constituent hypotheses associated with α_1 and α_2 are likely to be fruitful speculations in understanding the observed system behavior. More generally, the parameter space of the model can be cleaved into those parameters found by this procedure to be key, $\{\boldsymbol{\alpha}^K\}$ say, and those found to be redundant, $\{\boldsymbol{\alpha}^R\}$.

21.3.3 Predictive Task Specifications: Experience as a Substitute for Quantitative Observation

What we see, for our present purposes, is that the output of the model is required, as it were, to pass through a 'corridor' of constraints with 'hurdles' to be overcome (Figure 21.1c), and it either succeeds or fails for any given candidate parameterization. Passage through these constraints, in the spirit of Caswell's observations, can be thought of as the design task to be matched by the model. Specification of the task may be drawn—quintessentially—from the *experience* of an individual; from the expert field biologist, for example, who has walked round the estuary, or sailed across it, and witnessed the timing of the unsightly, rotting *Cladophora* biomass in each of the past five years or so. The potential for developing alternative approaches, caught in the word 'experience', yet passing by almost unnoticed in the earlier ASTM definition of model validation, can now be revealed, perhaps even realized. The notion of seeking only closeness of the match between observed and estimated history—in the *external* output space of the model—has been relaxed. Instead, the patterns of mere success and failure in performing the design task have been reflected back into the parameter space (of Figure 21.1d); and it will be from an analysis of these *internal* parametric properties that we shall construct our supplementary test of model (in)validation.

But at the core of what we are proposing is the freedom to draw upon the qualitative features of an individual's experience of the past—and imagination of future possibilities (Beck *et al.*, 2000)—in order to specify the set of design tasks to be performed by the model. A stakeholder, or a policy-maker, must have some idea of what he/she would not wish to have happen in the event of a newly synthesized chemical being released into the natural environment, for the first time, as part of a product of benefit to society. There is no history available for matching. Yet we would like to know, for example, whether candidate model A is a better designed tool than candidate model B for performing the task of predicting 'high-end' exposures to the toxic side-effects of the new chemical. In other words, we are seeking a novel form of quantitative assessment of the model's capacity to generate extrapolations beyond the confines of the conventional, formally observed record of the past. This we would like to place alongside the process of peer review of the composition of the model and the quantitative tests of the model's output performance in matching the formal record of past behavior.

21.3.4 Towards a Novel Statistic

The model is a refined map of the problem-specific science base. Each of its parameters resembles a tag, as it were, of the raw material assembled into the model's structure from any of the several conventional, primary disciplines. What might it therefore be about the model, we could enquire, i.e., which constituent parameter(s) might it be, that enables the model to generate a 'high-end' exposure or an exposure of 'no concern'? What might it be, more generally, that enables the model to perform its task? Which features in the map of the science base, we would want to ask, i.e., which members in the assembled composite of all the constituent, parameterized hypotheses, are key and which redundant? Indeed, what is the balance between the key and redundant parameters in achieving fulfillment of the specified task?

Now we have the prospect not only of thinking of such questions but also of answering them in a more quantitative fashion. For a well-designed (ill-designed) model could be gauged by the ratio of (key/redundant) parameters within its structure, a property notably *independent* of the size of the model. A model with 100 parameters in total, of which 30 are key to achieving the design task (and the remainder redundant), would be no better (or worse) as a predictive tool than an alternative containing 3 key parameters out of a total of 10. We would be able, therefore, to discriminate between the suitability of alternative models proposed as candidates for the given predictive task. Put another way, we could explore the relative suitability of a given model for performing alternative tasks. In the context of our predictive exposure example, which will be the subject of the case study to be presented below, we could ask whether the candidate model is a good or poor tool for predicting, on the one hand, high-end exposures and no-concern exposures, on the other. Thus, for instance, in Figure 21.2 six of the model's eight parameters are found to be key in the task of predicting high-end exposure concentrations while only two are critical to the task of predicting no-concern exposure concentrations. Were we to assert that a good design of a model would contain few redundant constituent hypotheses, clearly this particular model would be better suited to predicting high-end exposures. Quality of design is here not quite the same, however, as the familiar concept of parsimony of model parameterization as used in statistics. Parsimony should result in a model in which *only* those parameters demonstrably necessary for matching history are included. In the presently suggested measure of a good design—in performing a specified, predictive task—interest falls on enumerating the numbers of *both* apparently key and redundant parameters.

Yet a judgement about the quality of a model's design is not merely a matter of the relative numbers of key and redundant parameters. If fulfilment of the task is dependent upon many key parameters believed to be relatively well known, this will engender greater confidence in assigning the approval of a high-quality instrument to the given model. It is reassuring to know that effective performance of the predictive task is crucially dependent upon a good knowledge of the distance separating the receptor from the source of the contaminant, since such a distance ought to be easily and accurately measurable. To discover, in contrast, that

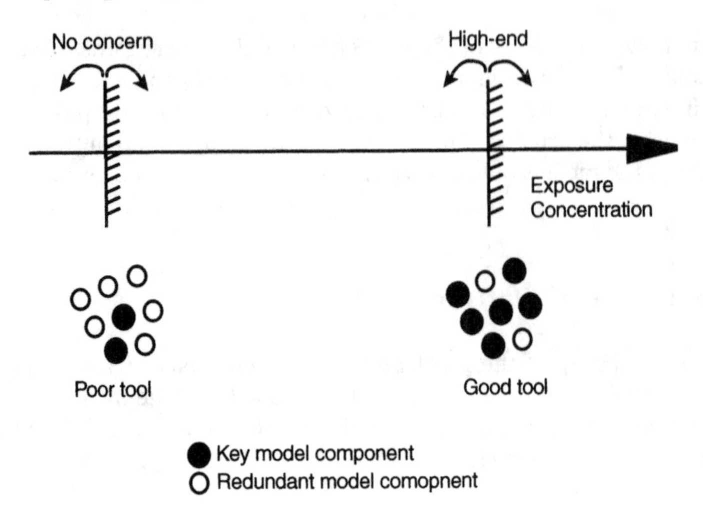

Figure 21.2 Illustration of the means of discriminating between ill- and well-designed models for a given predictive task; model components (or parameters) are indicative of the number of constituent hypotheses assembled in the model.

performing the task turns critically upon a handful of key but very uncertain parameters would be disquieting, to say the least. It would be troubling to learn that the task of predicting a no-concern exposure depends, above all else, on knowing precisely the rate of *in situ* microbial degradation of the contaminant, a parameter notoriously difficult to estimate (Blackburn, 1989).

We can develop further the ways in which the quality of the model might be judged more formally, but this will now best be achieved through the use of a specific case study.

21.4 ASSESSING THE QUALITY OF A MODEL FOR PREDICTIVE EXPOSURE ASSESSMENTS

Contamination of the subsurface land environment as a result of leakage from sites used to contain and store hazardous materials has been a predominant feature in the development of mathematical models of soil and groundwater systems over the past twenty years or so (Onishi *et al.*, 1990; National Research Council, 1990; Gee *et al.*, 1991; McLaughlin *et al.*, 1993). At a strategic level, the key questions to be answered are: what is the direction in which the plume of contaminants is moving; how fast is it moving; to what extent are the contaminant concentrations being attenuated with time and distance; what level of contaminant concentration will result at a particular receptor site; and, when contamination is forecast to be unacceptable, at which repositories will the greatest reduction in the risk of adverse exposure be achieved through expenditures on control and remediation measures? And herein lies the nub of the problem. There are many more sites where action might be taken to reduce risks of exposure than the funds to support all such actions. Accordingly, there is a need to rank the sites of potential action in terms of achieving the greatest reduction in the risk of exposure for a given sum of money. The process of arriving at such priorities may be termed a screening analysis, and as a part of this analysis the US Environmental Protection Agency (EPA) has developed a model for the propagation of contaminants from their source through the various media to receptor sites. In situations such as this, which are characterized by gross uncertainty, assessing the reliability of a model in performing the task of a screening analysis is especially important. The risks of ranking the sites for remedial action in an erroneous order are significant. The absence of a formal, conventional record of history derives here not from the sheer novelty of the substances involved but from the fact that there are so many storage sites with potential problems in general, yet so little empirical evidence of the actual problems at specific facilities.

21.4.1 The Model

The EPA's multimedia model, abbreviated here as EPAMMM (Salhotra *et al.*, 1990; Sharp-Hansen *et al.*, 1990), is a tool for predicting the transport and fate of contaminants released from a waste disposal facility into an environment composed of several media. Releases may be to the air or subsurface environment, the latter including both unsaturated and saturated zones, with the possibility of interception of the subsurface contaminant plume by a surface water system. The model contains seven modules: the landfill unit; the flow field in the unsaturated zone; the transport of solutes in this unsaturated zone; the transport of solutes in the saturated zone; the surface water system; an air emissions module; and dispersion of the contaminant in the atmospheric environment.

In the present case study, which is described in full elsewhere (Chen and Beck, 1998), only three of the above seven modules of EPAMMM will be used: flow in the unsaturated zone; transport of solute in the unsaturated zone; and transport of the solute in the saturated zone. In outline, therefore, the generic storage facility is supposed to be located above an unsaturated zone of the subsurface environment. Contaminated leachate from the facility infiltrates the ground, passing into the underlying unsaturated zone in the vertical direction (only). When the leachate reaches the saturated zone, it enters a horizontal flow of water driven by an input, laterally oriented recharge, whence it is subsequently transported with this flow in the saturated zone to the downstream receptor site. As the contaminated leachate passes through the unsaturated zone, its solute (considered simply as a single contaminant) undergoes attenuation and redistribution through dispersion, biodegradation (according to linear first-order kinetics), hydrolysis, and adsorption, this last being prescribed as a function of the concentration of organic matter in the unsaturated zone. Hydrolysis in the unsaturated zone is computed as a function of the bulk density and the distribution coefficient for contaminant in the dissolved and sorbed phases; it is assumed to occur in both phases with first-order kinetics that are a function of both temperature and pH. In the saturated zone, the solute likewise undergoes the same processes of attenuation. The receptor site is assumed to be at a 'worst-case' location, being directly downstream of the source, on the centre-line of the plume, and at the top of the saturated zone.

To summarize, the model requires the specification of more than 30 parameters ($\boldsymbol{\alpha}$) and, in its present application, the inputs $\mathbf{u}$ are assumed to be invariant with time, so that computational results are concerned merely with the steady-state solution of Equation (21.1) expressed as the value of the residual contaminant concentration at the receptor site, i.e., y.

21.4.2 The Tasks

A screening-level assessment of the risk of adverse exposure at the receptor site is concerned with knowledge of the probability that a particular contaminant concentration, say $\bar{y}$, will be exceeded. Within the setting of an RSA, therefore, 'acceptable behavior' from the given model will simply be defined as any candidate parameterization giving rise to y such that the following constraint is satisfied:

$$0 \leqslant y \leqslant \bar{y}. \tag{21.4}$$

Accordingly, 'unacceptable behavior', is equally simply defined as $y > \bar{y}$.

Computationally, we can implement the Monte Carlo simulation given the parent distributions of the parameters (as in Equation (21.3)), take the resulting distribution of y, and then separate behavior (B) from not-the-behavior ($\bar{B}$) for a *range* of different choices of $\bar{y}$. In particular, the performance of the model can be assessed in the context of different design tasks, for example, that of predicting a high-end exposure concentration at the receptor site (a high value of $\bar{y}$) versus that of predicting no-concern concentrations (a low value of $\bar{y}$), as illustrated earlier in Figure 21.2. Although these are hardly 'rich' task specifications, they will serve our present purposes, of revealing how the pattern of key and redundant parameters in the composition of the model changes with a changing definition of the task. What parameters rise to significance, we could ask, and which fall into insignificance (in terms of discriminating success from failure), as the single constraint in Figure 21.2 is moved from right to left, from one task to another? Just a binary cleavage of the parameter space into key and redundant features is perhaps unnecessarily crude in this context, however. Something more subtle may be attempted. For our interest lies in illuminating the roles of the various model parameters—be these anything on the *continuum* from absolutely critical to utterly redundant—in the ability of the model (EPAMMM) to discriminate the prediction of $y \leqslant \bar{y}$ from the prediction of $y > \bar{y}$ for different values of $\bar{y}$.

In sum, we would like to know whether EPAMMM is a good (reliable) model for predicting the entire range of exposures, or just the high-end exposures, or merely the mean exposures. What is it, in other words, that is most critical in the design of the model with respect to successful achievement of this particular task? For which of the model's parameters would the best possible knowledge be required in order to determine a particular percentile of the distribution of the contaminant concentration at the receptor site? Conversely, if the model is not judged to be reliable for fulfilling any of these tasks, we would like to know which of its parts are the least secure, for that given task.

21.4.3 Gauging the Quality of the Model's Design

Assessing the fate of a conservative, non-biodegradable contaminant (benzene) associated with a storage facility located in a sandy clay-loam soil will be the subject of this analysis (see also Chen and Beck, 1998). The 95th-percentile exposure concentration is deemed to be the upper bound on a tolerable high-end exposure and will therefore constitute one of the task definitions (conceptually, that to the right of Figure 21.2). The other task will be that of discriminating between concentrations of the contaminant at the receptor site above and below the 50th-percentile.

Before presenting our new measure of the quality of the model, let us first reflect on the information that will now be available from the RSA and how it might best be used. We have a model with some 30 or so parameters and we wish to employ the Kolmogorov–Smirnov statistics attaching to each of these in order to come to a summary judgement on whether the model is (relatively) ill- or well-suited to one among several task definitions. We wish in addition to have a measure with uniform properties maximally independent of the specific nature of a given task definition—as much, incidentally, as we would like a measure maximally independent of the size of alternative candidate models (although this latter will not be explored further herein). We do not, however, wish to use some standard value of the Kolmogorov–Smirnov statistic ($d_{m,n}$) to separate key from redundant parameters. What follows, then, is a tentative statistic q whose *distribution-like* properties are to be plotted for the purposes of assuring the quality of a model designed for a given predictive task (indexed by k):

$$q_k = \frac{d_{\max,k}(i)}{d^*} - 1. \tag{21.5}$$

Thus, for any parameter $\alpha_i, d_{\max,k}(i)$ can be determined as the maximum separation of the cumulative distributions of the m_k candidate behavior-giving values $\alpha_i(B_k)$ and the n_k not-behavior-giving values $\{\alpha_i(\bar{B}_k)\}$ for the kth task. d^* might be considered a value of the Kolmogorov–Smirnov statistic chosen to discriminate with a given degree of confidence between significant and insignificant such maximum separations ($d_{\max,k}(i)$) of the behavior-giving and not-the-behavior-giving distributions for the given parameter. Indeed d^* might be considered to fulfil the same role as $d_{m,n}$ previously. Yet, unlike $d_{m,n}$, d^* must be invariant and independent of the differing magnitudes of m and n, now denoted m_k and n_k, respectively, which arise from assessments against the multiple (k) task specifications.[3] These connotations of statistical confidence notwithstanding, d^* has here rather the function of a normalizing parameter, allowing different *distributions* of the model's 'parametric significances'—with respect to discriminating whether the task specification is matched or not—to be compared on a consistent basis, irrespective of the given task.

Figure 21.3 shows the frequency *distributions* of q_k, i.e., of q_1 and q_2, for assessing the performances of the model against the two predictive task specifications. The effect of d^*, chosen in this case to reflect a level of confidence of 0.001, is to scale the plot of the distribution of parametric significances so that 0.0 separates insignificance (redundancy) from significance. Once normalized as a plot of the relative frequency distributions for $(d_{\max,k}(i)/d^*) - 1$ (the parametric significances), the number of parameters in any model (indexed here through i) can to some extent be detached from the judgement on the quality of its performance. In general, if the distribution of the parametric significances were skewed towards the right, this would suggest that the model contains a relatively large number of parameters that are key in the performance of the specified task. If the distribution were skewed towards the left, the model might be said to be suffering from a preponderance of redundant parameters, relative to the task at hand.

Figure 21.3 shows that in performing both tasks, a majority of the constituent parameters of EPAMMM appear to be redundant, especially so in the case of predictions required to discriminate exposures above and below the 50th-percentile. At the same time, a small number of the model's parameters (13%) are critical to the performance of both tasks, notably more so in the case of the 95th-percentile task. In very general terms, we might be tempted to conclude from Figure 21.3 that EPAMMM is *better* suited to performing the task of predicting high-end exposures *relative* to the prediction of mean exposure concentrations.[4] Put in more familiar terms, we might say that EPAMMM is a better-designed tool for predicting high-end as opposed to mean exposure concentrations. This conclusion would be subject, of course, to the qualifying statement that the parameters associated with the highly positive values for the statistic in Figure 21.3 are relatively well known.

[3] The results of Chen and Beck (1998) show that $d_{m,n}$ does vary as a function of the varying m and n associated with varying numbers of random candidate parameterizations classified as behavior- and not-behavior-giving for differing task specifications.

[4] Note that this is not a statement of an absolute property but rather one of a more 'relativistic' character.

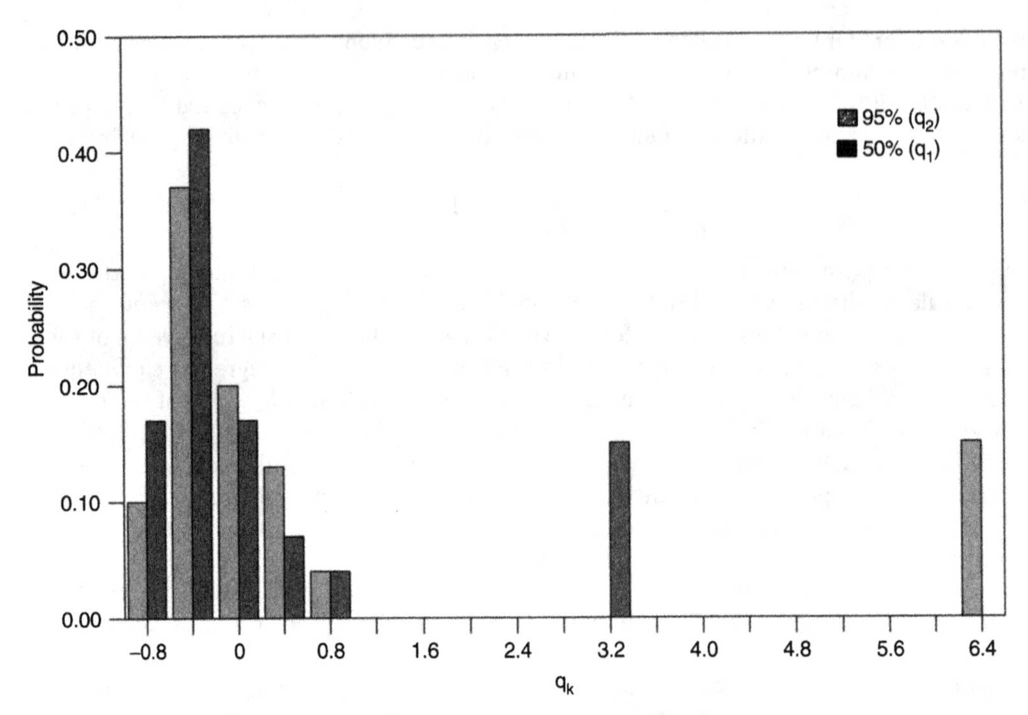

Figure 21.3 Comparative plot of distributions of the index q_k for assessing the quality of the model's design.

Having been tempted to draw such conclusions, however, it must be noted that the validity of computing and using the distribution of the above statistic (or index) has yet to be fully evaluated by much more extensive analyses. For instance, we should establish the legitimacy of using a single value for d^*, when $d_{m,n}$ varies significantly as a function of small magnitudes for either m_k or n_k for some of the assessments of the model's performance against the various task specifications. Similarly, judgements about the character of the relative frequency distribution of our statistic may be compromised in cases where the number of parameters in the model is very small. We note too that use of the Kolmogorov–Smirnov statistic has its limitations and that assessment of the model as a function of attributes of its *individual* parameters, as opposed to key and redundant *clusters* of parameters, should be interpreted with great care (Spear *et al.*, 1994). Nevertheless, here—in Figure 21.3—is a quantitative measure of the quality of a model conditioned upon how that model performs a task of prediction, including projection into utterly novel conditions, *not* upon how the model matches observed past behavior. This measure is cast in terms of the internal features of the model itself. What is more, we might imagine that a particular shape of the distribution of q_k could be attached to the concept of a well-designed model. For example, this might be a distribution in which there were neither too many redundant parameters (associated with the left tail of the distribution) nor a few excessively key parameters, affecting the distribution towards its right tail. In these respects, we argue that this statistic and its associated analysis have, in principle, great appeal; an appeal, moreover, tied closely to the practical purpose of decision-making for which the model was designed and developed.

21.4.4 Coming to a Judgement on the Trustworthiness of the Model

This examination of the quality of the model's design is one of several, all based on the application of some form of sensitivity analysis (Chen and Beck, 1998), none of which, crucially, requires access to observed data on the system's past behavior. The results of our analysis, together with the outcomes of matching history and peer reviews, were these to have been available, could have been marshalled and brought to bear on answering the question: given a predictive task, should we use the given model? Such simple questions are enormously difficult to answer, yet answered they must be. Actions must be taken.

In our case, the issue was one of establishing how well the Environmental Protection Agency's multimedia model (EPAMMM) performed as a tool for discriminating sites at risk from potential contamination of the sub-surface environment from those not at risk. EPAMMM is not all that complex. Acceptance of the test results of its performance, however, was by no means simple or straightforward. We found two things about the process of coming to a summary judgement. First, a test result frequently made no immediate sense until we had reasoned back from it to the starting point of the assumptions of the test conditions and their influence over the numerical computations and outputs of the model. Second, the circumstances under which the test result would have real import had to be qualified by many detailed footnotes, as it were. Our original report (Chen and Beck, 1998), perhaps rather like any legal document, does not make light reading; its arguments and interpretations are tortuous. In these reflections, we should detect nothing of surprise. Computer-based models are constructed because we cannot reason systematically, in our heads, through all the ramifications of the assumed web of interacting causal connections in the behavior of a system. In the end, however, whether we trust an outcome of the test of the model, and hence place our trust in the model itself, is a matter of whether we have a good mental image—a good understanding—of the inner workings of the mathematical and computational model. Thus was it ever.

21.5 CHALLENGING HIGH-LEVEL CONCEPTUAL INSIGHTS

Understanding something is achieved when information about this thing is assimilated into an appropriate mental model (MacFarlane, 1990). We believe we have understood the behaviour of Lake Erie, for example, when what is observed to happen can be faithfully reflected in terms of the simple, everyday, household concepts of beakers with stirrers, warm fluid rising towards the surface, switches that prevent the beaker's contents from mixing, and food for the growth of organisms. We believe we have understood the circulation of heat through the ocean when provided with insights through the image of simple switching mechanisms along the path of a conveyor belt (Weaver, 1995). Policy-makers and the public understand the implications of the increased release of carbon dioxide into the atmosphere and the associated possibility of a change in climate because of the analogy of the humble, but most familiar, greenhouse (familiar at least to those from northern temperate climates). It is through appeals to the commonplace, the mundane, and the familiar that we all have access to understanding and insight across the disciplines of science.

Of course, we know that things are not quite this simple. But until we have arrived at such a distillation of the essence of the matter—at such a macroscopic, high-level conceptual description—we are unable to make sense of all the information available. Yet what should we do if this 'making sense' is achieved through a defective mental model? After all, there are some well-known, perhaps even infamous, cases where sense was manufactured on the basis of a defective mental model, William Bowie's rejection of Alfred Wegener's theory of continental drift being one of them (Oreskes, 1998b). What should we do, for example, if the intuitive appeal of the ocean conveyor belt, whose realization has itself been enabled as much by the availability of very high-order computational models of ocean circulation, as by interpretation of the empirical observations, is misleading? How might we acquire any sense that the analogy of the conveyer belt is untrustworthy or wrong? How indeed would we provide evidence to support the generation of an alternative, candidate, high-level conceptual insight?

These questions call for the matching of history, but not in the same spirit as we would now understand from Konikow and Bredehoeft (1992). Macroscopic conceptual understanding and basic, microscopic theory are set in very different planes; on these our tentative descriptions of the 'truth' are resolved at very different levels. It is as though the scope of our enquiry is focused at very different powers of magnification. Whereas there may be observed current fields against which to match the behaviour of a high-order computational model of ocean circulation, we cannot literally observe the quantities of heat at given positions on the conveyor belt or the times at which the belt was switched on, slowed down, switched off, and so forth. We cannot *directly* reconcile the conceptual insight—vital though it is to understanding—with any empirical evidence, any more than we can easily come up with an insight very different from the ocean conveyor belt. But we might be able to undertake some useful form of evaluation provided interpretations of theory and observation can be derived at a more or less compatible power of resolution.

More crisply stated, let us suppose we have a high-order computational model, with a very high resolving power, i.e., with many state variables and parameters, denoted $[\mathbf{x}^{+n}, \boldsymbol{\alpha}^{+n}]$, yet we can only identify from the field observations models of a low order, denoted $[\mathbf{x}^{-n}, \boldsymbol{\alpha}^{-n}]$ for the sake of dramatic contrast. It is clear we cannot embellish the latter in order to achieve the detail of the former. It is just as clear, however, that we could seek to reduce the order of the former in order to compare a distillate of its essence with the latter. This, of course, will readily be recognized as a problem of sensitivity analysis, akin to the ideas of 'meta-modelling' discussed in Kleijnen (1987), for example. In the case study outlined below, taken from Young *et al.* (1996), the analysis of sensitivity is referred to as a dominant-mode analysis: the goal is to approximate the dynamic behavior of the given high-order model by identifying the small number of dynamic modes that appear to dominate its response to perturbations in the input variables. In other words, the low-order approximation (or 'meta-model') is in the form of an input–output time-series model, not a regression relationship, and is derived from a single, impulsive perturbation applied through an input variable, not from a number of trials with the high-order model for different, candidate combinations of values for its parameters. Having thus gone to an abstract representation, the manner of then re-extracting some kind of quasi-conceptual interpretation from the so identified low-order model is discussed at length in Young (1998).

The movement of carbon-bearing substances in the global environment can be represented by a set of 26 ordinary differential equations of the form of Equation (21.1a) (Enting and Lassey, 1993). Although not all that complicated, this must serve the role of a high-order computational model for the present purposes. Time-series observations are available over the period 1840–1990 for the single input of carbon-bearing fossil fuel substances (u) and the single output of atmospheric CO_2 concentration (y). In simple terms, it appears that the relationship between u and y can be succinctly identified from the empirical record as two continuously stirred tank reactors in series, one with a time constant of about 4.5 years, the other of some 115 years (Young *et al.*, 1996). That is to say, for this *data-based* low-order model (LOM), whose structure is parameterized in this manner, say $\hat{\boldsymbol{\alpha}}^{-n}(o)$ (i.e., two compartments in series with given time constants and steady-state gains), a notional impulse of fossil fuel carbon from anthropogenic sources would provoke something of a classical bell-shaped transient response in the atmospheric CO_2 concentration before returning (over a matter of centuries) to its pre-perturbation level. If the same notional input perturbation $\hat{u}$ is applied to the nonlinear high-order model of Enting and Lassey (at some equilibrium condition) and the resulting simulated response in atmospheric CO_2 recorded (as $\hat{y}$), an alternative *theory-based* LOM can similarly be identified from $[\hat{u}, \hat{y}]$ and parametrized as, say $\hat{\boldsymbol{\alpha}}^{-n}(m)$. This latter is found to be a linear system of four compartments acting in parallel, one of which is a pure integrator, the others having time constants of roughly 3.5, 16, and 459 years. When distilled down to its essence in these particular terms, it appears the theoretical, nonlinear model of the global carbon cycle exhibits linear properties, but the structure of this behavior is fundamentally *not* in agreement with the conceptual structure identifiable from the observed record. What is more, the atmospheric concentration of CO_2 simulated by the high-order model does not return to its pre-disturbance level once perturbed by a notional impulsive input of fossil fuel carbon; it remains, for ever, at a higher steady state (Young *et al.*, 1996).

When set on much the *same* plane, in a relatively coarse-grained representation of the system's behavior, the two 'interpretations' of this case study, i.e., the high-level conceptual descriptions (the one based on the observations, $\hat{\boldsymbol{\alpha}}^{-n}(o)$, the other on theory, $\hat{\boldsymbol{\alpha}}^{-n}(m)$, are inconsistent. That giving four compartments acting in parallel, with four parallel paths from the emission ultimately to the atmosphere, is in fact ambiguous. The identified behavior could equally well have been expressed as other structural arrangements, of compartments in series and with feedbacks (Young *et al.*, 1996). Such results are thought-provoking, not least when they confront us with somewhat surprising possibilities for conceptual interpretation. In this study of the global carbon cycle, the derived interpretation of the observed record (in contrast to that of the basic nonlinear simulation model) suggests that, once converted from its fossil form and emitted into the atmosphere, carbon does not behave in a conservative manner.

The analysis of the global carbon cycle has established a principle: a means of mapping the dominant modes of behavior of a high-order model and the strictly identifiable modes of observed behavior onto a common plane in which—through the lens of our enquiry—both can be brought into sharp, compatible focus. And from there, a constructive discourse might flow; one that has the possibility of corroborating or refuting the candidate high-level conceptual insight (the mental model), perhaps even provoking conception of an alternative.

21.6 CONCLUSIONS

Validation, in the classical sense of matching the performance of the model with a set of observed data, is getting no easier. Our models are becoming more complex. The order of their (internal) state–parameter vector $[\mathbf{x}, \boldsymbol{\alpha}]$ in a given field of study is tending to increase at a rate faster than the order of access to their (external) input–output properties $[\mathbf{u}, \mathbf{y}]$ through advances in instruments for observation. And even if this were not so, there are few, if any, methods of system identification that can be applied successfully to the problem of reconciling anything but the smallest of models with the available observations, copious though these may be. Furthermore, much of the basis for using a mathematical model in the analysis of environmental systems rests on the need to make extrapolations from past conditions into the possibly substantially altered circumstances of the future. In predicting the fate of novel substances released into the environment, no history is available to be matched, by definition. Collectively, we hold a very strong stake in the proper development, evaluation, and use of models; we are not free to walk away from such problems. Efforts can, of course, be redoubled in order to facilitate a better scope for peer review of a model and the matching of history with that model, where possible, in principle. Yet we can also augment the definition of what it might mean to assure the quality of a model to be used in the setting of policy forecasting.

Herein, therefore, we have sought to project the debate on model validation into a domain somewhat broader than the classical notion of demonstrating the match of the model with observed history (Konikow and Bredehoeft, 1992; Beck *et al.*, 1997). In particular, we have introduced the ideas of: (i) judging the trustworthiness of the model according to the quality of its design in performing a given task; (ii) opening out the terms in which the task can be specified, for example, as described by qualitative expert experience (as opposed to time-series of observations), as drawn from the terms of the regulatory setting, or as imagined as an unpalatable future by a stakeholder; and (iii) basing our judgement on properties of the *internal* attributes of the model—numbers of key and redundant parameters—as opposed to features associated with its output responses, i.e., its *external* attributes. Our approach has been inspired by the original work of Hornberger, Spear, and Young on a form of regionalized sensitivity analysis. From this we have proposed the possibility of employing distributions of a statistic, a normalized form of a Kolmogorov–Smirnov statistic, to discriminate between whether a model is well or ill suited to performing a given (predictive) task. A salient point is the relatively large measure of independence of the new statistic from the size of the model under scrutiny.

It is far too early to speculate on the eventual success of this approach, however. Our provisional conclusions are that it has worked in a prototypical case study of contaminant exposure assessment; the statistic appears to be sufficiently sensitive, in that it exhibits sufficient variation in response to relatively bland, somewhat weakly differentiated task specifications; but it may be hard to achieve consensus, in due course, on what constitutes a 'good design' of model and whether the shape of a distribution curve can be used to inform such a judgement.

If we assume the approach has promise, it would bring more varied forms of evidence—other than those of matching history and peer review—to bear on the makings of an answer to the essential question: given a predictive task, should we use the given model? The formalities of the process of coming to such a summary judgement might then have to be placed on a better organized footing, a patchy, preliminary sketch of which has been begun elsewhere (e.g., Beck *et al.*, 1995).

ACKNOWLEDGMENTS

This work has been supported in part by the Visiting Scientists and Engineers Program of the US Environmental Protection Agency (project title 'The Analysis of Uncertainty in Environmental Simulation'). M. B. Beck is currently Visiting Professor in the Department of Civil Engineering at the Imperial College of Science, Technology and Medicine, London.

22

Fortune and Future of Sensitivity Analysis

Andrea Saltelli

European Commission, Joint Research Centre, Ispra, Italy

22.1 INTRODUCTION: SENSITIVITY ANALYSIS AS AN INGREDIENT OF MODELLING

Sensitivity Analysis (SA) is, in the opinion of most, an important element of modelling. Kolb, quoted in Rabitz (1989), notes that theoretical methods are sufficiently advanced that it is intellectually dishonest to perform modelling without SA, while Furbringer (1996) argues '[Why] sensitivity analysis for modellers: Would you go to an orthopaedist who didn't use X-ray?' In spite of apparent consensus, SA has been used parsimoniously in the scientific community. A paradoxical proof of this statement is the common occurrence of articles devoted to Uncertainty Analysis (UA) and SA of a given model. In our opinion, any model should undergo UA and SA as a matter of course, prior to and during its use in different applications. SA should be implicitly assumed as part of any statement based on or involving a model.

Rabitz (1989) presents SA as a fundamental ingredient for model-building and a key tool in the understanding of a complex physical process. According to Rabitz, SA helps to analyze the contents of the model and to interface it with observational data. It helps to identify which factors are critically important, how they are interrelated, and especially how they, at a given level of description of the system (e.g. quantum mechanical potentials), influence the behavior at another, possibly higher, level (e.g. macroscopic rate constants). Rabitz points out another important use of SA: the antithetical process of reducing full detailed models to their essential or lumped structures.

While Rabitz refers to SA as a 'tool' for the modelling process (either in the building or application phases), in an article in *Science* entitled 'Verification, validation and confirmation of numerical models in the earth sciences', Oreskes *et al.*, 1994), attribute it a different meaning. SA is not viewed as a tool to build or improve a model but represents one of the

Sensitivity Analysis. Edited by A. Saltelli *et al.*

possible licit uses that can be made of the model itself. According to Oreskes, *et al.*, natural systems are never closed and models put forward as descriptions of these are never unique. Hence, models can never be 'verified' or 'validated', but only 'confirmed' or 'corroborated' by the demonstration of agreement (non-contradiction) between observation and prediction. Since confirmation is inherently partial, models are qualified by a heuristic value: 'Models are representations, useful for guiding further study, but not susceptible to proof'. In the opinion of Oreskes *et al.*, 'Models can corroborate a hypothesis [...]. Models can elucidate discrepancies in other models. Models can also be used for sensitivity analysis—for exploring "what if" questions—thereby illuminating which aspects of the system are most in need of further study, and where more empirical data are most needed.'

These comments suggest that the epistemological role of sensitivity analysis is varied and spans a considerable range of purposes, and justifies devoting an entire text to its tools. From the various chapters of this handbook, the diversity of applications is clear, as is also the variety of SA techniques that are available. SA and modelling merge and should not be separated when one desires to show that a given model is 'relevant' to a task; in that case the relevance of the factors may coincide with the sensitivity measures (see also Chapter 21).

22.2 THE FORTUNE

Perhaps the oldest applications of SA are in control theory (Tomovic and Vukobratovic, 1972) and in the resolution of the inverse problem (Rabitz, 1989; Turanyi, 1990a). With time, SA has also gained ground in the field of risk analysis, performance assessment, and analysis of environmental impact. Although other fields have been exposed to this discipline (see applications in JSCS, 1997; RESS, 1997; CPC, 1999), we would contend that SA is largely and happily ignored by most of its potential users.

So far, a combination of causes has limited the effective use of SA. Overconfidence in measurement accuracy and underestimation of uncertainty in the context of experimental activities has been discussed in Henrion and Fischhoff (1986). They remark that experimentalists systematically underestimate the experimental error, and show a convincing plot on the determination of *c*, the speed of light. In each chronologically consecutive new determination of *c*, its value was found to lie outside the uncertainty bound assigned in the previous determination. We suspect that a similar overconfidence might be present in the uncertainty linked to the use of models.

Other causes of the limited use of SA are of both a cultural and of a practical nature. Among the former, one might mention a problem of terminology. As pointed out in Chapter 1, SA means various things in different user communities, leading to a variety of different names being adopted in the literature to refer to the discipline. Helton and Burmaster (1996) note that many authors are not aware of the extensive work done by others in the area because of the proliferation of names for the concepts involved. It will be hard to convince potential users of the benefits of the various SA tools unless this terminological barrier is overcome.

Another cultural problem in the development of SA is perhaps a problem of 'cultural hegemony'. This is the predominance in the literature of the term SA used in conjunction with the concept of local derivatives of output versus input. The only review article on the subject published in *Science* (Rabitz, 1989) discussed only local approaches to SA. A single volume of the *Journal of Geophysical Research* (in 1997) contains as many as five articles where SA is performed either via an elementary OAT approach or via a derivative-based method (see Saltelli, 1999). For this reason, many user communities argue that they do not need SA, or that they have already performed it with uninteresting results.

Another difficulty for SA is when there are too many outputs to look at. When the models predict several variables, as in the case of temporally and spatially dependent outputs, the presentation and interpretation of the results of a sensitivity study may become problematic. Analyses and plots generated for each output need to be summarized. A strategy to overcome this difficulty is to base the analysis on 'model use' rather than 'model output'. For example, if the purpose of the model is to investigate the location of a polluting activity or the feasibility of its emplacement, rather than looking at pollution levels at each specific location on a map, one might consider the total area where a pollution threshold is exceeded, or the probability that it is exceeded anywhere in the region. In general, prior to an analysis, one should clarify which statement(s) the model is meant to support or disprove, and this should be taken as the output of interest in SA.

A further example, which touches on the difficulty of implementing SA on large, computationally expensive models, is when a complex, first-principle model is developed to either confirm or disprove simpler 'parametrized' models, i.e. models that make extensive use of empirical simplification whereby processes are summarized by phenomenological parameters. One such model could, for instance, provide a detailed description of the atmospheric physics in a limited area (few kilometers). Given that the model in this case must make statements about the possible discrepancy between first-principle and parametrized models, a sensitivity analysis could in this case have as target (objective) function some integral measure of discrepancy between the two models for some representative global output of the model (e.g., height of the mixing layer or intensity of the atmospheric turbulence as measured by the amount of turbulent kinetic energy, say). Such a discrepancy should be measured (averaged) over some finite domain in the space of all input factors.

Another way of dealing with excessive information is by using descriptive statistics to summarize the information. For instance, matrices containing $S_{T_i}(x_j)$, sensitivity indices for different output variables x_j, could be analysed via multivariate statistical methods. Principal-component analysis has been used in local SA studies (Chapter 5; Turanyi, 1990a).

Another common practical difficulty arises when it is impossible to sample some of the input factors, e.g. if the input comes from other models, or from databases. Correlated inputs can also cause difficulties for some of the SA techniques (importance measures were used in Chapter 14). Further, input factors might be defined on nominal scales.

One last difficulty is the computational cost of complex models, although this barrier to SA is an ever-receeding one. For systems with many uncertain input factors, SA could include a preliminary screening phase. A possible approach is to use a global screening method such as those of Morris or Andres (Chapters 2 and 4) to reduce the number of factors in order to make quantitative methods affordable.

Difficulties arise also in the use of 'global' methods. Before total sensitivity indices were introduced, frequently a global sensitivity study (e.g. using regression analysis or ordinary FAST) was completed, leaving an impression of laborious inconclusiveness when relevant fractions of output variation were left unaccounted for.

This happened because:

- regression coefficients only worked for linear systems;
- classical FAST only worked for additive models

Modellers were faced instead with inputs at different levels of uncertainties, often covering orders of magnitude. They needed a global SA method where the entire space of existence of the input factors would be covered and all the factors could vary simultaneously.

Mostly they needed an analysis of sensitivity that was informative, manageable, and not too expensive.

Rigorous tools capable of global quantitative analysis before now have not been available. Worthy exceptions to this lack of adequacy should nevertheless be mentioned.

The Morris method thoroughly described in this volume (Chapters 2, 4, 18 and 19) is one such exception, since it is model-independent, informative, and computationally inexpensive. We shall not recall the merits of this method here, which has been used quite extensively by several investigators.

Monte Carlo filtering (Rose *et al.*, 1991) is the process of rejecting sets of model simulations that fail to meet some pre-specified criteria of model performance (acceptable behavior). This process can be presented as an objective method for model calibration: the subsets of model parameters that generate acceptable model simulations can be regarded as equally satisfactory model calibrations. Rose *et al.* (1991) select the subset of factors for the analysis using a local SA method. More radically, Fedra *et al.* (1981) presents Monte Carlo filtering as an alternative to the concept of local calibration. The analyst should refrain from searching for the optimal solution but rest with the plausible ones.

Generalized sensitivity analysis is a term used by Hornberger and Spear (1981) to indicate the use of the outcome from a Monte Carlo filtering experiment for SA purposes (see also Chapter 21; Young *et al.*, 1996). The model output from the Monte Carlo run is categorized into two subsets. For a given input variable X, two subsets of possible values are identified: those that lead to acceptable behavior, subset X_a, and those that lead to unacceptable behavior, subset X_b. The sets of model inputs that lead to acceptable behavior are then statistically compared with those that do not. A statistical test of hypothesis is then applied to check if the two subsets are samples from the same statistical distribution. An input variable is regarded as important when the generated sample distribution functions are statistically different. In our view, this approach has a limitation, in that it takes into account only the output variation along the acceptable–unacceptable direction, while it ignores the variations of the output within the class of the acceptable values. In other words, an influential parameter could escape such an analysis only because it drives variation within the acceptable range.

With the exceptions mentioned above, the combination of factors reviewed in this section has determined the fortune of SA. While unquestionably several model-user communities would substantially improve their work by employing SA, convincing these users is hard—even more so if they have previously performed SA without satisfying their curiosity.

22.3 THE FUTURE

It would be appealing to be able to translate the contents of the previous section into a straightforward paradigm for the use of SA in modelling. The existence of a variety of sources of variation, many of which are linked to choices of the modeller, calls for objective model assessment procedures. Unfortunately the number of settings where models are applied is infinite, as are the possible uses of SA to improve them.

We may go back to Rosen's formalization of the modelling practice, mentioned in Chapters 1 (Figure 1.1) and 12. The link between a model (driven by a formal entailment structure) and the world (driven by different classes of causality, such as efficient or material causes) is the process of 'encoding' (from world to model) and 'decoding' (from model to world). As we said in Chapter 1, encoding and decoding are not themselves 'entailed' by anything, i.e. they are not amenable to a formal process of causality. Encoding and decoding are the object of the modeller craftsmanship. SA enters in this craftsmanship, especially in ensuring consistency between the two processes. SA should help to defend against dubious encoding, where the model is forced to mimic reality by 'special' values for the input factors that do not withstand scrutiny. In this way, one would reduce the chance of an arbitrary decoding, and, possibly, along the way, make the entire process more transparent.

In the present book, we have argued in favor of the use of SA prior and during model identification and calibration settings. In particular, before a given data acquisition activity that entails laboratory experiments or field data is undertaken, the candidate models put forward to describe the system should undergo SA (see Chapter 15). The convenience of this approach is in the low cost of the computational experiments with respect to the physical ones. Further, this analysis will tell the investigator whether, and for which values of the data still to be gathered, there is a possibility to discriminate between competing models. Note that in doing this, we have substituted model identification (which would imply that a given model is true, i.e. 'validated') by model discrimination, which implies that one model has been corroborated at the expense of another that has been falsified.

Model complexity and model corroboration should not be presented as conflicting, i.e. the more complex a model, the less likely is its corroboration. Increasing computer capacity and epistemological awareness suggest that model relevance becomes a prescription for the correct use of models.

A case where, nevertheless, we would dare a bolder prescription is in the use of predictive models in policy issues. Whenever a model is used to guide, influence, or inform different stakeholders (including policy-makers), our prescription would be to exclude every factor that—whether certain or uncertain—did not impact the model variation from the analysis. Here we mean that the fact that a given (set of) model(s) has been proven consistent with available evidence (even using SA paraphernalia), does not imply that the same model can be used in the policy related context. The model could still be at the stage that has been called—with some understatement—'parameter-rich'. The fact that such richness purportedly reflects actual knowledge on the causal relationships between variables should not be confused with the capacity of the model to make sustained predictions. If we run a SA at the same settings (e.g. space and time scales) where the prediction is sought, we are likely to find that only a subset of factors is capable of driving model variations. In our opinion, only those factors should be brought to hearing. In this approach, for any given task demanding a model, we do maintain causality in modelling (unlike Young *et al.*, 1996), but only to a level consistent with both observation and the task. In other words, with Oreskes *et al.* (1994), we believe that the task commands the choice of the model and that the latter cannot be defined without the former. The stakeholder should hence be confronted with the set of relevant inputs (with their uncertainty), models, and predictions. The information content and its quality should be seen with the ensemble of factors, models, and predictions. In this context, notions such as previous model performance and model importance based on size and complexity should be seen as irrelevant. Limpidity (transparency and relevance for Beck *et al.*, 1997) would be achieved by a process of model simplification whose objective is ultimately to tune the degree of complexity of the model to the questions being addressed.

22.4 CONCLUSIONS

The Danish writer Peter Høeg (1995) states in a novel: 'That is what we meant by science. That both question and answer are tied up with uncertainty, and that they are painful. But that there is no way around them. And that you hide nothing; instead, everything is brought out into the open'. Uncertainty is not an accident of the scientific method, but its substance.

Models as heuristic tools designed for a scientific task must be proven capable of dealing with uncertainty. Especially when the model is used to drive a choice or a decision, the importance of the associated uncertainties should be quantified, and the relevance of the model ensured.

Our working experience leads us to suspect, with Fedra *et al.* (1981), that, generally speaking, the scientific community overestimates the merit of punctual calibration in the context of scientific use of models. Very often the use of a model is forced within an optimization context, where the issue is to find some least-squares solution to estimate the input factors that disagree least with observation. Any model must include false and/or unrealistic assumptions in order to be of practical use (Morton, 1993), and these will not be made explicit in the calibration process. We see with empathy any attempt to map the input factor space into the prediction space, including alternative model structures, or the bootstrapping of the modelling process, and any reverse process, such as the mapping of the plausible outcomes back into the input space. SA could be an element of this alternative approach to 'calibration'.

Examples of instrumental use of models can be found in the literature, especially when models are used for making decisions having a large social and economic impact. Thus, it is not surprising to meet cynical opinions about models. An example is in *The Economist* (1998), where one reads that 'based largely on an economic model [...] completing K2R4 [a nuclear reactor] in 2002 has a 50% chance of being "least cost"'. Given that the model was used to contradict a panel of experts on the opportunity to build the aforementioned reactor, *The Economist* comments: 'Cynics say that models can be made to conclude anything provided that suitable assumptions are fed into them.' In the spirit of the approach defended in this book, it would be highly instructive to look at what factors were determining the variation around the 'least-cost' region. The outcome of this analysis could then provide experts with additional insight.

References

Adriaanse, A. (1993). *Environmental Policy Performance Indicators.* Sdu Uitgeverij Koninginnegracht, The Hague.

Allen, B. C., Covington, T. R. and Clewell, H. J. (1996). Investigation of the impact of pharmacokinetic variability and uncertainty on risks predicted with a pharmacokinetic Model for Chloroform. *Toxicology* **111**, 289–303.

Allen, D. M. (1971). *The prediction sum of squares as a criterion for selecting predictor Variables.* Report No. 23, Department of Statistics, University of Kentucky, Lexington.

American Society for Testing and Materials (1984). *Standard Practice for Evaluating Environmental Fate Models of Chemicals.* American Society for Testing and Materials. Standard E 978-84, Philadelphia.

Andres, T. H. (1987). Statistical sampling strategies. In: *Proceedings of Uncertainty Analysis for Performance Assessments of Radioactive Waste Disposal Systems.* NEA Workshop, Seattle, February 24–26, 1987.

Andres, T. H. (1997). Sampling methods and sensitivity analysis for large parameter sets. *J. Statist. Comput.* **57**, 77–110.

Andres, T. H. and Hajas, W. C. (1993). Using iterated fractional factorial design to screen parameters in sensitivity analysis of a probabilistic risk assessment model. In: *Proceedings of the Joint International Conference on Mathematical Methods and Supercomputing in Nuclear Applications, Karlsruhe, Germany, 19–23 April 1993* (ed. H. Küsters, E. Stein and W. Werner), Vol. 2, 328–337.

Archer, G., Saltelli, A. and Sobol', I. M. (1997). Sensitivity measures, ANOVA like techniques and the use of bootstrap. *J. Statist. Comput. Simul.* **58**, 99–120.

Atherton, R.W., Schainker, R. B. and Ducot, E. R. (1975). On the statistical sensitivity analysis of models for chemical kinetics. *AIChE J.* **21**, 441–448.

Ayers, G. P., Cainey, J. M., Granek, H. and Leck, C. (1996). Dimethylsulfide oxidation and the ratio of methanesulfonate to non-sea-salt sulfate in the marine aerosol. *J. Atmos. Chem.* **25**, 307–325.

Bair, E. S. (1994). Model *(In)*validation—A view from the courtroom. *Ground Water,* **32**, 530–531.

Bakun, A. and Parrish, H. (1982). Turbulence, transport and pelagic fish in the California and Peru Current sytems. *CalCOFI Rep.* **23**, 99–112.

Balogh, B. (1991). *Chain Reaction: Expert Debate and Public Participation in American Commercial Nuclear Power, 1945–1975.* Cambridge University Press.

Banks, J. (1998). *Handbook of Simulation; Principles, Methodology, Advances, Applications and Practice* Wiley Interscience, New York.

Barry, T. M. (1996). Recommendations on the testing and use of pseudo-random number generators used in Monte Carlo analysis for risk assessment. *Risk Anal.* **16**, 93–105.

Bates, T. S., Calhoun, J. A. and Quinn, P. K. (1992). Variations in the methanesulfonate to sulfate molar ratio in submicrometer marine areosol particles over the South Pacific Ocean. *J. Geophysi. Res.* **97**, 9859–9865.

Baumgartner, T. R., Soutra, A. and Ferreira-Bartrina, V. (1992). Reconstruction of the history of Pacific sardine and northern anchovy populations over the past two millennia from sediments of the Santa Barbara Basin. *CalCOFI Rep.* **33**, 24–40.

Beck, M. B. (1987). Water quality modeling: a review of the analysis of uncertainty. *Water Resources Res.* **23**, 1393–1442.

Beck, M. B. (1994). Understanding uncertain environmental systems. In: *Predictability and Nonlinear Modelling in Natural Sciences and Economics* (ed. J. Grasman and G. van Straten), pp. 294–311. Kluwer, Dordrecht.

Beck, M. B. and Halfon, E. (1991). Uncertainty, identifiability and the propagation of prediction errors: a case study of Lake Ontario. *J Forecasting*, **10**, 135–161.

Beck, M. B. (1998). Forecasting environmental change: a manifesto. Report, Warnell School of Forest Resources, University of Georgia, Athens.

Beck, M. B. Chen, J. and Osidele, O. O. (2000). Random search and the reachability of target futures. In: *Forecasting Environmental Change: A Manifesto* (ed. M. B. Beck, (to appear).

Beck, M. B., Kleissen, F. M. and Wheater, H. S. (1990). Identifying flow paths in models of surface water acidification. *Revi. Geophys.* **28**, 207–230.

Beck, M. B., Mulkey, L. A., Barnwell, T. O. and Ravetz, J. R. (1995). Model validation for predictive exposure assessments. In: *Proceedings of 1995 International Environmental Conference*, pp. 973–980. TAPPI, Atlanta, GA.

Beck, M. B., Ravetz, J. R., Mulkey, L. A. and Barnwell, T. O. (1997). On the problem of model validation for predictive exposure assessments. *Stochastic Hydrol. Hydraulics*, **11**, 229–254.

Beckman, R. J. and McKay, M. D. (1987). Monte Carlo estimation under different distributions using the same simulation. *Technometrics* **29**, 153–160.

Bell, D. (1995). Risk, return and utility. *Manag. Sci.* **41**, 23–30.

Bell, W. R. (1989). Comment to 'Sensitivity analysis of seasonal adjustments: empirical case studies', by J. B. Carlin and A. P. Dempster. *J. Am. Statist. Associ.* **84**, 22–24.

Bell, W. R. and Hillmer, S. C. (1984). Issues involved with the seasonal adjustment of economic time series. *J. Business Econo. Statist.* **2**, 291–320.

Bell, W. R. and Otto, M. C. (1993). Bayesian assessment of uncertainty in seasonal adjustment with sampling error present. SRD Research Report RR-92/12, US. Bureau of the Census, Washington, DC.

Belsley, D. A., Kuh, E. and Welsh, R. E. (1980). *Regression Diagnostics: Identifying Influential Data and Sources of Collinearity* Wiley, New York.

Berger, J. (1984). The robust Bayesian viewpoint (with discussion). In: *Robustness of Bayesian Analyses* (ed. J. Kadane). North-Holland, Amsterdam.

Berger, J. (1985). *Statistical Decision Theory and Bayesian Analysis*, 2nd edn. Springer-Verlag, New York.

Berger, J. (1990). Robust Bayesian analysis: sensitivity to the prior. *J. Statist. Plann. Infer.* **25**, 303–328.

Berger, J. (ed.) (1994a). Special issue on Bayesian Analysis. *J. Statisti. Planni. Infer.* **40**.

Berger, J. (1994b). An overview of robust Bayesian analysis (with discussion). *Test*, **3**, 5–124.

Berger, J., Betrò, B., Moreno, E., Pericchi, L., Ruggeri, F., Salinetti, G. and Wasserman, L. (eds) (1996). *Bayesian Robustness*. Lecture Notes IMS, Hayward.

Betrò, B. and Guglielmi, A. (1996). Numerical robust Bayesian analysis under generalised moment conditions. In: *Bayesian Robustness* (ed. J. Berger, J., B. Betrò E. Moreno, L. Pericchi, F. Ruggeri, G. Salinetti and L. Wasserman. Lecture Notes IMS, Hayward.

Betrò, B., Meczarski. M. and Ruggeri, F. (1994). Robust Bayesian analysis under generalized moment conditions. *J. Statist. Plann. Infer.* **41**, 257–266.

Bettonvil, B. (1990). *Detection of Important Factors by Sequential Bifurcation*. Tilburg University Press, Tilburg.

Bettonvil, B. and Kleijnen, J. P. C. (1997). Searching for important factors in simulation models with many factors: sequential bifurcation. *Eur. J. Oper. Res.* **96**, 180–194.

Beven, K. J. (1989). Changing ideas in hydrology—The case of physically-based models. *J Hydrol.* **105**, 157–172.

Beven, K. J. (1993). Prophecy, reality and uncertainty in distributed hydrological modelling. *Adv. Water Resources* **16**, 41–51.

Beven, K. J. (1996). Equifinality and uncertainty in geomorphological modelling In: *The Scientific Nature of Geomorphology* (ed. B. L. Rhoads and C. E. Thorn), pp. 289–313. Wiley, Chichester.

Beven, K. J. and Binley, A. M. (1992). The future of distributed models: model calibration and uncertainty prediction. *Hydrol. Process.* **6**, 279–298.

Bielza, M. C., Martín. J., Müller, P. and Ríos Insua, D. (1998). Approximating the nondominated set in continuous problems. Manuscript.

Bishop, W. P. and Hollister, C. D. (1974). Seabed disposal—where to look. *Nucl. Technol.* **24**, 425–443.

Blackburn, J. W. (1989). Is there an 'Uncertainty Principle' in microbial Waste Treatment? In: *Biotreatment of Agricultural Wastewater* (ed. M. Hurtley), pp. 149–161 CRC Press, Boca Raton, FL.

Blower, S. M. and Dowlatabadi, H. (1994). Sensitivity and uncertainty analysis of complex models of disease transmission: an HIV model, as an example. *Int. Statist. Rev.* **62**, 229–243.

Bolado, R., Alonso, A. and Moya, J. A. (1998). Using MAYDAY 1.2 to perform uncertainty and sensitivity analysis of a system model. In: *Proceedings of the Second International Conference on Sensitivity Analysis of Model Output (SAMO98)* (ed. K. Chan, S. Tarantola and F. Campolongo), pp. 37–40.

Bonano, E. J. and Apostolakis, G. E. (1991). Theoretical foundations and practical issues for using expert judgments in uncertainty analysis of high-level radioactive waste disposal. *Radioactive Waste Manag. and Nucl. Fuel Cycle* **16**, 137–159.

Bonano, E. J., Hora, S. C., Keeney, R. L. and von Winterfeld, D. (1990). *Elicitation and use of expert judgment in performance assessment for high-level radioactive waste repositories.* NUREG/CR-5411 SAND89-1821. Sandia National Laboratories, Albuquerque, NM.

Borchers, J. W. (ed.) (1998). *Land Subsidence Case Studies and Current Research; Proceedings of the Dr. J. F. Poland Symposium on Land Subsidence.* Special Publication No. 8, Association of Engineering Geologists, USA.

Box, G. E. P. and Draper, N. R. (1987). *Empirical Model-Building and Response Surfaces.* Wiley, New York.

Box, G. E. P., Hillmer, S. C. and Tiao, G. C. (1978a). Analysis and modeling of seasonal time series. In: *Seasonal Analysis of Time Series* (ed. A. Zellner), pp. 309–334. US Department of Commerce, Bureau of the Census, Washington, DC.

Box, G. E. P., Hunter, W. G. and Hunter, J. S. (1978b). *Statistics for Experimenters. An Introduction to Design, Data Analysis and Model Building.* Wiley, New York.

Box, G. E. P. and Jenkins, G. M (1970). *Time Series Analysis: Forecasting and Control.* Holden Day, San Francisco.

Box, G. E. P. and Meyer, R. D. (1986). An analysis for unreplicated fractional factorials. *Technometrics* **28**, 11–18.

Box, G. E. P. and Tiao, G. C. (1973). *Bayesian Inference in Statistical Analysis.* Addison-Wesley, Reading, MA.

Bratley, P. and Fox, B. L. (1988). ALGORITHM 659 Implementing Sobol's quasirandom sequence generator. *ACM Trans. Math. Software* **14**, 88–100.

Bredehoeft, J. D. and Konikow, L. F. (1993). Ground-water models: validate or invalidate. *Ground Water*, **31**, 178–179.

Breeding, R. J., Amos, C. N., Brown, T. D., Gorham, E. D., Gregory, J. J., Harper, F. T., Murfin, W. and Payne, A. C. (1992a). *Evaluation of severe accident risks: quantification of major input parameters: experts' determination of structural response issues.* NUREG/CR-4551, SAND86-1309, Vol. 2, Part 3, Rev. 1. Sandia National Laboratories, Albuquerque, NM.

Breeding, R. J., Helton, J. C., Gorham, E. D. and Harper, F. T. (1992b). Summary description of the methods used in the probabilistic risk assessments for NUREG-1150. *Nucl. Engng. Design* **135**, 1–27.

Breshears, D. D., Kirchner, T. B. and Whicker, F. W. (1992). Containment transport through agro-ecosystems: assessing relative importance of environmental, physiological, and management factors. *Ecol. Applic.* **2**, 285–297.

Burman, J. P. (1980). Seasonal adjustment by signal extraction. *J. R. Statist. Soc., Ser. A* **143**, 321–337.

Bushenkov, V. A., Chernyky, O. L., Kamenev, G. K and Lotov, A. V. (1995). Multidimensional images given by mappings: construction and visualization. *Patt. Recog. Image Anal.* **51**, 35–56.

Butler, J. L., Smith, P.E. and Lo, N. C.-H. (1993). The effect of natural variability of life-history factors on anchovy and sardine population growth. *CalCOFI Rep.* **34**, 104–111.

Bysveen, S., Kjelaas, A. G., Lereim, J. and Marthinsen, T. (1990). Experience from application of probabilistic methods in offshore field activities. In: *Proceedings of the Ninth International Conference on Offshore Mechanics and Arctic Engineering.* Saga Petroleum AS, Norway.

Byun, D. W. (1990). On the analytical solutions of flux-profile relationships for the atmospheric surface layer. *J. Appl. Meteorol.* **29**, 652–657.

Byun, D. W. (1991). Determination of similarity functions of the resistance laws for the planetary boundary layer using surface-layer similarity functions. *Boundary Layer Meteor.* **57**, 17–48.

Cambridge Course for Industry. (1998). *Dependence Modelling and Risk Management*. University of Cambridge.

Campolongo, F. and Saltelli, A. (1997). Sensitivity analysis of an environmental model; a worked application of different analysis methods. *Reliab. Engng. and Syst. Safety* **52**, 49–69.

Campolongo, F., Saltelli A., Jensen, N.R., Wilson, J. and Hjorth, J. (1999a). The role of multiphase chemistry in the oxidation of dimethylsulphide (DMS). A latitude dependent analysis. *J. Atmos. Chem.* **32**, 327–356.

Campolongo, F., Tarantola, S. and Saltelli, A. (1999b). Tackling quantitatively large dimensionality problems. *Comput. Phys. Communi.* **117**, 75–85.

Capaldo, K. P. and Pandis, S. N. (1997). Dimethylsulfide chemistry in the remote marine atmosphere: Evaluation and sensitivity analysis of available mechanisms. *J. Geophys. Res.* **102**, 23251–23267.

Carlin, J. B. and Dempster, A. P. (1989) Sensitivity analysis of seasonal adjustments: empirical case studies. *J. Am. Statist. Assoc.* **84**, 6–20.

Carlotti, G., Fioretto, D., Palmieri, L., Socino, G., Verdini, L. and Verona, E. (1991). Brillouin scattering by surface acoustic modes for elastic characterization of ZnO films. *IEEE Trans. Ultrason., Ferroelectr. Freq. Control.*, **38**, 56–60.

Carrera, J., Mousavi, S. F., Unusoff, E. J., Sánchez-Villa, X. and Galarza, G. (1993). A discussion on validation of hydrogeological models. *Reliab. Engng. Syst. Safety* **42**, 201–216.

Caswell, H. (1976). The validation problem. In: *Systems Analysis and Simulation in Ecology*, Vol IV. (ed. B. C. Patten), pp. 313–325. Academic Press New York, pp 313–325.

Caswell, H. (1989). *Matrix Populations Models: Construction, Analysis and Interpretation*. Sinauer, Sunderland, MA.

Cawlfield, J. D., Boateng, S., Piggott, J. and Wu, M. C. (1997). Probabilistic sensitivity measures applied to numerical models of flow and transport. In: *J. Statist. Comput. Simul.* **57**, 353–364.

Cawlfield, J. and Sitar, N. (1988). Stochastic finite element analysis of groundwater flow using the first-order reliability method. In: *Consequences of Spatial Variability in Aquifer Properties* (ed. A. Peck *et al.*). IAHS Publication No. 175.

Cawlfield, J. and Wu, M. C. (1993). Probabilistic sensitivity analysis for one-dimensional reactive transport in porous media. *Water Resources Res.* **29**(3), 661–671.

Chan, K., Saltelli, A. and Tarantola, S. (2000) Winding stairs—a sampling tool to compute sensitivity indices. *Statist. Comput.* **10**, 187–196.

Chan, M. S. (1996). The consequences of uncertainty for the prediction of the effects of schistosomiasis control programmes. *Epidemiol. Infect.* **117**, 537–550.

Chang, F. J. and Delleur, J.W. (1992). Systematic parameter estimation of watershed acidification model. *Hydrogeol. Processes* **6**, 29–44.

Chang, J. S., Brost, R. A., Isaksen, I. S. A., Middleton, P. B., Stockwell, W. R., and Walcek, C. J. (1987). A 3-dimensional Eulerian acid deposition model: physical concepts and formulation. *J. Geophys. Res.* **92**, 14681–14700.

Chang, J. S., Middleton, P. B., Stockwell, W. R., Walcek, C. J., Pleim, J. E., Lansford, H. H., Madronich, S., Binkowski, F. S., Seaman, N. L., and Stauffer, D. R. (1990). The Regional Acid Deposition Model and Engineering Model. SOS/T Report 4. In: *The National Acid Deposition Assessment Program: State of Science and Technology*, Vol. 1. National Acid Precipitation Assessment Program, Washington, DC.

Charlson, R. J., Lovelock, J. E., Andreae, M. O. and Warren, S. G. (1987). Sulfur phytoplankton, atmospheric sulfur, cloud albedo and climate. *Nature* **326**, 655–661.

Charlson, R. J., Schwartz, S. E., Hales, J. M., Cess, R. D., Coakley, J. A., Jr, Hansen, J. E. and Hofmann, D. J. (1992). Climate forcing by anthropogenic aerosols. *Science* **255**, 423–430.

Chatfield, C. (1993). Model uncertainty, data mining and statistical inference. *J. R. Statist. Soc. A* **158**, 419–466.

Chen, J. and Beck, M. B. (1998). Quality assurance of multi-media model for predictive screening tasks. Report EPA/600/R-98/106, US Environmental Protection Agency, Washington, DC.

Chuang, D.T. (1984). Further theory of stable decisions. In: *Robustness of Bayesian Analyses* (ed. J. Kadane). North-Holland, Amsterdam.

Clemen, R. T. and Winkler, R. L. (1999). Combining probability distributions from experts in risk analysis. *Risk Anal.* **19**, 187–203.

Cleveland, W. S. (1979). Robust locally-weighted regression and smoothing scatterplots. *J. Am. Statist. Assoc.* **74**, 829–836.

Cleveland, W. S. (1993). *Visualizing Data.* AT&T Bell Laboratories, Murray Hill, NJ.

Conover, W. J. (1980). *Practical Nonparametric Statistics*, 2nd ed. Wiley, New York.

Conover, W. J. and Iman, R. L. (1981). Rank transformations as a bridge between parametric and nonparametric statistics. *Am. Statist.* **35**, 124–129.

Cook, D. R. and Weisberg, S. (1982). *Residuals and Influence in Regression.* Chapman & Hall, New York.

Cook, I. and Unwin, S. D. (1986). Controlling principles for prior probability assignments in nuclear risk assessment. *Nucl. Sci. Engng.* **94**, 107–119.

Cooke, R. M. (1991). *Experts in Uncertainty, Opinion and Subjective Probability in Science.* Oxford University Press, Oxford.

Cooke, R. M. (1995). *UNICORN: Methods and Code for Uncertainty Analysis*, AEA Technology, Cheshire.

Cooke, R. M. and van Noortwijk, J. M. (1997). Uncertainty analysis of inundation probabilities; phase 2 of project uncertainty analysis of inundation probabilities. Technical Report, Delft University of Technology, Faculty of Mathematics and Computer Science.

Cooke, R. M. and van Noortwijk, J. M. (1999). Local probabilistic sensitivity measures for comparing FORM and Monte Carlo calculations illustrated with dike reliability calculations. *Comput. Phys. Commun.* **117**, 86–98.

Cornell, C. A. (1972). First order analysis of model and parameter uncertainty. In: *Proceedings of International Symposium on Uncertainties in Hydrologic and water resource Systems.* University of Arizona, Tucson.

Cotter, S. C. (1979). A screening design for factorial experiments with interactions. *Biometrika* **66**, 317–320.

Cox, D. C. (1982). An analytical method for uncertainty analysis of nonlinear output functions, with application to fault-tree analysis. *IEEE Trans. Reliab.* **R-31**, 265–268.

CPC (1999). Special Issue on Sensitivity Analysis. *Comput. Phys. Commun.* **117** (1–2).

Crystal Ball (1996). *Crystal Ball, version 4.0 Users Manual.* Decisioneering Inc., Aurora, CO.

Cukier, R. I., Fortuin, C. M., Schuler, K. E., Petschek, A. G. and Schaibly, J. H. (1973). Study of the sensitivity of coupled reaction systems to uncertainties in rate coefficients. I Theory. *J. Chem. Phys.* **59**, 3873–3878.

Cukier, R. I., Levine, H. B. and Schuler, K. E. (1978). Nonlinear sensitivity analysis of multiparameter model systems. *J. Comput. Phys.* **26**, 1–42.

Cukier, R. I., Schaibly, J. H. and Shuler, K. E. (1975). Study of the sensitivity of coupled reaction systems to uncertainties in rate coefficients. III. Analysis of the approximations. *J. Chem. Phys.* **63**, 1140–1149.

Cullen, A. C. and Frey, H. C. (1998). *Probabilistic Techniques in Exposure Assessment: A Handbook for Dealing with Variability and Uncertainty in Models and Inputs.* Plenum, New York.

Daniel, C. (1958). On varying one factor at a time. *Biometrics* **14**, 430–431.

Daniel, C. (1973). One-at-a-time-plans. *J. Am. Statist. Assoc.* **68**, 353–360.

Daniel, C. and Wood, F. S. (1980). *Fitting Equations to Data*, 2nd edn. Wiley, New York.

David, H. A. (1970). *Order Statistics.* Wiley, New York.

Davis, P. J. and Rabinowitz, P. (1984). *Methods of Numerical Integration*, 2nd edn. Academic Press: New York.

De Wit, M. S. (1997). Identification of the important parameters in thermal building simulation models. *J. Statist. Comput. Simul.* **57**, 305–320.

Demiralp, M. and Rabitz, H. (1981). Chemical kinetic functional sensitivity analysis: elementary sensitivities. *J. Chem. Phys.* **74**, 3362–3375.

Dempster, A. P. (1975). A subjectivist look at robustness. *Bull. Int. Statist. Inst.* **46**, 346–374.

Dennis, B., Desharnais, R. A., Cushing, J. M. and Costantino, R. F. (1995). Nonlinear demographic dynamics. Mathematical models, statistical methods, and biological experiments. *Ecol. Monogr.* **65**, 261–281.

Dennis, R. L., Arnold, J. R., Tonnesen, G. S., and Li, Y.-H. (1999). A new response surface approach for interpreting Eulerian air quality model sensitivities. *Comput. Phys. Commun.* **117**, 99–112.

Dennis, R. L., Byun, D. W., Novak, J. H., Galluppi, K. J., Coates, C. J., and Vouk, M. A. (1996). The next generation of integrated air quality modeling: EPA's Models-3. *Atmos. Environ.* **30**, 1925–1938.

Der Kiureghian, A. (1985). Unpublished class notes from CE 285, Structural Reliability and Risk Analysis, University of California, Berkeley.

Der Kiureghian, A. and Liu, P. L. (1986). Structural reliability under incomplete probability information. *J. Engng. Mech. ASCE* **112**(1), 85–104.

Der Kiureghian, A. and Taylor, R. (1983). Numerical methods in structural reliability. *Fourth International Conference on Applications of Statistics and Probability in Soil and Structural Engineering.* Universita di Firenze (Italy).

DeVries, T. J. and Pearcy, W. G. (1982). Fish debris in sediments of the upwelling zone off central Peru: a late Quaternary record. *Deep Sea Res.* **28**, 87–109.

Diaconis, P. and Freedman, D. (1986). On the consistency of Bayes estimates. *Ann. Statist.* **14**, 1–67.

Diaconis, P. and Shahshahani, M. (1984). On nonlinear functions of linear combinations, *SIAM J. Sci. Statist. Comput.* **5**, 175–191.

Dickerson, T. L., Macewicz, B. J. and Hunter, J. R. (1992). Spawning frequency and bath fecundity of chub mackerel, *Scomber japonicus*, during 1985. *CalCOFI Rep.* **33**, 130–140.

Dickinson, R. P. and Gelinas, R. J. (1976). Sensitivity analysis of ordinary differential equation systems—A direct method. *J. Comput. Phys.* **21**, 123–143.

Ditlevsen, O. and Madsen, H. O. (1996). *Structural Reliability Methods.* Wiley, Chichester. EXCEL (1995). EXCEL for Windows 95. Microsoft Corporation.

Domenico, P. A. and Schwartz, F. W. (1998) *Physical and Chemical Hydrogeology*, 2nd edn. Wiley, New York. Fetter, C.W. (1994). *Applied Hydrogeology*, 3rd edn. Macmillan, New York.

Donahue, N. M., Dubey, M. K., Mohrschladt, R., Demerjian, K. L., and Anderson, J. G. (1997). High-pressure flow study of the reactions OH + NOX = HONOX: errors in the falloff region. *J. Geophys. Res.* **102**, 6159–6168.

Draper, D. (1995). Assessment and propagation of model uncertainty (with discussion). *J. R. Statist. Soc. Ser. B* **57**, 45–97.

Draper, D. (1997). Model uncertainty in 'stochastic' and 'deterministic' systems. In: *Proceedings of the 12th International Workshop on Statistical Modeling* (ed. C. Minder and H. Friedl). *Schriftenreihe der Österreichischen Statistichen Gesellschaft* **5**, 43–59.

Draper, D., Pereira, A., Prado, P., Saltelli, A., Cheal, R., Eguilior, S., Mendes, B. and Tarantola, S. (1999). Scenario and parametric uncertainty in GESAMAC: A methodological study in nuclear waste disposal risk assessment. *Comput. Phys. Commun.* **117**, 142–155.

Draper, N. R and Smith, H. (1981). *Applied Regression Analysis*, 2nd edn. Wiley, New York.

Dunker, A. M. (1981). Efficient calculation of sensitivity coefficients for complex atmospheric models. *Atmos. Environ.* **15**, 1155–1161.

Dunker, A. M. (1984). The decoupled direct method for calculating sensitivity coefficients in chemical kinetics. *J. Chem. Phys.* **81**, 2385–2393.

Edelson, D. and Thomas, V. M. (1981). Sensitivity analysis of oscillating reactions. 1. The period of the Oregonator. *J Phys. Chem.* **85**, 1555–1558.

EEA (1996). *Atmospheric Emission Inventory Guidebook by EMEP/CORINAIR*. The European Environment Agency, Copenhagen.

Efron, B. and Stein C. (1981). The jackknife estimate of variance. *Ann. Statist.* **9**, 586–596.

Efron, B. and Tibshirani, R. J. (1993). *Introduction to the Bootstrap.* Chapman & Hall, London.

Eguilior, S., Prado, P. (1998). *GTMCHEM Computer Code: Description and Application to the Level* E/G Test Case (GESAMAC Project). CIEMAT, Madrid.

Eigbe, U., Beck, M. B., Wheater, H. S. and Hirano, F. (1998). Kalman filtering in groundwater flow modelling: problems and prospects. *Stochastic Hydrol. Hydrauli.* **12**, 15–32.

Eisenhart, C. (1964). The meaning of 'least' in least squares. *J. Washington Acad. Sci.* **54**, 24–33.

Enting, I. G. and Lassey, K. R. (1993). Projections of future CO_2. Technical Paper 27, Division of Atmospheric Research, CSIRO, Melbourne.

EPRI (Electric Power Research Institute), (1989). *Probabilistic seismic hazard evaluations at nuclear plant site in the Central and Eastern United States: resolution of the Charleston earthquake issue.* EPRI-NP-6395D. Electric Power Research Institute, Palo Alto, CA.

Eurostat (1999). *Towards Environmental Pressure Indicators for the EU.* Eurostat, Luxembourg.

Evensen, G. (1994). Inverse methods and data assimilation in nonlinear ocean models. *Physica*, **D77**, 108–129.

Falls, A. H., McRae, G. J. and Seinfeld, J. H. (1979). Sensitivity and uncertainty of reaction mechanisms for photochemical air pollution. *Int. J. Chem. Kinet.* **11**, 1137–1162.

Fedra, K. G., Van Straten, M. and Beck, M. B. (1981). Uncertainty and arbitrariness in ecosystems modeling: a lake modeling example. *Ecol. Modelling* **13**, 87–110.

Feller, W. (1971). *An Introduction to Probability Theory and Its Applications*, Vol. II, 2nd edn. Wiley, New York.

Fetter, C.W. (1994). Applied Hydrogeology, 3rd edn., Macmillan Publishing Co.

Fisher, R. A. (1935). *The Design of Experiments.* Oliver & Boyd, Edinburgh.

Fishman, G. S. (1996). *Monte Carlo: Concepts, Algorithms, and Applications.* Springer-Verlag, New York.

Freer, J., Beven, K. and Ambroise, B. (1996). Bayesian estimation of uncertainty in runoff prediction and the value of data: an application of the GLUE approach. *Water Resources Res.* **32**, 2161–2173.

French, S. and Ríos Insua, D. (2000) *Statistical DecisionTheory.* Arnold, London.

Frenklach, M. (1984). Modeling. In: *Combustion Chemistry* (ed. W. C. Gardiner, Jr). Springer, New York.

Friedman, J. and Stuetzle, W. (1981). Projection pursuit regression. *J. Am. Statist. Assoc.* **76**, 580–619.

Fuller, W. A. (1996). *Introduction to Statistical Time Series*, 2nd edn., Wiley, New York.

Funtowicz, S. O. and Ravetz, J. (1990). *Uncertainty and Quality in Science for Policy.* Kluwer, Dordrecht.

Fürbringer, J.-M. (1996). Sensitivity analysis for modelers. *Air Infiltration Rev.* **17**(4).

Gao, D., Stockwell, W. R., and Milford, J. B. (1995). First-order sensitivity and uncertainty analysis for a regional-scale gas-phase chemical mechanism. *J. Geophys. Res.* **100**, 23153–23166.

Gao, D., Stockwell, W. R., and Milford, J. B. (1996). Global uncertainty analysis of a regional-scale gas-phase chemical mechanism. *J. Geophys. Res.* **101**, 9107–9119.

Gardner, R. H., O'Neill, R.V. and Mankin, J. B. (1981). Comparison of sensitivity analysis based on a stream ecosystem model. *Ecol. Modelling* **12**, 173–190.

Gee, G.W., Kincaid, C. T., Lenhard, R. J. and Simmons, C. S, (1991). Recent studies of flow and transport in the vadose zone. *Revi. Geophys.* **29**, 227–239.

Geweke, J. (1996). Monte Carlo simulation and numerical integration. In: *Handbook of Computational Economics*, Vol. 1 (ed. H. M. Amman, D. A. Kendrick and J. Rust). Elsevier Science, Amsterdam.

Gilks, W. R., Richardson, S. and Spiegelhalter, D. J. (1996). Introducing Markov chain Monte Carlo. In: *Markov Chain Monte Carlo in Practice* (ed. A. Gilks, S. Richardson and D. J. Spiegelhalter). Chapman & Hall, London.

Girón, J. and Ríos, S. (1980). Quasi-Bayesian behaviour: a more realistic approach to decision making? In: *Bayesian Statistics I* (ed. J. M. Bernardo, M. H. De Groot, D.V. Lindley and A. F. M. Smith). University Press, Valencia.

Girosi, F. and Poggio, T. (1989). Representation properties of networks: Kolmogorov's theorem is irrelevant. *Neural Comput.* **1**, 465–469.

Goldsmith, C. H. (1998). Sensitivity analysis. In: *Encyclopedia of Biostatistics* (ed. P. Armitage and T. Colton). Wiley, New York.

Gómez, V. and Maravall, A. (1994). Estimation, prediction, and interpolation for non-stationary time series with the Kalman filter. *J. Am. Statist. Assoc.* **89**, 611–624.

Gómez, V. and Maravall, A. (1996). Programs SEATS and TRAMO: instructions for the User. Working Paper 9628, Bank of Spain.

Goossens, L. H. J., Harrison, J. D., Kraan, B. C. P., Cooke, R. M., Harper, F. T., and Hora, S. C. (1997). *Probabilistic accident consequence Uncertainty analysis: uncertainty assessment for internal dosimetry.* NUREG/CR-6571, EUR 16773, SAND98-0119, Vol. 1, Prepared for the U.S. Nuclear Regulatory Commission and the Commission of the European Communities.

Greenhouse, J. and Wasserman, L. (1996). A practical robust method for Bayesian model selection: a case study in the analysis of clinical trials. In: *Bayesian Robustness* (ed. J. Berger, B. Betró, E. Moreno, L. Pericchi, F. Ruggeri, G. Salinetti and L. Wasserman). Lecture Notes IMS, Hayward.

Grell, A. G., Dudhia, J., and Stauffer, D. R. (1994). A Description of the fifth-generation Penn State / NCAR Mesoscale Model (MM5). NCAR Technical Note NCAR/TN-398+STR, National Center for Atmospheric Research, Boulder, CO.

Gustafson, P. and Wasserman, L. (1995). Local sensitivity diagnostics for Bayesian inference. *Ann. Statist.* **23**, 2153–2167.

Gwo, J. P., Toran, L. E. Morris, M. D. and Wilson, G. V. (1996). Subsurface stormflow modeling with sensitivity analysis using a Latin-hypercube sampling technique. *Groundwater* **34**, 811–818.

Haldane, J. B. S. (1953). Animal populations and their regulation. *New Biol.* **15**, 9–24.

Hamby, D. M. (1994). A review of techniques for parameter sensitivity analysis of environmental models. *Environ. Monitoring Assess.* **32** ,125–154.

Hamby, D. M. (1995). A comparison of sensitivity analysis techniques. *Health Phys.* **68**, 195–204.

Hamed, M. M., Bedient, P. and Conte, J. (1996). Numerical stochastic analysis of groundwater contaminant transport and plume containment. *J. Contaminant Hydrol.* **24** (1), 1–23.

Hamed, M. M., Conte, J. and Bedient, P. B. (1995). Probabilistic screening tool for ground-water contamination assessment. *J. Environ. Engng., ASCE* **121** (11), 767–775.

Hanna, S., Chang, J., and Fernau, M. (1998). Monte Carlo estimates of uncertainties in predictions by a photochemical grid model (UAM-IV) due to uncertainties in input variables. *Atmos. Environ.* **32**, 3619–3628.

Hannan, E. J. (1980). The estimation of the order of an ARMA process. *Ann. Statist.* **8**, 1071–1081.

Hardouin Duparc, O., Sanz Velasco, E. and Velasco, V. R. (1984). Elastic surface waves in crystals with overlayers: cubic symmetry. *Phys. Rev. B* **30**, 2042–2048

Harper, F. T., Amos, C. N., Breeding, R. J., Brown, T. D., Gregory, J. J., Gorham, E. D., Murfin, W., Payne, A. C. and Rightly, G. S. (1991). *Evaluation of severe accident risks: quantification of major input parameters: experts' determination of containment loads and molten core-concrete issues.* NUREG/CR-4551, SAND86-1309, Vol. 2, Part 2, Rev. 1. Sandia National Laboratories, Albuquerque, NM.

Harper, F. T., Breeding, R. J., Brown, T. D., Gregory, J. J., Payne, A. C., Gorham, E. D. and Amos, C. N. (1990). *Evaluation of Severe Accident Risks: Quantification of Major Input Parameters: Experts' Determination of In-Vessel Issues.* NUREG/CR-4551, SAND86-1309. Vol. 2, Part 1, Rev. 1. Albuquerque, NM: Sandia National Laboratories.

Harper, F. T., Amos, C. N., Boyd, G., Breeding, R. J. Brown, T. D., Gorham, E. D., Gregory, J. J., Helton, J. C., Jow, H.-N. and Payne, A. C. (1992). *Evaluation of Severe Accident Risks: Quantification of Major Input Parameters: Experts' Determination of Source Term Issues.* NUREG/CR-4551, SAND86-1309. Vol. 2, Part 4, Rev. 1. Albuquerque, NM: Sandia National Laboratories.

Harter, H. L. (1983). Least squares. In: *Encyclopedia of Statistical Sciences* (ed. S. Kotz and N. L. Johnson), Vol. 4, pp. 593–598. Wiley, New York.

Harvey, A. C. (1989). *Forecasting, Structural Time Series Models and the Kalman Filter.* Cambridge University Press, Cambridge.

Hasofer, A. M. and Lind, N. (1974). An exact invariant first-order reliability format. *J. Engng. Mech., ASCE* **100**, 111–121.

Hedges, R. M. and Rabitz, H. (1985). Parametric sensitivity of system stability in chemical dynamics. *J. Chem. Phys.* **82**, 3674–3684.

Helton, J. C. (1993). Uncertainty and sensitivity analysis techniques for use in performance assessment for radioactive waste disposal. *Reliab. Engng. Syst. Safety* **42**, 327–367.

Helton, J. C. (1994). Treatment of uncertainty in performance assessments for complex systems. *Risk Anal.* **14**, 483–511.

Helton, J. C. (1996). Probability, conditional probability and complementary cumulative distribution functions in performance assessment for radioactive waste disposal. *Reliab. Engng. Syst. Safety* **54**, 145–163.

Helton, J. C. (1997). Uncertainty and sensitivity analysis in the presence of stochastic and subjective uncertainty. *J. Statist. Comput. and Simul.* **57**, 3–76.

Helton, J. C. (1999). Uncertainty and sensitivity analysis in performance assessment for the Waste Isolation Pilot Plant. *Comput. Phys. Commun.* **117**, 156–180.

Helton, J. C. and Breeding, R. J. (1993). Calculation of reactor accident safety goals. *Reliab. Engng. Syst. Safety* **39**, 129–158.

Helton, J. C., Anderson, D. R., Jow, H.-N., Marietta, M. G. and Basabilvazo, G. (1999). Performance assessment in support of the 1996 Compliance Certification Application for the Waste Isolation Pilot Plant. *Risk Anal.* **19**, 959–986.

Helton, J. C., Bean, J. E., Berglund, J. W., Davis, F. J., Economy, K., Garner, J. W., Johnson, J. D., MacKinnon, R. J., Miller, J., O'Brien, D. G., Ramsey, J. L., Schreiber, J. D., Shinta, A., Smith, L. N., Stoelzel, D. M., Stockman, C. and Vaughn, P. (1998). *Uncertainty and sensitivity analysis results obtained in the 1996 Performance Assessment for the Waste Isolation Pilot Plant.* SAND98-0365. Sandia National Laboratories, Albuquerque, NM.

Helton, J. C., Bean, J. E., Butcher, B. M., Garner, J. W., Schreiber, J. D., Swift, P. N. and Vaughn, P. (1996). Uncertainty and sensitivity analysis for gas and brine migration at the Waste Isolation Pilot Plant: fully consolidated shaft. *Nucl. Sci. Engng.* **122**, 1–31.

Helton, J. C. and Burmaster, D. E. (1996). Treatment of aleatory and epistemic uncertainty in performance assessments for complex systems. Guest Editorial, Special Issue of *Reliab. Engng. Syst. Safety* **54**, 91–94.

Helton, J. C., Garner, J. W., McCurley, R. D. and Rudeen, D. K. (1991). Sensitivity analysis techniques and results for the performance assessment at the waste isolation pilot plant. Sandia National Laboratories Report SAND90-7103.

Helton, J. C., Iman, R. L., Johnson, J. D. and Leigh, C. D. (1989). Uncertainty and sensitivity analysis of a dry containment test problem for the MAEROS aerosol model. *Nucl. Sci. Engng.* **102**, 22–42.

Helton, J. C. and Iuzzolino, H. J. (1993). Construction of complementary cumulative distribution functions for comparison with the EPA release limits for radioactive waste disposal. *Reliab. Engng. Syst. Safety* **40**, 277–293.

Helton, J. C., Johnson, J. D., McKay, M. D., Shiver, A. W. and Sprung, J. L. (1995a). Robustness of an uncertainty and sensitivity analysis of early exposure results with the MACCS reactor accident consequence model. *Reliab. Engng. Syst. Safety* **48**, 129–148.

Helton, J. C., Johnson, J. D., Shiver, A. W. and Sprung, J. L. (1995b). Uncertainty and sensitivity analysis of early exposure results with the MACCS reactor accident consequence model. *Reliab. Engng. Syst. Safety* **48**, 91–127.

Henrion, M. and Fischhoff, B. (1986). Assessing uncertainty in physical constants. *Am. J. Phys.* **54**, 791–799.

Hill, T. L. (1987). *Statistical Mechanics: Principles and Selected Applications.* Dover, New York.

Hillmer, S. C. and Tiao, G. C. (1982). An ARIMA-model-based approach to seasonal adjustment. *J. Am. Statist. Assoc.* **77**, 63–70.

Ho, T.-S. and Rabitz, H. (1993). Inversion of experimental data to extract intermolecular and intramolecular potentials. *J. Phys. Chem.* **97**, 13447–13456.

Homma, T. and Saltelli, A. (1994). Global sensitivity analysis of nonlinear models. Importance measures and Sobol' sensitivity indices. JRC Rep. EUR 16052 EN, Luxemburg.

Homma, T. and Saltelli, A. (1995). Sensitivity analysis of model output. Performance of the Sobol' quasi random sequence generator for the integration of the modified Hora and Iman importance measure. *J. Nucl. Sci. Technol.* **32**, 1164–1173.

Homma, T. and Saltelli, A. (1996). Importance measures in global sensitivity analysis of model output. *Reliab. Engng. Syst. Safety* **52**, 1–17.

Hora, S. C. and Iman, R. L. (1986). A comparison of maximum/bounding and Bayesian/Monte Carlo for fault tree uncertainty analysis. Technical Report: *SAND85-2839*, Sandia National Laboratories, Albuquerque, NM.

Hora, S. C. and Iman, R. L. (1989). Expert opinion in risk analysis: the NUREG-1150 methodology. *Nucl. Sci. and Engng.* **102**, 323–331.
Hornberger, G. M. and Spear, R. C. (1980). Eutrophication in Peel Inlet, I. Problem-defining behaviour and a mathematical model for the phosphorus scenario. *Water Res.* **14**, 29–42.
Hornberger G. M. and Spear, R. C. (1981). An approach to the preliminary analysis of environmental systems. *J. Environ. Management* **12**, 7–18.
Houghton, J. T., Meira Filho, L. G., Callander, B. A., Harris, N., Kattenberg, A. and Maskell, K. (eds) (1996). *Climate Change 1995 The Science of Climate Change, Contribution of WGI to the Second Assessment Report of the Intergovernmental Panel on Climate Change.* IPCC, Cambridge.
Høeg, P. (1995). *Borderliners*, p. 19. McClelland-Bantam, Toronto.
Hua Xia, Jiang, J. G., Wei Zhang, Chen, K. J., Zhang, X. K., Carlotti, G., Fioretto, D. and Socino, G. (1992). Elastic properties of a—Ge : H/SiO_2 heterostructures. *Solid State Commun.* **84**, 987–989.
Huber, P. (1981a). *Robust Statistics.* Wiley, New York.
Huber, P. (1981b). Projection pursuit. *Ann. Statist.* **13**, 435–525.
Hwang, J.-T. (1982). On the proper usage of sensitivities of chemical kinetics. Models to the uncertainties in rate coefficients. *Proc. Natl Sci. Council B ROC* **6**, 270–278.
Hwang, J.-T. (1983). Sensitivity analysis in chemical kinetics by the method of polynomial approximations. *Int. J. Chem. Kinet.* **15**, 959–987.
Hwang, J. T. (1985). A computational algorithm for the polynomial approximation method of sensitivity analysis in chemical kinetics. *J. Chinese Chem. Soc.* **32**, 253–261.
Hwang, J.-T. (1988). Computer modeling of chemical reactions. In: *Chemical Kinetics of Small Organic Radicals.* Vol. II: *Correlation and Calculation Methods* (ed. Z. B. Alfassi), pp. 149–189. CRC Press, Boca Raton, FL.
Ibrekk, H. and Morgan, M. G. (1987). Graphical communication of uncertain quantities to non-technical people. *Risk Anal.* **7**, 519–529.
ICI (1997). *Environmental Burden: The ICI Approach—A New Method to Evaluate the Potential Environmental Impact of Wastes and Emissions.* Imperial Chemical Industries, London.
Iman, R. L. (1992). Uncertainty and sensitivity analysis for computer modeling applications. *Reliab. Technol.* **28**, 153–168.
Iman, R. L. and Conover, W. J. (1979). The use of the rank transform in regression. *Technometrics* **21**, 499–509.
Iman, R. L. and Conover, W. J. (1980). Small sample sensitivity analysis techniques for computer models with an application to risk assessment. *Commun. Statist. A* **9**, 1749–1842.
Iman, R. L. and Conover, W. J. (1982). A distribution-free approach to inducing rank correlation among input variables. *Commun. Statist. Simul. Comput. B* **11**, 311–334.
Iman, R. L. and Davenport, J. M. (1980). Rank correlation plots for use with correlated input variables in simulation studies. SAND80-1903. Sandia National Laboratories, Albuquerque, NM.
Iman, R. L. and Davenport, J. M. (1982). Rank correlation plots for use with correlated input variables. *Commun. Statist.: Simul. Comput. B* **11**, 335–360.
Iman, R. L. and Shortencarier, M. J. (1984). A Fortran 77 program and user's guide for the generation of Latin hypercube and random samples for use with computer models. SAND83-2365. Sandia National Laboratories, Albuquerque, NM.
Iman, R. L. and Helton, J. C. (1988). A comparison of uncertainty and sensitivity analysis techniques for computer models. *Risk Anal.* **8**, 71–90.
Iman R. L. and Helton, J. C. (1991). The repeatability of uncertainty and sensitivity analyses for complex probabilistic risk assessments. *Risk Anal.* **11**, 591–606.
Iman, R. L., Helton, J. C. and Campbell, J. E. (1981a). An approach to sensitivity analysis of computer models, Part 1. Introduction, input variable selection and preliminary variable assessment. *J. Quality Technol.* **13**, 174–183.
Iman, R. L., Helton, J. C. and Campbell, J. E. (1981b). An approach to sensitivity analysis of computer models, Part 2. Ranking of input variables, response surface validation, distribution effect and technique synopsis. *J. Quality Technol.* **13**, 232–240.

Iman, R. L. and Hora, S. C. (1990). A robust measure of uncertainty importance for use in fault tree system analysis. *Risk Anal.* **10**, 401–406.

Iman, R. L., Davenport, J. M., Frost, E. L. and Shortencarier, M. J. (1980). Stepwise regression with PRESS and rank regression (program and user's guide). SAND79-1472. Sandia National Laboratories, Albuquerque, NM.

Iman, R. L., Shortencarier, M. J. and Johnson, J. D. (1985). A FORTRAN 77 program and user's guide for the calculation of partial correlation and standardized regression coefficients. NUREG/CR-4122, SAND85-0044. Sandia National Laboratories, Albuquerque, NM.

Ishigami, T. and Homma, T. (1989). An importance quantification technique in uncertainty analysis for computer models. Japan Atomic Energy Research Institute Report JAERI-M 89–111.

Ishigami, T. and Homma, T. (1990). An importance quantification technique in uncertainty analysis for computer models. In: *Proceedings of the ISUMA '90, First International Symposium on Uncertainty Modelling and Analysis, University of Maryland, USA, December 3–5, 1990*, pp. 398–403.

Jacoby, J. E. and Harrison, S. (1962). Multi-variable experimentation and simulation models. *Naval Res. Logistic Q.* **9**, 121–136.

Jakeman, A. J. and Hornberger, G. M. (1993). How much complexity is warranted in a rainfall–Runoff model? *Water Resources Res.* **29**, 2637–2649.

Jakeman, A. J., Post, D. A. and Beck, M. B. (1994). From data and theory to environmental model: the case of rainfall runoff. *Environmetrics*, **5**, 297–314.

Jang, J. C. C., Jeffries, H. E., Byun, D.W. and Pleim, J. E. (1995). Sensitivity of ozone to model grid resolution—I: Application of the high-resolution Regional Acid Deposition Model. *Atmos. Environ.* **29**, 3085–3100.

Jang, Y. S., Sitar, N. and Der Kiureghian, A. (1994). Reliability analysis of contaminant transport in saturated porous media. *Water Resources Res.* **30** (8), 2435–2448.

Jansen, M. J. W. (1996a) Winding stairs sample analysis program WINDINGS 2.0. Technical Report. Agricultural University of Wageningen (private communication).

Jansen, M. J. W. (1996b). WINDINGS: uncertainty analysis with winding stairs samples. Poster, Workshop on Assessment of Environmental Impact: Reliability of Process Models, Edinburgh, 18–19 March 1996.

Jansen, M. J. W. (1999). Analysis of variance designs for model output. *Comput. Phys. Commun.* **117**, 35–43.

Jansen, M. J. W., Rossing, W. A. H. and Daamen, R. A. (1994). Monte Carlo estimation of uncertainty contributions from several independent multivariate sources. In: *Predictability and Nonlinear Modelling in Natural Sciences and Economics* (eds. J. Gasman and G. van Straten) pp. 334–343. Kluwer Academic Publishers, Dordrecht.

Janssen, P. H. M., Slob, W. and Rotmans, J. (1990). Uncertainty analysis and sensitivity analysis: an inventory of ideas, methods and techniques from the literature. RIVM Report, 958805001, RIVM, Bilthoven, The Netherlands.

Jeffries, H. E. and Tonnesen, G. T. (1994). A comparison of two photochemical reaction mechanisms using mass balance and process analysis. *Atmos. Environ.* **28**, 2991–3003.

Jennrich, R. I. (1969). Asymptotic properties of non-linear least squares estimators. *Ann. Math. Statist.* **40**, 633-643.

Jesinghaus, J. (1999). *A European System of Environmental Pressure Indices. First Volume of the Environmental Pressure Indices Handbook.* Available on: http://esl.jrc.it/envind/theory/Handb.htm

JSCS (1997). Special Issue on Sensitivity Analysis. *J. Statist. Comput. Simul.* **57** (1–4).

Kadane, J. and Chuang, D. T. (1978). Stable decision problems. *Ann. Statist.* **6**, 1095–1110.

Kadane, J. and Srinivasan, C. (1994). Bayes decision problems and stability. Technical Report, Department of Statistics, University of Kentucky.

Kadane, J., Srinivasan, C. and Viele, K. (1997). Bayes decision problems and stability. Manuscript.

Karanikas, J. M., Sooryakumar, R. and Phillips, J. M. (1989). Dispersion of elastic waves in supported CaF_2 films. *J. Appl. Phys.* **65**, 3407–3410.

Kass, R. E. and Raftery, A. E. (1995). Bayes factors. *J. Am. Statist. Assoc.* **90**, 773–795.

Kass, R. E. and Wasserman, L. W. (1995). A reference Bayesian test for nested hypothesis and its relationship to the Schwarz criterion. *J. Am. Statist. Assoc.* **90**, 928–934.
Keeney, R. L. and von Winterfeldt, D. (1991). Eliciting probabilities from experts in complex technical problems. *IEEE Trans. Engng. Manag.* **38**, 191–201.
Keeney, R. L. and von Winterfeldt, D. (1994). Managing nuclear waste from power plants. *Risk Anal.* **14**, 107–130.
Keesman, K. J. and van Straten, G. (1990). Set-membership approach to identification and prediction of lake eutrophication. *Water Resources Res.* **26**, 2643–2652.
Kendall, M. and Stuart, A. (1979). *The Advanced Theory of Statistics*, Vol. 2, 4th edn. MacMillan, New York.
Kendall, M. G. and Stuart, A. (1961). *The Advanced Theory of Statistics*. Griffin, London.
Kleijnen, J. P. C. (1987). *Statistical Tools for Simulation Practitioners*. Marcel Dekker, New York.
Kleijnen, J. P. C. (1998). Experimental design for sensitivity analysis, optimization and validation of simulation models. In: *Handbook of Simulation—Principles, Methodology, Advances, Applications, and Practice* (ed. J. Banks). Wiley, New York.
Kleijnen, J. P. C. and Helton, J. C. (1999a). Statistical analyses of scatterplots to identify important factors in large-scale simulations, 1. review and comparison of techniques. *Reliab. Engng. Syst. Safety* **65**, 147–185.
Kleijnen, J. P. C. and Helton, J. C. (1999b). Statistical analyses of scatterplots to identify important factors in large-scale simulations, 2: robustness of techniques. *Reliab. Engng. Syst. Safety* **65**, 187–197.
Kleijnen, J. P. C. and Helton, J. C. (1999c). Statistical analyses of scatterplots to identify important factors in large-scale simulations. SAND98-2202. Sandia National Laboratories, Albuquerque, NM.
Kleissen, F. M., Beck, M. B. and Wheater, H. S. (1990). The identifiability of conceptual hydrochemical models. *Water Resources Res.* **26**, 2979–2992.
Klepper, O. and Hendrix, E. M. T. (1994). A method for robust calibration of ecological models under different types of uncertainty. *Ecological Model.* **74**, 161–182.
Klepper, O., Scholten, H. and van de Kamer, J. P. G. (1991). Prediction uncertainty in an ecological model of the Oosterschelde Estuary. *J. Forecasting* **10**, 191–209.
Koda, M. (1982). Sensitivity analysis of the atmospheric diffusion equation. *Atmos. Environ.* **16**, 2595–2601.
Koda, M., McRae, G. J. and Seinfeld, J. H. (1979). Automatic sensitivity analysis of kinetic mechanisms. *Int. J. Chem. Kinet.* **11**, 427–444.
Koga, S. and Tanaka, H. (1993). Numerical study of the oxidation process of Dimethylsulfide in the marine atmosphere. *J. Atmos. Chem.* **17**, 201–228.
Konikov, L. F. and Bredehoeft, J. D. (1992). Groundwater models cannot be validated. *Adv. Water Res.* **15**, 75–83.
Kramer, M. A., Calo, J. M. and Rabitz, H. (1981). An improved computational method for sensitivity analysis: Green's function method with 'AIM'. *Appl. Math. Modelling* **5**, 432–441.
Kramer, M. A., Rabitz, H. and Calo, J. M. (1984). Parametric scaling of mathematical models. *Appl. Math. Modelling* **8**, 341–350.
Krzykacz, B. (1990). SAMOS: a computer program for the derivation of empirical sensitivity measures of results from large computer models. Technical Report GRS-A-1700, Gesellschaft für Reaktor Sicherheit (GRS) MbH, Garching, Germany.
L'Ecuyer, P. (1998). Random number generation. In: *Handbook of Simulation* (ed. J. Banks), pp. 93–137. Wiley, New York.
Lam, S. H. and Goussis, D. A. (1988). Understanding complex chemical kinetics with computational singular perturbation. In: *Proceedings of 22nd Symposium (International) on Combustion*, pp. 931–941.
Lee, S., Hillebrands, B., Stegeman, G. I., Cheng, H., Potts, J. E. and Nizzoli, F. (1988). Elastic properties of epitaxial ZnSe(001) films on GaAs measured by Brillouin spectroscopy. *J. Appl. Phys.* **63**, 1914–1916.
Levin, S. A. and Goodyear, C. P. (1980). Analysis of an age-structured fishery model. *J. Math. Biol.* **9**, 245–274.
Li, Y.-H., Dennis, R. L., Tonnesen, G. S., Pleim, J. E. and Byun, D. W. (1998). Effects of uncertainty in meteorology inputs on ozone concentration, ozone production efficiency and ozone sensitivity to emissions reductions in the Regional Acid Deposition Model. In: *Preprints of the AMS–A&WMA*

10th Joint Conference on the Applications of Air Pollution Meteorology. American Meteorological Society, Phoenix, AZ.

Liu, P.-L., Lin H.-Z. and Der Kiureghian, A. (1989). CALREL User Manual. University of California, Berkeley, Report UCB/SEMM-89/18, Department of Civil Engineering (Last revised in 1992).

Lorentz, G. G., Golitschek, M. V. and Makovoz, Y. (1996). *Constructive Approximation*. Springer-Verlag, New York.

Luckman, P., Der Kiureghian, A. and Sitar, N. (1987). Use of stochastic stability analysis for bayesian back-calculation of pore pressures acting in a cut slope at failure. In: *Proceedings of 5th International Conference on Applications of Statistics and Probability in Soil and Structural Engineering*. Vancouver, BC.

Ma, J. Z. and Ackerman, E. (1993). Parameter sensitivity of a model of viral epidemics simulated with Monte Carlo techniques. II. Durations and peaks. *Int. J. Biomed. Comput.* **32**, 255–268.

Ma, J. Z., Ackerman, E. and Yang, J.-J. (1993). "Parameter Sensitivity of a Model of Viral Epidemics Simulated with Monte Carlo Techniques. I. Illness Attack Rates." *International Journal of Biomedical Computing* **32**(3–4): 237–253.

Maas, U. and Pope, S. B. (1992). Simplifying chemical kinetics: Intrinsic low-dimensional manifolds in composition space. *Combust. Flame* **88**, 239–264.

Maas, U. and Pope, S. B. (1994). Laminar flame calculations using simplified chemical kinetics based on intrinsic low-dimensional manifolds. In: *Proceedings of 25th Symposium (International) on Combustion*, pp. 1349–1356.

MacDonald, R. C. and Campbell, J. E. (1986). Valuation of supplemental and enhanced oil recovery projects with risk analysis. *J. Petrol. Technol.* **38**, 57–69.

MacFarlane, A. G. J. (1990). Interactive computing: A revolutionary medium for teaching and design. *Comput. Control Engng J.* **1**, 149–158.

Madsen, H. O. (1988). PROBAN: theoretical manual for external release. A.S. Veritas Research Report 88–2005.

Madsen, H. O., Krenk, S. and Lind, N. C. (1986). *Methods of Structural Safety*. Prentice-Hall, Englewood Cliffs, NJ.

Maravall, A. (1996). Unobserved components in economic time series. In: *Handbook of Applied Econometrics*, Vol. 1 (ed. H. Pesaran, P. Schmidt and M. Wickens). Blackwell, Oxford.

Maravall, A. and Planas, C. (1999). Estimation errors and the specification of unobserved component models. *J. Econometr.* **92**(2), 325–353.

Martin, D. and Yohai, V. (1986). Influence functionals for time series. *Ann. Statist.* **14**, 781–818.

Martín, J. and Ríos Insua, D. (1996). Local sensitivity analysis in Bayesian decision theory. In: *Bayesian Robustness* (ed. J. Berger, B. Betró, E. Moreno, L. Pericchi, F. Ruggeri, G. Salinetti, and L. Wasserman). Lecture Notes IMS, Hayward.

MathSoft (1998). *S-Plus 5 for Unix: Guide to Statistics*. Data Analysis Products Division, Mathsoft, Seattle.

Matsuda, H., Wada, T., Takeuchi, Y. and Matsumiya, Y. (1992). Model analysis of the effects of environmental fluctuation on the species replacement pattern of pelagic fishes under interspecific competition. *Res. Popul. Ecol.* **34**, 309–319.

Mauro, C. A. and Smith, D. E. (1982). The performance of two-stage group screening in factor screening experiments. *Technometrics* **24**, 325–330.

McCombie, C. and McKinley, I. (1993). Validation: another perspective. *Ground Water* **31**, 530–531.

McCulloch, R. (1989). Local model influence. *J. Ame. Statist. Assoc.* **84**, 473–478.

McKay, M. D. (1995). Evaluating prediction uncertainty. Technical Report NUREG/CR-6311, US Nuclear Regulatory Commission and Los Alamos National Laboratory.

McKay, M. D. (1996). Variance-based methods for assessing uncertainty importance in NUREG-1150 analyses. LA-UR-96-2695, Los Alamos National Laboratory.

McKay, M. D. (1997). Nonparametric variance-based methods of assessing uncertainty importance. *Reliab. Engng. Syst. Safety* **57**, 267–279.

McKay, M. D. and Beckman, R. J. (1994). A procedure for assessing uncertainty in models. In: *Proceedings of PSAM-II* (ed. G. E. Apostolakis and J. S. Wu), Vol. 1, pp. 13–18.

McKay, M. D., Beckman, R. J. and Conover, W. J. (1979). A comparison of three methods of selecting values of input variables in the analysis of output from a computer code. *Technometrics* **21**, 239–245.

McKone, T. E. (1994). Uncertainty and variability in human exposures to soil contaminants through home-grown food: a Monte Carlo assessment. *Risk Anal.* **14**, 449–463.

McLaughlin, D. B. (1995). Recent advances in hydrologic data assimilation. In: *US National Report to IUGG (1991–94). Rev. Geophys.* Suppl. pp. 977–984.

McLaughlin, D. B., Kinzelbach, W. and Ghassemi, F. (1993). Modelling subsurface flow and transport. In: *Modelling Change in Environmental Systems* (ed. A. J. Jakeman, M. B. Beck and M. J. McAleer), pp. 133–161. Wiley, Chichester.

McRae, G. J., Tilden, J. W. and Seinfeld, J. H. (1982). Global sensitivity analysis—a computational implementation of the Fourier amplitude sensitivity test (FAST). *Comput. Chem. Engng.* **6**, 15–25.

Meeuwissen, A. M. H. and Cooke, R. M. (1994). Tree dependent random variables. Report 94-28, Technische Universiteit Delft, Faculteit der Technische Wiskunde en Informatica.

Mellouki, A., Jourdain, J. L. and Le Bras, G. (1988). Discharge flow study of the $CH_3S + NO_2$ reaction mechanism using $Cl + CH_3SH$ as the CH_3S source. *Chem. Phys. Lett.* **23**, 231–236.

Meyer, M. A. and Booker, J. M. (1991). *Eliciting and Analyzing Expert Judgment: A Practical Guide.* Academic Press, New York.

Miller, D. and Frenklach, M. (1983). Sensitivity analysis and parameter estimation in dynamic modelling of chemical kinetics. *Int. J. Chem. Kinet.* **15**, 677–696.

Morris, M. D. (1987). Two stage factor screening procedures using multiple grouping assignments. *Commun. Statist. Theory Meth.* **16**, 3051–3067.

Morris, M. D. (1991). Factorial sampling plans for preliminary computational experiments. *Technometrics,* **33**, 161–174.

Morton, A. (1993). Mathematical models: questions of trustworthiness. *Br. J. Phil. Sci.* **44**, 659–674.

Mosleh, A., Bier, V. M. and Apostolakis, G. (1988). A critique of current practice for the use of expert opinions in probabilistic risk assessments. *Reliab. Engng. Syst. Safety* **20**, 63–85.

Myers, R. H. (1971). *Response Surface Methodology.* Allyn and Bacon, Boston.

Myers, R. H. (1990). *Classical and Modern Regression with Applications,* 2nd edn. Duxbury Press, Boston.

Myers, R. H. and Montgomery, D. C. (1995). *Response Surface Methodology—Process and Product Optimization Using Designed Experiments.* Wiley, New York.

National Research Council (1990). *Ground Water Models. Scientific and Regulatory Applications.* National Academy Press, Washington, DC.

Nau, R., Ríos Insua, D. and Martín, J. (1998) Incompleteness of preferences and the foundations of robust Bayesian statistics. Manuscript.

NEA PSAG User's Group (1989). *PSACOIN Level E Intercomparison* (ed. B. W. Goodwin, J. M. Laurens, J. E. Sinclair, D. A. Galson and E. Sartori). Nuclear Energy Agency, Organization for Economic Cooperation and Development, Paris.

NEA PSAG User's Group (1990). *PSACOIN Level 1A Intercomparison* (ed. A. Nies, J. M. Laurens, D. A. Galson and S. Webster). Nuclear Energy Agency, Organization for Economic Cooperation and Development, Paris.

Neter, J. and Wasserman, W. (1974). *Applied Linear Statistical Models.* Richard D. Irwin, Homewood, IL.

NRC (National Research Council) (1992). *Combining Information: Statistical Issues and Opportunities for Research.* National Academy Press, Washington, DC.

NRC (National Research Council) (1996). *The Waste Isolation Pilot Plant: A Potential Solution for the Disposal of Transuranic Waste* (Committee on the Waste Isolation Pilot Plant, Board on Radioactive Waste Management, Commission on Geosciences, Environment, and Resources, National Research Council). National Academy Press, Washington, DC.

NRW State Environment Agency (1997). *Identification of Relevant Industrial Sources of Dioxins and Furans in Europe.* Publication of the LandesUmweltAmt—North-Rhine-Westphalia State Environment Agency, Essen.

Odum, E., (1971). *Fundamentals of Ecology,* 3rd edn. Saunders, Philadelphia.

Onishi, Y., Shuyler, L. and Cohen, Y. (1990). Multimedia modeling of toxic chemicals. In: *Proceedings of the International Symposium on Water Quality Modeling of Agricultural Non-Point Source*, part 2 (ed. D. G. DeCoursey), pp. 479–502. Report ARS-81, Agricultural Research Service, US Department of Agriculture.

Oreskes, N. (1998a). Evaluation (not validation) of quantitative models for assessing the effects of environmental lead exposure. *Environ. Health Perspect.* **106**, (Suppl. 6), 1453–1460.

Oreskes, N. (1998b). When good models are false: sobering lessons from the history of science. In: *Proceedings of Sensitivity Analysis of Model Output, Second International Symposium, Venice, Italy*, EUR17758EN, Luxembourg, pp. 211–213.

Oreskes, N., Shrader-Frechette, K. and Belitz, K. (1994). Verification, validation, and confirmation of numerical models in the earth sciences. *Science* **263**, 641–646.

Ortiz, N. R., Wheeler, T. A., Breeding, R. J., Hora, S. C., Meyer, M. A. and Keeney, R. L. (1991). Use of expert judgment in NUREG-1150. *Nucl. Engng. Design* **126**, 313–331.

Owen, A. B. (1992). A central limit theorem for latin hypercube sampling. *J. R. Statist. Soc. Ser. B* **54**, 541–551.

Owen, A. B. (1997). Scrambled net variance for integrals of smooth functions. *Ann. Statist.* **25**, 1541–1562.

Øvreberg, O., Damsleth, E. and Haldersen, H. H. (1992). Putting error bars on reservoir engineering forecasts. *J. Petrol. Technol.* **44**, 732–738.

Park, R. A., Scavia, D. and Clesceri, N. L. (1975). CLEANER: the Lake George model. In: *Ecological Modeling in a Resource Management Framework* (ed. C. S. Russell), pp. 49–81. Resources for the Future, Washington, DC.

Parker, D. (1985). Learning logic. Working Paper 47, Center for Computational Research in Economics and Management Science, Massachusetts Institute of Technology.

Patel, M. S. (1962). Group-screening with more than two stages. *Technometrics* **4**, 209–217.

Patel, M. S. and Ottieno, J. A. M. (1984). Optimum two stage group-screening designs with equal prior probabilities and no errors in decisions. *Commun. Statist. Theory Meth.* **13**, 1147–1159.

Payne, A. C., Daniel, S. L., Whitehead, D. W., Sype, T. T., Dingman, S. E. and Shaffer, C. J. (1992). Analysis of the LaSalle Unit 2 nuclear power plant: risk methods integration and evaluation program (RMIEP). NUREG/CR-4832, SAND92-0537, Vols 1–3. Sandia National Laboratories, Albuquerque, NM.

Peña, D. (1990). Influential observations in time series. *J. Business Econ. Statist.* **8**, 235–241.

Pereira, A. de C. (1989). Some developments of safety analysis with respect to the geological disposal of high-level nuclear waste. SKI Technical Report 89 : 06, Stockholm.

Pierce, D. A. (1978). Seasonal adjustment when both deterministic and stochastic seasonality are present. In: *Seasonal Analysis of Economic Time Series* (ed. A. Zellner), pp. 242–269. US Department of Commerce, Bureau of the Census, Washington, DC.

Pierce, T. H. and Cukier, R. I. (1981). Global nonlinear sensitivity analysis using Walsh functions. *J. Comput. Phys.* **41**, 427–443.

Piggott, J. and Cawlfield, J. (1996). Probabilistic sensitivity analysis of one-dimensional contaminant transport in the vadose zone. *J. Contaminant Transport* **24**, 97–115.

Plackett, R. L. and Burman, J. P. (1946). The design of optimum multifactorial experiments. *Biometrika* **33**, 305–325.

PLG (Pickard, Lowe, and Garrick, Inc.) (1983). Seabrook Station probabilistic safety assessment. Prepared for Public Service Company of New Hampshire and Yankee Atomic Electric Company, Newport Beach, CA.

PLG (Pickard, Lowe, and Garrick, Inc., Westinghouse Electric Corporation, and Fauske & Associates, Inc.) (1982). Indian Point Probabilistic safety study. Prepared for the Power Authority of the State of New York and Consolidated Edison Company of New York, Inc., Newport Beach, CA.

Poggio, T. and Girosi, F. (1990). Networks for approximation and learning. *Proc. IEEE* **78**, 1481–1497.

Poland, J. F. (ed.) (1984). *Guidebook to Studies of Land Subsidence due to Ground-Water Withdrawal*. UNESCO.

Porter, K. G. (1996). Integrating the microbial loop and the classic food chain into a realistic planktonic food web. In: *Food Webs: Integration of Patterns and Dynamics* (ed. G. A. Polis and K. Winemiller), pp. 51–59. Chapman Hall, New York.

Porter, K. G., Saunders, P. A., Haberyan, K. A., Macubbin, A. E., Jacobsen, T. R. and Hodson, R. E. (1996). Annual cycle of autotrophic and heterotrophic production in a small, monomictic Piedmont lake (Lake Oglethorpe): analog for the effects of climatic warming on dimictic lakes. *Limnol. Oceanogr.* **41**, 1041–1051.

Prado, P. (1992). User's manual for the GTM-1 code. EUR 13925 EN Joint Research Centre.

Prado, P., Eguilior, S. and Saltelli, A. (1998). Level E/G test-case specifications (GESAMAC project). CIEMAT, Madrid (CIEMAT/DIAE/550/55900/04/08).

Prado, P., Homma, T. and Saltelli, A. (1991). Radionuclide migration in the geosphere: a 1D advective and dispersive transport module for use in probabilistic system assessment codes. *Radioactive Waste Manag. Nucl. Fuel Cycle* **16**, 49.

Press, W. H., Teukolsky, S. A., Vetterlings, W. T. and Flannery, B. P. (1992). *Numerical Recipes: The Art of Scientific Computing (FORTRAN Version)*, 2nd edn. Cambridge University Press, Cambridge.

Price, P. S., Su, S. H., Harrington, J. R. and Keenan, R. E. (1996). Uncertainty and variation of indirect exposure assessments: an analysis of exposure to tetrachlorodibenzene-*p*-dioxin from a beef consumption pathway. *Risk Anal.* **16**, 263–277.

Puolamaa, M., Kaplas, M. and Reinikainen, T. (1996). *Index of Environmental Friendliness—A Methodological Study.* Statistics Finland.

Quade, D. (1989). Partial correlation. In: *Encyclopedia of Statistical Sciences* (ed. S. Kotz and N. L. Johnson), Supplement Volume, pp. 117–120. Wiley, New York.

Rabitz, H. (1989). System analysis at molecular scale. *Science* **246**, 221–226.

Rabitz, H., Kramer, M. and Dacol, D. (1983). Sensitivity analysis in chemical kinetics. *Annu. Rev. Phys. Chem.* **34**, 419–461.

Rabitz, H., Alis, O., Shorter, J. and Shim, K. (1999) Efficient input–output model representations. *Comput. Phys. Commun.* **117**, 11–20.

Rabitz, H. and Shim, K. (1999). Multicomponent semiconductor material discovery guided by a generalized correlated function expansion. *J. Chem. Phys.* **111**, 10640–10651.

Rabitz, H. and Smooke, M. D. (1988). Scaling relations and self-similarity conditions in strongly coupled dynamical systems. *J. Phys. Chem.* **92**, 1110–1119.

Radhakrishnan, K. (1990). Combustion kinetics and sensitivity analysis computations. In: *Numerical Approaches to Combustion Modelling* (ed. E. S. Oran, and J. P. Boris), pp. 83–128. *Progress in Astronautics and Aeronautics*, Vol. 135, AIAA, Washington, DC.

Raghavarao, D. (1971). *Constructions and Combinatorial Problems in Design of Experiments*, pp. 309–314. Wiley, London.

Rahni, N., Ramdani, N., Candau, Y. and Dalicieux, P. (1997). Application of group screening to dynamic building energy simulation models. *J. Statist. Comput. Simul.* **57**, 285–304.

Raj, D. (1968). *Sampling Theory*. McGraw-Hill, New York.

RamaRao, B. and Mishra, S. (1996). Adjoint sensitivity analysis for mathematical models of coupled nonlinear physical processes. In: *Calibration and Reliability in Groundater Modelling* (ed. K. Kovar and P. Van Der Heijde). IAHS Press, Institute of Hydrology, IAHS Publication No. 237.

Ray, A., Vassalli, I., Laverdet, G. and LeBras, G. (1996). Kinetics of the thermal decomposition of the CH_3SO_2 radical and its reaction with NO_2 at 1 Torr and 298 K. *J. Phys. Chem.* **100**, 8895–8900.

Rechard, R. P. (1999). Historical background on assessing the performance of the waste isolation pilot plant. SAND98-2708. Sandia National Laboratories, Albuquerque, NM.

Reda, A. L. L. (1996). Simulation and control of stormwater impacts on river water quality. PhD dissertation, Imperial College of Science, Technology, and Medicine, London.

Reda, A. L. L. and Beck, M. B. (1997). Ranking strategies for stormwater management under uncertainty: sensitivity analysis. *Water Sci. Technol.* **36**, 357–371.

Remedio, J. M., Saltelli, A., Hjorth, J. and Wilson, J. (1994). KIM. A chemical kinetic model of the OH-initiated oxidation of DMS in air: a Monte Carlo analysis of the latitude effect. EUR Report 16045 EN, Luxembourg.

RESS (1997). Special Issue on Sensitivity Analysis. *Reliab. Engng. Syst. Safety* **57**, 1.

Riani, M. (1998). Weights and robustness of model-based seasonal adjustment. *J. Forecasting* **17**, 19–34.

Ríos Insua, D. (1990) *Sensitivity Analysis in Multiobjective Decision Making.* Springer-Verlag, New York.

Ríos Insua, D. and Ruggeri, F. (2000). *Robust Bayesian Analysis*. Lecture Notes in Statistics, Springer-Verlag, New York.

Ríos Insua, S., Martín, J., Ríos Insua, D. and Ruggeri, F. (1999). Bayesian forecasting for accident proneness evaluation. *Scandinavian Actuarial Journal* **2**, 134, 146.

Rose, K. A., Smith, E. P., Gardner, R. H., Brenkert, A. L. and Bartell, S. M. (1991). Parameter sensitivities, Monte Carlo filtering, and model forecasting under uncertainty. *J. Forecasting* **10**, 117–133.

Rosen, R. (1991). *A Comprehensive Inquiry into the Nature, Origin, and Fabrication of Life*. Columbia University Press, New York.

RSD (1990) Survey conducted by the Research Unit for Societal Development, University of Mannheim.

Ruggeri, F. (1990). Posterior ranges of functions of parameters under priors with specified quantiles. *Commun. Statist. Theory Meth.* **19**, 127–144.

Ruggeri, F. (1994). Local and global sensitivity under some classes of priors. In: *Recent Advances in Statistics and Probability* (ed. J. P. Vilaplana and M. L. Puri), VSP, Ah Zeist.

Ruggeri, F. and Sivaganesan, S. (2000). On a global sensitivity measure for Bayesian inference. To appear in Sankhya, A., Quaderno IAMI CNR–IAMI, Milano.

Ruggeri, F. and Wasserman, L. (1993). Infinitesimal sensitivity of posterior distributions. *Can. J. Statist.* **21**, 195–203.

Ruggeri, F. and Wasserman, L. (1995). Density based classes of priors: infinitesimal properties and approximations. *J. Statist. Plann. Infer.* **46**, 311–324.

Russell, A. G. and Dennis, R. L. (2000). NARSTO critical review of photochemical models and modeling. *Atmos. Environ.* **34**, 2283–2324.

Sacks J., Schiller, S. B. and Welch, W. J. (1989b). Designs for computer experiments. *Technometrics* **31**, 41–47.

Sacks J., Welch, W. J., Mitchell, T. J. and Wynn, H. P. (1989a). Design and analysis of computer experiments. *Statist. Sci.* **4**, 409–435.

Sacks, J. and Ylvisaker, D. (1984). Some model robust designs in regression. *Ann. Statist.* **12**, 1324–1348.

Sacks, J. and Ylvisaker, D. (1985). Model robust design in regression: Bayes theory. In: *Proceedings of the Berkeley Conference in Honour of Jerzy Neyman and Jack Kiefer* (ed. L. M. Le Cam and R. A. Olshen), Vol. 2, pp. 667–679. Wadsworth, Monterey, CA.

Salhotra, A. M., Sharp-Hansen, S. and Allison, T. (1990). Multimedia exposure assessment model (MULTIMED) for evaluating the land disposal of Wastes—model theory. Report (Contract Nos 68-03-3513 and 68-03-6304), Environmental Research Laboratory, US Environmental Protection Agency, Athens, GA.

Salinetti, G. (1994). Stability of Bayesian decisions. *J. Statist. Plann. Infer.* **40**, 313–320.

Saltelli, A. (1999). Sensitivity analysis. Could better methods be used? *J. Geophys. Res.* **104**, 3789–3793.

Saltelli, A., Andres, T. H. and Homma T. (1995). Sensitivity analysis of model output. Performance of the iterated fractional factorial design (IFFD) method. *Comput. Statist. Data Anal.* **20**, 387–407.

Saltelli, A., Andres, T. H. and Homma, T. (1993). Sensitivity analysis of model output: an investigation of new techniques. *Comput. Statist. Data Anal.* **15**, 211–238.

Saltelli, A. and Bolado, R. (1998). An alternative way to compute Fourier amplitude sensitivity test (FAST). *Comput. Statist. Data Anal.* **26**, 445–460.

Saltelli, A. and Hjorth, J. (1995). Uncertainty and sensitivity analyses of OH-initiated dimethyl sulphide (DMS) oxidation kinetics. *J. Atmos. Chem.* **21**, 187–221.

Saltelli, A. and Homma, T. (1992). Sensitivity analysis for model output; performance of black box techniques on three international benchmark exercises. *Comput. Statist. Data Anal.* **13**, 73–94.

Saltelli, A., Homma, T. and Andres, T. H. (1994). A new measure of importance in global sensitivity analysis of model output. In: *Proceedings of 6th Joint EPS/APS International Conference on Physics and Computing*.

Saltelli, A. and Marivoet, J. (1990). Non-parametric statistics in sensitivity analysis for model output: a comparison of selected techniques. *Reliab. Engng. Syst. Safety* **28**, 229–253.

Saltelli, A., Planas, C., Tarantola, S. and Campolongo, F. (1999a). The concept of sensitivity with applications to official statistics. *Res. Official Statist.*, **2**(2).

Saltelli, A., Prado, P. and Torres, C. (1989). ENRESA/JRC cooperation agreement in the field of nuclear waste management. 3383-88-03 TG ISP E. Final Report: 1 Results, 2 Annexes (SPI89.36/I, II).

Saltelli, A. and Scott, M. (1997). Guest Editorial: The role of sensitivity analysis in the corroboration of models and its links to model structural and parametric uncertainty. *Reliab. Engng. Syst. Safety* **57**, 1–4.
Saltelli, A. and Sobol', I. M. (1995). About the use of rank transformation in sensitivity analysis of model output. *Reliabi. Engng. Syst. Safety* **50**, 225–239.
Saltelli, A., Tarantola, S. and Campolongo, F. (2000). Sensitivity analysis as an ingredient of modelling. *Statistical Science.* **15** (4), 377–395.
Saltelli, A., Tarantola, S. and Chan, K. (1998). Presenting the results from model based studies to decision makers: can sensitivity analysis be a defogging agent? *Risk Anal.* **18**, 799–803.
Saltelli, A., Tarantola, S. and Chan, K. (1999b). A quantitative, model independent method for global sensitivity analysis of model output. *Technometrics* **41**, 39–56.
Šaltenis, V. and Dzemyda, G. (1982). Structure analysis of extremal problems using an approximation of characteristics. *Optimal Decision Theory* **8**, 124–138.
Saltzman, E. S., Savoie, D. L., Prospero, J. M. and Zika, R. G. (1996). Methane sulfonic acid and non-sea-salt sulfate in pacific air: regional and seasonal variations, *J. Atmos. Chem.* **4**, 227–240.
SAMO, (1998). *Proceedings of the Second International Symposium on Sensitivity Analysis of Model Output (SAMO98)* (ed. K. Chan, S. Tarantola and F. Campolongo). Venice, Italy, April 19–22, 1998. EUR Report 17758 EN, Luxembourg.
Sanchez, M. A. and Blower, S. M. (1997). Uncertainty and sensitivity analysis of the basic reproductive rate. Tuberculosis as an example. *Am. J. Epidemiol.* **145**, 1127–1137.
Sargent, D. J. and Carlin, B. P. (1996). Robust Bayesian design and analysis of clinical trials via prior partitioning. In: *Bayesian Robustness* (ed. J. Berger, B. Betró, E. Moreno, L. Pericchi, F. Ruggeri, G. Salinetti and L. Wasserman. Lecture Notes IMS, Hayward.
Satterthwaite, F. E. (1959). Random balance experimentation (with discussion). *Technometrics* **1**, 111–137, 157–192.
Savage, L. J. (1954). *The Foundations of Statistics.* Wiley, New York.
Savoie, D. L. and Prospero, J. M. (1989). Comparison of oceanic and continental sources of non-sea-salt sulphate over the pacific ocean. *Nature* **339**, 685–689.
Scatz, G. (1989). The analytical representation of electronic potential-energy surfaces. *Rev. Mod. Phys.* **61**, 669–688.
Schaibly, J. H. and Shuler, K. E. (1973). Study of the sensitivity of coupled reaction systems to uncertainties in rate coefficients. Part II, Applications. *J. Chem. Phys.* **59**, 3879–3888.
Schanz, R. W. and Salhotra, A. (1992). Evaluation of the Rackwitz–Fiessler uncertainty analysis method for environmental fate and transport models. *Water Resources Res.* **28** (4), 1071–1079.
Scheffe, H. (1959). *The Analysis of Variance.* Wiley, New York.
Schrader-Frechette, K. S. (1994). *Burying Uncertainty: Risk and the Case Against Geological Disposal of Nuclear Waste.* University of California Press, Berkeley.
Schrader-Frechette, K. (1995). Hard ecology, soft ecology, and ecosystem integrity. In: *Perspectives on Ecological Integrity* (ed. L. Westra and J. Lemons), pp. 125–145. Kluwer, Dordrecht.
Seber, G. A. F. (1977). *Linear Regression Analysis.* Wiley, New York.
Seidenfeld, T., Schervish, M. and Kadane, J. (1996). A representation of partially ordered preferences. *Ann. Statis.* **20**, 1737–1767.
Seinfeld, J. H. and Pandis, S. (1998). *Atmospheric Chemistry and Physics: From Air Pollution to Climate Change.* Wiley, New York.
Shackleton, L. Y. (1986). Fossil pilchard and anchovy scales—indicators of past fish populations off Namibia. In: *Proceedings of International Symposium on Long Term Changes in Marine Fish Populations, Vigo, Spain.* (ed.) T. Wyatt and M. G. Larraneta, pp. 55–68.
Shackleton, L. Y. (1987). A comparative study of fossil fish scales from three upwelling regions. *S. A. J. Mar. Sci.* **5**, 79–84.
Sharp-Hansen, S., Travers, C. and Allison, T. (1990), Subtitle D landfill application manual for the multimedia exposure assessment model (MULTIMED). Report (Contract No. 68-03-3513), Environmental Research Laboratory, US Environmental Protection Agency, Athens, GA.

Sheng, G., Elzas, M. S., Ren,T. I. and Cronhjorth, B. T. (1993). Model validation: a systemic and systematic approach. *Reliab. Engng. Syst. Safety* **42**, 247–259.

Shim, K. and Rabitz, H. (1998). Independent and correlated composition behavior of the energy band gaps for the $Ga_\alpha In_{1-\alpha}P_\beta As_{1-\beta}$ and $Ga_\alpha In_{1-\alpha}P_\beta Sb_\gamma As_{1-\beta-\gamma}$ alloys. *Phys. Rev.* B**58**, 1940–1946.

Shortencarier, M. J. and Helton, J. C. (1999). A FORTRAN 77 program and user's guide for the statistical analyses of scatterplots to identify important factors in large-scale simulations. SAND99-1058. Sandia National Laboratories, Albuquerque, NM.

Shorter, J. A., Ip, P. C. and Rabitz, H. (1999). An efficient chemistry solver using high dimensional model representations. *J. Phys. Chem.* A**103**, 7192–7198.

Shorter, J. A., Ip, P. C. and Rabitz, H. (2000). Radiation transport simulation by means of a fully equivalent operational model. *Geophys. Res. Lett.* (submitted)

Shreider, Y. (1967). *The Monte Carlo Method.* Pergamon Press, Oxford.

Silverman, B.W. (1986). *Density Estimation for Statistics and Data Analysis.* Chapman & Hall, New York.

Sitar, N., Cawlfield, J. and Der Kiureghian, A. (1987). First-order reliability approach to stochastic analysis of subsurface flow and contaminant transport. *Water Resources Res.* **23** (5), 794–804.

Skumanich, M. and Rabitz, H. (1982). Feature sensitivity analysis in chemical kinetics. *Comm. J. Mol. Sci.* **2**, 79–92.

Smidts, O. F. and Devooght, J. (1997). A variational method for determining uncertain parameters and geometry in hydrogeology. *Reliabi. Engng Syst. Safety* **57**, 5–19.

Smith, P.E., Chyan-Huei Lo, N. and Butler, J. (1992). Life-stage duration and survival factors as related to interdecadal population variability in Pacific sardine. *CalCOFI Rep.* **33**, 41–49.

Sobol', I. M. (1967). On the distribution of points in a cube and the approximate evaluation of integrals. *USSR Computat. Maths. Math. Phys.* **7**, 86–112.

Sobol', I. M. (1976). Uniformly distributed sequences with an addition uniform property. *USSR Comput. Maths. Math. Phys.* **16**, 236–242.

Sobol', I. M. (1990a). Quasi-Monte Carlo methods. *Prog. Nucl. Energy* **24**, 55–61.

Sobol', I. M. (1990b). Sensitivity estimates for non-linear mathematical models. *Matematicheskoe Modelirovanie* **2**, 112–118 (in Russion). [Transl. Sensitivity analysis for non-linear mathematical models. *Math. Modelling Comput. Exp.* **1**, 407–414 (1993).]

Sobol', I. M. (1992). An efficient approach to multicriteria optimum design problems. *Surv. Maths. Indust.* **1**, 259–281.

Sobol', I. M. (1993). Sensitivity analysis for non linear mathematical models. *Math. Model. Comput. Exp.* **1**, 407–414.

Sobol', I. M. (1994). *A Primer for the Monte Carlo Method.* CRC Press, Boca Raton, FL.

Sobol', I. M. and Levitan, Y. L. (1999). On the use of variance reducing multipliers in Monte Carlo computations of a global sensitivity index. *Comput. Phys. Commun.* **117**, 52–61.

Sobol', I. M. and Shukhman, B. V. (1995). Integration with quasirandom sequences: numerical experience. *Int. J. Mod. Phys. Ser. C.* **6**, 263–275.

Sobol', I. M., Turchaninov, V. I., Levitan, Y. L. and Shukhman, B. V. (1992). *Quasirandom Sequence Generators.* Keldysh Institute of Applied Mathematics, Russian Academy of Sciences, Moscow.

Soutar, A. and Isaacs, J. D. (1969). History of fish populations inferred from fish scales in anaerobic sediments off California. *CalCOFI Rep.* **13**, 63–70.

Soutar, A. and Isaacs, J. D. (1974). Abundance of pelagic fish during the 19th and 20th centuries as recorded in anaerobic sediments of the Californias. *Fish. Bull.* **72**, 257–273.

Spear, R. C., Grieb, T. M. and Shang, N. (1994). Parameter uncertainty and interaction in complex environmental models. *Water Resources Res.* **30**, pp 3159–3169.

Spear, R. C. and Hornberger, G. M. (1980). Eutrofication in Peel Inlet, II, identification of critical uncertainty via generalised sensitivity analysis. *Water Res.* **14**, 43–49.

SPSS (1997). *SPSS for Windows, 8.00.*

Srinivasan, C. and Truszczynska, H. (1990). On the ranges of posterior quantities. Technical Report, Department of Statistics, University of Kentucky.

Srivastava, J. N. (1975). Designs for searching non-negligible effects. In: *A Survey of Statistical Designs and Linear Models* (ed. J. N. Srivastava), pp. 507–519. North-Holland, Amsterdam.

Stein, M. (1987). Large sample properties of simulations using Latin hypercube sampling. *Technometrics* **29**, 143–151.

Steinberg, H. A. (1963). Generalized quota sampling. *Nucl. Sci. Engng.* **15**, 142–145.

Stigter, J. D. and Beck, M. B. (1994). A new approach to the identification of model structure. *Environmetrics* **5**, pp 315–333.

Stockwell, W. R., Middleton, P., Chang, J. S. and Tang, X. (1990). The second generation Regional Acid Deposition Model chemical mechanism for regional air quality modeling. *J. Geophys. Res.* **95**, 16343–16367.

Stoker, T. (1986). Consistent estimation of scaled coefficients. *Econometrica* **54**, 1491–1481.

Stone, C. J. (1982). Optimal global rate of convergence for nonparametric regression. *Ann. Statist.* **10**, 1040–1053.

Stone, C. J. (1985). Additive regression and other nonparametric models. *Ann. Statist.* **13**, 689–705.

Tarantola, S. (1998). Analyzing the efficiency of quantitative measures of importance: the improved FAST. Proceedings of SAMO 98: Second International Symposium on Sensitivity Analysis of Model Output, Venice, 19–22 April 1998, EUR report 17758 EN.

Tesche, T. W. and McNally, D. E. (1995). *Assessment of UAM-IV Model Performance for Three St. Louis Ozone Episodes.* Alphine Geophysics, LLC, Covington, KY.

The Economist (1998). More fall-out from Chernobyl. 27 June, p. 98.

Thorne, M. C. (1993). The use of expert opinion in formulating conceptual models of underground disposal systems and the treatment of associated bias. *Reliab. Engng. Syst. Safety* **42**, 161–180.

Tikhonov, A. and Arsenin, V. (1977). *The Solution of Ill-Posed Problems.* Wiley, New York.

Tilden, J. W., Costanza, V., McRae, G. J. and Seinfeld, J. H. (1980). Sensitivity analysis of chemically reacting systems. *Chemical Physics, Springer Series*, **18**, 69–91.

Tomlin, A. S., Li, G., Rabitz, H. and Tóth, J. (1994). A general analysis of approximate nonlinear lumping in chemical kinetics. II. Constrained lumping. *J. Chem. Phys.* **101**, 1188–1201.

Tomlin, A. S., Turányi, T. and Pilling, M. J. (1997). Mathematical tools for the construction, investigation and reduction of combustion mechanisms. In: *Low-Temperature Combustion and Autoignition* (ed. M. J. Pilling), pp. 293–437. *Comprehensive Chemical Kinetics*, Elsevier, Amsterdam.

Tomovic, R. and Vukobratovic, M. (1972). *General Sensitivity Theory.* Elsevier, New York.

Tong, H. (1990). *Non-linear Series: A Dynamical Systems Approach.* Oxford University Press, Oxford.

Tonnesen, G. S. and Dennis, R. L. (2000). Analysis of radical propagation efficiency to assess ozone sensitivity to hydrocarbons and NO_x. Part 1: Local indicators of instantaneous odd oxygen production sensitivity. *J. Geophys. Res.* **105**, (D7) 9213–9225.

Tonnesen, G. S. and Jeffries, H. E. (1994). Inhibition of odd oxygen production in the Carbon Bond 4 and the Generic Reaction Set mechanisms. *Atmos. Environ.* **28**, 1339–1349.

Turányi, T. (1990a). Sensitivity analysis of complex kinetic systems. Tools and applications. *J. Math. Chem.* **5**, 203–248.

Turányi, T. (1990b). Reduction of large reaction mechanisms. *New J. Chem.* **14**, 795–803.

Turányi, T. (1990c). KINAL: A program package for kinetic analysis of reaction mechanisms. *Comput. Chem.* **14**, 253–254.

Turányi, T. (1994). Parameterization of reaction mechanisms using orthonormal polynomials. *Comput. Chem.* **18**, 45–54.

Turányi, T., Bérces, T. and Vajda, S. (1989). Reaction rate analysis of complex kinetic systems. *Int. J. Chem. Kinet.* **21**, 83–99.

Turányi, T., Tomlin, A. S. and Pilling, M. J. (1993). On the error of the quasi-steady state approximation. *J. Phys. Chem.* **97**, 163–172.

US DOE (US Department of Energy) (1996). *Title 40 CFR Part 191 Compliance Certification Application for the Waste Isolation Pilot Plant.* DOE/CAO-1996-2184. US Department of Energy, Carlsbad Area Office, Carlsbad, NM.

US GLOBEC Executive Summary, http://www.usglobec.berkeley.edu/usglobec.

US NRC (Nuclear Regulatory Commission) (1990–91). *Severe accident risks: an assessment for five U.S. nuclear power plants.* NUREG-1150. US Nuclear Regulatory Commission, Washington, DC.

USEPA (1991). *Guideline for Regulatory Application of the Urban Airshed Model.* US Environmental Protection Agency (USEPA), Office of Air Quality Planning and Standards (OAQPS), Research Triangle Park, NC.

USEPA (1997a). National ambient air quality standards for ozone. 62 FR 38856.

USEPA (1997b). *Regulatory Impact Analyses for the Particulate Matter and Ozone National Ambient Air Quality Standards and Proposed Regional Haze Rule.* USEPA OAQPS, Research Triangle Park, NC.

Vajda, S. and Turányi, T. (1986). Principal component analysis for reducing the Edelson-Field-Noyes model of the Belousov–Zhabotinsky reaction. *J. Phys. Chem.* **90**, 1664–1670.

Vajda, S., Valkó, P. and Turányi, T. (1985). Principal component analysis of kinetic models. *Int. J. Chem. Kinet.* **17**, 55–81.

van Asselt, M. B. A. and Rotmans, J. (1996). Uncertainty in perspective. *Global Environ. Change* **6**, 121–157.

van Noortwijk, J. M., Vrouwenvelder, A. C. W. M., Calle, E. O. F., and Slijkhuis, K. A. H. (1999). Probability of dike failure due to uplifting and piping: an uncertainty analysis. In: *Proceedings of European Safety and Reliability Conference '99 (ESREL 00)*, TUM Munich-Garching, Germany.

Wagner, H. M. (1995). Global sensitivity analysis. *Oper. Res.* **43**, 948–969.

Walley, P. (1991). *Statistical Reasoning with Imprecise Probabilities.* Chapman and Hall, London.

Wasserman, L. (1992). Recent methodological advances in robust Bayesian inference. In: *Bayesian Statistics IV* (ed. J. M. Bernardo, J. O. Berger, A. P. Dawid and A. F. M. Smith). Oxford University Press, Oxford.

Watson, G. S. (1961). A study of the group screening method. *Technometrics* **3**, 371–388.

Weaver, A. J. (1995). Driving the ocean conveyor. *Nature*, **378**, 135–136.

Wegman, E. J. (1990). Hyperdimensional Data Analysis Using Parallel Coordinates, JASA, vol. 90, No 411, pp. 664–675.

Weisberg, S. (1985). *Applied Linear Regression* 2nd edn. Wiley, New York.

Weiss, R. (1996). An approach to Bayesian sensitivity analysis. *J. R. Statist. Soc.*, **58**, 739–750.

Welch, W. J. (1983). A mean squared error criterion for the design of experiments. *Biometrika* **70**, 205–213.

Welch, W. J., Buck, R. J., Sacks J., Wynn, H. P., Mitchell, T. J. and Morris, M. D. (1992). Screening, predicting, and computer experiments. *Technometrics* **34**, 15–47.

White, H. (1992). *Artificial Neural Networks: Approximation and Learning Theory*, Blackwell, Cambridge, MA.

Whiting, W. B., Tong, T.-M. and Reed, M. E. (1993). Effect of uncertainties in thermodynamic data and model parameters on calculated process performance. *Indust. Engng. Chem. Res.* **32**, 1367–1371.

Woods, J. D. (1999). Virtual ecology. *1995 Inaugural Lecture*, Imperial College Press, London.

Wu, M. C. and Cawlfield, J. (1992). Probabilistic sensitivity and modeling of two-dimensional transport in porous media. *Stochastic Hydrol. Hydraul.* **6**, 103–121.

Xiang, Y. and Mishra, S. (1997). Probabilistic multiphase flow modeling using the limit-state method. *Groundwater* **35**(5) 820–824.

Young, P. C. (1978). General theory of modelling badly defined systems. In: *Modelling, Identification and Control in Environmental Systems* (ed. G. C. Vansteenkiste), pp 103–135. North-Holland, Amsterdam.

Young, P. C. (1998). Data-based mechanistic modelling of environmental, ecological, economic and engineering systems. *Environm. Model. Software* **13**, 105–122.

Young, P. C. (1999). Data-based mechanistic modelling, generalised sensitivity and dominant mode analysis. *Comput. Phys. Communi.* **117**, 113–129.

Young, P. C. and Beven, K. J. (1994). Data-based mechanistic modelling and the rainfall-flow nonlinearity. *Environmetrics* **5**, 335–363.

Young, P. C., Hornberger, G. M. and Spear, R. C. (1978). Modelling badly defined systems—Some further thoughts. In: *Proceedings of SIMSIG Simulation Conference Australian National University, Canberra*, pp. 24–32.

Young, P. C., Parkinson, S. and Lees, M. (1996). Simplicity out of complexity: Occam's razor revisited. *J. Appl. Statist.* **23**, 165–210.

Zaldívar, J. M., Kourti, N., Villacastin, C., Strozzi, F. and Campolongo, F. (1998). Analysing dynamics of complex ecologies from natural recordings: an application to fish population models. JRC–ISIS Technical Note I.98.199.

Appendix

Software for Sensitivity Analysis—A Brief Review

Karen Chan

European Commission, Joint Research Centre, Ispra, Italy

E. Marian Scott

University of Glasgow, UK

Terry Andres

AECL Whiteshell Labs, Pinawa, Manitoba, Canada

A.1 INTRODUCTION

This appendix provides an overview of some specialized software available for performing sensitivity analysis. Other general-purpose (commercial or otherwise) software such as MS Excel and SPSS can also perform SA, but these will not be described in this appendix. Instead, we concentrate on those that have been developed specifically for implementing the SA techniques described in this book. This chapter is organized into three main sections: the first section gives descriptions of software that are mentioned in the methodological chapters, the next section provides information about some software not discussed in the methodological and application chapters, and the final section contains some generic algorithms that may be of interest to readers.

A.2 SOFTWARE FOR SENSITIVITY ANALYSIS

This section provides descriptions of some software previously described in the methodology and application chapters of this book. The particular packages included are

Sensitivity Analysis. Edited by A. Saltelli *et al.*

Table A.1 Specialized SA software packages.

Name	Methodologies
SAMPLE2	Sampling-based
CALREL	FORM and SORM
PROBAN	FORM and SORM
CHEMKIN-II	Local
CHEMKIN-III	Local
KINAL	Local
KINALC	Local
PREP/SPOP	Variance-based
SIMLAB	Variance-based
UNICORN	Graphical

summarized in Table A.1. The first package (SAMPLE2) does not actually perform SA, but instead provides a means of creating the necessary input based on sampling strategies for Monte Carlo-based SA. Some of these strategies are described in Chapters 3 and 6. The next two packages (CALREL and PROBAN) are used in Chapter 7 and the next four are for local methods, in particular for kinetic models. Finally, the PREP/SPOP package is a general SA package available in two versions. One is written in FORTRAN code and is less-up-to-date, but it is freely available on the web. The other version, currently being developed, is more user-friendly, and will facilitate more up-to-date sampling schemes such as LP_{τ} sequences and FAST sampling, and SA techniques such as FAST (described in Chapter 8). Finally, the UNICORN package presented in Chapter 11 was designed for use in reliability/failure studies. Special features include conditional sampling and cobweb plots.

SAMPLE2

Author/developer: T. H. Andres, AECL (Canada)

Availability: http://www.jrc.org/isis/services/services/samo_group/softw.asp

Version: N/A

Platform: Most operating systems

Programming language: FORTRAN code

Memory requirement: N/A

Brief description

SAMPLE2 generates samples from a sample space consisting of all possible combinations of parameter values that can occur in a model, and is a new version of SAMPLE — developed in 1989 — to implement a variety of the sampling techniques (Andres, 1987). It provides the user with the capability of generating a wide variety of composite experimental designs with both local and global properties. These include both the sample techniques (simple random, fractional factorial, Latin hypercube, combined fractional factorial/Latin hypercube, and differential analysis) and the subsequent transformation (rounding, discretization, and importance transformation). SAMPLE2 does not provide distribution-specific capability.

SAMPLE2 has a simple menu-driven interface, and can be used freely and disseminated to other people as a complete package of files, but AECL maintains all rights to the code.

CALREL

Author/developer: Armen Der Kiureghian and Pei-Ling Liu

Availability: NISEE/Computer Applications, Earthquake Engineering Research Centre, 379 Davis Hall, University of California, Berkeley, CA 94720, USA

Version: N/A

Platform: VAX, PC

Programming language: FORTRAN77

Memory requirement: 640 K of RAM

Brief description

CALREL is designed for performing structural reliability analysis, focused on computing failure probabilities of components of a structural system, or series reliability problems. Users can investigate a problem with a wide range of input assumptions such as means, variance, correlation, and distribution types. It offers a choice of first-order and second-order reliability methods (FORM and SORM), and also Monte Carlo simulations. The first-order analysis provides sensitivity information with respect to uncertain variables (with an option to make any of these uncertain variables a deterministic one) and distribution parameters (such as means and standard deviations). The user manual is published in Liu *et al.* (1989).

The authors would like to express their sincere thanks to Professor Jeff Cawlfield for providing information about this software.

PROBAN

Author/developer: Drt Norske Veritas AS

Availability: Drt Norske Veritas AS, Veritasveien 1, N-1322 Høvik, Norway

Version: 4.3 (1998)

Platform: NT, DEC, IBM, HP, SGI, SUN

Programming language: N/A

Memory requirement: 4 MB on NT

Brief description

PROBAN is a general-purpose program, which offers reliability, distribution, and sensitivity analysis. The program can be utilized for calculating reliability of structures and small mechanical systems. It computes FORM and SORM probabilities of failure, and can also perform Monte Carlo simulation. Sensitivity analysis is carried out to determine which variables influence the failure probability. Its user-friendly graphical interface makes modelling simple and display of results easy.

The authors would like to express their sincere thanks to Professor Jeff Cawlfield for providing information about this software and Mr Lars Tvedt of Drt Norske Veritas AS for reviewing the content.

KINAL

Author/developer:	Tamas Turanyi
Availability:	http://chem.leeds.ac.uk/Combustion/Combustion.html
Version:	2.1
Platform:	Platform – independent
Programming language:	FORTRAN77 source code
Memory requirement:	640 kbyte

Brief description

KINAL is a package for the analysis of kinetic mechanisms on personal computers. It consists of five programs, called DIFF, SENS, PROC, ROPA, and YRED. They require similar input data and use common subroutines. DIFF solves stiff differential equations and SENS computes the local concentration sensitivity matrix. The program PROC generates the rate sensitivity matrix or the quasi-stationary sensitivity matrix from concentration data, or uses a matrix computed by SENS and extracts the kinetic information inherent in sensitivity matrices by principal-component analysis. ROPA calculates the contribution of reaction steps to the production rate of species or calculates the lifetime of species. Finally, YRED provides hints for the elimination of species from the reaction mechanism. DIFFDAT, a utility program to KINAL, is an interactive program for the easy creation of data files. KINALC can be easily modified to handle not only chemical kinetic differential equations, but also ordinary differential equations of any other type. More details can be found in Turanyi (1990c).

The authors would like to express their sincere thanks to Dr Tamas Turanyi for providing information about this software.

CHEMKIN-II

Author/developer:	R. J. Kee, F. M. Rupley, J. A. Miller
Availability:	Distributed free by Sandia till 1997
Version:	Regularly updated till 1997
Platform:	Platform – independent
Programming language:	FORTRAN77 source code
Memory requirements:	Problem–dependent

Brief description

CHEMKIN-II is a software package to facilitate the solution of problems involving elementary gas-phase chemical kinetics. It provides a flexible and powerful tool for incorporating complex chemical kinetics into simulations of fluid dynamics. It consists of an interpreter of user-specified symbolic chemical reaction mechanisms, and libraries of chemical kinetic, thermodynamic, and transport subroutines. Databases of thermodynamic and transport properties, applications codes, mathematical software, and examples of usage in the form of UNIX shell scripts are also provided. Current gas-phase applications provide modeling capabilities for shock tube, premixed flame, stirred reactor, and equilibrium systems. All simulation programs calculate local sensitivities.

The authors would like to express their sincere thanks to Dr Tamas Turanyi for providing information about this software.

The CHEMKIN-III Collection

Author/developer: Robert J. Kee *et al.*

Availability: www.ReactionDesign.com

Version: Release 3.03

Platform: Win32, HP, SUN, SGI

Programming language: Mainly FORTRAN77

Memory requirements: Problem–dependent

Brief description

The features of CHEMKIN-II (see above) were extended with plasma chemistry and heterogeneous reactions simulation programs. Graphical interfaces are now available, and the documentation was updated.

The authors would like to express their sincere thanks to Dr Tamas Turanyi for providing information about this software.

KINALC

Author/developer: Tamas Turanyi

Availability: http://chem.leeds.ac.uk/Combustion/Combustion.html

Version: 1.3

Platform: Platform–independent

Programming language: FORTRAN77 source code

Memory requirements: Problem–dependent

Brief description

KINALC is a postprocessor to any CHEMKIN-based simulation program. It has been interfaced to the programs of the CHEMKIN package (SENKIN, PREMIX, PSR, SHOCK, and

EQLIB) and also to the RUN1DL package. KINALC carries out three types of analysis: processing sensitivity analysis results, extracting information from reaction rates and stoichiometry, and providing kinetic information about the species. KINALC can extract the important pieces of information from the sensitivity results dumped by the simulation programs. It can also calculate the sensitivity of objective functions, formed from the concentrations of several species. The program can suggest a list of rate-limiting steps. Principal-component analysis of the sensitivity matrix is also available. This eigenvector–eigenvalue analysis shows which parameters have to be changed simultaneously for a maximum change in the concentration of one or several species. The principal-component analysis groups the parameters on the basis of their effect on the model output. This information is useful for uncertainty analysis, parameter estimation, experimental design, and mechanism reduction. MECHMOD is an accompanying utility program for the transformation of reaction mechanisms. CHEMKIN, KINALC, and MECHMOD are specifically designed for the study of gas-phase reaction systems.

The authors would like to express their sincere thanks to Dr Tamas Turanyi for providing information about this software.

PREP/SPOP

Author / developer: A. Saltelli and T. Homma

Availability: OECD-NEA (http://www.nea.fr/)

Version: N/A

Platform: Unix

Programming language: FORTRAN77

Memory requirement: N/A

Brief description

The PREP (statistical PRE Processor) and SPOP (Statistical POst Processor) utilities are designed to assist users in the implementation of Monte Carlo-based uncertainty and sensitivity analyses, and are freely available from the above web site.

PREP has two main features. The first is the generation of samples in the domain of the input factors. Three sampling strategies are available in PREP: simple random, Latin hypercube and quasi-random. An established technique is implemented in PREP to impose rank correlations on two or more input factors at a prescribed level (Iman and Conover, 1982). Available distributions for the input factors are: piecewise-uniform, log-uniform, saw-tooth, normal, log-normal and Weibull. Mathematical relations between input factors and other variables can be specified. The second feature is the automatic inclusion of a number of FORTRAN subroutines that can be employed by users in their program to evaluate the model.

The model is executed with all the sample points generated by PREP and a file containing the model outputs is produced, which is used by SPOP to perform SA.

SPOP performs various types of uncertainty and sensitivity analyses on the model outputs. For uncertainty analysis, SPOP provides characterizations of the output distribution through means, confidence bounds, percentiles, histograms, and downward cumulative plots. Non-parametric techniques are also implemented (i.e. Tchebycheff and Kolmogorov confidence bounds) in order to cope with possible model nonlinearities. In the similar fashion, SPOP provides both linear estimators (e.g. Pearson correlation coefficient) and non-parametric ones (e.g. Spearman coefficient) as sensitivity measures. The user can compare the different ranking of importance produced by these estimators.

Other sensitivity measures available in SPOP include standard (rank) regression coefficients, partial (rank) correlation coefficient, and the importance measure (Hora and Iman, 1986) following a computational scheme suggested in Ishigami and Homma (1989). Some statistical tests are also implemented in SPOP, such as the Smirnov test, Cramer–von Mises test, Mann–Whitney test, and two-sample *t* test.

An upgraded version of PREP/SPOP (SIMLAB) has been developed (see below).

SIMLAB

Author/developer:	POLIS SCaRL Via della Signora, 3, 20122 Milano-ITALY
Availability:	Joint Research Centre, ISIS
	Contact: Andrea.Saltelli@jrc.it
Version:	1.0
Platform:	Windows 95/NT, UNIX
Programming language:	C++
Memory requirement:	32 Mb

Brief description

SIMLAB software is a menu-driven version of the PREP-SPOP software. This software is composed of three modules: Pre-processor, Model specification and Post-processor modules.

The Pre-processor module allows users to specify distributions for input factors from a wide range of distributions such as normal, log-normal, piecewise (includes uniform), log-uniform, Weibull, exponential, gamma, beta, and triangular. Correlation between factors can be imposed by using the dependence trees with copula (Meeuwissen and Cooke, 1994) or the Iman-Conover method (Iman and Conover, 1982).

The sampling techniques supported by the Pre-processor module are simple random, Latin hypercube, quasi-random LP_τ, Morris, and extended FAST. The user can also construct a sample by choosing any specified sequence of points. For each factor, descriptive statistics and information related to correlation characteristics are estimated. The sampling distribution of each factor and the correlation characteristics amongst the factors can be visualized, respectively, by using histograms and the cobweb plot or scatter plots.

The model specification module allows users to specify their model in two ways:

1. The user can link SIMLAB to an external model via executable files or statistical and mathematical packages, or
2. The user can define a model within SIMLAB by using a simple equation editor.

The module will perform model evaluation at each sample point, and yield a set of model outputs.

The post-processor module processes the model outputs to perform uncertainty and sensitivity analysis. SA techniques include Pearson correlation coefficient, Spearman rank coefficient; standard (rank) regression coefficient; partial (rank) correlation coefficient; Morris indices; extended FAST (first and total) indices. For the extended FAST, the post-processor module also incorporates the computation of sensitivity indices by groups; that is, factors are partitioned into groups according to (known) similar characteristics. For the uncertainly analysis, this module provides means, confidence bounds, percentiles, histograms and downward cumulative plots of the model outputs. Non-parametric techniques are also implemented (i.e., Tchebycheff and Kolmogorov confidence bounds) in order to cope with possible model non-linearities. The authors are grateful to Nicla Giglioli and Iacopo Cerrani for reviewing information about this software.

UNICORN

Author/developer:	Roger M. Cooke, J. R. van Dorp, and V. Kritchallo, Delft University of Technology
Availability:	AEA Technology, Thomson House, Risely, Warrington, Cheshire WA3 6AT, UK
Version:	R4.4 (July 98)
Platform:	PC
Programming language:	Turbo Pascal
Memory requirement:	Standard MS DOS/Windows

Brief description

UNICORN is a program for performing uncertainty analysis with up to 250 random variables; however, the limit may be much higher, depending on the hardware. The model is typically coded within UNICORN (though UNICORN may also be linked to external programs; see below). The main interdependent design choices are random sampling, marginal distributions represented by pre-calculated quantiles, dependence represented by rank correlation trees, and mathematical modelling support.

Users can specify their model by four different means: (1) via the FORMULA, the user enters random variables and a set of user-defined functions using UNICORN's internal parser; (2) via the FAULT TREES, the user enters basic events with uncertain failure frequencies and a fault tree; then UNICORN extracts the minimal cut sets; (3) via the DUALNET, the user provides uncertain supplies, demands, transport costs, and upper and lower transport bounds via a directed graph, then UNICORN computes flows and marginal costs for each arc in the graph, and the costs of the optimal solution; and (4) EXTERNALLY, where the user enters random variables, and a set of resulting sample is exported for external processing. The results, appropriately formatted, may be read back into UNICORN for further analysis.

Special features include conditional sampling, enabling previous samples to be called by the formula parser, indicator functions for random sets, and generation of sample files via cobweb conditionalisation. The method 'dependence tree with copula' is used for the specification of cross-correlation between the uncertain variables (Meeuwissen and Cooke, 1994). This operation is performed after specifying a model (i.e., marginal distributions and mathematical operations to be performed on the input variables). Hence, the same model can be processed with different dependence structures.

The GRAPHICS module is another component of UNICORN, and can handle at least 15 000 samples. Different plot types can be produced, namely cobweb plots, marginal plots, cumulative plots and scatterplots. The cobweb plots give a picture of the joint distribution of the percentiles of up to 20 variables. (See Chapter 11.)

The software is free for used for educational purposes, and can be obtained by contacting the author.

The authors would like to express their sincere thanks to Prof. Roger Cooke for reviewing information about this software.

A.3 OTHER SENSITIVITY ANALYSIS SOFTWARE

This section describes some commercially available software that can also perform SA. For example, Crystal Ball and @Risk are add-ons to commercial spreadsheet software that have been developed mainly for risk and uncertainty analysis, and are usually used by decision makers in the business and financial sectors. The following three packages (MAYDAY, SUSA-PC, and UNCSAM) have been developed by academics and government/organizational researchers, and hence are more applications-driven. For example, MAYDAY was developed for used in a safety study of a high-level nuclear waste repository, UNCSAM has been developed to deal with a wide range of environment models, and to be used in risk-assessment studies.

@Risk

Author/developer:	Palisade Corporation
Availability:	Palisade Europe, The Software Centre, East Way, Lee Mill Industrial Estate, Ivybridge, Devon PL21 9GE, UK
Version:	3.5
Platform:	IBM PC-compatible 386 or higher; MS Windows 95, NT or 3.x; MS Excel 4.0 or higher; Lotus 1-2-3 for Windows 4.0 or 5.0
Programming language:	N/A
Memory requirement:	Minimum 4 MB of RAM

Brief description

@RISK is a add-in to spreadsheet software to perform risk analysis. Uncertainty specifications of each input variable are assigned using the menu. There are 34 distributions to choose from, and the sampling scheme can either be Monte Carlo or Latin hypercube. Correlated input variables can be specified by means of a correlation matrix.

@RISK for Windows has two additional advanced analyses: Sensitivity Analysis and Scenario Analysis. Specifically, there are two methods for performing SA to choose from: stepwise regression and rank correlation coefficients. Results of SA are presented in the form of Tornado charts, with longer bars at the top representing the most significant input variables.

@RISK is also available for Lotus 1-2-3 on DOS and for Macintosh Excel, and in French, German, and Spanish.

Crystal Ball

Author/developer: Decisioneering Inc.

Availability: Decisioneering 1515 Arapahoe St., Suite 1311, Denver, CO 80202, USA; http://www.decisioneering.com

Version: 4.0

Platform: Macintosh (requires MS Excel 4.0 or 5), MicroSoft Windows (require MS Excel 4.0+)

Programming language: N/A

Memory requirement: At least 16 MB of RAM

Brief description

Crystal Ball is a spread sheet add-on that performs uncertainty analysis and creates graphic output for sensitivity analysis. In particular, it performs Monte Carlo forecasting and risk analysis. It uses either a simple MC or Latin hypercube sampling scheme. The user defines a mathematical model using familiar spreadsheet tools, then assigns distributions for input variables (from any of 12 or 16 mathematical distributions, including normal, triangular, Poisson, binomial, log-normal, uniform, exponential, geometric, Weibull, beta, and hypergeometric).

Choosing the Sensitivity Analysis option in the Run Preference dialog box can give a simple SA. Either rank correlations or contributions to variance can be used to rank input variables. The sensitivity chart displays the rankings as a bar chart. Raw data generated from a simulation can be exported to a new spreadsheet or file for further analysis or graphing.

MAYDAY

Author/developer: R. Bolado, A. Alonso, and J. A. Moya

Availability: Departamento de Ingegnería Nuclear, Escuela Técnica Superior de Ingenieros Industriales, Universidad Politécnica de Madrid, José Gutiérrez Abascal, 2-28006 Madrid, Spain

Version: 1.2

Platform: DOS/Windows, MAC/OS, UNIX, VMS

Programming language: C

Memory requirement: Unknown

Brief description

MAYDAY (Bolado *et al.*, 1998) consists of four modules: 0-var, 1-var, N-var, and a sampling strategy module. The 0-var module contains statistics and graphical tools to perform uncertainty analysis; these include summary statistics such as the mean, the variance, the skewness coefficient, kurtosis, and order statistics; confidence limits for these statistics; histograms, the empirical distribution function, its complementary curve, and Kolmogorov confidence limits. Also included in this module are some goodness-of-fit tests such as Kolmogorov, chi-square, Lilliefors, and the Shapiro–Wilk test for normality.

MAYDAY has a series of tools designed to apply variance-reduction techniques. These tools are proportional sampling and Neyman's stratified sampling, importance sampling, and Latin hypercube sampling.

MAYDAY includes in the 1-var module various sensitivity measures, such as Pearson correlation coefficient, Spearman rank correlation coefficient, partial (rank) correlation coefficients, standardized (rank) regression coefficients, R-squared, and the Fourier amplitude sensitivity test (FAST). The Wilcoxon statistic, the two-sample and k-sample Smirnov tests, the t-test, and the Kruskal–Wallis and Cramér–von Mises statistics can also be found in this module, to identify relationships between specific regions of an input variable on the output variable.

The N-var module can perform SA when several input factors are grouped into a set by taking into account possible relationships among them. Also included in this module is the classical FAST technique.

SUSA-PC

Author/developer:	Martina Kloos, Eduard Hofer, and Bernhard Krzykacz
Availability:	Gesellschaft für Anlagen- und Reaktorsicherheit (GRS)mbH, Forschungsgelände, 85748 Garching, Germany
Version:	3.1
Platform:	Microsoft Windows 95 (or NT), Microsoft EXCEL 7.0, FORTRAN compiler
Programming language:	The menu-driven user interface is a macro written in Visual Basics for Applications, Microsoft EXCEL 7.0. The executable programs performing the statistical data analysis are written in FORTRAN.
Memory requirements:	About 3.5 MB

Brief description

The main features include a wide choice of input distributions: (log-)histogram, polygonal line (customized), discrete, (log-)uniform, (log-)triangular, (log-)normal, beta, Weibull, gamma, exponential, chi-square, F, and extreme-value distributions of type I and type II, as well as the capability to handle truncated distributions and to derive distributions with specified quantiles. Plots of the selected densities, cumulative distribution functions and complementary cdfs can also be obtained.

Users can specify parameter dependence via ordinary correlation (Pearson), rank correlation (Spearman), induced rank correlation, complete dependence, conditional distribution, inequalities or functional relationships between uncertain parameters, and proportions (e.g. sums add up to 1). Scatter plots of specified dependence relationship are available. A choice of sampling schemes is available: simple random sampling and two options of Latin hypercube sampling. Output includes the design matrix and sample correlation coefficients on raw data as well as on their ranks.

Model specification is automatically prepared in the form of a FORTRAN skeleton for inclusion of model code statements. Ready-to-use model code can be included as a subroutine from a file.

Uncertainty analysis can be performed on scalar output as well as on one-dimensional time (space) model output. Output includes quantiles, mean, standard deviation, statistical tolerance limits, correlations between model results, all sample time (space) histories in one plot, as well as quantiles, mean, standard deviation over time (space), and cdf and scatter plot of scalar model output as well as of one-dimensional model output at selected points.

Sensitivity measures include correlation coefficients, partial correlation coefficients, standardized regression coefficients (stepwise), R^2, and correlation ratios for raw and rank-transformed data. Graphical output includes plots of sensitivity measures over time (space) or as bar charts for scalar model output, and scatterplots at selected points of time (space).

SUSA-PC software comes with a User's Guide and a Tutorial.

The authors would like to express their sincere thanks to Dr Eduard Hofer for providing information about this software.

UNCSAM

Author/developer:	P. H. M. Janssen and P. S. C. Heuberger
Availability:	National Institute of Public Health and Environmental Protection (RIVM-CWM), PO Box 1, 3720 BA Bilthoven, The Netherlands
Version:	1.1
Platform:	MS-DOS (mathematical co-processor recommended)
Programming language:	FORTRAN 77, embedded in ANSI-C environment
Memory requirement:	3 MB of RAM

Brief description

The first version of UNCSAM was available in 1992, and the main features included a variety of probability distributions, such as (log-)uniform, (log-)normal, (log-)triangular, exponential, logistic, Weibull, and beta, two sampling schemes (simple random and Latin hypercube), incorporation of correlated variables, and confidence limits for various summary statistics, in particular empirical cumulative distribution function of the output; the sensitivity measures used are regression and correlation analysis, and their rank version.

The authors would like to express their sincere thanks to Dr Peter Janssen for reviewing information about this software.

A.4 GENERIC ALGORITHMS

In this section, we include a number of generic algorithms, mainly written in FORTRAN, that are used to generate sample of inputs. We have included FAST, Sobol' LP_τ and Latin hypercube sampling.

A.4.1 FAST Sampling

The following is a piece of code to generate FAST samples using the search curve transformation described in Chapter 8 (Section 8.4.1):

```
C
C  AN ALGORITHM FOR SELECTING A SET OF FREQUENCIES
C
C  KI=NELGRP(L)!POPULATION OF GROUP OF INTEREST
C  KCI=NVARS-KI!POPULATION OF THE COMPLEM. GROUP
C  OMCIMAX=OMI/(2.*MCI)
C
      CALL SETFREQ(NVARS,KCI,OMCIMAX,OMGRCI,OTMP)
C
C  OMGRCI(I), I=1,KCI CONTAINS A SET OF FREQUENCIES
C  TO BE USED FOR THE COMPLEMENTARY GROUP
C
C  SETTING THE VECTOR OF FREQUENCIES 'OM' FOR
C  THE 'NVAR' FACTORS
C
     DO  I=1,KI      !FOR THE GROUP OF INTEREST
        OM (GR(L,I))=OMI
     ENDDO
     II=1       ! FOR THE COMPL GROUP
     DO NG=1,NGRPS
       IF(NG.NE.L) THEN
         DO  I=1,NELGRP(NG)
           OM (GR(NG,I))=OMGRCI(II)
           II=II+1
         ENDDO
       ENDIF
     ENDDO
C
     DO J=1,NVARS
       FI (J)=RAN1(ISEED) *2*PI
     END DO
C
     CALL PARCOMB2(OM,FI,MATSAM,IRANK,INDX,ARRIN)
```

The following subroutine is used to select a set of frequencies, as described in Section 8.4.4.

```
     SUBROUTINE SETFREQ(NVAR,KCI,OMCIMAX,OMGRCI,OTMP)
     IMPLICIT REAL (A-H,P-Z)
     INTEGER OMGRCI,OMCIMAX,OTMP
     DIMENSION OMGRCI(NVAR),OTMP(NVAR)
C
     IF (OMCIMAX.LT.(KCI)) THEN
       INFD=OMCIMAX
       ELSE
         INFD=KCI
     ENDIF
C
     ISTEP=INT(REAL(OMCIMAX-1)/real(INFD-1))
     IF (OMCIMAX.EQ.1)  ISTEP=0
     JJ=1
```

```
      DO I=1,INFD
        OTMP (I)=JJ
        JJ=JJ+ISTEP
      ENDDO
      DO J=1,KCI
        JJ=MOD(J-1,INFD)+1
        OMGRCI(J)=OTMP(JJ)
      ENDDO
      RETURN
      END
      Subroutine CURVE(OM,FI,X)
C  NRUN=number of runs
C  NVAR=number of factors
C  X=input matrix of nrun rows and nvar columns
C  OM=a vector of ωi
C  FI=a vector of φi
C
      DO J=1,NRUN
        S=PI*(2*J-NRUN-1)/NRUN
        DO I=1,NVAR
          X(j,i)=0.5+ASIN(SIN(OM(I)*S+FI(I)))/PI
        ENDDO
      ENDDO
C
      RETURN
      END
```

The authors would like to express their sincere thanks to Dr Stefano Tarantola for supplying the above codes.

A.4.2 Sobol' LP_τ

The following is a piece of code to generate the base matrix and the resampling matrix described in Section 8.3.4 using the LP_τ sequences (Section 8.5.1); The LPTAU subroutine (Sobol' *et al.*, 1992) is used, and is available from NEA Data Bank:

```
C
C  NSAM=sample size
C  NVAR=number of input factors
C  VECTOR=output of the LPTAU subroutine (dimension
C  must be less than 52
C  DATAMAT/SMATRIX=base matrix of dimension Nsam by
C  Nvar
C  RSMATRIX=Resampling matrix of the same dimension as
C  Smatrix
C
C
      DO I=1,NSAM
```

```
         CALL LPTAU(I,2*NVAR,VECTOR)
         DO J=1,NVAR
           DATAMAT(I,J)=VECTOR(J)
           SMATRIX(I,J)=DATAMAT(I,J)
         ENDDO
         K=1
         DO J=NVAR+1,2*NVAR
           RSMATRIX(I,K)=VECTOR(J)
           K=K+1
         ENDDO
       ENDDO
C
```

A.4.3 Latin Hypercube Sampling

The following piece of code generates a Latin hypercube sample with points that are centred within each interval:

```
C
C  NSAM=sample size
C  NVAR=Number of input factors
C  X=data matrix of dimension NSAM by NVAR
C  ICOLUMN is a matrix of dimension NVAR by
C  NSAM containing NSAM vectors of permutation of
C  distinct integers from 1 to NVAR
C
       DO I=1,NSAM
         DO J=1,NVAR
           X(I,J)=(FLOAT(ICOLUMN(J,I))-0.5)/NSAM
         ENDDO
       ENDDO
C
```

A.5 CONCLUSIONS

This appendix has briefly introduced some of the software available for performing sensitivity analysis. Clearly, the situation is very fluid, and this appendix can only cover those tools available at the time of writing—we expect there to be a continuing growth in software availability.

The common SA techniques are all provided within commercially available software. Simple random sampling and the correlation- and regression-based techniques are widely available, but the newer techniques (such as extended FAST and Sobol LP_{τ}) are still only available within more specialized software.

Within the modelling community, interface between the model code and the SA requirements has often been achieved by the programming in each case of the necessary tools, but as SA techniques become more widely known and used, there will be an increasing need for specific software tools.

It is our hope that as the use and demand for SA techniques as important modelling tools grows, we shall see a matching development in software availability.

Index

9 780470 743829